The Product Liability Handbook

Prevention, Risk, Consequence, and Forensics of Product Failure

Edited by **Sam Brown**

COMPETITIVE Manufacturing SERIES

VNR Van Nostrand Reinhold
New York

Library of Congress Catalog Number 90-35626
ISBN 0-442-21173-2

Printed in the United States of America

Van Nostrand Reinhold
115 Fifth Avenue
New York, New York 10003

Chapman and Hall
2–6 Boundary Row
London, SE1 8HN, England

Thomas Nelson Australia
102 Dodds Street
South Melbourne 3205
Victoria, Australia

Nelson Canada
1120 Birchmount Road
Scarborough, Ontario M1K 5G4, Canada

16 15 14 13 12 11 10 9 8 7 6 5 4 3 2 1

Library of Congress Cataloging-in-Publication Data

The Product liability handbook: prevention, risk, consequence, and forensics of product failure/edited by Sam Brown
p. cm.
Includes bibliographical references.
ISBN 0-442-21173-2
1. Products liability. 2. Standardization—Law and legislation. 3. Manufactures—Standards. 4. Quality assurance. 5. Engineering forensics. I. Brown, Sam,
1941–
K953.P764 1990
346.03'8—dc20
[342.638] 90-35626
CIP

To those who are responsible for making our work, our play, and our environment better as a result of the high quality and safe products for which they conceive, design, engineer, manufacture, operate, maintain, inspect, regulate, insure, provide operational instruction, and provide a legal framework.

Contents

Appendixes

Preface

This book is about the interrelationship among the manufacturer, design engineer, and consumer in a legal and forensic context, and it is also about change—changes in legal concepts and in engineering and scientific methods that have come about to protect citizens and businesses from injury and economic loss associated with products. *Product liability* refers to a legal action by an injured party (plaintiff) who seeks compensation for personal injury and/or loss of property from a second party (the defendant) who allegedly was responsible for the resultant loss, in which a product was involved. *Forensic engineering* is a science that deals with the relation and application of engineering/scientific facts to legal problems and product liability in general.

Between 1965 and 1971, the number of product liability claims increased from approximately 100,000 to one million, and the awards from $12.5 billion in 1972 to $50 billion in 1975. In the last decade, claims have tripled. The percentage of cases in which juries rule in favor of plaintiffs also has increased substantially in recent years, as has the amount of the average award. Design engineers and manufacturers are confronted with this dilemma—how to freeze costs so as to retain a competitive price and still improve quality control so as to realistically withstand the onslaught of product liability. The users and sociopolitical and economic sectors of our society have a vested interest in this question.

The legal aspects of product liability have become so complex that subspecialties have developed within legal specialties. For example, one set of rules governs the manufacturer of products involved in environmental control, another set governs food and drugs, another governs toys, and still another governs the automotive and aircraft industries. In the past decade, this legal complexity has generated intense interest in the impact of product liability on different sectors of American society. This interest reflects the recognition that the courts have been cast in the roles of arbiters of product safety and quality and de facto writers of manufacturing and engineering codes and standards.

The reality that is emerging in our highly industrial society is that of the tremendous growth in information access to all sectors of our society. There is increased pressure to quickly identify and correct deficiencies in product design, material selection, material processing, manufacturing, operation, and maintenance and inspection. Likewise, there is a growing interest in the engineering and scientific community in better understanding the legal implications of its roles in the manufacture and use of the products that it designs and engineers. Because manufacturers, fabricators, and contractors have historically been the defendants in product liability suits, they have long appreciated the implications of product liability for their businesses. However, there has been expanded interest on their part to understand the interface between the engineering/scientific community and the uses of their products in order to improve product quality and minimize litigation.

Industrial and individual consumers have been the most visible promoters of product safety in recent decades. As a result of this interest in improving the safety and quality of products, preventing product failure, and minimizing litigation, people realize that the burden of product quality does not rest solely on the shoulders of *the manufacturer*, but that *the design engineer* and *the user*, the three principals or primary influencers of product quality, also share the responsibility. The following instrumentalities, however, also influence the safety of products, either directly or indirectly: (1) training programs and professional development, (2) government regulations, (3) associations and societies, (4) insurers, (5) the marketplace, and (6) the litigation process. Because each entity that influences product quality and safety operates and interacts in an ever more complex technical and legal context, no one entity can improve the quality and safety of products on its own.

This book presents an overview of the interrelationship and functions of the nine influencers of product safety and quality previously listed in both a litigious and a nonlitigious context. The book is intended for use by the following groups that influence the quality of industrial, structural, and consumer products:

- Design engineers
- Manufacturers
- Owners-users
- Educators
- Governmental agencies
- Insurance industry

- Professional, technical, industrial, and user organizations that prepare voluntary standards
- Attorneys and judges
- Legislators
- Market place

Because we have attempted to knit together a broad range of technical and legal topics for a diverse audience, we have not included a great many technical details that would be of interest to scientific or legal scholars. We have, however, provided numerous references to highly technical books and periodicals for those who wish to consult them.

Each chapter is intended to give readers a new perspective on the mechanics of forensic engineering and product liability and, more important, on the complex process of bringing a safe and high-quality product to the marketplace. The book takes readers through the legal and engineering aspects of product liability by (1) outlining fundamental principles, (2) providing illustrative examples, (3) offering valuable guidelines, and (4) describing pitfalls, misconceptions, and conflicting perceptions for the reader's information and assessment.

The first four chapters provide an overview of the legal bases that bring together the entities that influence product quality, safety, and liability. They also provide guidance concerning the positive steps that can be taken to minimize injury or damage related to engineering design, manufacture, and use. Chapter 1 outlines the legal framework and the three legal criteria—negligence, strict liability and implied warranty, and expressed warranty—for measuring deficient products. Chapter 2 reviews the interrelationship between the entities that influence product quality, safety, and liability. The role of the litigation process in influencing product quality and safety also is explored. Chapter 3 describes how voluntary and mandatory codes and standards can reflect many different philosophies and many degrees of technical sophistication and that they can affect all aspects of an industry. Chapter 4 discusses product liability from the manufacturer's perspective; therefore, the focus is the relationship of quality control, safety, and legal liability to the other entities that affect the manufacturer's production and product.

In Chapter 5, the emphasis shifts to the engineering framework for product liability. The chapter gives a general overview of the eight stages of an accident investigation and failure analysis. Each stage of the accident investigation and failure analysis is discussed in more detail in Chapters 6 through 11.

Chapter 6 describes the forensic engineer's initial activities—collecting, examining, and documenting physical, recorded, and observational evidence that characterizes some aspect of the incident. Chapter 7 reviews the sources the forensic engineer consults for background information: for example, the manufacturer's specifications, engineering design reports and drawings, and manuals; the service history of the product; codes and standards; current practices; and databases. Chapter 8 reviews the destructive and nondestructive techniques for detecting defects in materials and for measuring the characteristics of materials. (The chapter is supplemented by an appendix that correlates the basic testing methods with the formulas used to describe the behavior of materials and why they fail.) Chapter 9 describes the physical evidence of failure, which plays an important role in determining the mode of failure, which, in turn, can be correlated with the mechanism of failure—the loading and environmental conditions that played a role in the failure. Chapter 10 discusses the use of simulations, both experimental and theoretical, to evaluate product performance and human factors. Computer simulations are especially effective for evaluating large, expensive products with respect to the choice of materials, product design, and the effects of defects. Chapter 11 describes how the forensic engineer reconstructs the incident of failure to determine causality: that is, who is responsible for the failure. It also discusses the final stage of the accident investigation and failure analysis, the formal report, which reviews the evidence gathered during the investigation and the engineer's opinions and recommendations based on the evidence.

Chapter 12, the final chapter, discusses the working relationship between the forensic engineer and the attorney when preparing for product liability litigation. Issues such as the following are discussed: the attorney and expert preparing for deposition, the types of questions that attorneys ask expert witnesses, and advice for attorneys who deal with expert witnesses. This format gives the reader a "feel" for the adversarial aspects of preparing for trial.

Finally, the book contains numerous examples that are intended to illustrate the lessons learned (sometimes the hard way). All the examples represent actual cases; however, the names of people, companies, and places have been changed in some instances.

Acknowledgments

Many individuals and organizations contributed directly and indirectly to the preparation of this manuscript. Contributions are manifested directly in the data and lessons presented in those illustrations, tables, lists, and case histories that provide guidance based on years of experience, and indirectly through the legal and engineering information contained in the publications that we have cited.

The editor would like to thank those who have contributed through their correspondence and discussions by sharing their experiences, advice, and questions. Thanks are first to Nelson Blank (Trenam, Simmons, Kemker, Scharf, Barkin, Frye, & O'Neill, PC) and Stan Grossel (Hoffman-LaRoche, Inc.) for their thorough review of this manuscript; Professors T. F. Talbot (University of Alabama at Birmingham), Stan Bates (University of Florida), Richard Henry (University of Akron), John Vance (Texas A&M University), Gerry George (University of Southwestern Louisiana), Andre Bazerqui (Ecole Polytechnique, Canada), Alan Gardner (Imperial College of Science and Technology, UK), and A. Kobayashi and A. Emery (University of Washington), who provided a perspective of the role of the educator in failure and safety analysis; Judge Lucious D. Bunton for his comments; Frank Silvestri and John Massicot (New Orleans, LA–per Napoleanic Base Code), Sam Miller, Henry Giessel, Larry Rudman, Kerry Blair, William Bell, Dan Linebaugh, Randall Crane, Martin Welp, and Bob Doty for their insights into various aspects of product liability; Cecil Kirby, and Randy Rooker of NASA (Langley Research Center) and Wayne Frazier of NASA (Headquarters Safety Division) for their role in obtaining a wealth of data on safety related to hazardous release protection; Victor Perez, Frank Stewart, Dave Hicks, Walt VonRiesmann (Sandia National Lab), Bill Short (ICI Americas), Tom Mulcahy (Argonne National Lab), James DeRaimo (NASA Lewis Research Center), Gary Steck (WFI), Jim Hademenos, Ed Bravenec (Anderson & Associates), Quy Troung (The MW Kellogg Co.), Frank Jandrasi, C. Taul (ARCO), R. Sims (EXXON), C. Langner (Shell Research & Development), J. Wantanabe (Japan Steel Works), Rene Labbens (Framatome, France), Richard Gwaltney and Claud Pugh (Martin Marietta), and Charles Becht, Dave Clayton, Don Weintritt, Joe Lestingi, "Raj" Sundararajan, and M. R. Pamidi (Intertech Services, Inc.) for their perspectives regarding design review, the redesign cycle, and prevention; and the people at ANSI, ABET, the Trial Lawyers Association, the American Bar Association, the Smithsonian Institute, and particularly the American Society of Mechanical Engineers for their information and help.

Finally, special thanks are to Jo Monistere for her participation in editing, organizing, and researching the manuscript material, to Troy Brown for assistance in preparing the technical and graphics data, and to Jaime Knowles for her clerical assistance.

Contributors

Sam Brown, Ph.D., P.E.,	President—Quest Engineering Development Corp., Humble (Houston) Texas, and board member—Intertech Services, Inc.
David Fromson, J.D.,	Trial Lawyer and Author—Member of the New York and Maine Bar Associations, West Palm Beach, FL
Iain LeMay, Ph.D., P.E.,	President—Metallurgical Consulting Services, LTD. Saskatoon, Saskatchewan, Canada, and board member—Intertech Services, Inc.
Jerry Lyons, Ph.D., P.E.,	President—Yankee Ingenuity, Inc., St. Louis, MO, and board member—Intertech Services, Inc.
Vytes Svalbonas, M.S.,	Vice President of Engineering, Mineral Processing Systems, Inc., York, PA
Justin Sweet, J.D.,	Professor of Law at the University of California at Berkeley, Berkeley, CA
Alvin Weinstein, Ph.D., J.D.,	Weinstein and Bohan & Associates, Portland, Maine; Adjunct Professor of Law at the Franklin Pierce Law Center, Concord, NH
Sam Zamrik, Ph.D.,	Professor in the College of Engineering at Pennsylvania State University, College Park, PA

1

The Legal Framework for Product Liability

The engineer's world is ruled by inviolate physical principles. It provides the comfort of predictable properties and behavior of materials. The engineer's reward is the satisfaction of modeling the world with bounded calculations that culminate in translating a design concept into a manufacturable, functional, and safe product. If the product fails during use, however, the legal system is highly likely to reevaluate the engineer's labors to establish whether the manufacturer is liable for damage or injuries caused by the failure.

But the legal system is not schooled in engineering; it does not have inviolate physical principles or predictable properties. Worse yet, the legal real world asks questions about the product that usually differ completely from those asked, and answered, by the engineer during the design process. The legal system is intrusive; it is rarely, if ever, quantitative. It espouses qualitative phrases that somehow are to be the basis for a jury's or a judge's decision about the engineer's design, the manufacturer's methods, or the user's operation.

As ephemeral and confusing as the legal system often appears to be, it is axiomatic that the courts are empowered to resolve the disputes that arise when a person complains that a product did not live up to expectations or caused physical injury or property damage because it was not properly designed. Such claims are inevitable, and discretion is usually the better part of valor. Thus, engineers should assume responsibility for understanding the legal system as it relates to the products they create or certify as safe for use under prescribed conditions for two reasons. First, they will be better able to respond to criticisms of their previous efforts. Second, in the future they will at least be more sensitive to the requirements of the legal system and, at best, actively incorporate those requirements into their designs from initial concept to final product. If the legal indicia are incorporated into the design of a product, fewer questions are likely to be raised about the efficacy of that product.

For the attorney (as well as the engineering designer and administrator, manufacturer-manager, user, insurer, and regulator) who is unfamiliar with the engineering and legal aspects of product liability, it is important to understand the legal and engineering fundamentals in addition to the lessons learned from past product liability, investigations of accidents, and failure analyses. We will therefore begin by exploring what seem to be the shifting sands and less-than-stable footing of the legal system as that system attempts to assess liability for structural failures.

ELEMENTS OF A PRODUCT LIABILITY SUIT

Consider first some of the terminology and the essential elements of a product liability lawsuit. *Product liability* is the descriptive term for this species of litigation: the complaining party (*the plaintiff*) alleges that a deficiency in the product caused personal injury, property damage, or both and wants to be fully compensated for the loss by one or more parties (*the defendants*), who are usually suppliers of materials, designers, manufacturers, contractors, assemblers, distributors, or sellers.

To prevail in the suit, the plaintiff must establish all of the following during the first phase of a trial: (1) the product was defective, (2) the product was defective when it left the defendant's hands, (3) the defect caused the harm to the plaintiff, and (4) this harm can appropriately be assigned to the defect. If the plaintiff fails to establish all four elements, he or she will lose the lawsuit.

If the plaintiff contends that the manufacturer was negligent, he or she must prove a fifth element: The manufacturer's conduct was unreasonable. Next, we will consider these five elements of product liability.

The concept of defect is the most elusive and most important one involved in product liability. Obviously, the plaintiff argues that the product is substandard in some way and that this substandard condition or defect was responsible for causing the injury or damage. Thus, the jury needs a measure or standard with which to determine that the product was, in fact, substandard or defective.

Defects usually fall into two broad categories: *production* errors and *design* errors. Defects caused by production errors are easily described because the product does not meet the manufacturer's, processor's, or fabricator's own standards. The following are examples of production defects: a hole is drilled out of alignment, the threaded length of a stud is shorter than that specified by the manufacturer, a weld did not penetrate completely. A manufacturer's own standards encompass all governmental or industrial standards adopted for the product. These standards can include those promulgated by engineering societies or voluntary consensus groups such as the American National Standards Institute and the American Society for Testing and Materials.

After establishing that a production defect exists—that the manufacturer failed to meet its own standards—all the plaintiff must do is demonstrate that the defect caused harm: the third element in the plaintiff's case. Clearly, the second element—that the defect occurred while the product was in the manufacturer's hands—does not need to be proved explicitly. The fourth element, which we will discuss in more detail later, usually is not an important consideration when a production defect is alleged.

Allegations of design defects created the battleground of product liability. In contrast to a production defect, which is caused by the manufacturer's failure to meet its own standards, the standards for judging whether a design defect exists are external to the manufacturer, and the courts generally impose those standards on the basis of relevant legal theories.

When a design defect is alleged, although the product has met all the manufacturing, governmental, and industrial standards, the plaintiff is contending that the product is defective because it failed to meet a separate and distinct legal standard. For example, the plaintiff can argue that the product should have been made of a material with better properties, it should have had a guard in an area where one was not provided, it should have had a two-button hand control rather than a foot switch, or it should have been made stronger or differently. In each instance, the plaintiff will argue that the identified shortcoming meets the test of a defect under a legal standard even though the product met the manufacturer's standards and all relevant governmental or industry standards.

LEGAL STANDARDS FOR ESTABLISHING THE PRESENCE OF A DEFECT

A product is measured for defect according to three basic standards or legal theories:

- *Negligence,* which tests the defendant's conduct.
- *Strict liability* and *implied warranty of merchantability,* which test the quality of the product.
- *Express warranty* and *misrepresentation,* which test the performance of the product against the explicit representations made by the manufacturer or seller on behalf of the product.

Negligence

Negligence has been defined as "conduct which involves an unreasonably great risk of causing damage" (Terry 1915) and, alternatively, as conduct that "falls below the standard established by law for the protection of others against unreasonably great risk of harm" (*Restatement* [*Second*] *of Torts* 1965). To make out a cause of action in negligence, the plaintiff does not need to establish that the defendant either intended harm or acted recklessly. The test is to measure the defendant's conduct against a norm.

The norm established by the courts is the "reasonable person" test. The law declares that defendants have been negligent if they (1) failed to conform to a standard of conduct necessary to protect others against unreasonable risks (every person's legal duty), (2) breached that legal duty, (3) caused injury to the plaintiff as a result, and (4) the causal relationship between their conduct and the resulting injury was a reasonably close one—sometimes called "proximate cause" (Eginton 1970).

Early in the development of negligence law, the courts did not hold the manufacturer liable unless some contractual relationship existed between the manufacturer and the injured party—a relationship known as "privity." The usual rule was "No privity, no duty," a common law rule that dated back to an English decision in 1842 called *Winterbottom v. Wright.* Winterbottom, the driver of a mail coach that broke down and injured him, sued Wright, who had the contract with the Crown of England for ensuring the safety of the coaches. Winterbottom claimed that Wright had failed in his contractual duty and caused his injury. The court denied recovery on the ground that no privity existed between them: that is, Winterbottom was not a party to be protected under the contract. The court, in denying recovery, declared that "if the plaintiff can sue, then every passenger, or even every person passing along the road who was injured by the upsetting of the coach, might bring a similar action." The court concluded: "It is no doubt a hardship upon the plaintiff to be without a remedy but by that consideration we ought not to be influenced." For some time thereafter, and even when the manufacturer could have foreseen injury to others, its duty was limited only to those people with whom it had contractual privity. To overcome the obvious inequities of this doctrine, a decisional law developed, under which a manufacturer of dangerous products such as poisons or explosives would be liable to users even when no contractual relationship existed between them. The rationale was that the manufacturer could have foreseen the hazardous consequences and therefore should be held liable.

Excluding situations involving dangerous products, the theory of privity continued until the case of *MacPherson v. Buick Motor Company* in 1916, when Justice Cardozo, speaking for the Court of Appeals in New York, swept away the privity requirement. MacPherson had purchased a Buick from a dealer; a wheel collapsed, causing his injury. MacPherson sued both the dealer and manufacturer. In the lower court, he emphasized that Buick had failed to inspect the wheel, which had been made by another manufacturer. On appeal, the court enunciated, for the first time, the broad determination that a motor vehicle was a dangerous instrument and thus privity was not required.

Although the *MacPherson* decision was originally limited to cases involving automobiles, the privity rule eroded from that time forward. Today, the rule is virtually nonexistent. In its place emerged the basic rule that if a product is inherently dangerous and is later found to be defective, the manufacturer is liable for injuries to the user, who can sue, independently of any contractual relationship, provided (1) the defect caused the injury, (2) reasonable means existed to prevent the defect, and (3) the user used the article in a reasonable manner.

The *MacPherson* rule was embodied in the *Restatement* [*Second*] *of Torts* (1965). In recent years, the rule has been extended to include bystanders (Flynn 1972).

Because all human activity involves an element of risk, a defendant's conduct is deemed negligent only when it falls below what a "reasonable" person would have done under similar circumstances. One could argue that such a standard is too vague for courts to use in everyday litigation; in fact, however, it is a flexible tool for judging behavior. For example, a bottle of soda pop that explodes because of a flaw in the glass may be defective, but holding that the manufacturer was negligent does not necessarily follow. If the manufacturer took "reasonable care," as demonstrated by its quality control techniques, its conduct was not unreasonable, even if the product that came off the assembly line was in fact defective and injured the plaintiff. If the manufacturer's conduct was judged to be reasonable under the circumstances, the plaintiff would not recover for any injuries. Instead, the plaintiff would have to absorb the costs of the injuries, even though his or her own behavior was exemplary.

The flexibility of the negligence standard is achieved by permitting interplay among three factors. In every negligence action, the court must consider (1) the probability that harm would occur, (2) the gravity of the harm, and (3) the burden of taking precautions to protect against the harm—what the manufacturer might have done to minimize or eliminate the risk of harm. The more remote and improbable the risk that the defendant's conduct will cause harm, the less reason the defendant has to protect that risk. If, however, a remote risk can result in serious injury, the defendant must consider the serious implications of even an extremely low probability of harm. Finally, the probability that harm will occur and the gravity of that harm must be weighed against the burden of precaution, the cost of preventing harm. Thus, in the case of the exploding pop bottle, assuming that the plaintiff argued that more exacting inspection procedures existed at the time of manufacture, the difference in the cost of using more sophisticated inspection procedures must be weighed against the increased probability that those procedures would result in the discovery of flawed bottles. The difference also must be weighed against the seriousness of injuries caused by an exploding pop bottle. In other words, the negligence principle admits the existence of real risks associated with use of the product. The question that must be answered in each case is whether the risk was justified.

One could argue that the flawed pop bottle is actually an example of a production defect, because no responsible manufacturer would want to use a flawed bottle and thus the bottle did not meet the manufacturer's own standards. Even so, the manufacturer need not be judged as negligent if expecting it to meet that standard is judged to be unreasonable.

The same allegation of defect could be raised in the context of a design defect. If flawed glass bottles occur with measurable frequency and no reasonable inspection techniques exist for detecting the flaws, the plaintiff could argue that plastic or metal containers should be used exclusively. In this scenario, using glass rather than readily available alternatives is an allegation of design defect. Again, however, the three factors of probability, gravity, and burden of precaution must be considered before the manufacturer can be judged as negligent for not using the alternate materials exclusively.

Whether the allegation involves a production defect or a design defect, the negligence test focuses on the technology and information that were available to the manufacturer at the time of production. Note, however, that a plaintiff can allege a cause of action based on what a reasonable manufacturer should have known at the time the defective product was produced. Thus, the reasonable manufacturer is expected to be aware of current industrywide standards and technology. An entire industry can be found negligent for continuing to apply unacceptable standards of manufacture (Hooper 1932) or for failing to make currently available technological improvements. *Marsh Wood Products Company v. Babcock and Wilcox* (1932) is an example of how this principle operates. The plaintiff, who was injured by an exploding boiler, alleged that the entire boiler manufacturing industry was negligent because it continued to use a hydrostatic test to establish the integrity of its boilers although a more sophisticated test was available. The court found that it was reasonable for a jury to determine that the new test, developed by a professor of metallurgy at the University of Wisconsin, was a feasible alternative to the hydrostatic test and had a greater probability of discovering flaws in the metal.

Although evidence that a manufacturer has met self-imposed standards, voluntary standards, or even statutory stands is significant, such evidence alone does not establish that the manufacturer is not negligent. These standards are often used as a floor, not as a ceiling, for establishing "acceptable," that is, nonnegligent conduct. In *Berkebile v. Brantley Helicopter Corp.* (1971), the defendant had met the standards set at the time by the Federal Aviation Administration (FAA) for engaging the autorotation control after engine failure. However, this fact did not constitute an absolute defense against liability. The court held that a jury could find that a reasonable manufacturer would design the controls to give the pilot more time after engine failure to activate the autorotation control than was required by the FAA.

The lesson is clear. Regardless of existing standards, either industrial or governmental, the final responsibility for reasonable behavior rests with the manufacturer.

Strict Liability and Implied Warranty

Product liability can also be based on a second theory, the concept of *strict liability*. In recent years, a social concept has evolved that the manufacturer should be held strictly liable, whether or not it is negligent, if it places a defective product in the marketplace. This theory developed from two parallel sources: the inability of the negligence theory to safeguard the injured person and the need to respond to the consumer protection movement.

In 1964 the concept of strict liability emerged as an alternate avenue for establishing a products liability action. The clearest expression of this concept is stated in Section 402A of the *Restatement [Second] of Torts* (1965):

1. One who sells any product in a defective condition *unreasonably dangerous* to the user or consumer or to his property is subject to liability for physical harm thereby caused to the ultimate user or consumer, or to his property if a) the seller is engaged in the business of selling such a product, and b) it is expected to and does reach the user or consumer without substantial change in the condition in which it is sold. (Italics added)
2. The rule stated in Subsection (1) applies though a) the seller has exercised all possible care in the preparation and sale of his product, and b) the user or consumer has not bought the product from or entered into any contractual relation with the seller.

Section 402A does not, however, make the manufacturer an insurer of all those who encounter its product (see *Helene Curtis Industries v. Pruitt,* 1967). Before the strict liability theory attaches and under the *Restatement (Second) of Torts,* a manufacturer will escape liability if it can demonstrate that the product was used in a manner that could not reasonably be foreseen, that is, that it was misused. Also, no liability attaches if the manufacturer can demonstrate that the plaintiff was aware of the danger, yet proceeded to use the product—a circumstance sometimes called "assumption of the risk." In any event, liability attaches only if the plaintiff can prove that the product was in a defective condition unreasonably dangerous and that the condition caused the injury.

The differences between the negligence and strict liability theories are significant and far-reaching. Fundamentally, when deciding whether a product is in a defective condition unreasonably dangerous, the focus in strict liability is on the product, not on the manufacturer's conduct. If nothing else, this shift from negligence to strict liability requires the inquiry to be directed to the product and its use rather than to what the manufacturer should or should not have done or perceived.

The case of the exploding pop bottle established that under the negligence theory, if a manufacturer could prove that its quality control techniques met the standard of reasonableness, it could not be found liable. Under the theory of strict liability, the argument that the manufacturer acted reasonably is no defense. If the product is, in fact, unreasonably dangerous (as shown by a risk/utility analysis) and has caused the plaintiff's injury, the manufacturer can be held liable. Liability will attach even if "the seller has exercised all possible care in the preparation and sale of his product." Arguing that better quality control procedures are prohibitively expensive will be of no avail to the defendant in a production-defect case.

The design-defect case does not provide a built-in internal standard for establishing defect. The product, by definition, meets the manufacturer's own internal standard. Under the strict liability theory, a balancing process clearly is necessary to help define an external standard for establishing a design defect. And that standard must be one of reasonableness—essentially the same type of standard defined for negligence, but one with a different focus.

A manufacturer will be liable under strict liability only if the product is unreasonably dangerous to the user or consumer. If a product is reasonably safe (i.e., not unreasonably dangerous), no liability will attach. Thus, the test under strict liability must distinguish between reasonable safety and unreasonable danger. This can be accomplished by balancing the risk of injury (the probability and gravity of harm) caused by the product against its usefulness to society, rather than against the reasonableness of the manufacturer's conduct, in assessing the burden of precaution, as in a negligence action.

When the test is the defendant's conduct, the resonableness standard is completely understandable. When, however, the product itself must be considered to establish defect (as it must be in strict liability), then the standard of "unreasonable danger" needs to be explained.

Because products are incapable of reasoning, the idea of the unreasonably dangerous product begins with the premise that all products present risks to the public. Some risks can be deemed "reasonable" for society to accept when balanced against the usefulness of the product—in other words, the importance of the functions the product performs outweighs the cost of making it safer. That is, a reasonable person who understood the product's potential for harm would conclude that marketing it in that condition was proper.

How does this notion differ from that of negligence? Strict liability theory, unlike negligence, is not concerned with the conduct that brought about the unreasonably dangerous condition. For example, assume that a manufacturer has acted reasonably in designing a product and has tested it adequately before marketing it. When the product is in actual use, however, certain dangers became apparent that the design and testing process failed to account for. Although it was reasonable not to anticipate those dangers, the design is in fact substandard. In this case, defendants cannot argue that they acted reasonably. If the plaintiff can claim that a reasonable person who was knowledgeable about the danger would not have marketed the product without altering the design or including additional warnings in the labeling or instructions, then the product is unreasonably dangerous.

McCormack v. Hankscraft Company (1967) illustrates how the standard of "reasonableness" in strict liability is determined after evaluating risk versus usefulness. Hankscraft, which manufactured steam vaporizers, placed on the market an inexpensive model in which the heating element was attached to a lift-off cap. One consumer placed the vaporizer on a stool in a child's bedroom. During the night, the child tripped over the electrical cord attached to the vaporizer. The device tipped over, the top came off, spilling near-boiling water, and the child received third-degree burns. An action brought against the manufacturer on the child's behalf claimed that the design of the vaporizer was unreasonably dangerous.

Using the balancing process for reasonableness under strict liability, one can identify the factors necessary to determine whether the design was unreasonably dangerous. The usefulness of a vaporizer is generally unquestioned. One can foresee, however, that steam vaporizers will be used in close proximity to children, and often in a darkened bedroom. Placing the device on a stool or chair to direct the steam toward the child also is a common and expected practice. Thus, the probability that a vaporizer will be inadvertently tipped over by either a child or adult also is foreseeable. Therefore, one must consider the implications of scalding water flowing from the overturned container.

Because the water in the vaporizer was hot enough to cause third-degree burns, one must ask whether an alternative design would eliminate the danger without seriously impairing the usefulness of the product or making it unduly expensive. Through expert testimony, the plaintiff established that a screw-on rather than a lift-off cap was both feasible and relatively inexpensive. The defendant countered that in addition to increasing the cost, the screw-on cap raised other design problems because of a possible buildup of pressure within the vaporizer. The plaintiff's experts responded that a small vent in the screw-on cap would alleviate that problem. Although small dribbles of water might flow from the vent if the vaporizer tipped over, the danger of severe burns would be significantly reduced. In considering the acceptability of the manufacturer's design, the court also took into account the fact that the user (the mother in this case) had no way of knowing that the water in the device was scalding hot and therefore could not realistically assess the danger. However, even assuming that she did understand the danger, the following question remains: Was she still likely to place the vaporizer near the child to enhance its effectiveness?

Note that the danger was assessed in the context of the environment in which the vaporizer would be used. Unlike negligence, where the emphasis is on the manufacturer's conduct, strict liability assesses the issue of unreasonable danger from the environmental vantage point or use by the consumer. Thus, the conclusion, through the litigation process, that a given design is, or is not, unreasonably dangerous is an expression of society's willingness to accept or reject that design. When the cost of a safer alternative outweighs the probable danger, then the product will be declared to be "reasonably safe" as is. When the risk posed by the product exceeds its inherent usefulness, the product will be declared to be unreasonably dangerous. In the *McCormack* case, the jury found that the vaporizer was defective because of the lift-off cap.

In describing the reasonableness test for strict liability, balancing risk versus utility is more appropriate than is balancing probability versus gravity versus burden of precaution, the test used for negligence. The element of risk is equivalent to the probability and gravity of the harm. But utility is a more accurate word with which to focus on the distinctive feature of strict liability, that is, assessing a product within its real-world use by real users. Because the manufacturer's conduct is not of concern, the utility of the product in the real environment of its use is the important factor, not necessarily the burden of precaution on the manufacturer.

Therefore, in balancing risk against utility to determine whether a product is unreasonably dangerous, an expanded set of measures is useful. Dean Wade (1965), a leading scholar of product liability, suggested seven considerations for this purpose (see Box 1-1). Although some of these indicia can be quantified and others require a subjective evaluation, all seven are required to decide whether a product contains an unreasonably dangerous defect.

BOX 1-1. Indexes for Determining Whether a Product Is Unreasonably Dangerous[a]

1. How useful and desirable is the product?
2. Are other, safer products available that will meet the same needs?
3. Is the product likely to cause injury, and is the injury likely to be severe?
4. How obvious is the danger?
5. How much knowledge about the danger does the public have, and what could it normally expect regarding the danger (particularly with respect to established products)?
6. Would care in using the product (i.e., following instructions or heeding warnings) avoid the injury?
7. Could the danger be eliminated without seriously impairing the usefulness of the product or making it unduly expensive?

[a]Derived from Wade (1965).

In one sense, the determination of defect and unreasonable danger is subjective because each product must be viewed in the context of its particular function and use. For example, if a product can be used in both domestic and industrial environments, different conclusions about defectiveness and unreasonable danger can be reached in the two situations. Thus, before the Wade indicia can be applied appropriately, the environment in which the product is used must be thoroughly understood.

Section 2-314 of the Uniform Commercial Code provides the user or consumer with a parallel guarantee of product quality, similar to that of strict liability:

> **1.** Unless excluded or modified, a warranty that the goods shall be merchantable is implied in a contract for their sale if a seller is a merchant with respect to goods of that kind . . . .
>
> **2.** Goods to be merchantable must be at least . . . fit for the ordinary purposes for which . . . [they] are used. . . .

The code has been adopted in all states but Louisiana. Although there is some variation, most courts agree that Section 2-314 provides the same consumer protection as does the strict liability requirement: A product must be reasonably safe. When two alternate theories of law seem to accomplish virtually the same purpose, the confusion that results is understandable. However, rather than delve into subtle legal complexities, it is sufficient to say that when proceeding under the Uniform Commercial Code (a statute designed primarily for commercial entrepreneurs and not directed specifically to consumers), the plaintiff must take special care to comply with all the requirements set forth in the complex provisions of the code. For example, if injured parties fail to notify a manufacturer about a defective product within a reasonable time after their injuries, they may be precluded from bringing a suit. In addition, the common practice among manufacturers of disclaiming the implied warranty of merchantability must receive special consideration. When plaintiffs proceed under the Uniform Commercial Code, they may be faced with an argument that the disclaimer prevents their recovery. Courts have differed sharply on whether a disclaimer between parties of unequal bargaining power (e.g., an ordinary consumer and a large manufacturer) is legally binding. The code is ambiguous on this subject. There is, however, no question that strict liability admits no disclaimers of any kind. When the courts permit a plaintiff to argue a breach of implied warranty as the legal theory, the standard for establishing a defect is the same as that under strict liability.

Express Warranty and Misrepresentation

The previous discussion has focused on legal theories that require the plaintiff to establish some form of product defect. Another theory for recovery, however, contains the notion that establishing the existence of a defect (that the product is unreasonably dangerous) is unnecessary. If a seller expressly warrants or represents a product as having certain characteristics or as performing in a certain manner, and the product fails to meet that warranty or representation, he or she will be held liable if the buyer can prove that an injury resulted from the failure of the product to meet the warranty.

Section 2-313 of the Uniform Commercial Code states that

1. Express warranties by the seller are created as follows:
   a) Any affirmation of fact or promise made by the seller to the buyer which relates to the goods and becomes part of the basis of the bargain creates an express warranty that the goods shall conform to the affirmation or promise.
   b) Any description of the goods which is made part of the basis of the bargain creates an express warranty that the goods shall conform to the description.
   c) Any sample or model which is made part of the basis of the bargain creates an express warranty that the whole of the goods shall conform to the sample or model.

2. It is not necessary to the creation of an express warranty that the seller use formal words such as "warrant" or "guarantee" or that he have a specific intention to make a warranty, but an affirmation merely of the value of the goods or a statement purporting to be merely the seller's opinion or commendation of the goods does not create a warranty.

Whether a certain statement is a warranty or a representation or is merely "puffing" is often a question. For example, at what point does representation of the qualities of a used car become a warranty rather than puffing? The answer depends on the type and age of the car and the nature and strength of the dealer's representations. Saying that a car is in "A-1 shape" may not mean much if it is seven years old and has 100,000 miles on its odometer. A stronger representation, such as "Factory New" applied to a demonstration model may qualify as a warranty. For instance, the courts have ruled that using the word "gentle" concerning a permanent wave (*Rogers v. Toni*), "blow-out proof" in relation to tires (*Goodyear Tire v. Hammond*), and the word "harmless" (*Simpson v. American Oil*) are not mere puffing but are express warranties. No hard-and-fast rules exist about where to draw the line; it is a matter of common sense.

The caution to manufacturers not to oversell the product is an important one. If a manufacturer's representations go beyond the capability of the product to perform, liability may result. This will be true even when the manufacturer honestly believes the representations and the belief is reasonable. *Crocker v. Winthrop Laboratories Division* is an excellent example. Winthrop developed Talwin—a fairly strong pain killer that was advertised to the medical profession as nonaddictive. Crocker, who had been injured in an industrial accident, took Talwin regularly, as prescribed by his physician, to relieve pain and became addicted to the drug. Later, he died after taking a toxic dose. Rather than declare Talwin to be an "unreasonably dangerous" drug, the court held the defendant to its promise that the drug was nonaddictive. Whether the company believed its own assertion when the product was marketed was irrelevant. Therefore, an express warranty is a form of absolute liability.

Section 2-315 of the Uniform Commercial Code expresses yet another theory of representation:

> Where the seller at the time of contracting has reason to know any particular purpose for which the goods are required and that the buyer is relying on the seller's skill or judgment to select or furnish suitable goods, there is unless excluded or modified under the next section an implied warranty that the goods shall be fit for such purpose.

For example, if a seller of power tools knows that a contractor relies on its judgment to select an electric drill that is suitable for heavy-duty, continuous service, the seller will be liable if the drill proves not to be suitable for the stated task. Note that the seller will be held liable even if the drill is not defective and performs well during light-duty service and even if the seller has not expressly warranted to the contractor that the drill will function effectively during heavy-duty, continuous service. The seller is therefore liable if it had "reason to know" about the particular purpose for which the drill would be used, knew that the contractor relied on the seller's judgment, and the drill failed to function for that purpose.

ADDITIONAL QUESTIONS THAT REQUIRE ANSWERS

Was the Product Defective When the Defendant Sold It?

Products often lead tortured lives. Those that meet the most demanding specifications can be used and abused to the point where they are clearly defective at the time of injury. Such defects cannot automatically be assigned to the manufacturer; *when* the defect was introduced into the product must be determined. Unless the plaintiff can establish that the defect existed when the product was in the manufacturer's hands, liability will not attach.

Of course, the mere fact that a defect did not become evident until later does not mean that the defect was not present when the product left the manufacturer's hands. Structural weaknesses and design problems often are not manifested until the product is used for a particular purpose. In each case, the facts must be examined carefully to determine whether the defect was attributable to the manufacturer or seller or whether it developed after leaving the manufacturer.

In our complex system of distribution, a product often goes through many hands before it reaches the ultimate user. Thus, the answer to the question "Who is the appropriate defendant?" is not an obvious one. In addition, a finished product need not begin and end in the same plant. The raw materials can be processed into a basic form such as steel by one manufacturer, converted into one component by another, assembled into a product containing a series of components by yet another, and distributed by still another party. Who, among all these participants, can be joined as defendants in a lawsuit? The simple answer is that under strict liability, all of them can be potentially liable.

In a negligence action, the defendant's conduct is at issue. For example, if a distributor or retailer simply sells or resells a packaged product, he or she has no duty to inspect the product for defects. Therefore, he or she is not negligent and cannot be found liable despite the existence of a defect.

Strict liability, on the other hand, focuses on the *sale* of a defective product. Therefore, anyone who *sells* the product, even though he or she did nothing more than take the packaged item off the shelf and sell it, can be liable under the theory of strict liability or implied warranty of merchantability.

The reasons for expanding the number of defendants to include every seller of a defective product across the manufacturing process as well as up and down the chain of distribution are set forth in Comment c, Section 402A, in the *Restatement (Second) of Torts*. In essence, the comment argues that the justification for imposing strict liability is that the seller, by marketing the product for use or consumption, has assumed a special responsibility toward any member of the consuming public who may be injured by the product—that in the case of products it needs and for which it is forced to rely on the seller, (1) the public has the right to and does expect reputable sellers to stand behind their goods, (2) public policy demands that the burden of accidental injuries caused by products intended for use or consumption be placed on those who market them and be treated as a cost of production against which liability insurance can be obtained, and (3) consumers of such products are entitled to maximum protection at the hands of someone, and the proper people to afford that protection are all those who market the products, wherever they happen to be in the distributive chain.

The only requirement for imposing strict liability or implied warranty is that the defendant is engaged in the business of selling. The occasional noncommercial sale will not admit the imposition of strict liability; however, such a seller must act reasonably (i.e., nonnegligently) to avoid liability for a defective product.

So far, case law has imposed strict liability on producers of raw materials; producers of component parts; assemblers and subassemblers; packagers of the final product; wholesalers, distributors, and intermediary; people who hold out a product as their own; and retailers. In each instance, however, the plaintiff must establish that the product was defective when it left the defendant's hands. When the defective product has been assembled from component parts supplied by a number of manufacturers, establishing the appropriate defendant is easier said than done. Establishing that the assembled product was defective can be fairly easy, but pinpointing the component that contained the defect can be extremely difficult if all or most parts failed during the event. The same problem exists with other members of the distributive chain. If a product can be tampered with and the defect was introduced after one member of the chain sold the product to another member, that member is not responsible for the defect. Therefore, adding names to the list of people who can be sued for strict liability is not a guarantee of recovery.

The defect does not need to manifest itself immediately. If the defect is a slumbering one that does not become evident until the product is used, but the fact that it was present at the time of manufacture or sale can be established, liability can attach. For example, a metallurgical flaw may not be manifested until the metal fails because of fatigue (crack growth); nevertheless, expert evidence may clearly identify the failure as the result of fatigue and pinpoint the structural deficiency of the metals to the date of the original processing. This would be sufficient to establish that the defect existed when the product left the hands of a defendant, in this instance, conceivably the producer of the base metal or the one who fabricated the component that failed from the metal.

Did the Defect Cause the Harm?

If a plaintiff can establish that the product was defective and that the defect existed when the product left the defendant's hands, the plaintiff must then prove that the defect was at least a contributing cause of the harm suffered. The standard legal formulation of causation is the "but for" test: for example, *But for* the presence of the defect, product failure, or malfunction, would the injury to the plaintiff have occurred? If the plaintiff cannot clearly establish that the defect was both necessary and sufficient to cause the injury, he or she will not prevail.

The "but for" test does not require the defect to be the sole, unequivocal cause of the injury; the defect only needs to be the probable cause. For example, in *Simpson v. Wellington Service Corp.*, the defendants drove their 137-ton rig on a street in violation of the statutory weight limitation. The plaintiff sought damages when pipes in his basement leaked because of a disconnected coupling. The defendant, on appeal, questioned whether the plaintiff's evidence was sufficient to show that the flooding was the result of the defendant's violation of the statute. The court stated that it was reasonable to conclude from the evidence

> that the great pressure in the street had caused a lowering of the connecting pipe in that part of its forty foot length that underlay the rig. An expert was not essential to demonstrate that the effect of downward movement on that part of the rigid pipe that was in the street would be an upward thrust of its other end inside the foundation wall on which it rested. It was also a reasonable layman's conclusion that all rigid pipes on the basement side of the coupling would as a result be under stress and that a fracture at some point in the system was likely.
>
> That the fracture occurred a number of hours after the time when the stress was created in the system did not make the cause conjectural. A delay between the distortion and the break was entirely consistent with initial slight adjustment in the position of the relatively rigid pipes which created tension throughout all of them, and the eventual release of the tension by a break occurring at the weakest point in the system.
>
> In this, jurors were concerned with the operation of established physical principles and circumstances to which they could apply their knowledge and experience. "The plaintiffs were not bound to exclude the operation of other possible causes. It is sufficient if the cause shown is made to appear the probable cause of the accident" (*Restatement* [*Second*] *of Torts,* Section 433, comment f).

Without commenting on the court's structural analysis, the important element to note in establishing causation is that all other possible causes of the failure or harm need not be negated. That the "defect" in question (the violation of the weight limitation in this case) was the probable cause is all that is necessary.

The mere presence of a defect in a product at the time of injury is not enough. The existence of a defect in a product may have had little or no bearing on the incident that caused the injury. The defect may have been dormant, and the cause of the incident may have been human failure and totally unrelated to the product. The facts in such situations are often extremely difficult to ferret out. After the injury, the product often appears to be defective. Thus, in *Barry v. Manglass,* in which motor mounts were found broken after an automobile accident, the following question arose: Did the mounts break before the accident and cause the driver to lose control, or did the driver lose control, causing the accident, and the motor mounts to break during the accident. If the latter occurred, then the fact that the mounts may have been defective was unrelated to the injury.

In other, similar situations, whether the defect played a causal role or became apparent because of the traumatic event, which was more probably caused by another factor, is often impossible to determine. That is why experts in the courtroom often clash in their assessment of what actually occurred.

Can the Harm Be Appropriately Assigned to the Defect?

The answer to this question has perhaps engendered more concern among manufacturers than anything else. Simply put, even if the product contains a defect, even if the defect was in the product when it left the manufacturer's hands, and even if the defect caused the harm, the manufacturer will argue that the plaintiff's actions were so bizarre or the risk was so remote (the plaintiff's actions could not reasonably be foreseen) that no liability should attach.

The phrase used by the courts to describe this policy question is *proximate cause.* Although the alleged defect may have played a role in the harm suffered by the plaintiff, the defect was not the proximate cause of the injury. In other words, the harm cannot be fairly assigned to the defect; therefore, imposing liability on the manufacturer is unfair when the injury resulted from the plaintiff's actions, which were *legally* unforeseeable by the manufacturer.

The issue here usually arises in the context of an allegation that the plaintiff misused the product. In *Ritter v. Narragansett Electric Company,* the product was a free-standing 30-inch kitchen stove. The plaintiff, a 4-year-old girl, was injured when she used the oven door as a step stool to look into a pot on top of the stove. The stove tipped forward, seriously injuring her. The expert for the plaintiff concluded that design of the stove was defective because the open door could not support 30 pounds without tipping over.

The issue, however, is *not* whether the design was defective. If the stove had tipped because the door was used as a shelf for a heavy roasting pan, the design would properly be the issue, because it could be argued that an intended use of the door was as a shelf for checking food during cooking, and hence could be foreseen by the manufacturer. The manufacturer argued, however, that using the door as a step stool was so unforeseeable (i.e., so remote a circumstance) that liability for the child's injury should not be assigned to the manufacturer.

Alternatively, a manufacturer can argue that the plaintiff misused the product. In an analogous case, *Magic Chef v. Sibley,* the Texas Court of Civil appeals said:

> It has been held by our Supreme Court that foreseeability is significant in product liability cases when the product is what it is intended and known to be, but injury is suffered because the product is misused . . . . (Citations omitted) A product is not "misused" merely because the manufacturer intended that it be used in a different manner; the manufacturer must show that the use which caused the injury was not *reasonably foreseeable.* (Italics added)

The question of misuse obviously centers on the plaintiff's behavior in using the product and whether this behavior was so aberrant as to bar recovery. The clear trend of the cases, however, is that the defendant must design a product to accommodate reasonably foreseeable misuse. Such use, though obviously not intended by the manufacturer, generally results from predictable, and thus reasonably foreseeable, behavior. Only outrageous misuse of a product will bar a plaintiff from recovery. What distinguishes outrageous or unforeseeable misuse from foreseeable misuse, for which the manufacturer will be liable, cannot be articulated because the courts consider each situation individually. Thus, the manufacturer is responsible for considering and designing for reasonably foreseeable misuse of the product, not only with respect to the features of the product but in communicating with the user as well.

THE ROLE OF THE USER'S BEHAVIOR

The preceding sections have focused primarily on the defendant's duty or responsibility for the design of a product. As discussed, if the plaintiff's behavior is categorized as outrageous misuse, the defendant will not be liable despite the existence of a defect. But two other forms of behavior, although perhaps not outrageous, can either bar recovery completely or reduce the defendant's liability. These are known as *contributory negligence* and *assumption of risk,* and are termed affirmative defenses that the manufacturer can raise in addition to or in place of the proximate cause issue.

The defendant alleged to be negligent can assert and must prove contributory negligence on the plaintiff's part. A defendant can only make such an allegation if a plaintiff alleges that the defendant is negligent. A defendant cannot assert contributory negligence when the plaintiff has based the case on the theory of either strict liability or implied or express warranty.

Recall that the test for a defendant's negligence is whether that person or company behaved reasonably with respect to the product. For contributory negligence, the same question is asked about the plaintiff's behavior with respect to use of the product. In many jurisdictions, if the defendant can establish that the plaintiff failed to act like a reasonable person would under the same or similar circumstances and that the plaintiff's behavior contributed to the injury, then the plaintiff is precluded from recovery, regardless of the extent of the defendant's negligence.

In an increasing number of jurisdictions, however, the courts have recognized the harshness of the all-or-nothing doctrine of contributory negligence. Instead, the emerging concept is one of *comparative negligence*: that is, the jury is asked to assign percentages to the relative contributions of the plaintiff and the defendant to the cause of the injury. In what is termed pure comparative negligence, the plaintiff can recover a portion of the damages even if the jury finds that the plaintiff was responsible for more than 50 percent of the damage. For example, if the plaintiff's fault is assessed at 70 percent, the plaintiff can recover 30 percent of the total damages awarded by the jury. In nonpure comparative negligence, plaintiffs can recover only if their fault is assessed at 50 percent or less. Again, the amount received is the total award multiplied by the defendant's percentage of fault.

Under the theories of strict liability and warranty, the affirmative defense characterizing the plaintiff's behavior is called assumption of the risk. If a plaintiff *voluntarily* and *unreasonably* assumes a *known* risk and that risk caused the injury, the plaintiff is either barred from recovery altogether or, as in comparative negligence, the jury is asked to assign percentages measuring each party's contribution to the cause of the incident.

The fundamental distinction between contributory negligence and assumption of the risk is the standard used to determine the existence of such behavior. In contributory negligence, the plaintiff's behavior is measured against an objective or community standard—what a reasonable person would have done under the same or similar circumstances. In contrast, assumption of the risk is measured by a subjective standard on the basis of what the plaintiff *actually knew* an instant before the incident.

In addition, assumption of the risk is governed by a three-part test. The plaintiff must be shown to have actually *known* about the risk before the injury and *voluntarily* and *unreasonably* encountered it. The defendant must prove all three elements to establish this defense.

In *Vernon v. Lake Motors,* Vernon's car was 8½ months old. One rainy morning, Vernon noticed that smoke was coming from the windshield area and that the windshield wipers would not shut off. The local dealership refused to repair the car under the warranty because Vernon had not bought it there. Later that morning, Mrs. Vernon decided to drive some 40 miles to Salt Lake City to take the car to the dealer, Lake Motors, who sold the car to the Vernons and to attend her granddaughter's dance recital. After checking the wipers again, she decided that the smoke was not too objectionable; the worst that could happen was that the fuse would blow, leaving her without wipers. About 30 miles from home, a fire started under the dashboard and completely destroyed the car. Lake Motors became one of the defendants.

Obviously, Mrs. Vernon *voluntarily* chose to drive the car. But whether her action was *reasonable* must be judged in context. If her sole reason for going to Salt Lake City had been to attend her granddaughter's recital, a jury might conclude that her action was unreasonable. The fact that the local dealer refused to make repairs would, however, argue against her act being unreasonable.

The third element, what Mrs. Vernon actually knew, required her to be aware that driving with smoking windshield wipers would result in a fire. According to her testimony, the risk she assumed was that of a blown fuse, which would leave her without windshield wipers, not the risk of destroying the car. Thus, the defense failed, even though her action could be characterized as voluntary and unreasonable. If the Vernons had filed the complaint as an action of negligence, one could argue that Mrs. Vernon's action was contributorily negligent, which would completely or partially defeat her claim for damages. Under strict liability or implied warranty, however, her action would be unlikely to be construed as assumption of the risk, thus permitting full recovery of damages. Additional illustrations of contributory negligence in the context of the six causes of failure will be discussed in Chapters 5 and 11.

THE ROLE OF THE EXPERT WITNESS

The role of the expert witness is pivotal in any attempt to prove or refute an allegation of defect. Earlier, we said that the plaintiff must prove, and the defendant must refute, the three elements of litigation: (1) the product was defective, (2) the defect was present when the product left the manufacturer's hands, and (3) the injury can be appropriately assigned to the defect. Each element has either a technological basis or requires the perceptions of a technologist who is familiar with the real world of how products function, how they behave, and how they are used. In addition, the technologist must be familiar with good design, production, and maintenance practices. In most cases, competent evidence can be introduced at a trial regarding these factors only through the testimony of a *qualified* technical expert.

But who is qualified? (This issue will be discussed in more detail in Chapters 2 and 12.) For example, if the issues of defect and causation hinge on whether a particular part failed because of fatigue (and it is highly probable that the failure precipitated the injury-producing event) or because of a single-load impact (in which case, the defendant will argue that the failure occurred as a result of the event that was induced by some other cause), only the testimony of an engineer or scientist who has specialized in materials, mechanics, or metallurgy is likely to meet the burden of proof.

The expert's role, however, should begin long before the trial. From the plaintiff's perspective, whether a product liability lawsuit is warranted is a decision that counsel can make only after the expert has collected and examined the appropriate evidence and understands the circumstances surrounding the event (Chapter 6). Once the lawsuit has been filed, counsel and the expert should plan the defense together. In most instances, the surface appearance of the tangible evidence does not speak unambiguously to the attorney. Even if the evidence is not ambiguous, only the expert can properly assess whether more detailed examination and testing are required to establish such issues as defect and causation with reasonable engineering or scientific certainty (Chapters 7–11).

The courts do not require experts to base their conclusion and opinions on an absolute sense of precision; they require those conclusions and opinions to be based on *reasonable certainty*. The expert's investigation, however, must be sufficiently thorough to permit that level of certainty in the testimony. The legal system is usually unaware that the final design of a product is the result of compromises among a series of complex tradeoffs—that the gap between the quantifiable aspects of a design and the uncertainties of the actual performance of a product in the real environment of use is closed by a safety factor, which focuses precisely on the inherent uncertainties in any design.

When a product is challenged as defective and unreasonably dangerous, the litigation often focuses on the band of uncertainty between the quantifiable parameters of the design and the actual performance of the product in the hands of users. It is unreasonable to expect experts to bring more certainty to their conclusions about defect and causation than existed in the original engineering of the product. To achieve substantive credibility, they must be able to communicate the underlying methodology of design to the court and the jury.

Expert witnesses must not only have a basic understanding of the product design process and the actual environment in which the product functions and is used, they also must understand the law well enough to make judgments about technical issues that reflect the proper perspective for advancing or refuting an opinion that a defect was unreasonably dangerous.

Although the technical expert may easily be able to trace the failure or malfunction of a product to either manufacture or design, this alone is usually not enough to establish defect and causation. Experts on both sides must weigh the general societal values reflected in considerations of risk and utility inherent in the determination of unreasonable danger according to the indicia set forth by Wade in 1965 as shown in Box 1-1.

If a production defect is alleged, the expert must (1) identify the flaw or flaws in relation to manufacturing or physical-property standards, (2) provide evidence that the failure or malfunction of the product is directly attributable to the flaw, (3) establish a relationship between the failure or malfunction and the expected standards of performance for the product, and (4) establish the causal link between the failure or malfunction and the injury.

Similarly, a case involving a design defect would require the experts to (1) identify the design flaw or flaws that led to the injury, (2) propose alternative design features that would have reduced the danger, (3) evaluate these alternatives in relation to the expected standards of performance for the product as well as evaluate the effect of the alternatives on the subsequent usefulness and cost of the product, (4) compare the product with other similar products, and (5) establish a causal link between the design deficiency and the injury.

The breadth of the expert's inquiries delineated by these issues implies a more expanded role than the role the attorney usually perceives. An expert who is called solely to testify that a proposed design alternative would or would not have prevented the injury is *not* fulfilling the expert's critical role. He or she must be capable of synthesizing all the diverse elements necessary to reach the ultimate legal conclusions (see Chapter 6).

The expert's function is to evaluate the close case and bring an *evaluative sense* to a difficult situation. To decide the liability issue, the jury should then choose between the competing evaluations of experts.

Thus, addressing the real probabilities and offering a sensitive evaluation of competing possibilities, rather than formulating unrealistic positions, is all the more crucial.

Juries are well equipped, as triers of fact, to evaluate human behavior when judging the validity of testimony. When questions center on the technological behavior of a product, however, juries must rely on the expert's interpretation of the evidence. This interpretation must be comprehensive enough to reveal how the experts reached their conclusions and must demonstrate their perceptions of the probable certainties and uncertainties. Without careful expositions of the process by which experts reach their conclusions, the evaluations of juries (and judges) will inevitably be superficial. Furthermore, neither juries nor judges will be able to probe the substantive questions that must be addressed when evaluating an unreasonably dangerous defect or establishing causation.

If an expert does not bring this comprehensive analysis to the litigation process or the court prevents its presentation, the legal system will be mired in an imperfect understanding of the design process and will have little hope of reaching conclusions regarding the defects that reflect the realities of the interaction among products, people, and the instrumentalities that influence them.

CONCLUSION

The law, through its principles, provides a matrix for resolving disputes. Of necessity, these principles can only articulate qualitative bench marks against which to measure and test factual disputes. The relative rigor with which engineers ply their craft often appears to be incapable of being influenced or judged by the nontechnically oriented legal system. The fact that the engineering community generally fails to define for the courts which engineers are qualified to assume responsibility for certifying the safety of products complicates both the legal and the technical process. Yet, we require an independent lay review of even complex technological issues.

Thus, the engineer's role is twofold. When a product is put on trial after an injury has occurred, the engineer, as an expert, must approach the question of whether a defect exists and the related legal issues as honestly and comprehensively as possible. More important, however, the engineer must prospectively incorporate the same sensitive balance between risk and utility when he or she puts pencil to paper and begins the design process (Weinstein et al. 1978).

2

High-quality Engineering Design: An Elusive Goal

Everyone wants to improve the quality of engineering design. High-quality design results in a product that functions in accordance with the user's reasonable expectations. Such a product is durable and serviceable commensurate with its price, it does not expose users or others to an unreasonable risk of harm, and it can be sold at a competitive price.

This chapter deals with legal controls over design at two levels: controls over those who create designs (e.g., claims against architects or engineers) and legal controls over the designs (e.g., claims against manufacturers of defective products). Although personal consumer products will be the main focus here, the observations will apply equally well to products such as commercial buildings, aeronautical vehicles, industrial equipment, and institutional consumer products.

Bear in mind that although engineering and designing a product are not usually synonymous, the emphasis here is a symbiosis of engineering and design, or engineering design.

The focus of training in engineering is the study of scientific principles (e.g., physics, chemistry, biology, mathematics) for purposes of application to ensure a system, structure, or product performs as efficiently and safely as intended by the design. The four basic engineering disciplines (electrical, mechanical, civil, and chemical), other disciplines such as metallurgical-materials engineering, and, more recently, biomechanical engineering evolved as the demand grew for more specialized applications of pure science.

Some engineering programs include departments or subdepartments specializing in the technical industry or in occupationally oriented subjects such as maritime, aeronautical, automotive, nuclear, agricultural, and industrial-manufacturing engineering that encourage students to specialize in an area of interest to a specific industry (which usually motivates such specialization with corporate grants).

In Europe and the Americas, however, graduate schools have traditionally encouraged students to specialize in an area covered in their undergraduate program: for example, fluid dynamics, heat transfer, soils, structural mechanics, microcircuitry, galvanic corrosion, or kinematics. In addition, because technology and its applications are changing rapidly, some large technical corporations encourage their employees to obtain baccalaureate and master's degrees in basic engineering disciplines.

The focus of training in design is the development of new functional or aesthetic concepts from existing technology. Typically, baccalaureate engineering programs teach design by illustrating the source or application of the scientific and engineering principles learned during classes and laboratory work. An undergraduate engineering program may contain some formal design courses, such as mechanical engineering design, based on Shigley's textbook (1977), which is used in the typical baccalaureate program in mechanical engineering. The focus of training in industrial design and architecture is on aesthetic concepts and on functional use by or interaction with individuals (human factors, or ergonomics).

For many products, aesthetics and functionalism are as important as is engineering for efficiency and safety. In some instances, only the engineering aspects of the product are regarded as important, whereas aesthetics and functionalism are viewed as having only marginal importance. A balance of all four characteristics should be the design engineer's objective when developing a product that will meet the needs of the manufacturer and user. The failure to achieve this balance can lead to product failure, injury, and economic loss.

Stairs and ladders, two products that are frequent sources of injury, illustrate these four aspects of engineering design. Stairs are usually designed, first, to be aesthetic and functional within the context of the style and space of a building. On the other hand, little attention is usually paid to the aesthetics of ladders. The function of both is to permit ascent and descent, but in a different manner. The human-factors aspect of functionalism or ergonomics related to stairs and ladders has been the object of study for centuries (see, for example, Archea 1983; Archea, Collins, and Stahal 1979; Carson et al. 1978; Rosen 1983). These studies indicate that human anatomy, human perception, and psychology require stairs and ladders to have treads of a certain height, width, slope, and so forth. These criteria translate into design criteria—principally, functional performance, rather than engineering-structural performance—that are imposed on the designer through codes and standards such as those in the Uniform Building Code or the Southern Building Code or those developed by the Building Officials and Code Administration or the American National Standards Institute.

The engineering of the stairs and ladders requires them to be strong and stable enough to perform safely under prescribed loads over a defined number of years (or subjected to inspection and maintenance over a prescribed interval with recertification). Obviously, the strongest stairs or ladders, designed without regard to human factors, are not necessarily the safest. The design of stairs may represent considerably more give and take between aesthetics, function, safety, and efficiency than does the design of a ladder, which requires greater attention to ergonomics and structural performance.

Structural performance sets a limit on the maximum stress allowed in the stair system when subjected to the maximum specified load that the side rails and individual treads will be subjected to over a specified period of time. For performance, the engineer must consider the strength and other characteristics of the materials to be used. Because the span and thickness of the treads influence levels of stress, the engineer sets about determining the necessary dimensions such as thickness.

Two causes of failure are generally associated with engineering design: faulty conceptualization and selection of inappropriate materials. Two causes of failure also are usually associated with the manufacture of a product: defective materials or defective methods of fabrication. Failure of a product during use, too, usually has two causes: improper operation and inadequate maintenance and inspection procedures. As discussed in Chapters 5 and 11, however, one organization may bear the responsibility for all six possible causes of failure.

INSTRUMENTALITIES THAT AFFECT THE QUALITY OF DESIGNERS

Box 2-1 lists the instrumentalities that can affect the engineer–designer, the manufacturer, the user, and, ultimately, the quality of engineering design. Because public and private accreditors, professional registration agencies, and continuing education act only along the margins, we will put these instrumentalities aside.

BOX 2-1. Instrumentalities that Affect the Quality of Products

A. Primary Influencers of New, Cost-effective, and Safe Products
Engineer/designer (service)
Manufacturer (product)
User

B. Causes of (Responsibility for) Product Failure
Defective design
Selection of improper materials
Defective materials
Improper fabrication or processing
Improper operation
Improper or no maintenance and inspection

C. Instrumentalities that Influence the Engineer/Designer, Manufacturer, and User Directly or Indirectly
Training
Government regulations
Technical associations and societies
Insurers
Litigation
Marketplace

C.1 Instrumentalities that Affect the Quality of Engineering Designers
Universities, colleges, professional schools
Professional certification or registration
Professional and technical societies
Liability insurance
Litigation regarding defective design
Professional services to manufacturers, users, and all of the above

C.2 Instrumentalities that Affect the Quality of Manufacturing & Fabrication
Training programs at all levels of the company
Government regulations and enforcement
Industry associations that fund research and development and develop voluntary codes and standards
Liability insurance
Litigation regarding defective design and processing
Rejection of inferior products by users

C.3 User Instrumentalities that Influence Design, Manufacture, and Maintenance
Awareness of experiences with products
Gathering of information
Requiring voluntary and statutory codes and standards
Membership in user groups and associations that gather information, fund research, and recommend the development of standards and development
Insurance coverage for injury and damage
Initiation of law suits for compensation regarding property damage and personal injury
Refusal to buy inferior products

Schools of Engineering and Design

Engineering schools provide engineering designers with their initial education—and often, their only formal education. These schools influence, if not determine, who will become an engineering designer, and they decide, within the realities thrust upon them by the market, who and how many students will be enrolled, how many students will be trained, which ones will receive the school's stamp of approval, and what degrees will be offered (e.g., associate, technology, baccalaureate, master's, or doctorate).

The Accreditation Board for Engineering and Technology (ABET) (1987a, 1987b) approves programs offered by engineering schools. The board's members are representatives of approximately 24 technical and professional engineering societies (ABET 1987c), including the National Council of Engineering Examiners (NCEE). Most engineering societies regard themselves as representatives of both individuals and industry, some represent only one industry, and others represent only technology (see Pletta 1984 for a historical overview of engineering societies).

The ABET-approved engineering programs leading to bachelor's and master's degrees consist of the four basic branches of engineering—mechanical, civil, electrical, and chemical, referred to as Group I disciplines—in addition to a number of specialized programs of study. ABET also approves associate and baccalaureate programs in engineering technology, which is defined as "in the occupational spectrum between the craftsman and the engineer" (ABET 1987c).

The members of the NCEE are representatives from state boards of registration for engineering practice. The organization provides testing in the Group I disciplines and in 10 other specialized disciplines referred to as Group II (ABET 1987c).

Generally speaking, no strong formal relationship exists between the ABET and the NCEE. But their relationship, as it does exist, has both advantages and disadvantages, as will be described below. Those who administer the registration laws *do,* however, influence educational institutions by usually allowing only graduates with baccalaureate degrees from approved schools to seek registration and by selecting the methods that determine who will be registered. In addition, because rigorous standards may discourage prospective students, registration policies can affect school enrollment. Finally, the content of the registration examination undoubtedly influences academic curricula. But their influence, though arguably more than marginal, is small compared to that of the schools.

Schools, through their power to grant admission, can significantly influence the quality and characteristics of designers, especially when the demand for designing positions is greater than the number of applicants to fill those positions. For example, only 14 percent of engineers in the United States obtain a master's degree, and only 2 percent obtain a doctoral degree. Also, of all the master's degrees earned in engineering, 60 percent are earned by foreign-born students, and of all the doctorates earned, 55 percent are earned by foreign-born students, which further illustrates the problem. But do the schools have the tools to select applicants who are most likely to become good designers, engineers, and engineering designers?

Intelligent people can and do debate the difficult question of what constitutes high-quality design. But predicting who will be a good designer is even more difficult. Of course, all engineers should be able to read, communicate, reason analytically, understand engineering principles, and be able to apply those principles. But how does one measure qualities such as the drive to excel, the passion to advance the state of the art, the willingness to test and experiment, and the common sense as well as scientific approach required to investigate the human factors and safety aspects of their engineering and design? And what about perseverance?

The selection process rarely uncovers these qualities or identifies their absence. In other words, the tools for predicting who will create high-quality designs are extremely crude. Imperfect though it is, however, the educational selection process could accomplish some sifting. On the assumption, thin though it may be, that those who select applicants can predict who will be a good designer, the quality of design could be improved by accepting fewer students. But educational and financial pressures minimize the likelihood that schools will reduce their enrollment to accomplish this sifting process. In any event, some argue that we need more, not fewer, engineers.

What could the curriculum of the engineering college or school do to aid the sifting process? Some engineering professionals argue for the development of professional engineering schools, similar to law and medical schools, that would offer three- or four-year programs beyond the general college or baccalaureate program. Such schools would enlist the participation of professional practitioners who are on the cutting edge of technical operations and would work in concert with theoretically oriented researchers and professional educators. The purpose of such schools would be to provide a more comprehensive educational program for the engineering designer than the baccalaureate program, which is considered too cursory in an era when even the fundamentals of science have expanded tremendously (Pletta 1984; S. J. Brown 1986b).

Another thought is that a more comprehensive educational program would better clarify the qualifications of an engineering designer who can certify the design and engineering of a product. (In fact, because of the impact of technology, other professionals such as lawyers, physicians, architects, and business professionals are earning baccalaureate degrees in engineering.) Still others argue that one can hardly expect professional education to end with a baccalaureate degree (S. J. Brown 1986b), and that the introduction of two-year associate degrees and four-year engineering technology degrees into college curricula confuse manufacturers, consumers, and the legal system regarding who is a qualified engineer. (Compare the relatively clear-cut progression of a medical or legal education with the graphically confusing progression of an engineering education illustrated in Fig. 2-1.)

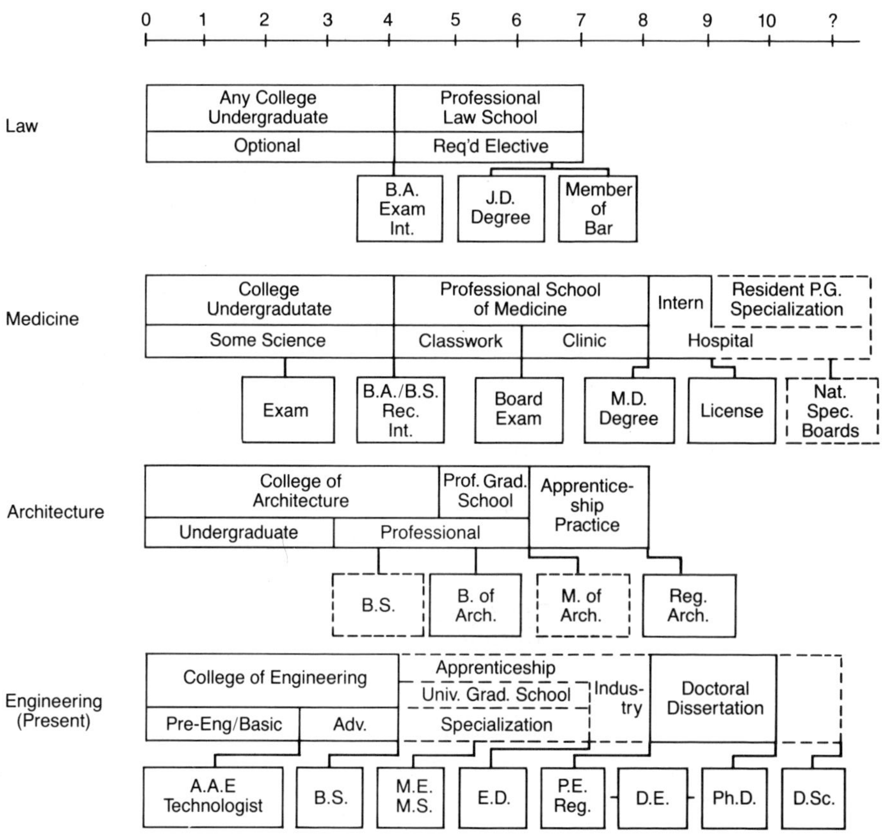

Figure 2-1. A comparison of professional practice-oriented education in law, medicine, architecture, and engineering. Solid lines = required education, dotted lines = optional education. (Accr. = accreditation, D.E. = doctor of engineering, Int. = internship, P.E. = professional engineer, Rec. = registration, and * = degree from an existing four- or five-year institution.)

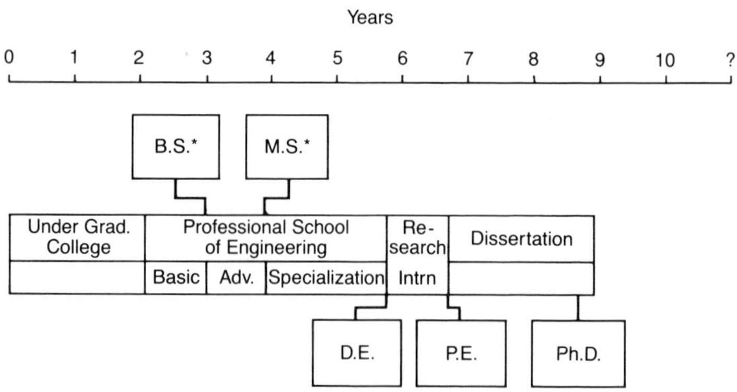

Figure 2-2. Proposed engineering school vs. graduate school training. (D.E. = doctor of engineering, P.E. = professional engineer, and * = degree from an existing four- or five-year institution.)

Chapter 12 discusses the roles that the technologist, engineer, and technobureaucrat play in industry, academia, and private practice. For readers who are unfamiliar with the engineering profession, we will show that the profession is similar to the more familiar medical profession, which has its paramedics, medical technologists, physicians, medical researchers, and nonmedical bureaucratic administrators and product developers. In general, however, engineers and their employers have blurred the distinction between who is and who is not a professional practitioner. For example, in a case in the Akron, Ohio, Common Pleas Court involving an improperly designed ladder foot, the manufacturer's designated "design engineer" had no formal training in engineering and had "borrowed" features from other existing designs, then modified them for production and economic purposes.

As shown in Figure 2-2, engineering programs could have a significant impact on improving engineering and design capabilities. As an extension of the present educational system, however, such programs would be geared toward influencing qualifications with respect to certification and being charged with responsibility "responsible charge."

Can the educational process itself improve the quality of engineering design? From an attorney's (or engineer's) viewpoint, one can make the following observations:

1. *Competent designers need many pragmatic skills.* Designers must appreciate and be able to predict and control costs. They must be able to forecast what users will need and want. They need to know what materials are best and what equipment is available.

And they need to keep up with the state of the art. They also need to have people-related skills, and to understand the legal requirements that affect their designs. Although some schools may be better at teaching these skills than others, most schools cannot or may not want to teach them in a baccalaureate program.

2. *Design skills and judgment are mastered best on the job.* Design skills improve and mature with trial and error, diverse industrial experience, experimentation, and continuous study. That is, skills are mastered best on the job. Because a baccalaureate program is a brief educational experience, the schools must leave the cutting edge to the workplace, to continuing education programs, and, possibly in the future, to professional schools.
3. *Design skills should be built upon training in scientific basics.* In special circumstances, experience can be a substitute for formal training in design at school. The trend in modern society, however, has been away from apprenticeship for professional training to institutional training in the basic theories and fundamentals, followed by a specified period of internship. Although the quality of an apprenticeship or internship is difficult to measure, such training is regarded as part of a professional's qualifications for registration (see Appendix A).
4. *Training in the scientific fundamentals is complemented not replaced by experience.* Do years of employment with a manufacturer qualify employees who lack formal training as engineers? Do these employees understand ergonomics, the fundamental principles of engineering design, modes of product failure, and methods of predicting the performance of a product? If the answer is yes, then how should these individuals be certified? Should they be certified at all? Is their expertise based on an understanding of and experience with the instrumentalities that influence their industry or occupation? (Does employment with a manufacturer of medical equipment or pharmaceuticals qualify a person as a physician?) As S. J. Brown (1986b, p. 1491) pointed out, "the fact is that many corporations award titles to their 'engineering' employees based upon internal performance without regard to education; whereas, others have specific guidelines."

Schools can, in a modest way, do some sifting. They can also encourage and subsidize important research and scholarship. But they are currently not a major force for improving the quality of engineering designers. The search for ways to improve designers must continue.

Registration

Registration laws should protect the public from incompetent engineering designers. Note the difference between getting rid of incompetent designers and developing truly great designers. Education and registration may play a limited role in getting rid of incompetent designers but virtually no role in developing truly great ones. The fact that a person holds a professional license *should* assure the public that the person has reached and, one hopes, maintained a minimal level of competence.

Registration laws do not certify "product" or "occupational" engineering such as a "warnings," "ladder," or "human factors," engineer. In a June 1988 issue of the *Louisiana Engineer*, the executive secretary for the State Board of Engineers (which recognizes the four basic branches of engineering in addition to five subbranches: metallurgy, mining, petroleum, industrial, and agriculture) stated in an inquiry that there is no board rule or law that would prohibit engineers from calling themselves an "XYZ" engineer. The State Board of Louisiana under Rule 911 states "that the board will not add branches to correspond to job titles or job functions such as 'XYZ' engineer." In other words, the board has determined that they will not provide a professional engineering registration for a branch for XYZ engineer.

Thus, in calling oneself an XYZ engineer, the individual (or his or her employer) assumes all responsibility for the qualifications to practice in the XYZ discipline. The response goes on to cite the rule governing registered engineers practicing in areas for which they are qualified.

But do the requirements for registration mimic the attributes of a competent designer? Applicants for registration must achieve a certain level of education, have a designated amount of professional training in the field, and pass an examination (Constance 1988; Hicks, Tyler, and Mueller 1986; also refer to the examination specifications listed in Appendix A). Examinations can crudely measure some attributes, such as mastery of scientific and engineering principles, but they cannot measure what makes a good designer. Passing grades can be and are drawn arbitrarily, often simply to limit the number of people registered.

Although the following statement may be heretical, the principal effects of the registration laws probably have been to increase the cost of being a professional and to limit the number of people who are registered. Because, historically, the purpose of the engineering registration laws was to require formal training in engineering (a baccalaureate degree) in addition to internship and an examination, the cost of engineering services has increased without much demonstrable improvement in quality. Why?

Although the ambit of registration laws seems to expand continually, many designers who are employed by manufacturers to design products *need not be registered*! The law (as well as voluntary and statutory codes and standards in most instances) does not require professional registration for engineers who design a product to conform with codes and standards. Nor is registration required to participate in developing the codes and standards, which are usually written by industry representatives rather than by unaffiliated professional engineers. Although manufacturers should unquestionably participate in writing codes and standards, why should designers, users, and other instrumentalities that affect good design be excluded?

Furthermore, should manufacturers and user-representatives be registered professional engineering designers? By not requiring a qualified designer to certify the drawings and specifications of a product before production, is the registration system permitting the manufacturer to make the ultimate decision about the sufficiency of the balance between adequate engineering design and production tradeoffs?

According to a study by the National Science Foundation (1987), 81 percent of the approximately 2.2 million engineers in the United States who have a baccalaureate or higher degree work in industry. The remaining 19 percent are employed either as academics or consultants. According to the U.S. Department of Commerce (1986), all 540,000 practitioners in law, 150,000 practitioners of dentistry, and all 530,000 practitioners of medicine are registered. The 350,000 registered engineers, on the other hand, represent only 21 percent of all graduates of ABET-approved schools of engineering.

Examples of product failures caused by design deficiencies appear throughout this book. Many of them were the result of a manufacturer's employee, designated as an engineering designer, who had no formal training in engineering, was unable to anticipate that the product would be subjected to fundamental loading conditions, or was unable to apply fundamental scientific principles properly. In many instances, the manufacturer did not have the product certified by a registered engineer; in some instances, an engineering evaluation was not performed.

In an ideal world free of economic struggles between professionals and politics, the registration process might be helpful. As it currently exists, however, it is not a major force for producing better designers. In addition to the weaknesses already described, the people who operate the registration system lack the resources to protect the public and the will to resist political pressure. The search for ways of improving design must continue.

Professional and Technical Societies

What about the influence of technical societies, such as:

American Society of Mechanical Engineers (ASME)
American Institute of Chemical Engineers (AIChE)
American Society of Civil Engineers (ASCE)
Institute of Electrical and Electronics Engineers (IEEE)
American Society of Metals (ASM)
Society of Engineering Mechanics (SEM)

and professional societies, such as the National Society of Professional Engineers (NSPE), and their counterparts in the architectural (e.g., American Institute of Architects) and allied industrial occupations (e.g., American Society of Bakery Engineers, The Fertilizer Institute, and American Welding Society)?

Technical and professional societies do not control who can practice; the registration laws control who can practice. Although they *do* influence the quality of designers through their professional programs and publications, their function goes beyond this obvious but by no means insignificant role. These societies—especially the technical societies—are clearing houses for informal information: informal because their power to suspend or discipline their members exerts little influence on the quality of designers, mostly because they rarely suspend or discipline a member. Pletta (1985a) discussed the role that some societies have played in addressing the issue of the engineer's ethical responsibility to report unsafe practices and methods of protecting the engineer from reprisals, particularly when the offending organization is the engineer's employer. According to Pletta (1985a) and Boisjoly (1987), engineers have an ethical responsibility to try to work first through established channels. Only if this fails should they seek recourse through appropriate regulatory or legal agencies.

Technical societies usually do not require their members to have an engineering degree or professional registration. Also, the law has placed significant limits on the power of professional societies to determine who will be admitted for membership or to establish the grounds for discipline, particularly if a member has engaged in "ungentlemanly" conduct such as competing with a fellow member on the basis of price (*U.S. v. American Society of Civil Engineers* 1977).[1] But independent consulting

---

[1]This case upheld the contention of the United States that the society violated its content decree when it suspended members who supplanted another member. The suspension was related to offering a price for certain work.

engineers are often hired because of their reputation within technical or professional societies (the original function of which was to provide a social milieu for members). Information is exchanged and reputations are developed by social and professional contacts made in the societies. Despite recommendations that are occasionally based on nonprofessional criteria such as school ties or friendship, professional reputation is the most important factor.

The societies can screen and help improve the quality of designers. However, despite the undoubtedly valuable role that they play in screening designers (not all designers are members of such groups, either by choice or because they do not fit in), the search for improving design must continue.

Insurers

Clients often demand that the professional have liability coverage. Therefore, liability insurance carriers *could* sift the qualified from the unqualified, at least among independent engineers, but they do not.

Policies are issued and premiums are charged largely on the basis of the volume of business, the type of services, the principal locality of professional practice, and, perhaps most important, the insurer-carrier's underwriting capacity. Although the terms under which services are performed, the claims made, and the design professional's education and training undoubtedly play a role in the carrier's decision to issue a policy, these considerations are not influential enough to allow the conclusion that insurance carriers play a significant role in sifting engineers.

Insurance carriers rarely have the resources to check professional qualifications carefully. Furthermore, they do not want to risk a lawsuit by canceling an existing policy (see *Spindle v. Travelers Insurance Co.* 1977). Unfortunately, modern society also is claims-dominated. One cannot assume that a professional who is sued, successfully or otherwise, is incompetent. Finally, any carrier who is willing to brave all these obstacles may well find that a competitor will insure the professional whose coverage was refused or canceled.

The Litigation Process

What role does the judicial system play in sifting designers? Together with or apart from professional insurers, the system may be in a position to screen designers. In this instance, courts are *not* viewed in their role of implementing public law standards, such as denying unlicensed designers the use of courts to collect for their services (Sweet 1989), or implementing design controls specified by public law in civil litigation. Civil litigation also sets design standards when it measures the performance of products of those against whom a claim has been made, such as those who design or manufacture products. Both topics will be discussed later in this chapter; here, the issue is whether civil litigation can be viewed as a vehicle for removing incompetent designers.

Suppose designers who have been sued, even successfully, in civil litigation can be found. Suppose a number of them are no longer designing. Some or all of them may have withdrawn because of inefficient business practices, fatigue, or boredom. But suppose liability claims forced them out. Should they have been forced out? Does civil litigation eliminate incompetent designers? Although data on this subject are impossible to obtain, the system is unlikely to work that way. What was said about insurance carriers is relevant in this context. When consulting engineers are denied insurance or when insurance is prohibitively expensive, do they continue to work without it or do they become employees? As will be discussed in the next section, market imperfections make it likely that clients will retain these engineers despite a malpractice claim against them, even a successful one.

A manufacturer may drop a product line or be forced out of business because it cannot get insurance after a series of claims or because it cannot make its product competitive. That does not necessarily mean, however, that the product *should* be removed from the market. In any event, the civil litigation process is unlikely to remove incompetent designers.

The Marketplace

What about the influence of the marketplace—in this country, a relatively free one? The market can reward the good players and ruthlessly eliminate the bad ones. Does it operate that way with designers?

A market requires information. Comparing professional services is not like comparing mass-produced pencils. When choosing a designer, clients have difficulty measuring and comparing the quality of the designers available.

A knowledgeable client can evaluate prospective designers by looking at their completed products or by contacting their former clients. Prospective clients who have the resources and the time can gather relevant information, make intelligent comparisons, and select a designer based on predictions about the quality of the designs he or she will produce. Note, however, how this changes if the client is selecting a designer for the normal design *and* site services incident to construction, such as reviewing shop drawings, giving interpretations, issuing certificates, and resolving disputes. Many other factors now enter the evaluation process.

Even in the best of circumstances, however, can a prospective client assume that a designer's previous clients are capable of judging the competence of designers in relation to the quality of their designs, their ability to deal with costs, and, in the case of architects and civil engineers, their abilities as administrators and as monitors of performance? Many clients cannot do this, and even if they can, the search costs usually make a market approach to engaging independent designers ineffective. Another problem is that clients often require the consulting engineer or firm to sign a confidentiality agreement regarding information about the design of a product or process.

Excluding highly specialized and skilled clients such as the aerospace, nuclear power, and automobile industries, market attributes make the market an unlikely vehicle for sifting qualified designers from the unqualified ones. Yet, despite all of its imperfections, the market at least operates in the real world where quality is best judged. Thus, enlisting it in the quest to improve design is important.

Another question is: If products fail in the market, as they do, can the market eliminate the people who designed those products if the product failure can be traced to poor design? If engineering designers lose their jobs, does that mean that they will no longer design? No. Because of deficient avenues of information and prohibitive search costs, incompetent engineering designers are rarely eliminated from the market. In addition, designers often do not have final approval over the form of the design that goes into the product, fabrication, or structure—especially if they are employees of the manufacturer, fabricator, or builder.

Sadly, although the instrumentalities mentioned in this section can act as levers for improving the quality of designers, their effect is marginal at best. With that depressing conclusion behind us, what is the influence of controls on the design itself? What instrumentalities affect the manufacturer of a product: that is, what controls directly affect design?

INSTRUMENTALITIES THAT AFFECT THE MANUFACTURER

Historical Background

The period from the beginning of industrialization until roughly the end of World War II was one of minimal legal control over design. Little direct state or federal regulation over products existed, with the exception of food and drugs. Even the passage of the Pure Food Act and the

Pure Drug Act in 1906 did not result in massive legal interventions into design.

Similarly, courts exercised only minimal control over design in their function of hearing claims by people who were injured by products. Professional designers such as architects and design engineers were protected by legal doctrines such as the requirement of privity between the injured party and the professional—that any liability that might exist terminated when the client accepted the project or the interval between execution of the design and injury to the user was too long. In other words, designers were insulated from having the courts judge their performance.

Similarly, legal rules protected manufacturers. A relationship of privity, usually a contractual one, had to exist between the claimant and the manufacturer before the claimant could sue. Claimants who surmounted this obstacle had to establish negligence by the manufacturer. Even if a claimant was aided by rules allowing the burden of establishing of negligence to be satisfied by inference, recovery against a manufacturer was extremely difficult.

Of course, some laws did control design. Registration laws for professional designers limited who could design improvements to real property. Similarly, local entities often exercised their power to regulate local land use by requiring a design to be signed by a registered design professional and to be approved by public authorities before a permit to improve or occupy real property could be issued. During this process, the design was reviewed, and if the local entity had the resources to review the design carefully, it eliminated designs that were clearly unsafe. (Unfortunately, current financial constraints make this process far less likely). Finally, as was mentioned earlier, products such as food and drugs were subject to some direct regulation.

On the whole, however, designers were not subject to legal restraints; the free market was the sole regulator of design quality. If a product was in demand and profits could be made, manufacturers would enter that market. If their products did not function as expected or proved to be unsafe, consumers or wholesalers would not buy them and they would be withdrawn from the market. [Today, the objective of economic "efficiency" causes concern about "externalities"—the true cost of a product is not reflected when it causes harm for which its manufacturer does not pay. (See Posner 1977)]

In the 1960s, public authorities such as legislators and judges who operated the control levers quickly pounced on the obvious defects of the market as regulator. Courts began to view consumers as victims of large aggregates of bargaining power (*Henningsen v. Bloomfield Motors, Inc.* 1960) and began holding manufacturers strictly liable for defective products that injured buyers or users (*Greenman v. Yuba Power Products* 1963). Similarly, legislators, both state and federal, saw the need for and the political advantage of regulation. The political strength of the consumer movement during the 1960s inspired legislation to protect the public: for example, the Occupational Safety and Health Act of 1970 (Public Law 91-596), which deals with injuries in the work place; the National Highway Act of 1970 (Public Law 91-605), which regulates automobile design; and the Consumer Product Safety Act of 1972 (Public Law 92-573), which gave authority to a federal agency to enact product safety standards. Keep in mind that a manufacturer or a user-consumer can transfer its losses to a legal entity consisting of one owner or numerous owners, an owner with modest assets or billions of dollars, an owner who is the only employee, or one with hundreds of thousands of employees. In general, issues surrounding the plight of an individual who seeks redress from a design engineer or manufacturer are referred to as consumer issues.

Four apparently unconnected events that occurred in the early 1980s demonstrate the complexity of design controls and the controversy surrounding them:

- In 1982 the United States Supreme Court, in *American Society of Mechanical Engineers v. Hydrolevel Corp.*, upheld a treble-damage antitrust (a conspiracy to restrain trade) award on behalf of a manufacturer against the ASME. Even though the ASME did not *itself* violate the law, it manifested that members of one of its standard-making committees had authority to set standards in a way that violated the antitrust laws.
- That same year, Professor Theresa M. Schwartz published the findings of her study of the Consumer Product Safety Commission, a regulatory agency created in 1972 to administer the Consumer Product Safety Act.
- In April 1983, the Federal Trade Commission (FTC) published a report recommending the regulation of standard-making organizations because of their potential power to reduce competition and injure consumers.[2]

---

[2]The modest controls suggested in the FTC report are unlikely to be enacted because the full commission probably will find that the controls would impose unnecessary costs upon these organizations.

With reference to eliminating competition, Schwartz described the "pool slide fiasco," in which the Consumer Products Safety Commission, acting on a manufacturer's petition, adopted mandatory standards for swimming-pool slides. After these standards, which probably were unnecessary, had been enacted, the number of companies that made swimming-pool slides dropped from seven to one—the petitioner!

- In 1983 Merrell Dow Pharmaceuticals, Inc., announced that it would no longer produce *Bendectin*—a drug widely used to control nausea and vomiting during pregnancy—saying that it could not afford to market the drug because of the product liability law.

In a manner similar to instrumentalities for designers, government and voluntary regulations affect the manufacturer and the quality of its product (see Box 2-1).

Product Codes and Standards

Product standards are technical prescriptions for material, production, design, installation, safety, or performance characteristics of products (see the FTC study 1983). These standards are usually developed by industry associations and technical societies, which create committees composed of people who are knowledgeable about specific products. In general, members of committees that write codes and standards are institutional representatives: that is, they represent the position or interests of their company in relation to competitors and, in some instances, represent instrumental organizations such as insurance companies and governmental regulatory agencies. Professional engineering designers rarely participate in these committees as individuals, and rarely are the members required to have technical credentials such as an engineering degree, a professional registration, or experience in design engineering.

The standard-writing committees make judgments about what characteristics are acceptable for a product; these judgments are then condensed into product standards. Although product certification will not be discussed here, note that certification provides a link between product information contained in a standard and a particular product. Certifiers test products to determine whether they comply with certain standards. The seal or label attached to the product or an "Approved by ________" statement attests to compliance with the standard. (In Chapter 3, we will explore the impact of codes and standards on the manufacturer with respect to design, testing, analysis, inspection, and maintenance.)

Design versus Performance Standards. An important distinction is often drawn between design and performance standards. Design standards state the requirements in terms of factors such as materials, production processes, and product dimensions. Performance standards, on the other hand, address requirements concerning actual use of the product: for example, strength (stress, strain, displacement), durability, conductivity, and flammability.

These standards take the following forms:

- Definitional standards, which establish precise, uniformly accepted meanings and nomenclature for materials, apparatuses, processes, and the like.
- Specification standards, which prescribe the use of specific materials, designs, production processes, performance levels, and testing methods.
- Interchangeability standards, which dictate the dimensions of products or equipment; they promote the uniformity, simplification, and compatibility of products.
- Classification standards, which allow competing products to be compared on the basis of their relative quality or performance grading.
- Recommended practice standards, which describe the instructions for installation, use, or maintenance to ensure safe or satisfactory product performance.
- Testing standards, which specify the apparatus, procedures, and testing conditions needed to determine whether a product meets the requirements of an applicable standard.
- Safety standards, which address the safety of a product.
- Protection standards, which come into play when the design engineer determines that (1) the consequences of a product failure are too severe or (2) the probability of failure is too great. (An example of complying with protection standards would be to add a fragment shield to a lawnmower, turbine, or pressure vessel or a containment vessel for a hazardous chemical process).

Mandatory versus Voluntary Standards. Mandatory standards are dictated by laws such as the Code of Federal Regulations. Voluntary standards, such as those developed by the American National Standards Institute (ANSI), are not. However, the distinction is often blurred. As a rule, private organizations make standards; thus, compliance with their standards, at least in theory, is voluntary. Yet their standards may be adopted by public authorities and become mandatory (e.g., building codes). Also, voluntary standards have a tendency to become mandatory—not in the sense that the law requires their use, but in the sense that failing to follow them may make marketing a particular product impossible.

Organizations that Make the Standards. According to the FTC's report, as of 1975 more than 400 private organizations in the United States either wrote standards or sponsored the development of standards (see ANSI table of organization in Appendix A). By 1975 these organizations had produced roughly 20,000 nongovernmental commercial standards. Most standard-making organizations are technical and trade association—often, engineering technical societies. Two major testing organizations, the American National Standards Institute and the American Society for Testing and Materials (ASTM), are devoted to developing standards (Hamilton 1978).

By and large, the private organizations are supported by the businesses affected by the standards. Standard-making organizations function through committees composed of individuals who are knowledgeable about the particular product studied. They rarely, however, impose professional requirements (registration or training in a school of design) on their members. Because standards typically involve highly technical subjects, industry usually provides direct or indirect financial support.

Some organizations that write codes and standards have engaged in concerted efforts to achieve a balanced membership on code-writing committees. For example, the ASME selects committee members on the basis of their employment or affiliation with a manufacturer, user, constructor, or insurer. In other words, most members of ASME's code-writing committee actually represent their employers. In addition, the ASME has no designation for an engineering design professional.

Users are becoming increasingly involved in the development of codes and standards, and they represent one of the major memberships in codes and standards committees of organizations such as the ASME. For example, users of pressure vessels and piping equipment—for which the ASME code provides design performance criteria—include the petrochemical, power, process, maritime, and aerospace industries.

Although most mandatory standards are written by private organizations, then adopted by public agencies, a current trend is to have public entities, especially federal agencies, make the standards. Similarly, the American National Standards Institute adopts the standards of industry or technical societies.

The Consumer Products Safety Commission example of regulatory instrumentalities. A high incidence of product failure or a catastrophic failure is often the driving force behind the promulgation of voluntary standards (Farr 1982) and mandatory standards (Suarez and Schupack 1988). In the early 1970s, the National Commission on Product Safety (NCPS) completed a three-year study on product safety and concluded that consumers were exposed to unsafe products to an excessive degree and that existing methods for reducing those health and safety hazards were ineffective. The NCPS found that 20 million people had been injured in accidents at home and at a cost of over $5.5 billion.

Governmental entities. As a result of the NCPS study, Congress passed the Consumer Product Safety Act of 1972, which created the Consumer Products Safety Commission and gave the commission the power to create mandatory standards. Schwartz's critical study (1983) of the commission provides the best glimpse of the relationship between a federal agency that has the power to mandate standards and the private standard-making organizations that the agency enlists to help perform its task. Because the trend is toward mandatory standards and because the commission is the most recently created mandatory regulating instrumentality, a discussion within the context of Schwartz's comprehensive study of the commission may be useful.

Difficulties Involved in Developing Standards. As Schwartz pointed out, the Consumer Product Safety Act gave the commission jurisdiction over an estimated 10,000 consumer products and more than a million producers and sellers of products, with the exception of food, drugs, automobiles, tobacco, and firearms. It also gave the commission a broad range of enforcement tools, such as the power to ban products, issue recalls, order refunds and repairs, seize and condemn products, and seek injunctive judicial relief to implement its orders.

Not surprisingly, strenuous opposition to the commission's broad powers developed in the business community. And when controversial regulations affect business, it is not unusual for the regulatory agency to be starved financially. The commission faced this problem in 1983. In 1981 the Congress eliminated some innovative aspects of the act, sought to force a cost-benefit analysis, gave preference to voluntary standards, and determined that the commission's power to set mandatory standards applied only to performance standards (Public Law 97-35 1981).

In 1983 the commission faced financial starvation again when Congress not only sharply limited its funds and personnel but also gave itself the power to impose a legislative veto—an act of doubtful validity (see *INS v. Chadha* 1983)—on regulations it disagreed with.

One innovative provision of the Consumer Product Safety Act gave individual users—through the petition process—greater rights than they had when dealing with other administrative agencies. Petitioners could, by court action if necessary, force the commission to issue mandatory standards. They also could participate in the early stages of developing those standards. Schwartz concluded that, unfortunately, the commission allowed petitions to control its agenda, which resulted in spending too much time on matters that were not sufficiently important. The petition experiment was dropped in 1981 (Public Law 97-35 1981; see Klayman's criticism of this action, 1982.)

The commission also was criticized by consumer groups for enacting only seven mandatory standards during its first 10 years of existence. As

Schwartz pointed out, however, an unusual "offerer" system was built into the act. If the commission wanted mandatory standards, it asked for bids from interested groups to develop those standards. Because of delays caused by this offerer process—in combination with the unrealistic deadlines required by the act, shortages of personnel and funds, and the requirement for input from the public—the average time spent developing a mandatory standard was four times that required by the act. Furthermore, the courts and Congress overturned some of the standards that the commission did adopt.

In addition, because of its lack of experience, the commission often embarked on quixotic ventures. (See, for example, the appendix in Schwartz where she describes the commission's standards for power mowers, a nightmare of mandatory standard making.)

Schwartz concluded that the commission did some of its jobs well, especially product bans and recalls, but that it was essentially in a no-win situation. The strong opposition of industry, the reluctance of Congress to give it much support, and the huge job it was given guaranteed that the commission would fail.

Schwartz's study demonstrated the difficulties involved in making standards and the resources needed to do so, especially when the standards are made by an uneasy consortium of consumer groups, technical consultants, academics, researchers, and manufacturers who are presided over by a modestly endowed public agency. Although its main concern was consumer safety, the commission also took into account technical feasibility, costs, and benefits of any proposed standards (*Cost-Benefit Analysis* 1982). Furthermore, its attempts to make standards affecting the workplace generated sharp disputes among scientists over acceptable standards regarding acceptable levels of noise, radiation, asbestos, and other environmental pollutants (see, for example, *American Textile Manufacturers Institute v. Donovan* 1981).

Advantages of Standards. Despite the unfortunate situation just described, those who make standards perform an indispensable function. The FTC report (1983) detailed some of the many advantages of standards. For example, they improve competition by perfecting markets; this is especially true of performance standards. They help provide a basis for comparison and establish consistent terminologies, measurements, and testing procedures. Their principal advantage is that they help the user, who must select, purchase, and use materials and equipment.

In addition, performance criteria indirectly set technical standards for the engineering designer by requiring analytic, testing, operational proof, or all three while giving latitude to innovative design concepts.

Most important, however, standards increase the flow of information, reduce the costs of purchase and production transactions, and promote the efficient allocation of resources for production and consumption. All this information comes at a low cost. Independent research and verification by users (purchasers) or public officials would be expensive. In addition, users may lack the technical competence to determine quality, even when they are provided with basic information.

Standards offer other direct advantages to users. When markets are characterized by low entry-and-exit costs, producers may be tempted to deceive users about the quality of their products. Standards can eliminate "fly-by-night" operations by providing users with an easy way of checking producers' claims about quality. Increasing users' confidence may strengthen overall demand for the product and lead to economies of scale. Standardizing durability and reliability may facilitate price–quality comparisons in competitive markets. Standards also have other advantages, but the few discussed here are enough to indicate that standards can serve a vital function at a relatively low cost.

Disadvantages of Standards. The awesome power that some standard-making organizations possess was demonstrated in *American Society of Mechanical Engineers v. Hydrolevel Corp.* (1982). That case and the FTC report (1983) demonstrate the obvious—unless checks and balances are incorporated into the standard-writing process, a standardizing organization can abuse its power (McLean 1985).

The ASME case involved a federal antitrust claim by Hydrolevel (a company that ultimately went out of business) against two of its competitors and the ASME. The two competitors had requested an ASME committee to determine whether Hydrolevel's low-water fuel cutoff, with its time-delay factor (one of its features), complied with the ASME's Boiler and Pressure Vessel Code. The committee concluded that Hydrolevel's product did not comply; this interpretation, although made in the committee's name, had actually been formulated by an employee of one of the two competitors.

The competitor whose employee had rigged the ASME's interpretation settled with Hydrolevel before the trial for $725,000; the other competitor settled for $75,000. The first trial resulted in a treble-damage award of more than $7 million against the ASME, and a divided U.S. Supreme Court upheld the award. The award was based on the ASME having created the appearance that its committee members, who were mainly interested in preventing Hydrolevel from penetrating their market, had the authority to set standards, which violated the antitrust laws. This theory justified the treble-damage award, although the ASME was *unaware* of the wrongful conduct and did *not* benefit from that conduct.

The dissenting judges emphasized the importance of the standard-making organizations, especially those that were nonprofit, and stated that the burden of wrongful conduct should be borne by the companies whose employees had rigged the system.

The second trial resulted in a finding that Hydrolevel suffered $1.1 million in damages, which the judge tripled to $3.3 million. The judge deducted the $800,000 that Hydrolevel had received in the first settlement and awarded a judgment of $2.5 million. Finally, he set a date for a hearing in which Hydrolevel would be allowed to establish its attorneys' fees, which were likely to be extensive (*The Wall Street Journal* June 27, 1983).

The ASME case illustrates several important points. First, standardization can be abused. Second, those who make standards must provide better controls over their operations. One reason for the reluctance of the FTC to regulate standard-making organizations is the likelihood that the ASME case will generate voluntary internal controls to prevent abuse.

The ASME case also raises questions. The facts in the case were extreme. What would the outcome be in less flagrant cases? And what effect would the case have on the willingness of individual volunteers to contribute their expertise and time—the core of standard making?

The FTC study cited other instances of abuse, which were mainly attempts by manufacturers to gain markets that were stifled by what the FTC staff believed to be the intransigence and self-interest of some standard-making organizations. In addition, the FTC study cited incidents of deception of consumers and instances where consumers were denied the opportunity to buy products that functioned as well as more expensive ones.

Potential abuses, however, should not distract our attention from the tremendous advantages of having standards. With that in mind, we can discuss how standards are implemented.

The laws that make standards mandatory usually give regulatory agencies the power to ban the use of products that do not meet the mandatory standards and to recall products that violate the standards. Agency or law enforcement representatives can obtain cease-and-desist orders, the violation of which can result in severe sanctions. For example, a public authority can refuse to issue permits for construction projects that do not meet the building codes.[3] Finally, marketing products

[3]In these sue-everyone days, issuance of a permit can generate a liability claim against the public entity issuing the permit (see *Garrett v. Holidays Inns* 1983).

such as drugs requires prior approval. For obvious reasons, requirements for prior approval are extremely rare.

Looking next at voluntary standards, products that do not meet those standards are unlikely to succeed in the marketplace, which is perhaps the most effective sanction of all. But courts also are involved in the implementation of standards, either through public law or private law.

Courts can implement public law, such as the Consumer Product Safety Act, by granting orders sought by the Consumer Product Safety Commission and providing sanctions for noncompliance. Similarly, they can implement regulations by the National Highway Safety Commission by fining manufacturers who do not comply with that commission's regulations or by enforcing recall orders issued by the commission. In this public law function, courts vindicate the interests of the general public, as expressed in laws and regulations.

When implementing private law, courts decide on the validity of claims by individuals against other individuals—these days, usually a manufacturer, public utility, a professional, or a public entity that claimants assert has caused them injury for which they should be compensated. (As will be discussed in the next section, courts not only implement standards made by others but make standards themselves as well.)

Even though the same judge can operate in a public and private law capacity, and even though decisions based on both types of law can be made in the same courtroom, a division exists between these two functions. When deciding a private law claim, courts can, if they choose, borrow from the public-law regulatory system. For example, public law prohibits drivers from driving over 55 miles per hour. Violating that law can result in a fine, imprisonment, or revocation of one's driver's license.

Suppose the driver who exceeds the speed limit causes an accident and someone is injured. The court can decide what weight, if any, to give to the violation of the law (Prosser 1971). On rare occasions, the legislature may indicate that the violation is conclusive on a civil claim. Usually, the violation will be regarded as negligence per se, unless it can be excused: for example, if the driver suffered a stroke and pushed his foot down on the accelerator. Neither judge nor jury need pass on the defendant's conduct to determine whether he or she has been negligent.

Sometimes the violation is less significant: that is, it simply may be evidence of negligence to be considered along with other evidence. Sometimes the violation cannot be considered at all.

A court's determination of how much weight it will assign to the violation can help or hinder the implementation of regulatory policy. The varying weight courts give the statutory violations also indicates, however, the great variety of laws, which often reflect dubious wisdom. Making it a crime to spit on the sidewalk is representative of the many bizarre laws that have been enacted in the United States, usually by local governments.

Thus far, we have discussed defendants who have not conformed with legal rules such as statutes or regulations. Suppose the defendant claims as a defense that it has performed in accordance with the standards required by law. What effect does compliance have on civil liability?

Compliance with a statute or regulation has less effect on the resolution of a civil case than does a violation. Public law rules set up minimum standards of conduct that may be unreasonable under certain circumstances. Suppose a builder knows that a requirement of a building code, if followed, will expose occupants to a high risk of personal harm. Complying with the code may bar any action for violating the law, but it may not exculpate the builder from either a claim of negligence or a warranty that the building would be a reasonably safe and habitable one. Also, a client may want more than is required by an existing code or standard.

What is the effect of nonmandatory standards, such as custom and usage in the profession or voluntary industry standards, on the civil claim? Again, compliance must be differentiated from noncompliance. Failure to comply with customary professional practices, as established by the expert testimony of professionals, is conclusive. With regard to products, a manufacturer's failure to make a product in accordance with customary industry practices will be evidence—even strong evidence—that the product is defective. Failure to comply with professional standards has a similar, and possibly even greater, effect.

The more difficult problems arise when the manufacturer has followed customary practices. Compliance is not conclusive as to the defectiveness of a product, but it can be looked at; some courts give compliance great weight and others give it less.

Courts are not completely confident about the standard makers. They also do not want an industry to be able to set legal standards. The fact that judges must face victims in the courtroom exacerbates its difficulty in deciding the claim.

Courts: Product Liability Law

In their private law role, courts pass judgments on claims by people who have suffered harm (the fifth instrumentality under manufacturers in Box 2-1) that their losses should be transferred to the party that has caused the harm. In doing so, the courts do not use standards created by others, they make standards themselves.

As we stated at the beginning of the chapter, we have dealt with legal control over design at two levels: control over the designers (e.g., claims against architects or engineers) and control over the designs (e.g., claims against manufacturers of defective products). The first type of control, which has caused fewer problems than the second type, is discussed only briefly. Generally, the law holds designers to the professional standard. Designers are not judged by the success of their design, but on a comparison with what they did and what other designers would have done (Sweet 1977).

As was indicated earlier, until the 1960s, users who were injured by products and sought recovery from the manufacturer faced many obstacles. The courts subsequently made it much easier for a victim to recover against a manufacturer, especially if that victim had suffered physical harm. Generally, the claimant had to show that the manufacturer had sold a product that was in "a defective condition unreasonably dangerous to the user or consumer" (*Restatement [Second] of Torts* 1965).

A manufacturing or construction defect could be shown by comparing the product that caused the harm with other similar products made by the manufacturer. If, among 100 units coming off the assembly line per hour, *one* unit caused physical harm, it may be easy to prove that the unit was defective by comparing it with the other units or to the design. But when the claim is based on a defective design, the unit cannot be compared with other units from the same assembly line. Instead, a dreary "battle" between experts precedes the jury's determination about the reasonableness of the design.

Courts in some states, led by California, believe that cases involving design defects, unless treated differently, would in effect require victims to prove negligence—the very obstacle the product liability revolution in the 1960s sought to eliminate. California courts hold the manufacturer liable if the product did not perform in accordance with the buyer's or user's reasonable expectations. Even if the product did perform in accordance with those expectations, the manufacturer will be liable unless it can show that its design was reasonable under the circumstances (*Barker v. Lull Engineering Co.* 1978; Diamond 1983). California has gone even further: It holds pharmaceutical companies liable to the extent of their share of the market, even if the victim cannot establish that the harm was caused by the company's drug (*Sindell v. Abbott Laboratories* 1980).

In expanding product liability law, courts have taken the essentially private law of torts and turned it into a system for compensating victims. Proponents of this transformation believe that manufacturers should pay for the human costs caused by their products. Manufacturers can pass these costs on to everyone who buys their product by including them, or the cost of insuring against risk, in the price of the product. In addition, to the extent that victims are not compensated by the manufacturer, the burden may fall upon the state.

Although this system of spreading risk was generally believed to be a relatively efficient and painless way of dealing with injuries, judicial expansion of manufacturers' liability has received a tremendous amount of criticism. The following points represent a formidable indictment of the courts as setters of design standards.

1. Each state determines these standards by comparing them with the national standards set by the federal government (the Code of Federal Regulations) through mandatory standards or by voluntary standards developed by associations such as the American National Standards Institute.
2. Cases are decided by lay jurors, who are often baffled by conflicting expert testimony. Judges are frequently confused about what qualifies a person to render an expert opinion in a product liability trial. And both jurors and judges are usually confused about the difference between technical expertise and occupational expertise. Trial attorneys exploit this naivete to disqualify highly qualified engineers and scientists because they have not been "an employee of a manufacturer of the subject product," are "not a product engineer," or are "not a member of some [trade] association or society." In addition, the courts commonly accept as technical experts people who have little or no formal technical engineering training and little training that is relevant to a product-failure analysis or an accident investigation, but who have expertise by virtue of their employment. Expert testimony requires an understanding of the basic engineering principles that precede the design of a product, not the other way around—unless trial and error are necessary requirements in the design cycle.
3. Large awards and enormous defense costs have caused a sharp increase in the cost of insurance. As a result, many manufacturers simply cannot afford to carry insurance. Thus, they must decide to market the product without insurance or not market the product at all.
4. There is a moral objection to liability that is not based on fault, especially by "hindsight judgments."
5. The law has begun to award large punitive damages against manufacturers if the manufacturers acted recklessly, although this is notoriously difficult to prove when a corporate milieu is domi-

nated by the motto "Don't look and you won't find" (*Grimshaw v. Ford Motor Co.* 1981). This open-ended jury determination can, however, be too great a risk and can force a useful product out of the market.

6. The cost of products will increase drastically as a result of the costs of insurance, overdesign, and elimination of marginal manufacturers from the market.
7. The jury verdict simply says the product is defective; it does not tell the manufacturer how to make the product properly. Although expert testimony (and in some instances, technical reports) often includes criticisms of a design and suggests alternatives, this is not always a precise way of telling the manufacturer what must be done to avoid liability.

The decision by Merrell Dow Pharmaceuticals to withdraw antiemetic *Bendectin* demonstrates the sledgehammer power of product liability law. According to an editorial in the *Wall Street Journal* in June 1983 ("Defects in the Courtroom"), there were no medical reasons for withdrawing the drug; it was on the list of approved drugs in all Western countries. Although a few animal tests had raised doubts about possible birth defects, the editorial noted that the scientific evidence for the safety of the drug when prescribed by a physician was overwhelming. Nevertheless, the product was withdrawn because of "misleading articles in the press and of litigiousness run amok" (p. 28).

Product liability may be *too* effective. Although it does not directly determine whether a product can be marketed, it can have a devastating effect on market decisions. For example, more than 300 lawsuits alleging birth defects have been brought against Merrell Dow. Two jury findings on behalf of claimants were reversed by trial judges after the drug had been withdrawn (*Mekdeki v. Merrell National Laboratory* 1983; *Oxendine v. Merrell Dow Pharmaceuticals, Inc.* 1986; the *Oxendine* decision was reinstated on appeal). Merrell Dow's total annual liability insurance premium, which covers all pharmaceuticals produced, was more than $10 million—almost as much as total sales of *Bendectin* of $13 million in the United States. The company said it simply could not afford to market the drug. Thus, society has been deprived of a useful product.

Numerous state statutes have been enacted to reduce product liability; some of these statutes have been found to be unconstitutional (*Heath v. Sears, Roebuck & Co.* 1983). Furthermore, continual attempts have been made to enact a federal law that deals with the portion of product liability law involving personal injury. Generally, legislative proposals to change the product liability law have evolved in state courts. In addition to providing for a uniform federal system, these bills, if enacted, would reduce the manufacturer's liability, require that workers' compensation payments reduce the amount of recovery, make compliance with mandatory standards a defense, limit punitive damages, and make other changes in current product liability law.

Sharp differences exist between regulation of engineering design by setting standards and regulation by using a private product liability lawsuit. Perhaps most important is how awesomely powerful the product liability process has become. In the long run, its sanctions can be far more severe, particularly when risk of personal harm is concerned, than the sanction of market failure. In product liability cases, juries ultimately determine the standards, often on the basis of conflicting testimony by experts. On the other hand, product standards are usually made by experts.

Perhaps the most crucial difference between the two methods of regulating engineering design is that in a product liability case, jurors often must look directly into the eyes of a person who has been ruined for life by a product produced by an apparently affluent manufacturer. Under these conditions, the fact that the jury will find the manufacturer liable, even it demonstrates that it did all it could be expected to do, is understandable.

Finally, whether the organizations that set standards are governmental or technical organizations, they are likely to place a heavier emphasis on practicality and cost than are juries or even judges.

CONCLUSION

The comparisons discussed in this chapter do not mean that one approach must be used and another must not be used. For example, to say that attempts to control designers are useless and should be abandoned does not take into account the fact that complex problems require multilevel solutions or the fact that institutions such as registration systems and schools are here to stay and may become more effective.

Similarly, the comparisons do not inevitably lead to the conclusion that a particular public or private approach should be discarded. Societal and political forces will dictate whether and how they will continue to operate. An effort must be made to focus more sharply on the qualifications required to monitor design and determine who should be accorded primary responsibility.

The following statements are premised on basic beliefs about the current capabilities and attributes of institutions and individuals.

- Control of engineering designers is, at present, too difficult to achieve because the available tools are either too crude or too complex. The interest groups of the "engineering profession" are fragmented, they have not set standards for professional certification of products, and they have not defined the criteria for a professional engineer. Thus, control of design is preferable to control of engineering designers.
- The state must exert some oversight regarding controls over engineering design, but the politics and rigidity that characterize the public approach tend to tip the scales toward private controls. In other words, private controls are preferable to public ones.
- Design standards tend to stifle creativity and become too rigid. Thus, performance or outcome codes are preferable to design standards.
- The marketplace, despite its admitted deficiencies, is preferable to rules imposed by law.
- The courts, in their function of deciding claims, are often emotionally biased or influenced—primarily by compassion for the victim, not the overall cost and benefit to society, but also by sympathy for businesses as victims of opportunistic attorneys and victims. The courts are often unable to grasp the technical concepts involved and sort through the complexities of product design, process, and user responsibilities. And they are often unable to determine what information is relevant and who is qualified to provide it. Thus, direct controls such as publicly or privately developed codes and standards are preferable to indirect controls imposed by the courts.

Finally, those who operate each instrumentality described in this chapter should (1) know the precise nature of the instrumentality's objective, (2) realize that other instrumentalities seek the same objective, (3) help other instrumentalities achieve goals that are compatible with its own objectives and resources, (4) recognize the strengths and limitations of its approach in relation to the approaches of others who seek the same objective, and (5) do its best to achieve its objectives after considering the direct and indirect costs imposed by its approach.

3

Legal Implications of Codes and Standards for Manufacturers

During the past decade, industry has witnessed substantial technological growth, reflecting a reach for higher pressures, larger flows, improved systems application, and new mechanical products. These developments generate expanded problems of "product liability," which in turn necessitate the updating of existing codes and standards and the development of new ones to accommodate emerging technical frontiers.

As Gee (1975) pointed out, these new developments "raise tough, new challenges for both the integrity and the ingenuity of corporate management" and afford the consumer what has been termed "social justice" (*Ketterer v. Armour & Company* 1912). Thus, corporate management, engineering designers, and production personnel must exercise the highest degree of correct decision making no matter where such decisions are made—at the board level, at the engineering design level, or at the production level.

To fill this need, the office of "risk manager" has been established. The function of this office is to oversee "those activities performed to prevent accidental loss . . . and [to create] the optimum conditions to meet those accidental losses that are not preventable or that occur despite preventive measures" (McCahill 1971).

Quality assurance and legal principles are interrelated, and neither yields to precise technical treatment. Therein lies the threshold to the problems we will discuss in this chapter.

Today's manufacturer, as a matter of practicality, must establish correct preventive measures to minimize the impact of potential or actual litigation. This has become acutely necessary because of the "widespread attractiveness of the theory that deep pockets are best suited to bear the risks of personal and property injuries in our modern industrial society" (Doelle 1974, p. 201).

The mood of the judiciary continues to match the action of legislators to protect the user. The late Justice Jackson of the U.S. Supreme Court declared in a product liability case that "a dependent society must exact greater care than in more simple days and must require from manufacturers and producers increased integrity and caution as the only protection of its safety and well being" (*Dalehitt v. U.S.*). As early as 1912, a federal court determined that "The remedies of the injured consumer . . . should rest upon the demands of social justice" (*Ketterer v. Armour & Co.*). This trend toward consumer protection has led to the use of disclaimer clauses for limited relief (see the Uniform Commercial Code (UCC) now adopted in 49 states; *Walsh v. Ford Motor Car Co.* 1969; "Emerging Product Liability Under Section 2-318 UCC" 1974).

From conceptual design through development, testing, the manufacturing process, procurement of vendor items, assembly, inspection, and packaging, the need for codes and standards to establish, and thereafter preserve, product integrity is paramount.

Chapter 2 presented an organizational overview of codes and standards. This chapter focuses on the legal implications of those standards for the manufacturer with respect to the in-house disciplines of design, testing, analysis, inspection, and maintenance. Although factors such as proper packaging, warnings, disclaimers, and seals of approval are properly related to product liability, they will be discussed in later chapters.

HISTORICAL DEVELOPMENT OF CODES AND STANDARDS

The concept of standard making is not new. Standardized weights date back to 3500 B.C. (Verman 1973). In Proverbs 11:1, "A false balance is an abomination to the Lord, but a just weight is His delight." Consider the following performance criteria and liabilities citation:

> If a structure collapses as a result of improper construction, the contractor will be required to: (1) repair damages at his expense; (2) reconstruct irreparable damages at his own expense; (3) compensate the owner for injury, and (4) be put to death if death of the owner occurs as a result of the collapse (Code of Hammurabi, 2124–2081 B.C.).

The guilds of the Middle Ages prescribed the manufacturing process in minute detail: for example, the number of rivets to go into a breast plate, or not permitting resin or tallow in candles. The processing of English wool filled 30 chapters in a code used in Barcelona in 1438. By the end of the eighteenth century, the French Academy of Science had developed the metric system. In 1798, Eli Whitney manufactured muskets from interchangeable parts under U.S. contract.

The first attempt to develop a standard screw thread occurred in 1864, "possibly the most important single engineering standard" (Rech 1956, p. 13). In a paper presented at the Franklin Institute, William Sellers urged the creation of a uniform American screw thread. Regrettably, this goal was achieved at a snail's pace. By 1904 a massive and disastrous fire leveled the downtown area of Baltimore, Maryland. Because the nozzles of their hoses did not match the thread of the Baltimore hydrants, the out-of-town fire fighters who responded were unable to help.

Codes and standards are principally intended for the user, not the engineering designer. They allow the user to measure performance and uniformity when purchasing a product or system.

A standard cannot be precisely defined. In essence, a standard establishes a level of quality, purity, or fitness. A more detailed definition is that a standard "is a constant uniform or invariable means, method or tool adopted by a group for use throughout an industry to reach a specified goal" (Lamb, Carrington, and Shields 1971, p. 76). The term "constant" does not mean "never changing"; it means that a word or symbol or method has a uniform meaning at a given time. The term itself connotes application in industry. Therefore, it should be distinguished from the type of standard that a manufacturer may establish "in house" to meet its own criteria. In this chapter, the general use of the term standard is applied: namely, a recommendation achieved through industry consensus.

Coles (1943, p. 123) proposed another definition:

> A standard is a model for a material object or a rule for a course of action established by authority, custom, or general consent by which others of a like nature may be identified, compared, or regulated, or which in itself presents the ideal or the "one best" for a particular purpose.

As was discussed in Chapter 2, a standard can originate from a number of sources. A standard prepared by a legislative body may take the form of a mandatory code or statute. A manufacturer can establish a voluntary standard for its own use; such a standard can cover a variety of steps within the plant, beginning with a product design, and proceeding through shipping and maintenance. A trade association or a similar nonprofit entity may develop a voluntary standard for use by members of a particular industry or by a number of unrelated industries (Peyton 1973).

Procedurally, after applying good engineering design practices to develop a prototype product, one seeks to meet the general standard. As Rockwell (1970) said: when an industry standard is available, it "should require only reasonable methods to verify compliance with the standard." If no general standard exists, one relies on the best engineering know-how in an effort to achieve one's objectives.

Associated with the term standard are such words as standardization, simplification, and certification. These are "uses" made of a particular standard. One may or may not label the results of a test as certification. In cases of self-certification, one reinforces what the law has already imposed—namely, that the warranty, either express or implied, suggests that the product is fit for its particular use and that it complies with the designated standard.

Another classification can be based on whether the standard measures performance, process, inspection, fabrication, or design. As was stated in Chapter 2, a *performance* standard prescribes a level of operation to be achieved for the purpose of safety or efficiency. Because performance codes and standards set limits on product performance (e.g., strength, deformation, wear) rather than on dimensions, materials, and the like, they tend to be less restrictive with respect to accommodating diverse design concepts. A *design* standard, on the other hand, establishes parameters that define the product, such as the materials and its dimensions. Design standards are generally used when interchangeability is desired. In any event, as John Kincaid, a government official, said in 1967:

> One solid fact is emerging with crystal clarity; for a mass society in an era of accelerating scientific and technical change, standards are the stabilizing, the protective factor. Without standards there would be technological chaos; without standards the user would be unprotected.

From the standpoint of product liability, standards that establish safety criteria assume great importance (see *Final Report of the National Commission on Product Safety* 1970). What was regarded as a risk years ago may not be regarded as a risk today. Consider the matter of electrical current. In 1889 Thomas Edison sought to protect the public against what he considered to be a major public danger—the use of alternating current. He believed that introducing such a system had "no element of permanency and every element of danger to life and property." Any engineer during that period who recommended the use of alternating current would be liable if its use resulted in the injury of a person, because such use would not be in accordance with the then-prevailing standards.

The Federal Trade Commission has long recognized that private standard-setting bodies are the major source of voluntary standards and that they will continue to be so in the near future (Interagency Committee on Standards Policy 1970). Safety standards establish specifications that are designed to ensure that regardless of grade, quality, or price, the article will not expose users to a foreseeable risk or injury. Safety standards are currently graded on a pass-or-fail scale, and are critically important to consumers (see "Liability of Certifiers of Products for Personal Injuries" 1970).

Guidelines for Organizing a Product Safety Program, a publication of the American National Standards Institute (ANSI), contains a detailed checklist concerning the elements of product safety and the management of safety controls (see, also, Rockwell 1972). The scope of modern standards is indicated by the five categories listed by ANSI: (1) definitions, terminology, symbols, and abbreviations, (2) materials, (3) performance characteristics, procedures, and methods of rating, (4) testing and analysis for size, weight, volume, and rating, and (5) practices (as methods, designations, or symbols indicative of certain industries), safety, health, and building construction. Although safety is specified only in the fifth category, implications for safety are evident in the other categories. The International Standards Organization has established many standards in the international field.

MANDATORY STANDARDS

In recent years, we have witnessed the enactment of an increasing number of laws and regulations concerning control of the environment, culminating in the passage of the National Environmental Act of 1969. All these efforts are directed toward the control of safety and health while outdoors. With the passage of the Occupational Safety and Health Act (OSHA) in 1970, the federal government assumed jurisdiction over the workplace, embracing more than 4 million employers and over 57 million employees. The stated purpose of the act is "to assure so far as possible every working man and woman in the nation safe and healthful working conditions and to preserve our human resources" (OSHA, Section 2[b]). OSHA permits each state to assume jurisdiction over local working conditions by submitting a state plan, providing the plan is "at least as effective as" the controls set up under OSHA (Forster 1971).

The purpose of the OSHA job and safety standards or regulations is to avoid conditions that have been proved by research and experience to be hazardous to personnel. These regulations apply equally to both employer and employee (OSHA, Sec. 5), as set forth in the "Industrial Standards" of the *Federal Register* (Part 1910, 1972).

Federal regulations issued by the Department of Labor (*Federal Register,* Title 29, Part 1910, 1972) deal with general industry standards for manufacturing (*Federal Register*), which for convenience can be classified into six major categories: (1) workplace, (2) machines and equipment, (3) materials, (4) employees, (5) power source, and (6) processes ("Work Place Standards" 1973).

A manufacturer is subject to all industrial standards; the critical impact of OSHA occurs when one manufacturer's component is incorporated into a piece of equipment or a system produced by another. In such a case, the question arises whether such a component, when used in such manner, comes within any of the prescribed OSHA regulations.

OSHA uses approximately 185 national consensus standards and relies on ANSI for new or revised standards (Peyton 1973). The original standards, effective on May 29, 1971, were derived from existing federal legislation and from the voluntary standards of ANSI and the National Fire Protection Association (NFPA). Literally thousands of standards are grounded in OSHA. These standards are not organized by industry, but instead are addressed to a particular safety problem or health hazard, and they are revised each year ("Factory" 1984).

Health conditions within a plant are the subject of OSHA's Target Health Hazards Program, which focuses on five substances that are hazardous in the workplace: asbestos, lead, silica, cotton dust, and carbon monoxide ("Target Health Hazards" 1983; see, also, *Federal Register* 1983). In June 1984, the U.S. Department of Labor reported that it would issue through OSHA "a long awaited standard . . . [that] would sharply reduce worker exposure to Ethylene Oxide which is used as a sterilizing agent in hospitals" (*Wall Street Journal* 1984, p. 2). Such rules become effective 60 days after publication in the Code of Federal Regulations. Hence, when attempting to follow proposed federal standards, the first task is to examine the code, which is indexed by subject (see Appendix C).

One serious industrial problem involves exposure to occupational noise (*Federal Register* 1972). The general rule states that permissible exposure to noise during an 8-hour day shall not exceed 90 decibels. OSHA is now struggling to revise this standard (NFPA Reporter 1973; *Noise: The Environmental Problem* 1973).

The OSHA regulation on noise provides that "when employees are subjected to sounds exceeding those listed in table G16, feasible administrative or engineering controls shall be utilized" (*Federal Register,* Part

1910, p. 95). Administrative control simply means removing the employee from the excessive noise for a given period. Engineering controls require a reduction of the sound at its source and the protection of personnel through innovative engineering design and innovative equipment—this would include not only hearing devices but practical solutions such as properly oiled machines, proper mounting of machines, and mufflers on noisy parts (*Federal Register* 1972).

Lacking a government standard, industry will develop its own standard (NFPA Recommended Standard, 1970, 1971, Project T2.7.3), influenced by other instrumentalities, for both national and international use (International Standards Organization/Technology Committee 1971). Of particular interest to the manufacturing, processing, and power industries are the standards pertinent to "power source" (see *Federal Register,* 1972, Part 1910; ASME Boiler and Vessel Pressure Code).

Currently, much valuable work in devising a standard for eventual consideration by OSHA is being done by ANSI under the cosponsorship of the National Safety Council and the National Machine Tool Builders Association. The scope of the ANSI B-11 Standards Committee's work on power presses has expanded to include all types of machine tools: mechanical and hydraulic power presses; power press brakes; shears; ironworkers; lathes; cold headers; drilling, milling, boring, grinding, sawing, and gear cutting machines; formers and benders; and automatic screw/bar and chucking machines. (The status of these proposed standards can be obtained from the Safety Director of the National Machine Tool Builders Association.) Note that when no standard exists for a specific machine tool under OSHA regulations, the compliance officer can cite any or all existing violations pursuant to Subpart O, Part 1910, of the *Federal Register* (1972).

Manufacturers, both at exhibits and in their advertising, emphasize that they produce and sell components that meet OSHA's safety, noise reduction, and health standards. Thus, one manufacturer proposes a fluidic, two-hand, non-tied-down control package for operating press clutches with a fluidic control requiring two signals to advance at the same time to permit the system to operate (*Hydraulics and Pneumatics* 1973). Another firm, seeking to eliminate the release of oil mist into its plant, has conceived of an air cylinder that is lubricated for life during manufacture. Still another, seeking to improve the safety of workers who use pneumatic controls, suggests double valves for safer power-press clutch-brake controls, two-hand control circuits to keep workers' hands out of dangerous areas, lockout valves to increase the safety of workers when servicing pneumatic components and mufflers to reduce sound levels, and directional control valves to reduce air pollution.

LITIGATION AND CODES AND STANDARDS

In a product liability case, the plaintiffs' attorneys usually join multiple theories in their complaints—negligence, warranty, and strict liability—so that if their proof based on one theory fails, they may prevail on the others. In addition, if they have the option, they will select as the site of a trial the state with laws that are most favorable to their particular theory. For example, if a plaintiff has contributed to the defendant's negligence and sues in New Jersey for breach of warranty, his or her action would be dismissed because, in that state, a plaintiff's contributory negligence is a defense in a warranty case. If, however, the plaintiff can bring the suit in New York, the suit may succeed because, in that state, contributory negligence is not a defense to an action based on a breach of warranty (*Codling v. Paglia, Blessington v. McCrory's Stores*).

The nature of the suit, whether based on negligence (tort) or warranty (contract), is directly related to recovery. Damages that are recoverable in tort may be more substantial than those recoverable in a contract (22 Am. Jr. 2d Damages Section 18, 1965).

One reason why industry standards have not been universally accepted in the courts or given great weight is the judiciary's awareness that members of an industry (for a variety of reasons, including cost) tend to keep safety and other requirements *to a minimum*. A standard cannot be used to prove that one was reasonably prudent when, in fact, the standard does not represent current technology (*Texas & Pacific Railroad v. Behymer* 1903). If a manufacturer failed to meet its own industry standards and claims to have taken due care, this defense may be rebutted if the plaintiff can show that the entire industry and its codes and standards are lax.

Some industries have no codes and standards and do not contemplate developing any unless pressured by a high incidence of failures, injuries, lawsuits, or adverse publicity—that is, economic losses. Some standard-writing committees collectively protect their industries from lawsuits by carefully limiting some incriminating safety requirements. In fact, some standards are written as advisory or "motherhood" documents that do not cite any design or performance criteria, but do admonish users to assess certain problems. Thus, a manufacturer's attorney may state that the product does not violate the code or standard—that, in fact, the code or standard does not either specifically address or require any product performance or design requirement.

In some instances, individual members of an industry association may conspire to defeat the association's codes and standards to gain an economic advantage over a competitor. Consider the following litigation that took place in 1989. The plaintiff was struck on the head by a component of a product that was described in the industry association's guidelines as "designed to be maintenance free, and other items require no attention either, under normal circumstances." The defendant–manufacturer (one of five in the United States) stated in its deposition that it disagreed with its industry association guideline and would seek to amend the part of the report that contained this statement. Although this situation could apply, and has applied, to a number of products, the component in question was the truck and finial attached to the top of a flagpole. The truck and finial can indeed be easily and safely attached to the flagpole. (For a more detailed discussion about the technical diversity of codes and standards criteria and their interrelationship with the technical community of professionals, manufacturers, insurers, users, regulators, and educators, see Chapter 6.)

Nevertheless, despite certain judicial criticism to the contrary, standards and safety codes are invaluable tools in the defense of a product liability case. When a proposed standard reflects the combined expertise of the industry and is written with a view to minimizing risk, the defendant–manufacturer's legal position is vastly improved. In this instance, the manufacturer will be able, if it complied with the proposed standard, to indicate that the requisite legal minimums have been met and, furthermore, that its actions were those of a reasonably prudent person acting under the same or similar circumstances. Of course, the jury normally decides whether the manufacturer, in fact, complied with the standard.

Although a product has been made in accordance with the state of the art, such defense is unavailable in a strict liability case. In many states, however, "compliance with the state of the art at the time the product was issued, has been made a statutory defense to a product liability suit" (*New York Bar Association Journal*, 1979, Note 31, p. 499).

Leaving aside the question of which state has the most favorable substantive law, a basic problem is whether the case is barred by the statute of limitations. Decisions vary in different states. If the suit is based on tort (negligence), the time proscribed by the statute of limitations usually begins from the date of the injury. This period varies from state to state or for the period prescribed in the *Restatement (Second) of Torts* (1965, Para. 402A). If the suit is based on contract (warranty), under the Uniform Commercial Code, the period begins on the date the breach occurred regardless of whether one had knowledge of the breach, and the suit must be initiated within four years (UCC, Sec. 1-272).

If the suit is based on the strict liability theory, the fault of the defendant is immaterial. The root question is whether the defendant had physical control of the product when the product was found to be defective. Although courts differ in their decisions, the prevailing view favors the period of statute of limitation used in tort actions (see, for example, *Brody v. Overlook Hospital* 1972). A review of selected cases will demonstrate the use of a "standard" in litigation.

In *Dimeo v. Minster Machine Co. Inc.*, a case heard in the U.S. District Court of Connecticut and affirmed in the U.S. Court of Appeals (Second Circuit), the court reviewed the use of standards. The plaintiff was injured when the stamping press he was operating recycled while he was removing a finished piece from the die. In his action against the manufacturer of the press, the plaintiff contended that a valve, which the defendant had *not* made, was defective and therefore the defendant was negligent in failing to detect the defect. The defendant claimed that the plaintiff's employer had allowed water and dirt to get into the valve and that the trial court had erred in excluding as evidence a National Safety Council manual dealing with the care and operation of power presses. The defendant had offered this manual to show that the plaintiff's employer had not followed a procedure that might have avoided the accident. The Court of Appeals, after commenting on the use of safety manuals to show that a defendant has or has not compiled with accepted standards of care, said that although admitting this manual into evidence might have been proper, the defendant had been allowed to introduce another manual containing the same information—that all that the defendant had lost was the "prestigious name and green symbol" of the National Safety Council. The defendant's appeal was denied.

In *Brandon v. Yale & Towne Mfg. Co.* (1963), liability was imposed on the manufacturer–supplier of a forklift because an operator was injured by a falling load. The forklift was not equipped with a canopy guard as recommended in the ANSI code that describes the accepted minimum safety standard for equipment used in the materials-handling business. (Interestingly enough, a canopy guard that is supposed to protect the driver can be dangerous if the forklift overturns. Unless the driver is wearing a safety belt, he can be ejected from the driver's seat and crushed by the canopy. In other words, both the canopy and the safety belt work together as a protective system.) Safety codes, safe practice standards, and manuals setting out standards in the trade were properly admitted into evidence. Both the plaintiff's and defendant's experts recognized the manuals as authorities on the subjects of safety precautions and wire, rope, cables, and clamps. The safety codes could not be excluded as treatises, they were admissible as objective standards of safe construction, and they were generally recognized and accepted as such in the particular construction industry (see *McComish v. DeSoi* 1963).

In the past, defendants had been unable to place into evidence the standard they had used when producing the allegedly defective component on the theory that this standard was hearsay evidence. Instead, they were required to furnish expert testimony in support of their position. In *Philip J. Nordstrom v. White Metal Rolling & Stamping Corp.* (1968), a leading case correcting this situation, the Supreme Court of Washington State heard an action against a manufacturer–seller of a ladder for injuries sustained by the buyer–plaintiff when the ladder collapsed. The defendant introduced the ANSI safety code for portable metal ladders, and attempted to show that its product had been made in accordance with the code. The court overruled previous decisions and held that standards published by private entities, which did not have the force of law, were nevertheless relevant, trustworthy, and necessary to prove the matter and therefore were admissible as evidence.

However, can a manufacturer of a ladder with a defective foot design (e.g., see Fig. 3-1) claim that its product complies with the ANSI standard even though the standard does not contain performance criteria addressed to the defective part? Defendants often claim that the product does comply—that it conforms to, or does not violate, the applicable standards (the minimum requirements for marketing a product). Safety has been held to be a proper objective of standardization (see *Roofire Alarm Co. v. Underwriters Laboratories Inc.* 1960).

A standard covering the subject of safety may validly include recommendations concerning (1) safety analysis, (2) composition of materials, (3) design, and (4) construction when these factors are deemed to be essential to safety, strength requirements, energy supply and control, reliability, durability, and product limitations, to mention only a few of the items that are permissible to include (Rockwell 1972). Safety standards are currently graded on a pass-or-fail scale (Federal Trade Commission 1972).

Thus, we see that the promulgation of a reasonable and properly formulated standard recommended by an association of manufacturers

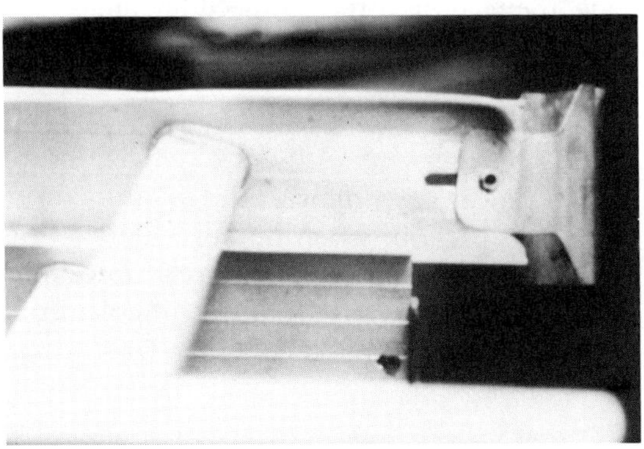

Figure 3-1. Poorly designed ladder foot. Note that the slot allows the thin sheet-metal foot to rest on uneven ground with only one foot to carry the ladder load. (Courtesy of QED)

can help establish the current state of the art and, in doing so, substantially aid the defendant–manufacturer. Although adhering to a standard will not automatically relieve a defendant of product liability, the combined technological wisdom of industry representatives will unquestionably be of substantial assistance.

To be effective, the standard in question must have been formally adopted when the product was designed and manufactured. Thus, in *Vroman v. Sears Roebuck & Co.* (1969), which involved an allegation of defective design, the plaintiff was originally allowed to introduce into evidence ANSI's safety specifications for power lawn mowers. This evidence was subsequently held to be improper, however, because the specifications had not been formally adopted until some two years after the mower in question had been designed and manufactured.

An industry standard that is inadequate and incomplete will be opposed on the ground that no industry should be allowed to control liability by establishing such patently self-serving standards (see *Safety: Can It Be Standardized?* 1961).

In many jurisdictions, proof that the defendant failed to meet mandatory federal, state, or local standards will be conclusive evidence of negligence or a defect under strict tort principles. Under certain circumstances, however, compliance with a mandatory standard may not relieve the defendant of liability. For example, in *LaGorga v. Kroger Co.* (1969), the manufacturer was found liable despite compliance with the Flammable Fabrics Act because the standard was insufficient. This principle has been written into Section 108(c) of the National Traffic and Motor Vehicle Safety Act, which reads as follows:

> Compliance with any Federal motor vehicle safety standard issued under this subchapter does not exempt any person from any liability under the common law.
>
> Since the fitness of a product is judged in relation to standards current at the time it is made, the manufacturer also need not anticipate liability for harm occurring at some future time when technology might allow improved construction or conditions affecting its safe use might change.

Thus, in another case involving a power lawn mower, the manufacturer was not liable to a user who injured his foot when he fell while cutting grass on a sloping grade. The court sustained objections to questions dealing with current design standards as evidence that the nine-year-old mower was defective (see *A Management Guide to Product Quality and Safety*).

METHODS OF ENSURING THAT STANDARDS ARE MET

Testing

Essentially, testing is undertaken to establish a rating or to verify a rating pursuant to a standard, if a standard is available. From a legal standpoint, a manufacturer is obligated to exercise due care in its manufacturing process. As proof that it used due care, it must be able to prove at trial that it made all reasonable tests and inspections.

After a product design has been approved and a prototype has been built, the first purpose of testing is to establish the satisfactory performance, safety, and reliability of the basic design (see *A Management Guide to Product Quality and Safety*).[1] Although some products, such as nuclear power plants, do not lend themselves to prototype testing by the manufacturer, the government usually sponsors prototype studies (called demonstration projects or plants) under its own auspices. Such studies use computer simulations and scale-model testing, and place the emphasis on applicable codes and standards, such as Section III of the ASME's boiler and pressure vessel code, "Nuclear Power Plant Components," which permit "design by analysis criteria," which are performance criteria.

Computer simulations have proved to be an invaluable complement to experimental testing. For example, such simulations permit the engineering designer to evaluate the effects of various changes in parameters such as material, geometry, and the like on the product before the final configurations are set for testing.

Because testing should be done during both the premanufacturing and the manufacturing processes, one must decide at the outset whether to use testing facilities within the manufacturing plant or to engage an independent laboratory (see Hazard 1968 concerning the use of experts and outside testing). For many fabrication processes, testing procedures such as statistical analysis (sampling techniques) are needed to determine what the proper test is (Johnson and Johnson 1973). An independent testing agency has been defined as a private organization that "autonomously tests and certifies products for consumer or manufacturer clients" (Rechlin 1970). For most manufacturers, certification by an independent testing agency serves a dual purpose: It ensures the safety and quality of the manufacturer's product and acts as a valuable advertising

[1]Reliability, the principal goal of testing, has been defined as "the probability that a product will operate successfully for a specific period of time and under specific conditions when used in the manner and for the purpose intended" (Scerbo 1974, p. 23).

tool. To the extent that the certifier is both independent and impartial, both goals are served effectively (Rechlin 1970). In some cases, of course, a customer's requirements determine whether a manufacturer will use in-house or independent testing. For example, the Coast Guard, when purchasing cylinders, sometimes requires the manufacturer to perform in-plant testing.

An independent testing agency may test a product for only one specific purpose. Underwriters Laboratories, Inc., for example, tests products only for safety. An independent third-party certifier ordinarily tests only a small sample of the product, and its liability to the public generally includes only "class defects": that is, a defect common to an entire line, such as a defective design. If a person is injured from some other cause, the manufacturer, not the third-party underwriter, is liable (*Tort Liability of Independent Agencies* 1968).

Because the independent testing laboratory is a seller of services—the testing and certification of products—not a seller of goods, it does not fall within Article 2 of the Uniform Commercial Code, which says that all sellers of goods who warrant their products are liable. We must caution, however, that the legal distinction between a seller of goods and a seller of services, for the purpose of imposing common law, with respect to strict liability appears to be disappearing and that today joint liability is and can be imposed on all defendants (see, for example, *Jackson v. Muhlenberg Hospital* 1970).

Regardless of who does the testing, the three elements required to spell out strict liability are those discussed in Chapter 1: (1) the product must be defective, (2) the defect must have been present before the product left the manufacturer's hands, and (3) the defect caused the injury. Thus, a manufacturer should consider the following question: "Would my testing and findings disclose the defect?"

A defective product can be defined, as Trayner did in a lecture, "as one that fails to match the average quality of like products. . . . [Thus,] the manufacturer is . . . liable for injuries resulting from deviations from the norm."

Finally, regardless of whether manufacturers test in-house or through an independent laboratory, they are legally required to make their own calculations or tests to determine the fitness of a product. If no standard is available, they must substitute their own best engineering know-how (*Product Liability Reports* 1974).

Quality Assurance

Qualified quality assurance personnel are a key factor in preventing product liability. The initial design of a product is critically important; it should be made in a manner that allows one to determine how the product might fail. Once that is known, one should be able to determine how to prevent or avoid such failures.

In most companies, the quality assurance manager's role includes inspection and testing. These two jobs are subdivided into receiving (including vendor controls), in-process, component, subassembly, assembly, and shipping. Quality assurance personnel should be familiar with both the quality of the product and relevant procedures. They also should understand the issues involved in the typical product liability suit, which include the design, test records, controls, analysis of components, photographs and data-recording devices, and correct measurements.

Design. The designer should consider the wide variety of possible environments to which the product may be exposed. As a representative of the automobile industry once said:

> The place to start to avoid recalls is right in the drafting room. . . . Defects occur in one of four ways—a flaw in the design (performance or ergonomics) that doesn't show up in testing . . . is the most serious type of defect because it will involve the largest number of actions; material selection; a flaw in the materials or parts sometimes bought from a outside supplier; a flaw in the manufacturing process, including sloppy workmanship or faulty tools. . . . ("Automobile Recalls" 1972, p. 39).

Therefore, in *Frey v. Chrysler Corporation* (1968), the court found that the manufacturer was liable for injuries a motorist sustained when the accelerator stuck because a guide bracket was too thick and a pull-back spring was too weak. One could ask: "Would testing have revealed these defects?"

Test Records. Test records should be maintained in a manner that reflects the experience developed during testing. From the manufacturing standpoint, recordkeeping that assures that vendor items can be traced is extremely important. These items should be traceable not only by batch, lot, or serial numbers but also to all processing, manufacturing, inspection, and testing processes and to shipping (to the first purchaser or customer, at least). Although keeping such records may be an economic burden, their importance cannot be overemphasized. Furthermore, all records referring to traceability should be retained because they provide evidence of the precautions taken during manufacturing, and may show that the defect occurred after the product left the manufacturer's hands.

Controls. Controls must be established to permit feedback about experiences in the field and circulation of testing data to interested personnel within the plant. Such controls provide evidence that when errors were discovered, corrective measures were taken promptly.

Analysis of Components. Whether one is dealing with a product or a purchased component, strict specifications should be established to achieve the desired standards of quality, reliability, and performance. This analysis should include data on the type of material used, the size and strength of the product, its anticipated use, appropriate instructions for its assembly, and procedures for inspecting it.

Most end products consist of components. Thus, every effort should be made to certify components purchased from outside suppliers. Is the physical strength and chemical composition of a component precisely as anticipated? If a component proves to have been defective when it left the supplier's hands, any liability to the manufacturer can be shifted to the supplier. Yet the manufacturer will be joined as a codefendant, with all the expense and inconvenience that attends such a position.

Today, the trend is to hold the manufacturer of the component liable under the strict liability theory. For example, in *Burbage v. Boiler Engineering & Supply Co., Inc.* (1967), involving a maintenance worker who was killed when a boiler exploded due to a defective valve, only the manufacturer of the valve was held liable because the manufacturer of the boiler could prove that it had incorporated the valve into the boiler without substantial change.

Fabrication. Current drawings also are important. In other words, is the manufacturer observing the most recent in-house and national voluntary and statutory standards for fabrication procedures and operator qualifications?

Inspection Equipment. Because high-quality testing enables a manufacturer to support warranty statements and permits the preparation of realistic instructions and warning labels, the obvious first goal of any testing procedure is to identify what to measure and what specific instrumentation is required. All this information should be recorded on a test data sheet. It has been said that, "If you can't measure it, you really don't know very much about it," to which can be added: "If you measure it, and you haven't properly calibrated the measuring equipment, then you won't pinpoint flaws that could later become the type of defect that causes injury."

Type and Frequency of Testing. Although much has been written about destructive and nondestructive testing, adhering to nondestructive techniques will do much to pinpoint flaws that later could cause injury. Such testing enables one to support warranty statements and to prepare realistic instructions and warning labels.

The frequency of testing, of course, will depend on many factors, including the complexity of the product. Because engineers understand the well-founded principles of statistical analysis, sampling methods, and reliability analysis, they know how much testing should be done. When testing is cut short, the reason is probably economic.

Photographs and Data-recording Devices. It seems self-evident that if one takes photographs of or videotapes the stages of testing, these records of success and failure can do much to verify that the product that was actually approved is the one that was not defective when it left the plant.

Correct Measurements. Measurement standards fall into two classes: those involving direct, accurate measurements such as temperature, and those that involve the destructive testing of a similar product such as testing of the chemical constituents of a plastic sheet. Mandatory statutory standards usually involve direct measurements (Mandel 1974).

Examination for Latent Defects

Inspection procedures ordinarily detect visual flaws. Detecting and preventing internal or latent defects, the most dangerous type, requires specific testing procedures. For example, a manufacturer of grinding wheels was held liable to workers who were injured when the wheels exploded due to variations in density and tensile strength (*Taylor v. The Carborundum Co.* 1974).

Figure 10–14a,b illustrates another grinding wheel failure. In this case, a consulting engineer's independent testing and analysis determined that because of fatigue, a crack propagated through the composite material, causing a catastrophic fracture when the size of the crack became critical. Because the manufacturer had not achieved total shielding of the disk, the consultant concluded that the manufacturer should have developed an inservice inspection procedure that would enable users to determine whether cracks of critical size were present.

Figure 3-2 illustrates a fabrication defect—misalignment of a drilled hole that caused excessive wall thinning of the bicycle steering fork.

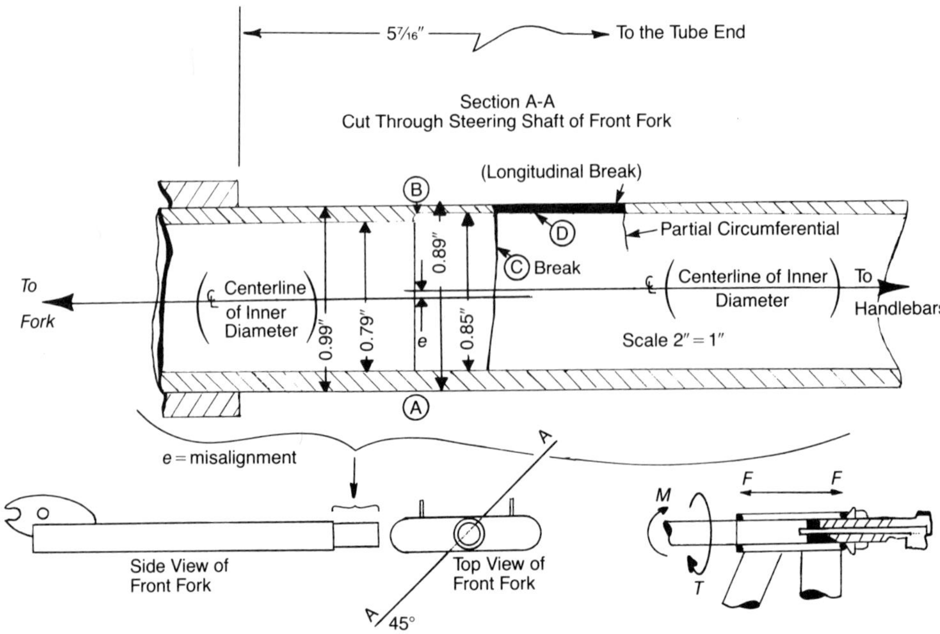

(a)

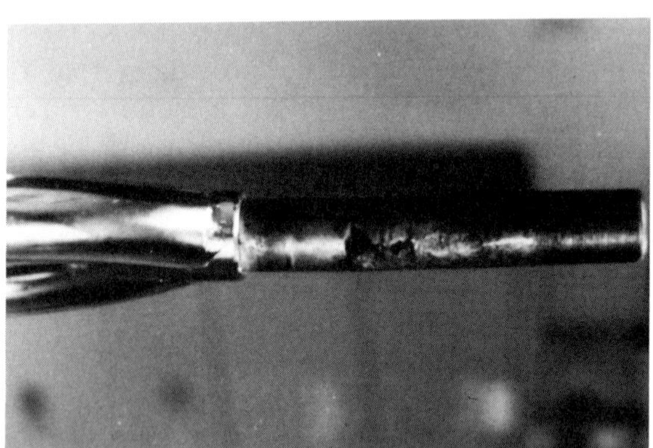

(b)

Figure 3-2. Illustration of a fracture in a steering tube caused by a misdrilled hole. (a) Cross-sectional schematic illustrating the break and the misdrilled hole. (Courtesy of QED) (b) Photograph of the fracture and the thinned tube wall that occurred during the fabrication process.

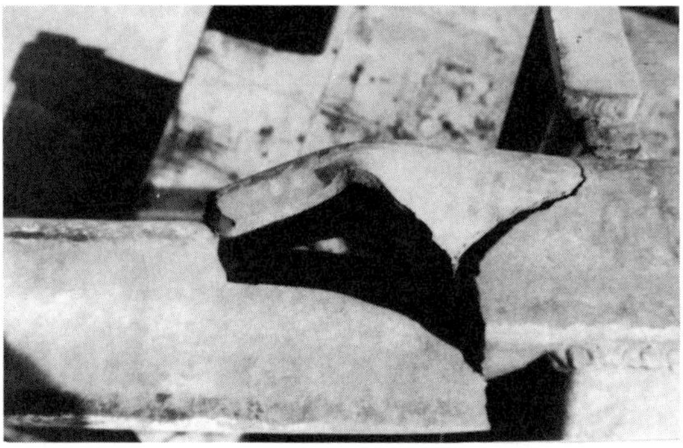

Figure 3-3. Failure analysis of a drilling derrick's square tube corners revealed longitudinal lamination and full-penetration defects that were introduced during milling.

Figure 3-3 shows a longitudinal defect in formed tubing caused by dimensions in excess of the manufacturer's specifications. Figure 3-4 illustrates a crack caused during the capping process at a bottling plant. Rather than develop an alternative inspection method to eliminate such events on the capping line, the bottler substituted plastic containers for glass bottles, which are not susceptible to explosive brittle fractures.

In *Sweeney v. Matthews* (1968), a nail manufacturer was liable to a carpenter-user when the head of a nail broke off during hammering and flew into the carpenter's eye. The court found that the nail had been improperly heat-treated and thus was brittle and weak. In the case illustrated in Figure 3-5, the straightness of the head and shank were not maintained during manufacturing.

Updated testing techniques for latent defects are extremely important. Numerous technical and professional societies provide forums through journals, magazines, conferences, and exhibits as important sources of information about the latest techniques for testing and inspection.

Figure 3-4. Crack defect in the neck of a glass bottle caused during the capping operation.

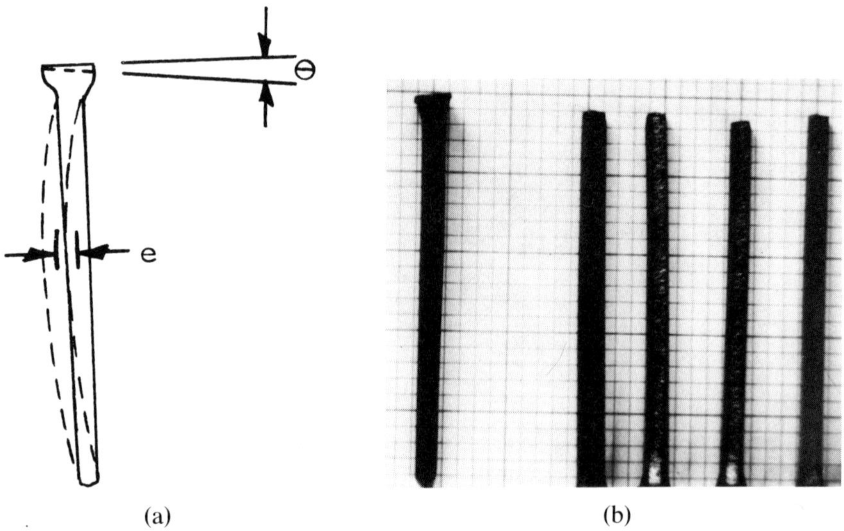

Figure 3-5. (a) Eccentric (e) and skewed (θ) concrete nails. (Courtesy of QED) (b) Photograph illustrating dimensional variation in concrete nails.

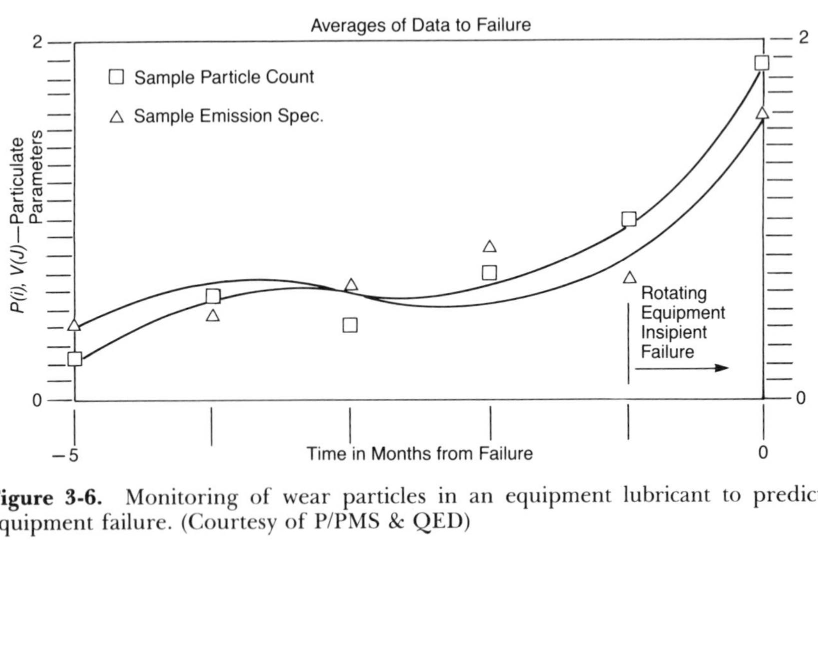

Figure 3-6. Monitoring of wear particles in an equipment lubricant to predict equipment failure. (Courtesy of P/PMS & QED)

Tests that can be used to measure materials and to detect defects are discussed in Chapter 8. Here, however, consider the following example of how to test or monitor mechanical parts for excessive wear. In an article that appeared in *Scientific American* in 1974, Seifert and Wescott explained that when parts of a machine wear, they shed tiny bits of metal. The authors described a new microscopic method of analyzing these metal particles in an effort to determine in advance when a machine would fail (see Fig. 3-6). With this method, they determined that the type of particle associated with wear is usually generated by rubbing or adhesive wear. These particles are produced in large quantities when one metal surface slides over another; for example, when a piston in an automobile engine slides in a cylinder.

Such sophisticated testing may not be required, but in a product liability case involving a latent defect, an attorney who wants to prove that the client has met the warranty would find such data extremely persuasive. In industry, such methods are routinely used for preventive maintenance in conjunction with equipment dynamics time-histogram equipment signaturing, a technique that is not new but has become more practical because of developments in microcircuitry (see *Lubrication and Maintenance,* published by Fluid Marketing, Corp. and Trucker, Co. as a source reference).

THE TECHNICAL PROFESSIONAL: LEGAL AND MANUFACTURING ASPECTS

Use of Technical Experts to Interpret Codes and Standards

When searching for answers about elements such as integrity of product design, the presence of concealed danger, defects in the composition of a product, or the adequacy of safety devices, the attorneys of both plaintiff and defendant need independent technical assistance. This assistance is usually provided by "experts" (Hazard 1968), who will need test evaluation data as the basis for their own investigation.

Cases involving design defects become a battleground for experts, and so-called defects in engineering design present the jury with difficult questions. A sampling of jury decisions reveals that the design fails to incorporate (1) properly applied engineering principles to anticipate modes of failure, (2) materials with appropriate properties, (3) correct specifications for a milling process, (4) specifications for the operator-user, (5) fabrication or welding data that are correct or appropriate for production of a safe product, and (6) instructions for correct maintenance, inspection, and documentation procedures (Society of Automotive Engineers 1970).

As was mentioned in Chapter 2, the engineering and scientific experts who participate in product liability cases can represent an extremely broad range of expertise: from general design engineering to specialization in engineering mechanics; metallurgy; biomechanics; or electrical, chemical, or civil engineering. Other important aspects of some product designs require specialists in the psychology of instructions, warnings, human behavior, and the like.

The relationship between the expert, the attorney, and the client will be explored further in Chapter 12. Here, however, we should mention that technical experts should be familiar with (or become familiar with) the codes and standards that affect their area of expertise. Although this may sound paradoxical to the nonengineer, *design codes* often reflect an attempt to provide a simplistic or greatly simplified procedure as a minimum criterion. Many codes, such as the ANSI standard for ladders, are geared for implementation by nonengineers. A manufacturer should seek the assistance of a technical expert who can apply the appropriate technology to the product. This will ensure that the product exceeds the codes and standards with respect to design, fabrication, and instructions for use, and consequently reduce the risk of failure.

Application of Legal Principles

The failure to prepare and execute procedures for design, testing, and other elements that would reveal defects amounts to a breach of one's duty to exercise ordinary care. How does one ascertain the degree of evaluation required to reduce the risk of liability? In each case, the facts will determine the ultimate result of litigation.

As a general legal principle, a manufacturer is required to conduct as many evaluations of its product as necessary during the manufacturing process and after the process is completed to ensure the safety of that product. If the manufacturer failed to make an evaluation that would have revealed a defect likely to cause injury, this failure constitutes negligence and the manufacturer is liable. If an evaluation is judged to be careless or inadequate, this, too, constitutes negligence.

Furthermore, the number and scope of the evaluations must be proportionate to the risk of harm the manufacturer could reasonably anticipate. Thus, if harm cannot be foreseen, an evaluation is usually not required. If the manufacturer can prove that the latest techniques, including accredited standards, were used to discover defects, its liability will be minimal. If the manufacturer developed detailed procedures that revealed no defect, the plaintiff will have difficulty proving that the product was defective (*Product Liability Reports,* Commerce Clearing House (CCH), p. 4756).

Assume that a customer furnishes the manufacturer with specifications for a certain product. Under these conditions, the manufacturer has no duty to evaluate the safety of the customer's design. In *Garrison v. Rohm & Haas* (CCH, para. 7127), a manufacturer of industrial equipment fabricated a dolly from a customer's specifications. When a worker was injured by a load that fell from the dolly, he sued the manufacturer. But the court held that the manufacturer had no duty to test the device once the manufacturer proved it had followed the customer's specifications.

Earlier, we discussed *Philip J. Nordstrom v. White Metal Rolling & Stamping Corp.* (1968), in which the plaintiff sued both the manufacturer and seller of a ladder that collapsed. This landmark decision confirmed the admissibility of an industry standard. During the trial, evidence was admitted concerning the ANSI Safety Code for portable metal ladders, and the manufacturer established that the ladder had been manufactured in accordance with the code. The court found that the code had been published before the litigation and could be properly admitted into evidence, thus relieving the manufacturer of liability. The court declared that a standard or code did not need to be written into the law and gave great weight to voluntary industry standards. This case is well worth reading because, during the trial, the defendant's experts gave detailed testimony about tests made on similar ladders to determine their strength. This evidence also was admissible.

We should emphasize, however, that when a code or standard specifically addresses a specific aspect of the design in question, the manufacturer's position is greatly enhanced. In contrast, if the manufacturer takes the position that design does not violate the applicable code and standard, but the code and standard does not address the issue in question, the manufacturer's position is precarious.

Testing to Identify Defects

This section begins with the general proposition that if a manufacturer failed to conduct a test that would have revealed a defect, this failure constitutes negligence and therefore the manufacturer is liable.

For example, a plaintiff who purchased special aluminum wheels for his car claimed that one wheel cracked when it struck a curb. As a result, the plaintiff lost control of the car, which collided with a stanchion. He attempted to show at the trial that the standard used to test the strength and quality of aluminum wheels was for the steel wheels used on most cars. The court refused this evidence. The court of appeals, however, reversed the original decision, holding that the plaintiff's evidence would have enabled the jury to determine whether the wheels were as safe as an ordinary consumer would expect them to be.

In *Mooring v. Fram Corp.*, the owner of a truck with a damaged engine sued a manufacturer of oil filters, claiming that the filter installed in the truck had been negligently made and that the manufacturer had failed to inspect and test the filter properly. The plaintiff introduced expert testimony that the return duct in the filter had separated from the lid on the filter. The cause, the expert said, was a buildup of pressure because of an inadequate number of perforations in the return duct; thus, the oil could not flow freely from the filter, then back into the engine through the return duct. The court held that the defendant had failed to conduct the proper tests and that the expert's testimony was admissible. (See, also, *International Harvester Co. v. Sharoff).*

The importance of the instrumentation used in testing is illustrated by the following examples. In *Crown Cork & Seal Co. Inc. v. Morton Pharmaceutical Inc.* (1969), a judgment on a counterclaim against a manufacturer of aerosol cans of a pet-grooming aid in favor of the purchaser was reversed. The purchaser had bought some 36,000 cans, 3000 of which leaked. The purchaser claimed that the cans were defective and sought recovery for negligence and breach of implied warranty of merchantability. Whether the leakage was the result of faulty manufacturing or of faulty crimping of the cans by the purchaser when installing the valve-cup assembly was in sharp dispute. The trial court had erred by refusing the manufacturer's evidence consisting of test results in which a newly developed micrometer was used to measure the depth of the crimp around the valve-cup assembly. The manufacturer contended that the evidence would show that (1) the depth of the crimp was improper, which resulted in a bad seal between the can and the assembly, and (2) the improper crimping was the primary cause of the leakage.

In the case of *Perfecting Service Co. v. Product Development & Sales Co.* (1963), an engineer's testimony about an experiment he had conducted with a fan mechanism installed in a particular type of automobile was properly excluded because he failed to provide evidence that the mechanism had been installed in the type of car the specialty company had intended. A metallurgy engineer's testimony about a laboratory test he conducted to determine the tensile strength of the metals used in the mechanism also was properly excluded because he was unable to establish whether the materials tested had been manufactured before or after the design had been modified.

In *Whitefield v. Earl E. Knox Co.* (1956), liability was imposed on both the manufacturer and the reconditioner of a diesel engine for injuries sustained by the owner's employee when the engine went out of control and the flywheel disintegrated. The company that reconditioned the engine was judged to be negligent because it failed to test the engine before placing it back in operation, and the manufacturer was judged to be negligent because it failed to inspect the engine before starting it.

In a leading case involving a latent defect (*Argo Industries Inc. v. Heat-Timer Corp.* 1972), a distributor of heating equipment sued a manufacturer of a heating-zone component. The court found that the component had been inadequately tested before shipping and, in addition, that the product should have been tested under actual field conditions before marketing. Because the defects were latent, they did not appear until the component had been installed and placed in operation for varying periods of time.

In *Maurizio D. Fortunato v. Ford Motor Co.* (1972), an owner of a Ford Mustang testified that he was injured when he lit his cigarette with the car's lighter and the car immediately burst into flames. The main issue was his contention that a defectively designed gas tank made the car unsafe. He offered proof that the metal used in the top part of the tank led to a small hole that allowed gasoline vapors to escape into the trunk and, eventually, into the passenger compartment. Ford contended that the plaintiff's explanation was physically impossible and introduced the testimony of an employee who had tested a similar car and was unable to generate enough heavy vapors to cause an explosive condition. The court stated that enough evidence had been developed to warrant sending the case to a jury, which later held in favor of the plaintiff.

THE PROPOSED FEDERAL UNIFORM PRODUCTS LIABILITY ACT

The current product liability crisis did not happen overnight. It reflects the erosive consequences of changing from the principles of common law to the strict liability doctrine. What the outer limits of this doctrine will be is unknown (see Epstein 1980).

More than 30 states have acknowledged the crisis by enacting remedial legislation, but these attempts have proved to be ineffective. Because an estimated 70 percent of goods produced leave their state of origin, a manufacturer cannot predict which law will apply if an injury results from use of its product.

In 1982 a decision in New Jersey (*Beshada v. Johns-Manville Products Corp.*) concluded that both the manufacturer and its distributor were liable whether or not they were aware of the defect or risk. The theory behind this decision, which dealt with "failure to warn," was grounded in the belief that the manufacturer and distributor are in the best position to allocate the cost of the injury. Therefore, by "imposing on the manufacturer the cost of failure to discover hazards, we create an incentive for them to invest more actively in safety research."

This problem of inequitable relief for comparable injuries has been under study within the executive and legislative branches of the federal government since 1976. The Department of Commerce established the Federal Interagency Task Force on Product Liability, which completed its initial report in 1977 (*Final Report*). On the basis of that report, the Department of Commerce prepared the Draft Uniform Product Liability Act for comment in 1979 (*Federal Register* 1979, Sec. 2996). A final version of the act was published in the *Federal Register* on October 31, 1979.

Some relief in reducing product liability is found in the Risk Retention Act, which became law in September 1981. This federal statute offers suppliers and sellers the option of forming their own cooperative insurance company without having to comply with the customary guarantees required under Public Law 97-45.

Efforts to introduce a uniform statute in the various states were unsuccessful. Meanwhile, various congressional committees conducted independent studies and eventually introduced bills in both the Senate and the House. (Senator Robert Kasten and Representative Henry A. Waxman authored these proposals in July 1984.)[2] Because recent administrations have favored some form of federal tort reform, such legislation is likely to be passed eventually.

One principal objection to the proposed federal statute is that the federal law would preempt all state product liability laws, although the statute would be interpreted by state courts. Both the Senate and House versions of the statute include provisions covering the following (compare H.R. 7284 and S 2631): preemption of state law, the manufacturer's liability (indicating the grounds for such liability), defective construction, design defects, failure to warn, liability of sellers other than manufacturers, failure to comply with federal standards (creating a rebuttable presumption that the product was unreasonably dangerous), evidence of postmanufacturing improvements, a statute of repose (25 years from the date of first delivery), and a statute of limitations requiring the claim to be filed within two years.

---

[2]For a review of the main arguments for and against such legislation, see the *American Bar Association Journal* (1984) and Atkeson and Neidich (1983).

THE PRINCIPALS AND CODES AND STANDARDS

As was stated in Chapter 2, the principals that influence product design are the design engineer, the manufacturer, and the user-consumer—each of which can be an individual or a giant multinational corporation with hundreds of thousands of employees. The principal influencers of product design are, at once, adversaries, cooperative, and accommodating. Voluntary codes and standards as well as governmental legislation are, or can be, their stage.

Users want codes and standards that will ensure that products are uniform, compatible, durable, and safe. And they want those products to be as inexpensive and foolproof as possible.

Engineering designers have a vested interest in having codes and standards oriented toward *performance* (durability and safety) and *function* (ergonomics, uniformity, and compatibility). Some of them view *design codes and standards* as a mechanism for reducing the risk to themselves. That is, codes and standards eliminate the element of decision making concerning analytic procedures.

Manufacturers view codes and standards as a vehicle for eliminating other manufacturers who would use practices that would cause product failures, injuries, or economic impacts that would be unfavorable to their industries. In other words, they view codes and standards as a vehicle for reducing liability by establishing minimum standards and, in many instances, by adopting design standards rather than performance standards. Because design standards usually provide an opportunity to use technicians rather than qualified professional engineers to certify a product, such standards are favored by less technically developed industries and manufacturers. Some manufacturers, like designers and users, become members of codes and standards committees as a defensive posture for their organization; that is, they do not want requirements that would have an unfavorable impact on their competitiveness.

As was mentioned in Chapter 2, some codes and standards committees have a balanced representation of engineering design professionals, manufacturers' representatives, and representatives of user organizations. Most organizations that develop voluntary codes and standards, however, are composed of manufacturers' employees.

The historical absence or discouragement of the engineering professional's and consumer-user's participation in the development process—and their inability to organize representation—has led users to resort to the courts as their primary instrument for affecting product quality.

Some of those interested in this issue regard balanced codes and standards committees as the best means of achieving effective codes and standards. As was pointed out in Chapter 2, however, no instrumentality, by itself, can improve the quality of design.

CONCLUSION

Just as the eighteenth century was characterized as the beginning of the Industrial Revolution, historians are likely to characterize the twentieth century as the beginning of the safety revolution. The strength of the consumer movement demands a greater degree of care when evaluating products. Thus, the valuable contribution of industry standards in helping manufacturers must not be underestimated. Because decisional law now reflects the aid given to manufacturers who comply with industry standards, committees that develop codes and standards should be encouraged to make the fullest useful possible of them.

More than any other factors, the "deep pocket theory" and liability—even when the defendant is not at fault, and all in the name of protecting the consumer—have led to excessive insurance costs and to uncertainty in the entire marketing process. The seriousness of the problem and the attempts to solve it to date may well herald a new era in which federal law will bring some order out of the chaos. Although such a law may preempt current state provisions, no constitutional barriers prevent the federal government from enacting a federal reform tort act.

4

Guidelines for Manufacturers

In the past, when a customer-user was displeased or angry because a product failed to function or function properly and caused injury or economic loss, the manufacturer or its sales agent might receive a phone call or a letter demanding restitution, compensation, or both, or the customer might dismiss the incident as an "act of God." In either case, the parties were more than likely to resolve the matter between themselves—happily or unhappily—on the basis of prevailing moral or ethical considerations or of political and economic advantage and power. Some companies and individuals were inclined to settle the matter through their attorneys.

Today, parties in conflict are more likely to rely on the courts to resolve their disputes, particularly when personal injury is involved. Law firms specializing in personal injury advertise on television to appeal to larger audiences. Some believe that these developments mean that justice is being democratized; others believe it is being trivialized. But democratic institutions always seem to be shifting to adjust to contending forces for the economic good of all.

As was discussed in Chapter 1, all that plaintiffs may have to prove is that a defective product was responsible for the injury or damage; they do not need to show that the manufacturer was careless in making the product. Although breach of warranty and negligence still play an important role in a number of product liability cases, the trend is toward strict liability (the concept of implied warranty, which says that if someone is injured, then someone will pay), and, for the most part, it is the manufacturer (the "deep pocket") who pays.

Today, judgments in product liability cases can exceed $10 million. This is a major reason why the insurance premiums of some manufacturers have increased as much as 1000 percent over the past decade. Even at these high rates, many insurance companies will not offer product liability insurance to certain industries or provide certain types of coverage because of the increased risk of suits. In part, this is the result of user awareness, which has increased because of the publicity the media gives to private and governmental investigations, product recalls, and major catastrophes.

To give readers a frame of reference concerning the costs of product liability and the combatants, a number of product failures discussed in this book resulted in court awards or settlements in excess of $20 million. The major economic loss was not attributable to personal injury but to loss of production and the cost of repairs experienced by the owner of the equipment—the manufacturer's customer. In some instances, the legal fees involved in the suit reached or exceeded $2 million and the costs of engineering/scientific analyses were approximately $200,000. In one instance, a law firm took more than 65 depositions, spending approximately $20,000 on the services of engineers or scientists. The damage in most cases was caused by a deficient design, unsatisfactory material, a defect, an improper fabrication process, misuse, or improper maintenance, that should have cost no more than $200 to $20,000 to prevent.

Some settlements involving personal injury may seem high until one examines the case further and realizes the amount of medical, rehabilitation, support systems, and personnel that the victim requires over many years. In many instances, the victim is left with only modest assistance or no assistance at all. In all the personal injury cases referred to in this book, the victims indicated that they would gladly return to the physical, mental, and financial condition they had been in before the incident.

Two causes of increased pressure on manufacturers are the increased sophistication of the user and the quantum leap in technology, the latter having given the public greater access to large databases and information retrieval systems. As a result of product liability lawsuits, manufacturers who attempt to reduce production costs at the expense of safety are eliminated from the marketplace because they threaten to stigmatize an entire industry.

To produce a safe, functional, and cost-effective product and to defend against allegations of a defective product, a manufacturing, fabrication, or processing organization must have an effective defense. But its defense can only be as effective as its offense—its knowledge of the instrumentalities that affect product liability and its ability to respond according to standard operating procedures and policies.

Within this context, eight basic topics are addressed in this chapter: the manufacturer's need for (1) a product liability loss assurance (PLLA) committee, (2) qualified engineering designers, (3) careful quality assurance procedures, (4) well-thought-out manuals, instructions, and labels, (5) well-prepared marketing procedures, (6) stringent procedures for packaging and shipping, (7) equally stringent procedures for handling customer complaints and claims, and (8) complete records.

Before we review these topics, consider Fogel's observations (1987) concerning project or contract management and administrative guidelines. Managers and engineers are increasingly embroiled in battles in which failure is defined as improper administration or management rather than the failure of a product or system to function as designed. These individuals are charged with failing to properly manage or administer one or all of the following: the predesign and design process, the bidding process, responsibility during the construction or fabrication phase, or the closeout phase of a project.

Engineers like to think of themselves as professionals who are skilled in applying scientific principles to achieve practical results. However, they often lose sight of the fact that in the process, they must manage their contracts to eliminate "failures" and the claims that result from them. A claim or lawsuit resulting from the failure to manage or administer properly is just as grievous as one resulting from a design error.

Contract administration is no less important than the design and construction process. Because most management failures result from errors of omission rather than errors of commission, it is important to prepare and follow through with a checklist. The following generic list can be modified to suit any project.

1. Check drawings for proper coordination between trades.
2. Check to make certain that the information on the drawings conforms to that written on the specification.
3. Inspect to see that parts, components, structures, and materials conform, specifically and uniquely, to the project.
4. Enter into written agreements with consultants that detail precisely what their responsibilities are.
5. Review the work performed by consultants for conformity with agreements.
6. Check the final program against the estimate or budget to determine its current validity.
7. Communicate properly with consultants so that a change made by one consultant that affects the work of another is properly accounted for.

8. Include owner-consumers in the communication process to permit them the privilege of determining in advance how their money will be spent.
9. Establish a reasonable time frame for review.
10. Establish a reasonable time for implementation. (Or, we bow to the owner's desire for a project by a fixed date and insert that date into the contract, knowing that duration specified is unreasonable or is possibly only at a cost that far exceeds the estimate.)
11. Specify the schedule, using one of many available network scheduling techniques.
12. Review the specifications to determine whether the items specified are still available or, for that matter, whether they have been available in recent years.
13. Check whether the information given to or by a sales representative or "sales engineer" is valid and conforms to the specification sheet prepared by the company that person represents.
14. Check the bids, quotations, or proposals for conformity with the contract documents.
15. Review the insurance requirements of the contract documents to assure compliance.
16. Review the schedule carefully. (Often the schedule is filed away with approval or disapproval although it may contain a specific contract reference to output.)
17. Establish and adhere to an orderly system of controlling or keeping track of the documentation that must pass through the manager's hands or office to assure completion of the project.
18. Retain the backup information or supporting data for the approval, modification, or rejection of requests for payment.
19. React in a timely manner to requests for clarification or interpretation of the project documents.
20. Accept the responsibilities that are written into the contract documents. (Some are especially lax in asserting the authority for coordinating the work of the various independent agents involved with the project.)
21. Monitor the progress of the project and report objectively to the owner as required.
22. Assume responsibility for discrepancies and omissions in the project documents.

23. Issue clarifications and instructions that will point out errors on one's part in a timely fashion and thereby eliminate or minimize one's exposure.

PRODUCT LIABILITY LOSS ASSURANCE COMMITTEE

The first thing a manufacturer must do when addressing product liability and safety is to have its chief operations officer establish a PLLA committee. Such a committee assures the implementation of an overall strategy and a procedure for coordinating product design, quality assurance, and the other six programs listed in the previous section.

The function of the committee is to assure that every department in the company is aware of the appropriate steps to take in a project and that every department will take action to prevent or minimize liability claims by establishing a safety program that will reduce the risks of hazards (Boley 1977; Environmental Protection Agency 1972; Green 1982; Hammer 1972; National Academy of Engineers 1970; Zajic and Hummelman 1978).

Hammer (1972) presented a model that an organization can use as a reference when forming a PLLA committee and carrying out a safety program. The manufacturer already may have a safety organization in place that performs some of the tasks outlined in Figure 4-1 (see also Fig. 4-5, p. 116) and pursues some of the objectives regarding product design and other subjects discussed in this chapter. New processes and safety developments identified by the committee can benefit the organization. The committee's function should be reviewed periodically, however, with respect to its potentially changing role. If the committee is accomplishing its objectives, it should be able to answer affirmatively the questions listed in Box 4-1.

PRODUCT DESIGN

The challenge for the design engineer is to create products that sell, have aesthetic appeal, are functional, operate safely in a prescribed manner, and are "foolproof." Recall that a product is defective if it is not reasonably fit for its intended use according to the warranty of fitness theory, which is found in contract law. Further, under the law of tort, the strict liability theory says that a product is defective if it is unreasonably dangerous.

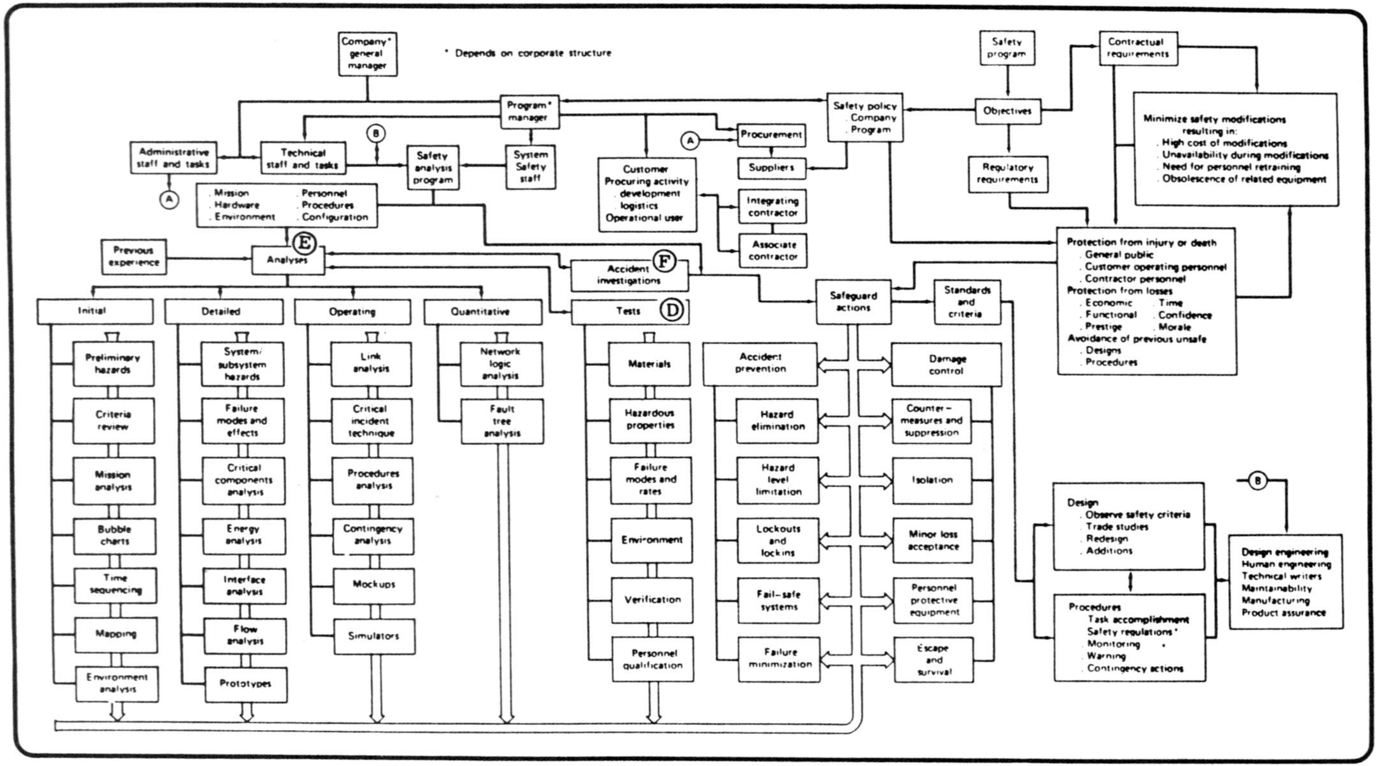

Figure 4-1. Safety program model. (From W. Hammer, *Handbook of System and Product Safety*, pp. 36, 49. Englewood Cliffs, NJ: Prentice-Hall, © 1972.) (See Fig. 4-5 for continuation.)

BOX 4-1. Questions that Management Should Ask the Product Liability Loss Assurance Committee

1. Are the quality assurance manuals up to date?
2. Are records maintained on the testing of product and process?
3. Are guidelines available for ensuring that the committee's mission will be carried out during design, development, production, tests, transportation, handling, operations, maintenance, and repair?
4. Are procedures in place for implementing, updating, and reviewing manuals, instructions, and labels?
5. Is a procedure for packaging and shipping in place?
6. Has a marketing policy or procedure been established?
7. Is a procedure or policy in place for handling customers' complaints or claims?
8. Does the product meet or exceed the standards and/or codes developed by industry associations, technical societies, or government entities?
9. Does the reliabilty of the product match or exceed that of other, comparable products on the market?
10. Do the instructions, labels, or brochures call the customer and end-user's attention to specific hazards? Do these materials specify the user's responsibilities and obligations for the product?
11. Do the safeguards designed into the product include misuse or other foreseeable problems? Are these safeguards active or passive? (See National Academy of Engineers 1970.)
12. Have adequate steps been taken to prevent the inherent dangers of the product to company employees as well as to customers? (See Boley 1977; *Chemical Engineering Progress Manual* 1987; Hammer 1972.)
13. Have steps been taken to monitor outside suppliers of components or parts to ensure that the product containing those components or parts will be safe?
14. Has the company conducted every reasonable test to assure the safety of the product?
15. Do the records reflect the company's safety endeavors?
16. Are databases on product performance (in general and by manufacturer) available from industry associations, technical societies, user associations, and governmental agencies? Is this information used to remedy problems in as expeditious a manner as possible?

BOX 4-1. *(continued)*

17. Have consequence and risk analyses been performed? What is the worst possible consequence that can be the result of the worst possible defect (based on current inspection methods and the safety features already built in)?
18. Has the committee recommended preventive or protective measures to eliminate or control hazards?
19. Have all operating and faulted (emergency, runaway, scram) reactions caused by operator or equipment malfunction been identified in fault-tree, risk, and consequence analyses? Does the equipment have proprietary or specialized uses? Has the user supplied performance requirements? Have these requirements been specified contractually and reviewed by the committee's engineering and legal representatives? (See Green 1982; Hammer 1972; Fig. 4-1.)
20. Have the histories of hazards, failures, and mishaps of existing systems been reviewed?
21. Does the committee keep management informed about its status, significant problems, deficiencies, and methods of improvement?
22. Does the committee participate in design preparations and reviews to ensure that incompatible or unsafe design, process, operation, and maintenance procedures are not incorporated?
23. Does the committee participate in product performance ratings (see Table 4-1) to ensure that safety is not compromised by changes in mission, hardware, configuration, procedures, and the like?
24. Does the company have an independent review procedure or audit for each major function in the company, including the PLLA committee?
25. Does the committee monitor reports of failures and incidents to determine the presence of discrepancies, deficiencies, or trends that might affect safety? Does it make suitable recommendations for corrective action?
26. Does the committee ensure that safety training programs are adequate to meet organizational needs? Does it initiate improvements for such training?
27. Does the committee investigate mishaps involving the product while the product is the company's responsibility?
28. Has the committee developed a plan for investigating mishaps involving the product?
29. Do the members of the PLLA committee serve on boards or committees that deal with industrial safety or related areas?

BOX 4-1. *(continued)*

30. Are protective equipment, warning devices, and emergency and rescue equipment required for the product? Is that equipment selected and properly identified?
31. Has information on hazards been disseminated through other organizations that might be affected?
32. Have corrective actions been taken to prevent the recurrence of mishaps caused by similar deficiencies or practices?
33. Have the effects on product safety of outside suppliers been determined?
34. Is a mechanism in place for preparing safety analysis programs required by customers or clients?
35. Does a mechanism exist for analyzing safety requirements, procedures, and milestones required by subcontractors?

One extremely important step the manufacturer must take in preparing to market—or to revise an existing product—is to retain or employ the services of a professional design engineer who is qualified by training and experience. Although designers and engineers are not necessarily synonymous, design engineers provide a safety net for the manufacturer because they use scientific principles to assess, both theoretically and experimentally, the safety and function of a product within an economic framework. Their academic degree and area of specialization (e.g., mechanical, civil, electrical, chemical) is dictated by the product, process, and use.

A design concept can come from a variety of sources—markets, competitors, inventors, housewives, and even a designer or engineer. No matter what the source, however, a qualified design engineer is trained to translate the concept into a functioning, manufacturable, safe, and marketable product. The manufacturer who copied features of an extension ladder produced by another company and manufactured a ladder with a dangerous foot and side-rail assembly illustrates what can happen when an engineer is not qualified to interpret borrowed concepts properly.

A product failure can often be traced to the fact that a manufacturer modified or extended a product with an excellent record beyond its original design to accommodate a user's request without asking an engineer to evaluate how the changed product would perform. Sometimes a revision based on a hunch succeeds, but in most instances of catastrophe, the means and methods were available to extend the "window" or "envelope" of product performance. For example, if the space shuttle O-rings were known to work only for 30°F to 1000°F, then to extend the "window" or "envelope" tests and analyses would be needed to determine whether they would work for 21°F to 1000°F.

If a user wants to modify a product, the manufacturer should make it clear to the user that no modifications are permitted without the manufacturer's approval. Consider an example in which a user who purchased four large-capacity pumps for its facility decided to fabricate its own replacement parts. The user changed the design, material, fabrication tolerances, use and maintenance from those originally specified without authorization from the manufacturer and without the benefit of an engineering analysis concerning how the changes would affect the risk of failure. The user's changes, particularly the material changes, led to failure caused by fatigue. Subsequently, an independent engineering analysis identified the mode and mechanism of the failure and the remedy (S. J. Brown 1989).

Many product failures are caused by the manufacturer's failure to have a product reviewed from "design through maintenance and inspection procedures" by an independent engineering consulting firm. Such a review is especially important if the firm is small or has no engineering staff of its own. This type of review, however, is also important for large organizations, which can become complacent about how they have always done things. In addition, having an independent design review or a manufacturing audit by an engineering consultant with broad experience in other industries is a good practice for large organizations, because the consultant can provide new, innovative, or different ways of solving problems and identify new, more effective techniques.

In some industries, independent in-house audits must be performed by different departments or divisions. For purposes of illustration, consider the following incident that occurred in 1987. A small manufacturer who had no previous experience in designing or fabricating pressure vessels identified a need for metallic sampling spheres in the chemical process industry and began to produce and advertise a unit that, it claimed, could operate at pressures up to 3500 pounds per square inch (psi) with a relief device. A customer's technician, who was using one of the spheres attached to a production line, was killed when a vapor cloud was released and the hydrocarbons subsequently ignited.

After the incident, a design engineer discovered that, according to standard engineering-design practices, the sphere should not have been operated above 1300 psi. The most widely quoted industry standard for pressure vessels and piping—the *ASME Boiler and Pressure Vessel Code*, Section VIII, Division 1 (1968 to present)—recommended a limit of 850 psi. In addition, the spherical vessel did not contain a vent line for the control of discharged vapors through the protective relief device, and it did not have a plate to indicate the maximum allowable working pressure (MAWP). The unit also had no permanently affixed serial numbers of traceability or a warning that if it was attached to a pressure system with a higher maximum allowable working pressure, it would rupture before the other system ruptured. Cases such as this are the main reason why product designers must establish procedures for product liability loss control that emphasize safe designs and coordination with the manufacturing departments.

Design reviews are a means of critiquing designs, analyses, engineering specifications, processes, and operating, maintenance, and other procedures as they potentially relate to safety. Hammer (1972) described the sequence and content of these reviews as follows.

1. A *concept review* considers requirements concerning mission, performance, environment, and other related factors. During this review, the approaches proposed for meeting these requirements, the advantages and disadvantages of each approach, tradeoff and feasibility studies, and laboratory investigations are discussed; expected problems for the proposed approach are pointed out; and methods of controlling problems are offered. (See Table 4-1 for an example of a product performance rating.)
2. A *definition review* addresses requirements and designs in relation to the ability of the proposed product or system to meet stipulated objectives.
3. A *preliminary design review* (PDR) is conducted on initial designs. It takes place either before or soon after the initial details of the design have been established. Safety personnel participate by providing an initial analysis of hazards (see Fig. 4-1) and other significant information that is available during the PDR.
4. A *critical design review* (CDR) is an extension and updating of the PDR as the design proceeds. Tests (especially of failures), progress in eliminating hazards, and provision of safeguards developed since the earlier reviews are covered during this review. In addition, subjects discussed during the PDR that resulted in requests for additional studies are generally discussed again at this time. Progress reports are presented if the design study is still under way. If the study has been completed, the findings are reported.
5. The *prototype* or *first-article configuration inspection* reviews the proposed design before the prototype is built. Updated or newly completed analyses, tests, and product performance ratings are

TABLE 4-1. Sample Product Performance Rating[a] of Wigit to Control Displacement

Selection Criteria	*Wigit Concepts by Group and Design Variation (V)*											
	I Hydraulic Group			*II Pneumatic*		*III Thermal Expansion (Active)*				*IV Thermal Expansion (Passive)*	*V Mechanical Group*	
	V1	*V2*	*V3*	*V1*	*V2*	*V1*	*V2*	*V3*	*V4*	*V1*	*V1*	*V2*
Reliability	3	3	2	2	3	3	3	3	2	1	2	2
In-service inspection	3	3	2	3	3	3	3	3	1	3	3	3
Maintenance	3	3	2	3	3	3	3	3	3	2	3	3
Replacement	—	—	1	2	—	3	3	2	3	3	3	3
Compatibility	3	3	3	3	3	2	2	3	2	2	2	2
Manufacture	2	3	3	3	2	3	3	3	2	3	3	3
Code acceptance	3	3	3	3	3	2	3	3	3	3	3	3
Design limits	2	3	3	3	3	2	3	3	3	1	2	2
Development	3	3	2	2	3	3	2	3	2	1	3	3
Cost	3	3	1	1	3	3	3	2	3	3	3	2
Human factors	3	2	3	2	2	1	2	2	3	3	2	3

[a]Rating: 1 = unsatisfactory; 2 = satisfactory; 3 = excellent.

presented, and subjects left open for determination at the CDR are discussed.

6. A *"flight" readiness review* is conducted to check the readiness of principal items and support equipment and is usually accomplished immediately before the "first flight" or initial production of a product to ensure that no outstanding problems remain uncorrected. Objectives also are considered.
7. A *preproduction review* assesses problems encountered during tests and inspections to determine which solutions should be incorporated into the production model. This is the last complete review of the product or system before its design is accepted.

Although designers, engineers, and design engineers are usually concerned primarily with the function and performance of a product, they must be equally concerned about the customer's safe use of the product. One method of promoting their participation in this concern is to place one or more design engineers on the PLLA committee. This will provide engineering with feedback from the marketing, legal, sales, and other departments about potential and actual problems. The consultant design engineer (if one is retained) should be knowledgeable about product liability from both a legal and an engineering perspective and should work with an engineering or legal representative on the committee in the course of providing design engineering services to the manufacturer.

If all dangers cannot be designed out of the product, then adequate protection devices and warning labels must be attached to the product and be described in all manuals and instructions. If incorporating all the necessary safety features into the product prevents the manufacturer from being competitive, then removing the product from the market may be the appropriate decision. Otherwise, a product liability settlement could wipe out all profits or even the company (see, for example, *Beshada v. Johns-Manville Products Corp.* 1982). As Hammer (1972) pointed out, the safety function of engineering can be delegated to one engineer or, as is often the case in large organizations, to many engineers. The engineers can have a degree in one of the four basic engineering disciplines or in a subdiscipline. The duties of engineers involved in a safety program are outlined in Box 4-2.

To reduce the chances of a lawsuit and increase the manufacturer's confidence, a product design checklist should be incorporated into the organization's standard operating procedures. The checklist is guided to a great extent by—or even based on—consideration of the six causes of product failure (a deficient design, improper use or selection of materials, defective materials, deficient manufacturing, fabrication, or process, improper or unusual operating conditions, and improper maintenance and inspection), the eight instrumentalities that affect the

BOX 4-2. The Duties of Engineers Involved in a Safety Program

Function or Occupation[a]	*Normal Duties*	*Safety Duties*
Research and development	Explore new technology that may have a positive impact on all aspects of the organizational mission and develop them for implementation.	Evaluate theoretical and hardware concepts throughout the organization from academia and the market. Perform trial tests and evaluation of new concepts under controlled conditions.

[a]These engineering assignments can be performed separately by separate individuals or departments or by one individual or department, depending on the size of the company.

BOX 4-2. *(continued)*

Function or Occupation[a]	*Normal Duties*	*Safety Duties*
Design and analysis	Prepare equipment and system concepts and configurations that will meet contractual specifications for performance and accomplishment of the corporate mission.	Conduct engineering design and analysis of the most appropriate and safest concepts, configurations, and procedures. Ensure that safety requirement in end-item specifications are met. Incorporate safety requirements for subcontractors and vendors in company specifications and drawings.
Human factors	Ensure optimal integration of human, machine, and environment.	Conduct analyses to ensure the well-being of personnel involved in operating, maintaining, repairing, testing, or performing other tasks in the proposed environment, especially to minimize fatigue and possible human error. Analyze procedures.
Reliability	Ensure that the equipment will operate successfully for the specified period under stipulated conditions.	Analyze failure modes and effect analyses. Perform tests on parts and assemblies to establish failure modes and rates of failure. Conduct special studies and tests. Review reports of trouble and failure with safety connotations.
Maintainability	Ensure that the hardware will accomplish its prescribed mission.	Ensure that the system or equipment operates safely by minimizing failures caused by fatigue through replacement of failed items by surveillance of possible degrading environments. Participate in analyzing proposed maintenance procedures regarding safety.

BOX 4-2. *(continued)*

Function or Occupation[a]	*Normal Duties*	*Safety Duties*
Testing	Conduct laboratory and field tests of parts, subassemblies, equipment, and systems to determine whether their performance meets contractual requirements.	Evaluate hardware and procedures to determine whether they operate safely, changes are required, or additional safeguards are necessary. Determine whether equipment has any dangerous energy levels or other characteristics or failure modes. Evaluate the effects of adverse environments on safety.
Product (field) support	Act as a liaison between the customer and the producer.	Assist the customer regarding safety problems encountered in the field. Serve as the major channel for feedback of field information on performance, hazards, mishaps, and near misses.
Production	Determine the most economical and best means of production in accordance with approved designs.	Ensure that designed safety is not degraded by poor workmanship or unauthorized changes in the production process.
Industrial safety	Ensure that company personnel are not injured and that company property is not damaged.	Ensure that personnel involved in system development, production, and operation have the necessary training to accomplish their assigned tasks safely. Certify workers for work on safety-critical operations, especially for test operations.

cost-effectiveness and safety of products (see Chapter 2), and various modes of failure (see Chapter 9). At a minimum, this checklist should contain the questions listed in Box 4-3, which the manufacturer asks of the independent engineering design organization or its own engineering design department.

The importance of having designers, engineers, and research personnel keep accurate, dated, and legible records cannot be overemphasized. The designer should keep written documentation of design ideas, design changes, and improvements for as long as the product is used. Testing procedures and results should also be part of the company's documentation. These data demonstrate that the manufacturer not only is conscientious but also is dedicated to producing a safe product. An example is a test program that documented the torque procedures of a bolted assembly on a large cooler. Figure 4-2 contains the engineer's conceptual sketch from the preliminary design report and a photograph from the shop-test report. Note in Figure 4-2(b) the presence of strain

BOX 4-3. A Sample Checklist of Questions the Manufacturer Should Ask of Engineering Designers

1. Has the product been designed to minimize injury if an accident does occur?
2. Has the product been designed for all foreseeable uses (human factors)? Does the operator require training?
3. Can the product cause an accident?
4. Are warnings located in a reasonable place and attached in a reasonable manner so that the consumer will be protected from inherent danger or damage?
5. Are safeguards designed into the product to prevent misuse or removal of the safeguards?
6. Has the product been designed for reliability? How reliable are the predictive methods used; that is, are they experimental or theoretical?
7. Has the environment (e.g., the opportunity for corrosion) been considered for all potential applications?
8. Does the product have a "design life" (will it function X number of years)? If the answer is yes, has a mechanism been identified to determine deterioration and decommission of the product before failure (via the user or manufacturer)?
9. Does the product meet or exceed established guidelines such as governmental statutory codes and regulations, industrial or com-

BOX 4-3. *(continued)*

mercial standards, and national voluntary codes and standards that reflect minimum safety requirements? If the applicable codes conflict, has a ruling or interpretation been requested and has the ruling or interpretation been conveyed to the PLLA committee?

10. Does the company use engineering drawings and related documentation to determine that the dimensions and tolerances meet those specified in standards developed by organizations such as ANSI Y14.5? Are the materials properly specified?

11. Are the latest engineering and scientific techniques used to develop and test products?

12. Has rough handling of the product during everyday use and shipping as well as overall wear and tear been considered?

13. Are the instructions for shipping, storage, assembly, installation, operation, and maintenance easy to understand? What will happen if the user does not perform the maintenance procedures?

14. Did the designer review all brochures, catalogs, and other advertising vehicles before they were published?

15. Has product failure been considered in cases where the product is used in conjunction with other components and systems?

16. Can the reliabilty of the product be improved if a different fabrication process is used?

17. Has the product been assessed for all possible modes of failure? Have critical (high-risk) modes of failure been evaluated adequately (in fault-tree, risk, and consequence analyses)?

18. Has the recent technical literature been evaluated with respect to accidents involving users (see, for example, Archea 1979; Carson *et al.* 1978; Rosen 1983; *Work Fatalities* 1977)?

19. Was more than one design assessed in a product performance evaluation?

20. Have all loadings (forces, temperature, environmental, etc.) of the product been considered in relation to performance: for example, operation, shop test, in situ tests, transportation loads, and emergency, faulted, and accident loads?

21. Has the engineering designer reviewed all data on product failure? (See Gangadharan and Brown 1977, for uses of failure data and failure analysis.)

22. Has a lawyer reviewed your quality assurance procedure as a preventive measure?

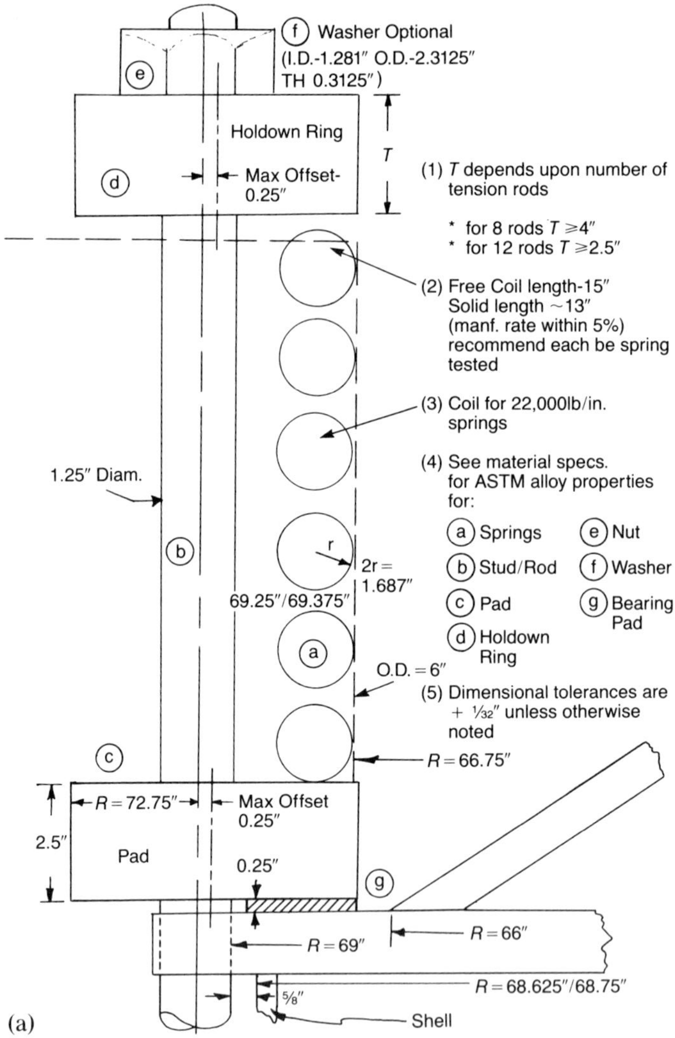

(a)

(b)

Figure 4-2. A heat-exchanger drum. (a) Engineer's conceptual drawing of a holddown ring and spring system for the drum. (b) Shop test and performance evaluation of the drum. (See S. J. Brown 1990.)

TABLE 4-2. Vessel and Spring Average Deflection Data (Houston Test)

Load Incre-ment	*Number of Nut Turns (average)*	*Total Stud Load Based On Spring Deflection (lb)*	*Spring Average Length (in.)*	*Stud #3 Load Based on Strain Gauge Reading (lb)*	*Shell Stress Based on Strain Gauge Reading (ksi)*
1	$1\frac{1}{2}$	0	14.951	25,462	2.100
2	$1\frac{3}{4}$	49,338	14.854	56,696	—
3	$1\frac{3}{4}$	110,496	14.735	103,083	4.200
4	$1\frac{3}{4}$	174,224	14.611	206,167	5.600
5	2	262,107	14.475	222,629[a]	7.0
6	2	334,055	14.342	335,021	4.9
7	$3\frac{1}{2}$	462,537	14.073	453,567	7.7
8	3	565,323	13.847	592,730	9.8
9		unload	14.70	—	—
10	$8\frac{3}{4}$	reload	14.25	—	—

[a]284,768 (based on average of 4 readings).

gauges on the bolt/rods for monitoring stress levels during the torquing or preloading operation of the 12 nuts. The relationships between the number of nut turns versus stud stress, spring compression, vessel stress, and nut sequence of torquing were documented in the test report. Nut sequencing of torquing refers to the pattern in which the nuts are to be tightened. This pattern affects changing distributions of loads in the system. An evaluation of a worker's implementation of the torque procedure with respect to ergonomics, clarity of instructions, effects of errors, human factors, and hazards was conducted as part of the design evaluation. Table 4-2 illustrates partial tabulated data from one in situ test.

QUALITY ASSURANCE

Chapter 3 discussed quality assurance in the context of liability. Here, the role of the quality assurance department is reviewed. Although companies vary in size, location, policies, the type of products they manufacture, and financial resources, the charter or scope of quality assurance is the same—to monitor production, ensure that the product meets the design requirements, and ensure that it can be used safely. Just as the product designer should be able to respond to a checklist of questions by the manufacturer or PLLA committee, the quality assurance department should be able to say yes in answer to the following questions.

- Does the department have a quality assurance policy manual?
- Does it have an interdepartmental coordination representative for the PLLA committee and other company functions?
- Do supply vendors provide products that are of the same quality as those of the purchaser or manufacturer? Are the vendors' control procedures the same as those of the purchaser or manufacturer?
- Does the department have procedures for testing products?
- Does it have packaging and shipping procedures?
- Does it have a program of corrective actions and analyses for materials and equipment that do not conform?

Coordination of Departments

Today, more than ever, coordination between the design engineering and manufacturing departments, divisions, or companies (if the manufacturer has retained an engineering design firm) is essential for the review of product designs to verify realistic tolerances as well as the accuracy and completeness of design data. These two entities should work together as a team to implement changes, record the data on these changes, and coordinate rejection of materials that do not conform to specifications. In addition, they should work together to determine what testing and production equipment is required to ensure a high-quality, economical product.

It is also extremely important to have the quality assurance representative (or designate) on the PLLA committee report to members of top management who have the authority to mediate interdepartmental disputes and can reflect the company's best interest. The quality assurance function should never be under the jurisdiction of the manufacturing department because this gives the impression that the fox is guarding the hen house.

The quality assurance manager should ask him- or herself and others in the organization the following question about product acceptability: "Would you allow yourself and those who are closest to you to use the product without any reservation?" If the answer is not an unqualified yes, the product should be rejected until the deficiencies are corrected.

Quality Assurance Manual

The quality assurance manual should spell out the company's product loss and quality objectives and the company's policy. It should also contain a list of people who are responsible for the various phases of quality assurance, including a description of procedures for implementation of these phases.

Control of Outside Suppliers

The quality assurance department must keep track of raw materials and finished parts obtained from outside sources for use in production. The following is a checklist for good quality control of outside suppliers.

- Check the supplier's operations and reputation before placing orders.
- Give the supplier all product specifications.
- Provide the supplier with objectives regarding quality and criteria for acceptance.
- Tell the supplier what procedures and gauges will be used to measure quality.
- Inform the supplier whether the design engineering personnel have developed any specifications for components and have prohibited any substitutions or changes without the written approval of all appropriate departments.
- Check the supplier's listing of product specifications and continue to monitor the level of quality to assure acceptable orders.
- Inspect all orders before accepting delivery.
- Station a full-time inspector in the supplier's plant or make certain that the supplier has its own inspector conduct periodic inspections to analyze in-process quality. (This item applies if a manufacturer does a large amount of business with one suppler.)

Testing the Product

A manufacturing company should keep complete and accurate records of tests and methods used to determine the properties of materials used in the product, the dimensions and chemical or metallurgic content of the product, and the results of performance tests. The company also should record the special processes used during production; for example, plating, welding, heat treating, painting, and stress relieving. Each process can play an important role in the integrity of the final product and should be carefully monitored in accordance with testing procedures, such as those of the American Society for Testing and Materials (ASTM), dependable statistical sampling techniques, and dependable methods of establishing reliability.

In addition, quality assurance procedures should be scrutinized by line inspectors in the field or by the shipping department, and an inspection of the products they review or sign off should be monitored periodically to help determine whether the manufacturer's examiners are achieving the objectives of quality assurance. Finally, all testing equipment and tools, whether personally or company owned, should be calibrated and certified periodically (e.g., four times a year) to ensure that they meet recognized standards of tolerance, such as those cited by the National Bureau of Standards. Calibration data should be readily visible on all testing equipment.

Rejection of Product, Material, or Process

When errors do occur, rejections should follow. Thus, the quality assurance group is responsible for recording its analysis of product that does not conform to specifications. The procedure the group should follow is as follows.

- Segregate rejected product or material from approved product.
- Identify all rejects by clearly marking them with a tag or another suitable form of identification that is highly visible (see the sample tags illustrated in Fig. 4-3).
- Reinspect product or material that has been reworked or repaired to ensure that it conforms to the testing specifications.
- Develop a process for recording and returning rejected components or raw materials to their suppliers.
- Record all discrepancies and dispositions to identify potential problems in design, manufacturing procedures, or processes. (Such records can reveal economic losses or losses of productivity; they also reflect a sound and responsible liability program.) It is important that company records reflect concern for safety over any monetary consideration.

Packaging and Shipping

Another responsibility of quality assurance personnel (and designers of the product) is to assess how the product is packed and shipped. The best design and manufacturing can be thwarted if the product arrives at the user's site in a visibly or invisibly altered condition that may contribute to failure.

To assure a safe and undamaged product, the manufacturer's quality assurance group should determine whether (1) packaging and shipping comply with the customer's purchase order, (2) the packaging is

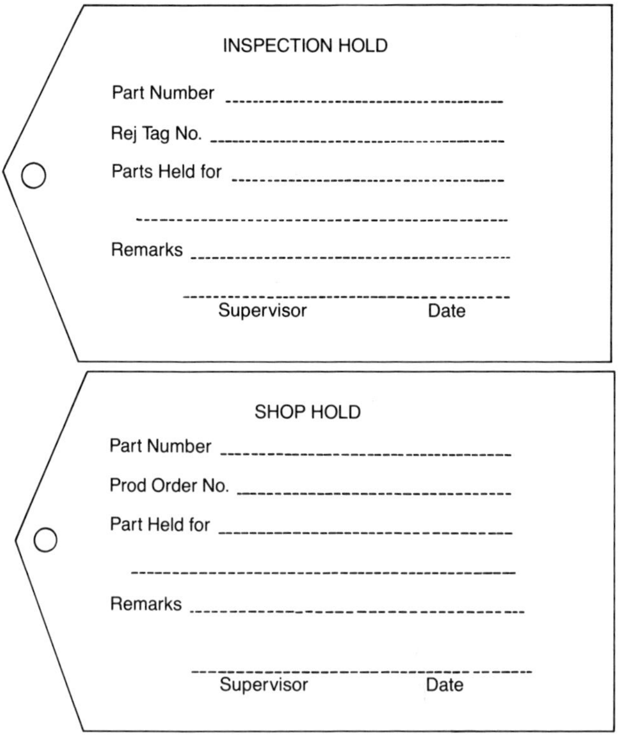

Figure 4-3. Illustration of an "inspection hold" tag and a "rejection" tag.

sufficient to prevent damage to the product, (3) the instructions, manuals, and labels are included in the package, (4) the proper warning labels are on the product, (5) the proper warnings are on the package and visible in all positions and meet all standards where required, i.e., add appropriate warning labels if the product is required to meet the "right to know" law, (6) records for locating, identifying, and tracing shipments are adequate, and (7) the method of transportation has been approved by the company's legal and engineering personnel.

MANUALS, INSTRUCTIONS, AND LABELS

No matter how safe a product is, the manufacturer can be held liable if the customer does not understand the purposes, limitations, and dangers of the product. Therefore, not only what the manufacturer says but how the manufacturer says it is important.

A booklet or manual of instructions should provide the customer with enough information to assemble, install, test, and use the product safely. The following list will serve as a guideline for the type of information that should be provided in an operation and maintenance manual:

- A complete list of parts, and a description of how to assemble or install the product.
- The maintenance procedures and periodic inspections required to keep the product functioning efficiently and safely.
- A list of high service areas and trouble spots, and instructions on how to service or correct them.
- The operational life of the product in terms of time or cycles of operation and loss of material caused by erosion or corrosion.
- The mechanism of recertification.
- Procedures or antidotes for coping with accidents or other emergencies.
- Information to the product sales personnel, representatives, distributors, and customers about the uses, limitations, and potential hazards of the product and the consequences of misusing it or failing to follow instructions.
- A warning that operators should read the instructions before trying to use the product.
- A warning against modifying the product or substituting parts or materials, unless approved by a responsible corporate manager who can accept liability on the manufacturer's behalf.

Another important point is that all parts used in the product must be documented so that in the case of injury or damage, any modifications of the product can be deciphered. The manufacturer's guarantee of quality and safety is based on the fact that the product contains parts the manufacturer has specified. Serializing each unit to keep track of any improvements, modifications, or recalls also is a good idea. Drawings and photos can be kept to identify any changes made to the original product.

When product labels are used, all possible uses and misuses must be considered when determining what information consumers need, and the wording on the label must be clear, simple, and concise. If the product is dangerous or incorporates hazardous materials, a highly visible warning label such as the one shown in Figure 4-4 must be used. In many cases, such a label may be required to meet state or federal regulations; furthermore, specific symbols, languages, or colors may be required. If symbols are used, they should be easily recognizable (Peters 1984).

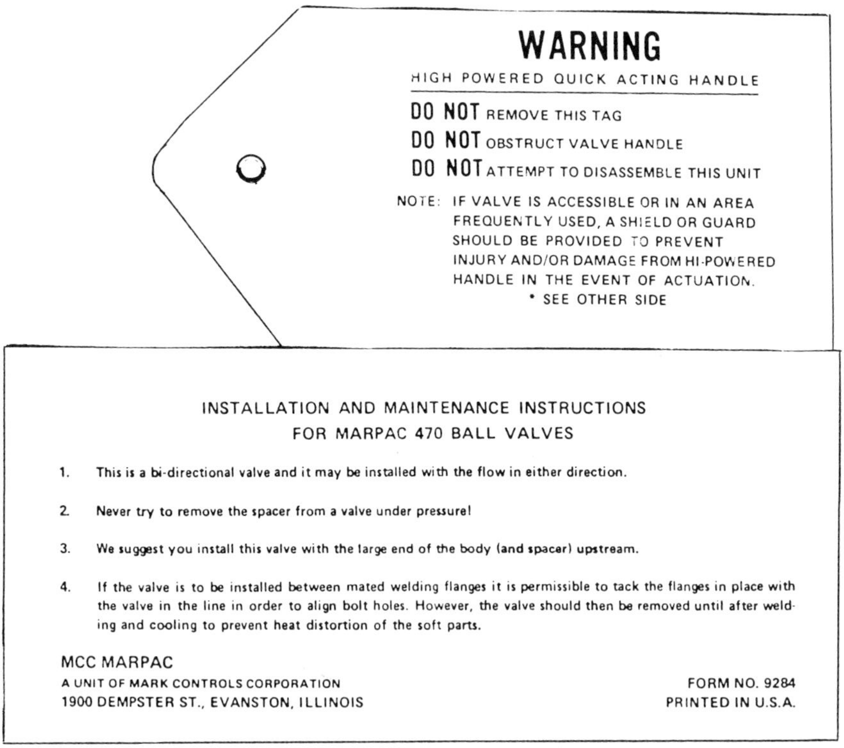

Figure 4-4. Illustration of a warning label and instructions for installation and maintenance.

Product designers, engineers, and manufacturing and quality assurance personnel should receive training or counseling, from a legal and ergonomics perspective, concerning the language and tone of the information used in manuals, instructions, or labels. Equally important, marketing, sales, and advertising personnel should receive training or counseling in the context of product liability with regard to sales brochures, advertising, and promotional materials. These policies will help protect the company against accusations of misstatement and misrepresentation. Having the PLLA committee review and approve *all* manuals, instructions, labels, and brochures is an extremely important policy.

MARKETING

The manufacturer's marketing organization must have a plan for controlling product liability losses. This begins with an understanding that every product carries a warranty and that the warranty does not necessarily have to be in writing or attached to the product. As was discussed in Chapter 1, the implied warranty of merchantability or fitness is an unwritten guarantee of product safety based on the theory that every product is safe and fit for its intended use. An implied warranty can extend to anything related to the advertising and selling of a product. A manufacturer can be held liable for any misleading or inaccurate representations, either written or verbal.

Because the company is responsible for what its representatives say about the product, the marketing organization, in conjunction with the PLLA committee, should develop a strategy for instructing the sales force, dealers, wholesalers, and distributors about the principles of restraint and common sense. If customers injure themselves by pushing the product beyond its safe limits because of a salesperson's overenthusiasm, the manufacturer may be liable for a breach of warranty of fitness. Disclaimers or limitations of warranty do not always help if bodily injury or property damage occurs as a result of product failure.

Most companies guarantee the quality, life expectancy, or use of a product. This promise can be written or oral and is an express warranty. Drawings, marketing literature, or models that mislead a customer and result in damaged property or injury can inadvertently contribute to liability. Therefore, words such as shockproof, fireproof, foolproof, or lifetime guarantee should be avoided. As was mentioned earlier, the manufacturer's legal department or counsel should review all marketing material for clarity and accuracy. This practice can help safeguard against a marketing error and demonstrate a company's commitment to good advertising policies.

PACKAGING AND SHIPPING

The fact that a product is the best designed one in the world may be meaningless unless it arrives undamaged and on time. To ensure product safety during shipping, the packaging must be suitable for the size, weight, and fragility of the product and must compensate for shelf life, environmental conditions, storage and stacking conditions, corrosiveness, volatility, and flammability. Companies that specialize in packaging and shipping can be of great assistance to a manufacturer during the development of standards or procedures for a product.

To assure that all necessary information has been supplied, the manufacturer should ask the packing and shipping organization the following questions.

- Are all the necessary instructions for assembly, service, and maintenance enclosed?
- Is a complete list of parts enclosed?
- Are all the appropriate warnings and instructions attached to the product?
- Are the handling instructions for the product and package clear, visible, and easy to understand?
- Is the shipping label complete, correct, and affixed in a suitable manner?
- Do the contents, customer shipping instructions, and product comply with the customer's order as agreed upon?
- Have all the federal, state, and foreign standards for shipping and packaging of the product been met?
- Are sensing devices required to detect possible damage to the product caused by changes in temperature, acceleration, humidity, and so forth during transportation?
- Has the design engineer considered transportation loads and storage loads?

Because keeping records of how, when, and who ships the manufacturer's packages is a good practice, the manufacturer should keep a list of carriers that have excellent reputations for dependable, safe, and timely deliveries at a price that meets the company's objectives. Other factors to consider are insurance coverage, 24-hour delivery, and the carrier's ability to trace and follow up on late pickups or overdue schedules.

COMPLAINTS AND CLAIMS

Even if a manufacturer has tested its product for every conceivable application, it may receive complaints for a number of reasons. For example, the operational experience of users obviously is not readily duplicated or substituted completely by experimental and simulated testing in a laboratory. Thus, some manufacturers conduct extensive, controlled, consumer testing or monitoring before they release a product for general production, sales, and marketing. Furthermore, each customer may have different operating procedures, safety practices, and so forth. Complaints are extremely important, however, because they reflect real-life situations and, over time, can provide valuable data leading to a better product. A company policy that recognizes this source of information for engineering design has a positive effect not only on company personnel but on customer relations as well.

Customer Complaints

A manufacturer should never ignore a complaint or suggestion from a customer, and it should never repair or replace a defective product without notifying the PLLA committee. A rapid response to a customer's complaint, no matter how insignificant it may seem, demonstrates the manufacturer's concern for its customers and may avoid a liability suit as well.

The company's procedure for handling each type of complaint should be coordinated by the PLLA committee, and each department should be involved in problem-solving activities and decisions. The committee should make certain that customers, dealers, sales representatives, and service personnel know how to report incidents of product malfunction or product misuse to the company's service department. The service department should then notify the PLLA committee immediately. If the service department responds to a complaint by telephone to expedite its concern, the call should always be followed by a letter that has been reviewed by the PLLA committee and the legal department.

When a complaint is received, the manufacturer should, if possible, try to visit the place where the product is used (particularly in the case of large products); retrieve the product or the relevant parts; review its own records concerning development and production of the product and records concerning sales and delivery; and review the user's maintenance and inspection records (for more in-depth information, see Chapters 5 to 7). This information can be helpful in three ways: (1) it will help the PLLA committee determine what action can be taken to improve the product and prevent a similar failure, accident, or injury in the future, (2) it can help the manufacturer maintain good public relations with the complainant as well as potential customers, and (3) it will help the legal department and insurance company establish a defense if the complaint develops into a claim or suit.

Gathering facts about an incident as soon as it is reported is another important task. Physical evidence should be examined, recorded, and photographed before it is disturbed or removed. Also, witnesses should be interviewed while facts are still fresh in their minds. Finally, these tasks should be completed as quickly as possible.

Customer Claims

How a company handles claims can either weaken or strengthen its future. When the manufacturer's service department first receives infor-

BOX 4-4. Guidelines for Handling Product Liability Claims

Which claims should receive special treatment?
All claims or potential claims resulting from allegedly defective products or services that involve circumstances that could possibly impose significant liability on the company: (1) personal injury or death, (2) damage or potential damage to other property, (3) labor costs and other expenses connected with repair or replacement of the product, and (4) loss of profits resulting from shutting down the plant or curtailing production.

To whom should claims be reported?
Claims should be reported to the company's legal department.

How should the claim be reported?
The initial report should be made by telephone as soon as it is received, and it should be confirmed with a detailed written report. The person who prepares the report should seek the guidance of legal counsel before and upon completing of the draft.

What information should be reported?
The report should contain all the facts available. Preliminary conclusions about the mode and cause of the failure and what or who is responsible should not be delayed because of the lack of essential information.

Personnel should be instructed as to how to handle questions from anyone regarding company problems. This should be part of company policy and standard operating procedure.

Generally, the following details should be provided (if additional pertinent details are available, they should be included).

1. The date and place of the event.
2. When and how the notice was first received.
3. The type of product involved.
4. The customer's name and the order number.
5. The nature of the claim.
6. The injuries or damages that were sustained.
7. The evidence, including background information such as eyewitness accounts, physical evidence, maintenance and inspection records, and operational history (including any misapplication or misuse), and photographs.

A formal accident investigation and a failure analysis should be performed, based on the preliminary investigation.

Discretion and judgement must be exercised in investigating claims. Moving cautiously and deliberately is desirable on some occasions, whereas prompt action commensurate with the seriousness of the situation will be required on other occasions. The legal department will provide guidance regarding the legal ramifications of the event.

BOX 4-4. *(continued)*

What should the person reporting the claim not do?
The person should *not* do the following.

- Express conclusions or opinions about the cause of the accident, either orally or in writing, to anyone unless requested to do so by the legal department.
- Communicate, either orally or in writing, information that could be construed as an admission that the product was defective (such an admission can result in loss of insurance as well as in a lawsuit).
- Take any action to correct "deficiencies" in the design, construction, or marking of a product unless the legal department and insurance company have given their approval.
- Make any formal statements without the prior approval of the legal department.
- Mention that the company has insurance protection against claims.
- Divulge information related to the claim to the claimant or its investigators or attorneys without consulting the legal department first.
- Offer to settle a claim.
- Fail to notify the legal department immediately if served with a summons.

mation about a claim for damage or injury caused by product failure, they should immediately contact the PLLA committee. At the same time, the manufacturer should notify its insurance company and forward a report of the claim to the claims department. The report must contain complete information about the product: the conditions surrounding the failure; defects; inherent dangers; design changes; critical conditions; testing data; other complaints received about the product, if any; and verification that the manufacturer made the product. This information will help the insurance company and the PLLA committee prepare a defense.

A manufacturer should never make improvements to a product after a claim is filed without first notifying the PLLA committee, counsel, and the insurance company. Such changes may have a detrimental effect on the outcome of the claim as well as on the outcome of a lawsuit.

The importance of a guideline for handling claims cannot be overemphasized. This guideline, containing instructions for the proper handling of product liability claims, must be supported and endorsed by top management and usually should be written by the company's legal department. It should be given to all departments for ready reference, it should be followed to the letter, and it should be constantly reviewed. The sample guidelines in Box 4-4 can serve as a starting point for the manufacturer who must develop guidelines tailored to its own products and company operations and based on the advice of its legal department or counsel.

RECORDKEEPING

The quality of the records a manufacturing company keeps can greatly reduce the risk of product liability suits. A manufacturer must never forget that a court can subpoena its records. If a product has been designed for safety, the company's records can substantiate these efforts. If the records are not complete or thorough, the company's defense could be in jeopardy.

Records should be accurate and easy to understand. The information should be signed, dated, and verified in a consistent manner by an authorized official of the company. The records that are usually most important are those developed by engineering, manufacturing, and quality assurance personnel. Design records should include formulas, engineering specifications, design testing data, changes in product design, and information about the reasons for selecting or rejecting materials. Manufacturing records should describe all production steps and processes. Quality assurance records should contain information about the configurations of the product, the identification marks on parts purchased from outside suppliers, and the results of inspections of raw materials used in the product. Other inspection records should include the results of manufacturing and process controls, descriptions of testing equipment and calibration methods, and information about nonconforming material and its disposition.

Other necessary data include final inspection records on materials and parts that were shipped with the product, how the product was packaged, how it was shipped, and where it was sent. In addition, the manufacturer should keep records of all customer complaints.

Finally, maintaining files on all written forms—for example, instructions for handling and assembling the product, operating guidelines, maintenance procedures, parts lists, warning labels, invoices, special instructions, brochures, vouchers, letters of special conditions, and advertisements—can help the manufacturer's product liability defense and demonstrate its concern for producing a safe product.

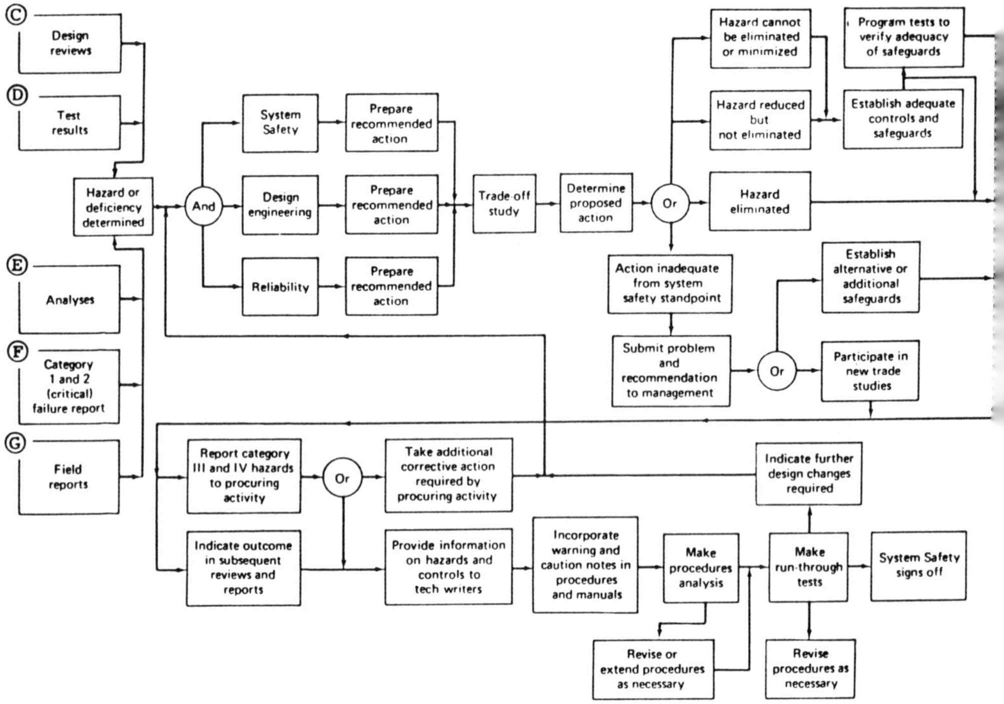

Figure 4-5. Hazard reporting and correction flow diagram. See also Fig. 4-1, p. 90. (From W. Hammer, *Handbook of System and Product Safety,* pp. 36, 49. Englewood Cliffs, NJ: Prentice-Hall, © 1972.)

CONCLUSION

Although the product liability laws differ in every state and country, product liability can be greatly reduced or limited if a manufacturer has made an effort—from the beginning of a new product design—to develop, implement, and enforce a policy of informed decision making and informed risk taking based on knowledge of the law, engineering design practices, and control of product quality.

The remaining chapters provide the manufacturer with more detailed guidance concerning the principles and procedures that are important when performing an accident investigation and a failure analysis—guidelines that can be used to reduce product failures, injuries or damages, and lawsuits.

5

The Accident Investigation and Failure Analysis: Engineering Framework

This chapter provides an overview of the procedures that an engineer or scientist follows when conducting an accident investigation and failure analysis. Thus, it serves as a transition between the legal and engineering aspects of product liability.

The engineer or scientist typically begins an accident investigation and failure analysis at the request of a principal such as the manufacturer or user. If the principal anticipates that litigation will be needed to determine the validity of the claim of responsibility and compensation for damages or injury, an attorney or law firm is usually retained to act on the client's behalf. At that point, the attorney or law firm usually contacts the engineer or scientist who will perform the investigation and analysis. The relationship between the attorney and the technical expert is an important one; thus, each needs to understand the professional objectives and capacities of the other. This need for understanding is the theme of this book, and increased understanding is its objective. (Throughout the remaining chapters, the word product will be used unless a specific structure, piece of equipment, or component is being discussed. Similarly, the term engineering investigator will include the scientist.)

The role of forensic engineers who perform accident investigations and failure analyses is not unlike the role of a medical forensic pathologist: that is, they must do an examination or autopsy to determine how and possibly who is responsible for a defect in or the fatal demise of a product, structure, piece of equipment, system, or component. Attorneys, on the other hand, are charged with the responsibility of demonstrating only that their client is not responsible. They do not necessarily need to demonstrate what caused the product failure and who is responsible. Keep in mind the dilemma of investigators who conduct an accident investigation: Can causality or lack of causality be demonstrated without an autopsy? No. Yet the client instructs them to do only what is necessary to define the client's position, either positively or negatively. Unfortunately, no direct relationship exists between the cost of an accident investigation and failure analysis and the cost of damages. For example, $25,000 worth of damages may require $100,000 worth of investigation, whereas an investigation involving millions of dollars worth of damages could conceivably cost $100,000 or less.

The term *product failure* can refer to the degeneration of a product or the presence of a flaw, defect, or deformation (variance from the design) that disrupts its operation or presents a possible safety hazard and requires repairs. The material in a product does not need to fail to contribute to damage or injury. For example, a poor functional design—one that fails to consider ergonomics—may precipitate, or be the "proximate cause" of, injury or damage.

The severity of a product failure can be classified in one of two ways: (1) as a *disruptive* or *catastrophic failure* that results in total loss of the product or requires major repairs, or (2) as a *nondisruptive failure* that requires only remedial or minor repairs. The purpose of an accident investigation is to evaluate the capacity of the product to perform its intended function. The failure analysis addresses the technical aspects of the mode of failure and may include simulations and reconstructions that address the mechanism of failure. The overall investigation brings together all the background information on the accident, the results of inspection, and data on the mode of failure, which may include the results of simulations or reconstruction that address the mechanism (the how) of failure. These data are then used to formulate a position on how the incident occurred, what caused it, and who was responsible, and to make recommendations based on what has been learned. The investigation can be conducted in-house at the manufacturer's or user's request, or it can be conducted as part of a product liability case at an attorney's request. Product liability litigation consists of the accident investigation or failure analysis and the legal basis for the complaint, which together affix responsibility to the defendant or plaintiff.

The legal aspects of an accident investigation involve at least two of the following principals who are involved in some way with the damaged or failed product: the design engineer (or architect), the manufacturer, the owner-operator or consumer, and a third party such as an employee, bystander, or service organization. Strictly speaking, the third party can be interpreted as an operator (intentionally or unintentionally). Any of these principals can be an individual, a large corporation, or somewhere in between. And, as was discussed in earlier chapters, the design engineer can be an employee of the manufacturer or owner. The designer, manufacturer, and the owner of the equipment can, in rare instances, be one corporation. Generally speaking, engineers or engineering firms that conduct accident investigations and failure analyses should not be employed directly by one of the litigants, nor should they have designed the product that failed. The obvious reasons for this general rule are the need for objectivity, the need to avoid being influenced by personal interest, and the need to convince the jury or judge of their objectivity.

Accident investigators typically face a number of pitfalls because of the types of clients and law firms they represent. Litigants and attorneys who are technically knowledgeable tend to steer a technical expert in the direction they would like him or her to go; however, they can help the expert build a strong engineering position. Personal injury attorneys, on the other hand, tend to favor only marginal technical investigations, if any, and lean toward expert opinions.

During a typical accident investigation, the engineering investigator

1. Collects background data on the incident
2. Documents and examines the failed product
3. Conducts nondestructive and destructive tests of materials in the failed product
4. Determines the mode and mechanism of failure
5. Performs simulations of the incident
6. Reconstructs the incident
7. Determines the cause or who is responsible
8. Draws conclusions and makes recommendations

Although the eight steps are listed in the typical order, they do not necessarily have to be taken in that order. Furthermore, certain steps may be left out: for example, simulations are not always necessary. Each step can involve one set of data or multiple sets, each of which presents a partial picture of the accident or failure and all of which must be included in the reconstruction process. A failure analysis usually involves steps 1, 2, and 3 and, occasionally, steps 1 through 5; it does not address the issue of causality.

Two critically important ingredients in an accident investigation and failure analysis are the expertise and money required to perform them. Both engineering expertise and legal expertise are needed to assess the technical and legal implications of the accident, and economics may determine whether the evidence can be developed sufficiently. Therefore, it is important for technical experts to understand that their service agreements and financial practices with product liability clients (usually a law firm) must be handled differently from those for a typical engineering design-related project.

An engineering investigator usually begins the accident investigation and failure analysis after a preliminary meeting with the client or a cursory review of material outlining what happened, who the principals are, and whether litigation is contemplated and after the client has issued a service agreement, which can be a work authorization or a purchase order.

In the following sections, it will become clear that engineering investigators must have a broad understanding of engineering principles, even though they may have specialized in a specific type of engineering (Vance and Gould 1985). It also will become evident that they must be able to recognize when and what analyses must be performed, what data must be gathered, and when other experts should be retained.

The following case provides an overview of the steps involved in an accident investigation and failure analysis. A well-known international manufacturer of heavy equipment we will call Highlift Manufacturing Co., introduced a foot-control pedal for its forklift that allowed the operator to control the direction and speed of the vehicle with both hands free (typically, one hand is required to shift into forward and reverse).

The foot pedal consisted of springs, levers, and set screws and resembled the brake pedal found in most automobiles, with the following exceptions: operators could turn the vehicle to the right or left, depending on which side of the pedal they pressed. In addition, increasing the pressure on the pedal on one side or the other put the vehicle into forward or reverse and also increased its speed. Thus, the spring system required to control the vehicle direction via the rocking motion of the pedal had to respond more quickly than the springs controlling acceleration. Otherwise, when attempting to select a gear, the vehicle might have accelerated when the operator was attempting to select a gear.

Several obvious conclusions can be drawn about the design of the pedal. First, operators were highly likely to operate the pedal incorrectly because of its complexity. Second, the pedal was highly likely to malfunction as a result of wear on the pivots, set screws, and springs. Most competitors criticized the design from the outset, and none introduced a similar pedal. Furthermore, governmental and technical organizations advised, through announcements, codes, and standards, that combining control and acceleration was dangerous and unacceptable.

After years of use, a unique characteristic of the pedal emerged. As it wore, many operators had difficulty changing the direction of the vehicle, hence they held their left foot on the brake while manipulating the control pedal with their right foot. Then they slowly released the brake to ensure they were going in the desired direction. This delayed selection of direction became especially dangerous when the forklift was used on loading docks, which are usually about four or five feet above the ground; a number of operators were injured or killed when the forklift fell off a loading dock. These incidents resulted in lawsuits against Highlift Manufacturing Co.

Around the time the multifunction control pedal was introduced, voluntary and statutory codes began requiring installation of an overhead roll cage on all forklifts to protect operators from objects overhead. This technical requirement complicated the pedal issue. Studies by industry and government-sponsored agencies found that if the vehicle overturned, the operator was likely to be crushed by the roll cage. Thus, these organizations recommended that forklifts should be equipped with safety belts. (An illustration of a forklift equipped with a roll cage and safety belt is shown in Fig. 5-1.) Because operators of forklifts equipped with the multifunction control pedal and a roll cage but no safety belt were at greater than normal risk of being crushed if the vehicle fell off a loading dock, the pedal was eventually withdrawn from the market.

In one case involving a Highlift truck, the consulting engineering firm retained by the estate of a fatally injured operator conducted an accident investigation and failure analysis involving the following steps:

1. *Background research*
- A visit to the site of the accident to examine maintenance records and interview other forklift operators.
- A survey of other companies that had purchased Highlift trucks to assess their experience with the multifunction pedal.
- A review of the literature published by truck manufacturing associations and by technical societies regarding the requirements for multifunction pedals.
- A review of voluntary and statutory standards concerning accidents involving forklifts and, in particular, those involving multifunction pedals.
- A review of experimental and theoretical studies performed

by both profit and nonprofit organizations and with and without manufacturer's funds.

- A review of studies on ergonomics in relation to multifunction controls.
- A review of legal cases regarding Highlift and other manufacturers of multifunction controls.
- A review of Highlift's specification drawings; specification sheets; shop, operator, and repair manuals; advertisements; and memoranda.

2. *Documentation of the accident*
    - An examination of the forklift involved in the accident.
    - The taking of photographs and measurements of the site and the vehicle.
3. *Assessment of the mechanism of failure*—What caused the vehicle to fall off the dock?
4. *Testing*
    - Experimental simulations of how the multifunction pedal functioned in an actual vehicle and in the laboratory after removal from the vehicle.
    - Theoretical simulations of the dynamics of the vehicle falling off the edge of the loading dock and testing of the response time of the multifunction pedal when the gears were switched from forward to reverse and back during actual operation.
5. *Reconstruction of the accident*—A reconstruction of the sequence of events leading up to and including the accident.
6. *Assessment of responsibility*—Who was responsible for the accident?

After completing the investigation, the investigator reached the following conclusions: the multifunction control pedal was (1) confusing from an ergonomic (functional) standpoint and (2) deficient from an engineering design standpoint—that is, as the mechanical linkages were subjected to wear, shifting from forward to reverse became sluggish and, eventually, impossible. When the pedal worked correctly, operators could change direction quickly by "punching" the opposite side of the pedal. When the pedal was worn, however, they could punch the forward side after going into reverse and discover, to their surprise, that the vehicle continued in reverse. This posed a serious problem for operators working near the edge of a loading dock.

An interesting aspect of this case was that the plaintiff's technical expert was *prohibited from testifying* about the results of his investigation. Although he had a master's degree and a doctorate in mechanical and

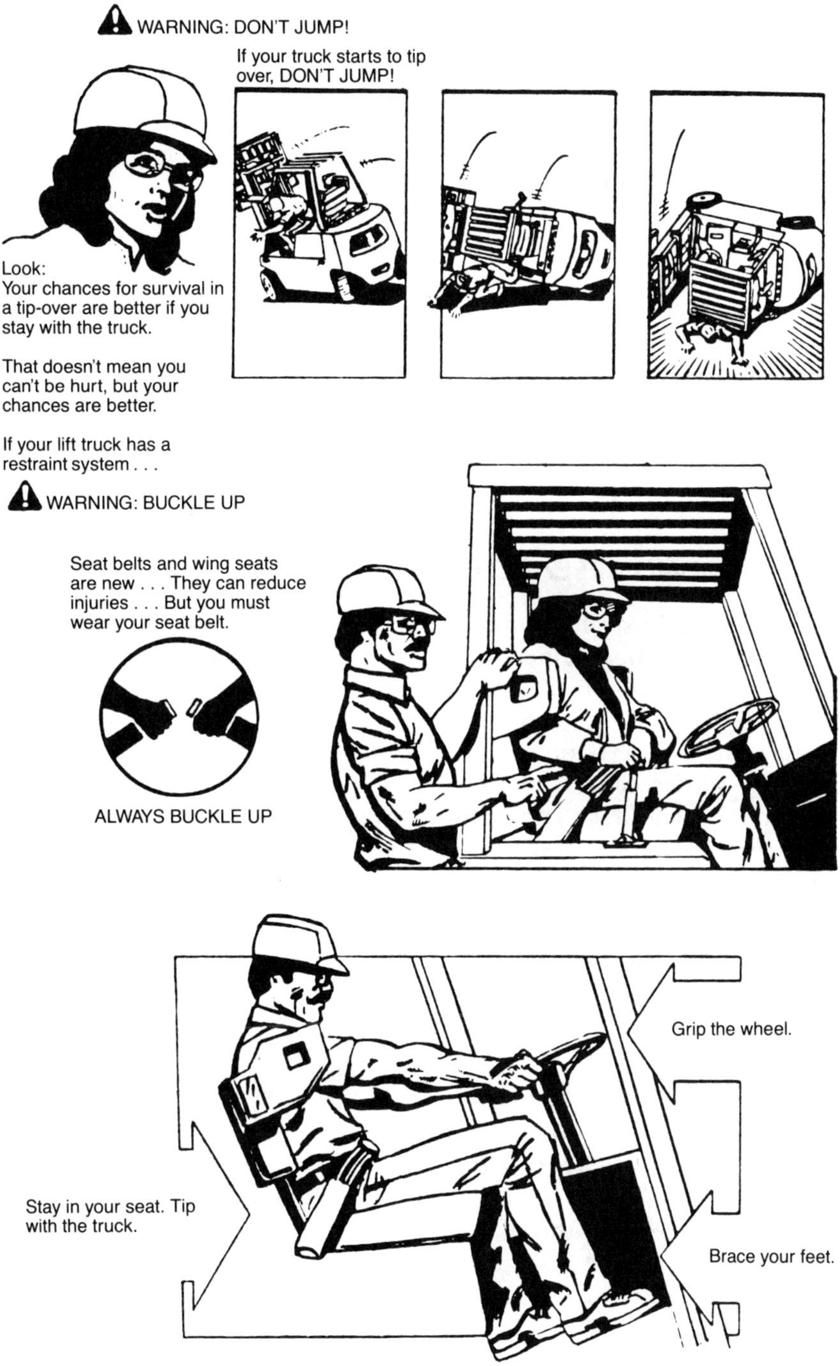

Figure 5-1. A warning for operators of forklift trucks. Note the roll cage in the roof.

engineering mechanics, was a university instructor in component design, had 22 years of industrial experience, was a registered professional engineer, had received professional recognition, and had ample experience with codes and standards, the defense attorneys convinced the judge that such a thing as a "forklift multifunction pedal engineer" existed and that the plaintiff's technical expert was not one, and the defense won the trial.

The remainder of this chapter provides a brief overview of the eight steps involved in an accident investigation and failure analysis. Chapters 6 through 11 will discuss each step in detail. This chapter also introduces some of the issues the engineering investigator should keep in mind regarding product liability litigation while conducting the investigation. Chapter 12 is devoted exclusively to litigation.

BACKGROUND RESEARCH

As this section illustrates, numerous sources of background information are available to the engineering investigator. Background information can be classified into two categories: (1) documentation of physical and observational evidence, and (2) collection of documentary evidence. These sources range from on-site investigations and eyewitness accounts to case histories cited in the literature and current practices cited by associations, manufacturers, and publications. Information the investigator receives before visiting the site will guide the selection of the necessary equipment.

Documenting the Incident

It is important for a qualified forensic engineering expert to visit the scene of the product failure as soon after the event as possible. In the early stages of an accident investigation, it may be difficult to determine the mode, mechanism, and cause of the failure. Therefore, any clue can be important in reconstructing the accident. Also, if litigation is involved, it may be years before depositions are taken and the case goes to trial. Even if the failed product is unavailable or the incident occurred sometime in the past, the standard practice is to visit the scene of the failure. If the product is small enough, it should then be shipped to the investigator's office or laboratory. If it is large, it should be stored in a secured building or area.

Investigators often complain that the evidence was moved before they or other qualified professionals were able to document it adequately, especially with respect to the failed product. In some instances, the investigator is contacted months or years after the incident occurred. In such situations, the investigator should inquire whether anyone who was present during the incident or immediately afterward took photographs or collected other documentation.

Eyewitness Accounts. Eyewitnesses who can describe the sequence of events before, during, or after the accident or failure should not be overlooked. Eyewitnesses include people who are involved in the legal complaint as well as bystanders. Depositions should not be used as a substitute for discussions or interviews with witnesses about the technical aspects of the accident. When possible, the investigator should debrief and interrogate eyewitnesses at the scene of the accident and encourage them to reenact as well as articulate what happened. Because an eyewitness's documentation of a failure may be influenced by the trauma of the event, physiological and psychological effects, and personal motives, the data he or she provides carry less weight than the physical evidence. Nevertheless, the information may lead to the discovery of new evidence.

Measurement and Visual Documentation. When visiting the site, the investigator should bring appropriate portable equipment for taking physical measurements, photographs, and videotapes. Standard measuring devices for investigating structural failures include a measuring tape, a micrometer, a hardness tester, a dye penetrant, a strobe or digital rpm (revolutions per minute) meter, a scale, and a magnet.

In some instances, closeup photographs of the surfaces of the failed part in situ may be required. Thus, the investigator should have a complement of high-quality photographic lenses and equipment with which to document the orientation of the failed components as well as surface features. Before collecting the pieces of a failed component, it is a good practice to label the pieces, mark their location on a grid (relative to their position before the accident), photograph the piece in situ with its label, and weigh each piece. This information can be extremely useful if the accident involved an explosion, the rupture of a vessel, or a failure of rotating equipment, and estimates of the energy or velocities of the fragments is of interest. If possible, closeup photographs of the surface of a fracture should be taken immediately after the failure. Finally, all parts and pieces should be carefully packaged, transported, and stored under secure conditions.

Videotaping (with color and sound) the wreckage at the site of the accident is another good practice. The scene should be viewed from several angles indicating the general layout and the relative positions of parts and pieces. If the failure involved adjacent or relative motion or conditions, the investigator may want to return when the scene has returned to normal to videotape additional data for comparison with the initial record. Photographs and videotapes should be made with the expectation that they may be used in a report or as part of a videotaped presentation to the jury and judge.

Data from the User's Recording Devices. In some industries and technical areas, recording instruments are used to monitor the parameters of the equipment or structure. Examples of these devices are the flight recorders used in aircraft and the recorders used in power or processing facilities to measure vital parameters such as pressure, temperature, and flow rates. In addition, many facilities use videotaping equipment to record the activities of personnel and equipment. Obviously, such data can be extremely useful to the investigator.

Documentary Evidence

Product Specifications. A manufacturer of a finished product has its engineering designer prepare specifications for its own processing equipment. In addition, the manufacturer may have specifications for and from a host of suppliers, vendors, and subcontractors who are responsible for various subsystems or parts of the finished product. These specifications can range from those regarding design, materials, processing, and fabrication to those concerning operational and inspection procedures. Furthermore, a complex maze of warranties between purchaser, vendor, and subcontractor may be involved.

The investigator must evaluate these specifications to determine whether the product was capable of operating safely under the specified conditions or whether the conditions were more severe than those specified. The investigator also must determine the implications of the warranties (the attorney can be of help here). He or she also should review warnings and disclaimers, if any are cited.

Service and Maintenance History. The service and maintenance history of the product before the failure is another valuable source of information. Unfortunately, the availability of complete service records varies greatly, depending on the type of product and its applications. In instances where complete service records are unavailable, the investigator should examine the data that are available on repairs, inspections, and maintenance.

Codes and Standards. Reviewing the applicable codes and standards can be helpful during an investigation. The investigator should be circumspect when conducting this review, however, and do so from the perspective that codes and standards are no substitute for good engineering design and usually are minimum criteria self-imposed by the manufacturer or industry. As was pointed out in earlier chapters, the product should comply with the codes and standards, but compliance does not mean the product is free of defects.

A review of current practices in design, analysis, manufacture, and inspection may be an important part of the accident investigation. Codes and standards may have become obsolete, or they may not exist for certain products. For example, high-pressure systems have been designed, fabricated, and operated for decades without the benefit of a recognized code (see S. J. Brown 1984 for a history of the high-pressure industry). The ASME recently formed a High Pressure Systems Committee to develop codes and standards for systems operating at pressures of 3000 psi to 250,000 psi. Most design practices in the high-pressure area, particularly with respect to protection from energy release, are based on empirical formulas derived from physical data obtained from the field, tests, and experience.

Case Histories. Case histories of similar failures or accidents can be a source of corroboration for the mode and cause of failure of the product being investigated, and they may shed light on previously overlooked sources of evidence. Case histories are cataloged according to the mode and cause of failure (see Boyer, American Society of Metals 1979; Rider *et al.* 1975).

Public and Private Records. Records from previous litigation, governmental agencies, manufacturers, insurance companies, technical societies, and industrial associations are good sources of data on product failures and accidents. Consultants who specialize in design, analysis, testing, and manufacturing of the type of equipment involved in the accident are an additional source of information about past failures.

Manufacturing associations can provide information on design, analysis, and testing practices based on consensus. Examples are the Expansion Joint Manufacturers Association, the Tire and Rim Manufacturers Association, and the Industrial Truck Manufacturers Association. (See Appendix C for a list of members of the American National Standards Institute.)

Statistical data obtained from databases can be used to evaluate parameters such as the frequency of accidents or failures by type of equipment, mode of failure, cause of failure, types of materials, fabrication techniques, inspection methods, and maintenance methods. Data that confirm a high incidence of failure in the past under conditions similar to those present during the current accident can be damning indeed. On the positive side, however, such data can identify the need for improvement in certain areas (S. J. Brown 1976; Gangadharan and Brown 1977; Phillips and Warwick 1968).[1]

LABORATORY EXAMINATION OF THE FAILED PRODUCT

If structural or material failure precipitates an accident, a failure analysis is usually required. This analysis, which is performed as part of the engineering investigation, differs from a failure analysis conducted in conjunction with a product liability case. In a product liability analysis, the protocol is more formal, and more detailed records must be kept.

Figure 5-2 shows a failure report form used in the United Kingdom to document failure. This form is filled out at the site of the accident to the extent possible. A laboratory form should be prepared for each item involved in the failure (see Colangelo and Thornton 1981). Colangelo and Thornton (p. 255) provided the following good piece of advice about conducting a laboratory examination: "Do nothing right away. Contemplate the article [first] and determine a plan of action."

Although dimensions should be measured in the field to the extent possible before the equipment or parts are transported to the laboratory, precise measurements can only be obtained in the laboratory in some instances. This laboratory work is especially important when the specifications contain exact tolerances and dimensions and then wear, corrosion, or deformation may have affected these "built-in" parameters.

Maintaining the security of all parts and pieces involved in the accident is essential to ensure that no inadvertent damage occurs to them. As Vander Voort (1978) pointed out, neophyte failure analysts often have an overpowering urge to see whether the failed parts fit together and to rush to the nearest machine shop and cut up the pieces before obtaining all possible information and before photographing all details. Both actions can result in the loss of valuable information.

The most important and, usually, first step in a failure analysis is to examine the items involved in the accident visually. This examination can answer such questions as the following.

---

[1]In their excellent survey on pressure vessels, Phillips and Warwick found that cracks and defects were the cause of 89 percent of failures: 35 percent of the defects were present before the vessels left the manufacturer and 15 percent were the result of fatigue. Welds were the most likely location of cracks and defects.

- Where are the origins of the failure?
- Where was the stress that contributed to the failure concentrated?
- What was the orientation and magnitude of the stress?
- Were contaminants present?
- What initiated the cracks and in what direction were the cracks propagated?
- In what sequence did the failure occur?
- What was the mode of the failure?
- What was the mechanism?
- Did any imperfections contribute to the failure?

The next step is the laboratory analysis, the purpose of which is to determine the mode of failure. Laboratory tests provide information about parameters such as the rate and magnitude of loading, cyclic loading, temperature, and environment—the type of information that can be useful in reconstructing the accident. The phase of the investigation is analogous to an autopsy performed by a medical pathologist. The investigator seeks physical evidence that will help in reconstructing the incident and identify the cause (i.e., who or what is responsible). Tests that are commonly used to evaluate structural failures can be either destructive or nondestructive.

Some laboratory examinations and tests are performed on parts that have not undergone material failure. For example, the product may have failed to function correctly. Thus, the investigator may have to determine the structural qualities or features that should or should not have been present for the product to function properly; for example, deformation, tolerances, surface finish, friction, heat transfer or transmission, wear, corrosion, and irradiation.

In some instances, the investigator must perform controlled parametric tests on a structure or piece of equipment that is similar to the failed product. These and other types of simulations are discussed briefly in a later section and in more detail in Chapter 10.

Two methods of testing that have received considerable attention in failure analyses are metallography and fractography. Metallography is the study of metals and alloys, especially with microscopic techniques. However, macroscopic examination is included in this discipline.[2] A me-

[2]McCall and French (1978) provided an excellent collection of articles on the fundamentals of metallography and failure analysis and their use in industries such as aerospace, armaments, mining, and heavy equipment.

ASSOCIATED OFFICES TECHNICAL COMMITTEE

Survey of Pressure Vessel Failures **Failure in Service** **Failure Prior to Commissioning**

Report of Failure

---

1. Company issuing report

2. References of item

3. Description of item

4. Overall parameters of item
(length, diameter, thickness, etc.)

5. Age of item

6. Manufacturer

7. Design
- (a) Design code or specification
- (b) Materials used
- (c) Design condition (pressure, temperature, etc.) *None Shop Site
- (d) Heat treatment

8. Actual working conditions
- (a) Contents
- (b) Max. working pressure
- (c) Pressure cycling
- (d) Max. or min. working temperature
- (e) Temperature cycling
- (f) Any unusual conditions

9. Failure
- (a) Date
- (b) Nature of failure
- (c) Size of failure
- (d) Position of failure
- (e) Original thickness at failure
- (f) Reason for failure
- (g) Repairs carried out

Figure 5-2. Failure report form. Used in the United Kingdom to document failure.

(h) Failure discovered by	*Catastrophic failure visual inspection, hydraulic test, nondestructive test, leakage other reason (specify) ______________
(i) When failure is found prior to commissioning at what stage was it found	
(j) History of item relevant to the failure (e.g., details of crack growth or of increase of deformation)	

Signed____________Date_____

*Please strike out where not applicable.

Figure 5-2. (*continued*)

tallographic examination provides information about the material in question with respect to the following.

- Its method of manufacture
- The effects of use (e.g., corrosion, oxidation, and severe hardening)
- The characteristics of cracks
- The presence of abnormalities and defects
- Incorrect chemical composition
- Data or clues about the effects of processing
- The loads the material was subjected to
- In some cases, the history and magnitude of those loadings

Although the term fractography was first used in 1944 to describe the study of the characteristics of fracture surface features, the method has been used for hundreds of years. Before the 1940s, fractography was accomplished with the naked eye or with low magnification lenses. The last 50 years have witnessed the development of optical microscopic techniques and, more recently, the transmission electron microscope (TEM) and the scanning electron microscope (SEM).

DETERMINATION OF THE MODE AND MECHANISM OF FAILURE

LeMay (1985) has divided the modes of failure into two major categories: failure caused by overload and environmentally assisted failure of materials. This approach takes into account that the failure of the material in the part or product is influenced by how the load is applied to the product and by the characteristics of the surrounding environment.

According to Hodge and Phillip (1971), the modes of product failure can be classified as follows:

Plastic collapse	Brittle fracture
Excessive deformation	Creep
Fatigue	Corrosion
Instability (elastic or inelastic)	

Although others (e.g., Collins 1981) have devised more complex classifications, Hodge's classification system, which has been adopted in many engineering design codes (see, for example, Secs. III and VIII, ASME Boiler and Pressure Vessel Code), is the system used in this book.

The following factors, referred to as the *mechanisms* of failure (or failure-mode parameters), define the modes of failure.

- Force/displacement—load deflection or character of loading
- Time—history of when the failure is initiated, was impending, and occurred
- Dimensions—geometric characteristics of the structure
- Temperature—thermal state of the product or structure
- Environment—chemical or physical
- Material state—properties or characteristics of the material

As Seeley and Smith (1963) point out, a product made of ductile steel and one made of a brittle material such as concrete can fail in different ways—that is, yielding (plastic deformation) may constitute structural damage in the steel, whereas fracture (separation) may constitute such damage in concrete. Both ductile and brittle behavior depend not only on the material itself but on the conditions to which the material is subjected as well. Thus, strictly speaking, a material is designated as ductile or brittle relative to other materials or conditions, or qualified with respect to the six parameters of failure just listed. For example, at very low temperatures, steel can fail in a brittle manner similar to concrete or glass. The terms ductile and brittle, however, are convenient for designating the relative ease with which a material either yields markedly or fractures before appreciable yielding occurs under conditions that are relatively favorable to yielding.

The type of action that constitutes structural damage to a product subjected to repeated loads may be different from the structural damage to a product made of the same material but subjected to static (gradually applied) loads. Under repeated loads, a product is likely to fail due to the progressive growth of cracks and fractures even though the material is classified as ductile, especially if the product changes abruptly in the section where the stress is concentrated. Conversely, under static loads, the mode of failure is likely to be yielding; however, a sudden catastrophic brittle fracture can occur.

The action within steel that leads to its failure at elevated temperatures, such as can occur in boilers and chemical and nuclear reactors, can be different from the action that leads to failure at room temperatures, even though both actions are classed as yielding. In general, yielding that occurs at elevated temperatures is dependent on time, whereas yielding that occurs at ordinary temperatures is almost independent of time.

Gamma radiation, moisture in the air, or a corrosive medium such as chlorine may affect the action that constitutes structural damage to a product and therefore limits the maximum loads the product can sustain. Galvanic corrosion is an electrochemical corrosion that occurs when two dissimilar metals are in contact by means of fasteners, welds, or interference fit (i.e., wedges, snaps, or clips). An electric current produces the corrosion and the more current, the greater the rate of corrosion. Corrosion can be inhibited by the material selected, sacrificial anodes, induced electrical polarity, and coatings. The basic concepts of the behavior of materials—stress–strain, load–stress, and crack growth versus load cycles—are reviewed in Appendix B.

The commonly used variables that define the modes of failure are listed in Table 5-1. Note that these variables generally define the mechanistic organization of specific topics that are given unified treatment in engineering, usually because of the convenience of some analytical divisions. To design a structure to prevent failure or to evaluate a failure to determine a remedy, one must understand the modes and mechanisms of failure.

The investigator must be able to identify the presence of fractures by examining the surface markings. Modes of failure such as fatigue, cracks, brittle fracture, ductile rupture, and creep have certain characteristic surface features. Even stress can be evident on the surface. The evidence on the surface alone, however, rarely allows one to determine both the mode and cause of failure. Therefore, in addition to the back-

TABLE 5-1. Factors that Define the Modes of Failure

Factors (Variables or Functions)	*Failure-Mode Parameters*			
1. Material:	Elastic (linear)	Plastic (load-dependent)	Creep (slow, time-dependent)	Fracture (slow and/or rapid crack growth)
2. Load:	Static	Cyclic	Dynamic	
3. Geometry:	Stable	Neutral	Unstable	
4. Temperature:	Low	Ambient	Elevated	
5. Time:	None	Short term	Long term	
6. Environment:	None	Material degeneration (e.g., corrosion)	Structural degeneration (e.g., erosion)	

ground information and the results of the examination already discussed, the investigator may have to perform experimental or theoretical simulations of some or all conditions that may have played a role in the sequence of events leading up to the accident. A thorough knowledge of applied mechanics and mechanical metallurgy is desirable when assessing what tests are required to conclusively document the mode or mechanism of failure. (Some of these testing methods will be explored in Chapter 10.)

Neophyte attorneys who are handling one of their first product liability cases often have the urge to rush out and look for an "engineering expert" concerning the part, component, or equipment that failed or was involved in the accident. For example, if a pump impeller fails, they look for an impeller expert; if the failure is related to a control pedal on a forklift, they look for a forklift expert. In general, however, engineering investigators who have a background in engineering science, mechanics, and materials are best suited to conduct the preliminary stages of the investigation (Vance and Gould 1985). In other words, these people are best suited to identify the important parameters involved in the failure or accident and determine whether specialists in fields such as metallurgy-materials, vibration, dynamics, or ergonomics are needed.

Sometimes, one encounters so-called professional investigators, the "product experts," who have no formal technical training but cite, as the basis for their technical credentials, their occupational experience and membership in trade, industry, and occupational organizations that have no technical or professional requirements. Because industry practices are an important source of information, people who lack technical training but have experience in a particular industry can provide useful information. If the court's decision is to be an informed one, however, judges and attorneys must understand this person's role in developing evidence versus the role of the trained and experienced professional engineer, who uses experimental or theoretical methods or both to identify the mode of failure, reconstruct the sequence of events, and identify the causes of failure. Readers who are interested in basic engineering texts on materials and design should refer to Chu *et al.*'s overview (1983) of fundamental engineering mechanics, Shigley's (1977) text on mechanical design, and Dieter's (1976) overview of mechanical metallurgy.

SIMULATIONS

Having gathered background information, conducted a laboratory examination, and arrived at some preliminary conclusions about the mode and mechanism of failure, the investigator may decide that simulations will clarify the mechanism of the failure as well as its mode and cause. Simulations can be either experimental or theoretical. Experimental simulations involve parametric tests versus full simulation, scale, prototype, and in situ. Theoretical simulations consist of classical, numerical, and/or parametric versus full simulation. Theoretical simulations are often selected because they are more economical than experimental tests. For example, testing a large structure such as a bridge, building, ship, or manufacturing plant can be prohibitively expensive. On the other hand, if the nature of a failure is complex, the mathematics used in a theoretical simulation may be uncertain; thus, an experimental simulation may be more accurate. The advantage of theoretical simulations is that many parameters can be evaluated, whereas only one model or specimen may be available for an experimental simulation in many instances.

Simulations are used to demonstrate that a hypothetical scenario will yield the same results as those that were produced during the actual accident. Although the investigator may perform parametric tests in some cases to isolate the effects of critical parameters, a full-scale or prototype simulation involving all the parameters during the accident is the most desirable test.

Tests on scale models are desirable because they are economical, especially when large structures costing billions of dollars are to be simulated. Of course, there are disadvantages or potential problems in justifying scaling effects. Simplistic examples are the poor special effects that are characteristic of grade B movies; such simulations can be detected readily even by the untrained eye.

Tests performed in situ on the component or structure that was actually involved in the accident are advantageous because they provide accurate data. In situ testing is often used to obtain data on determinable structural responses such as stiffness, flexibility, damping, and mass for input into a computer model. In situ tests eliminate some uncertainties concerning the accuracy with which the computer simulation reflects the actual structure or boundary conditions. Once the in situ data are incorporated into the computer model, the investigator can perform a variety of parametric and full-scale simulations.

Theoretical simulations are usually divided into two types: classical and numerical. The term classical refers to solutions ranging from simple formulas to special functions obtained by solving the governing differential and constitutive equations. Numerical methods such as the popular "finite-element method" (S. J. Brown 1983a), the finite-difference method, and the numerical-integration method take advantage of digital computer technology to solve complex structural problems. Simulations based on numerical methods can readily illustrate graphically, on screen or on paper, a variety of parameters such as deformation, stress, temperature, velocity, and acceleration.

As was mentioned earlier, the theoretical modelings are only as good as their material characterizations. Few theoretical methods consider the microstructural behavior of materials. Generally speaking, a continuum approach is used in theoretical predictions about the development of structural or fluid responses.

The finite-element technique has been used to predict elastic, plastic, creep, and fracture material characterizations, both solid and fluid, over a wide range of strain rates for thermal, vibration, and dynamic forces (S. J. Brown 1986a; Zienkiewicz 1977). The use of theoretical fracture mechanics as a tool in design and analysis is relatively new, and the method was introduced into engineering curricula only recently. Theoretical fracture mechanics incorporates the concept of a "flawed" material—the existence of a crack, which is the most frequent cause of catastrophic failure—into its theory. (Some basic concepts of linear fracture mechanics are presented in Chapter 8 and Appendix B; see also, Labbens 1985 for a historical perspective and Garwood and Harrison 1985 for a discussion on the use of yielding fracture mechanics in failure analyses).

The motion of large structures such as bridges, ships, and buildings as well as biomechanics and missile impact can be simulated readily by computer. Finally, illustrating the mechanism of failure to juries, courts, and attorneys is often accomplished best by showing videotapes of either the experimental sequences or the results of computerized analysis.

RECONSTRUCTION OF THE INCIDENT

One extremely challenging aspect of an accident investigation is the reconstruction of the sequence of events leading up to the incident involving a product that failed to serve its intended purpose and resulted in damage or personal injury. This part of the investigation brings together all elements gathered from the background research, the examination, the evaluation of the mechanism or mode of failure, and any theoretical and experimental simulations. The result provides a basis for developing a statement about what caused the failure or accident.

An accident reconstruction is usually motivated by a desire to (1) determine how the incident occurred and what caused it, (2) improve or correct the product, the design, the manufacturer's procedures, or the user's operation, (3) determine who was responsible if loss, injury, or damage was involved, or (4) accomplish all three. At this phase of the investigation, the intent is not to affix causality or responsibility within a legal framework.

One can think of the reconstruction process as analogous to attempting to correlate several jigsaw puzzles of the same subject when the pieces of the puzzles are not interchangeable and some pieces are missing. For example, the field investigation of wreckage provides many pieces of the puzzle but usually not all the pieces. Eyewitness reports, recorded data, databases, inspections, and simulations of the accident represent independent puzzles. In other words, the investigator must try to determine which parts of each set of data overlap with parts in other sets, and look to the various sources of information available to fill in the missing parts of the overall picture.

DETERMINATION OF CAUSALITY AND RESPONSIBILITY

A generally accepted practice in the engineering community is to assign the cause of a failure or the responsibility for the failure to one of the following: a deficient design, inappropriate materials, defective materials, improper manufacturing and processing, improper or unusual operating conditions, or lax or inappropriate maintenance and inspection procedures.

The investigator formulates a technical position about the cause of the accident by combining the scientific facts used to reconstruct the accident and establish legal responsibility. Recall that a product is measured for defects against three basic standards or legal theories: negligence, strict liability and implied warranty, and express warranty and misrepresentation. Thus, engineering investigators should cooperate with attorneys by reviewing applicable codes and standards, warranties, specifications, current practices, the state of the art, and any relevant implied or explicit agreements. They should not overlook two other resources when determining whether their conclusions about the cause of the accident are sound. One resource is peer review—the independent evaluation of the investigator's report by one or more independent experts. The other is blind testing—one or more independent reviews of the report, from which information and conclusions, indicating which party or viewpoint is represented, have been deleted.

Each case history described below involved a failure analysis. In some instances, the failure analyses were part of an accident investigation. Each case involved some aspect of design, processing or fabrication, or the user's operation. Together, the histories indicate the diversity in the modes of failure discussed in previous sections.

Failure of a Pipe Liner

Figure 5-3(a) is a photograph of a longitudinal cross section of a pipe containing a liner that failed. Figure 5-3(b) is a graphic representation of the pipe and the thin 10-gauge liner attached concentrically within the pipe. The pipeline and liner contain a gaseous flow field of approximately 200 feet per second. In this example, the flow of gas induced vibration of the liner, which caused cracking in the liner. Fatigue occurred at the point where the liner was welded to the wall of the pipe. The finite-element theoretical simulation shown in Figure 5-3(c) and (d) indicates a first fundamental *in vacuo* frequency of 66.5 Hertz (Hz). An in situ instrumented test of the liner in the flow field indicated a critical frequency of vibration of approximately 7 Hz. The purpose of these tests was to correlate the theoretical vibration response of the liner *in vacuo* (no fluid) versus the critical vibrational response in the actual field-of-flow (with fluid) and the suitability of using the theoretically derived frequency *in vacuo* as a specification for fabricating the liner.

Flow-induced vibration was the mechanism of failure (excitation or loading), and the fluid caused the coupling effects between the liner and the pipe wall. The sensitivity of the liner to the added mass of fluid or gas is demonstrated by the fact that even accelerometer weight influenced the frequency of vibration of the liner *in vacuo*. For example, a

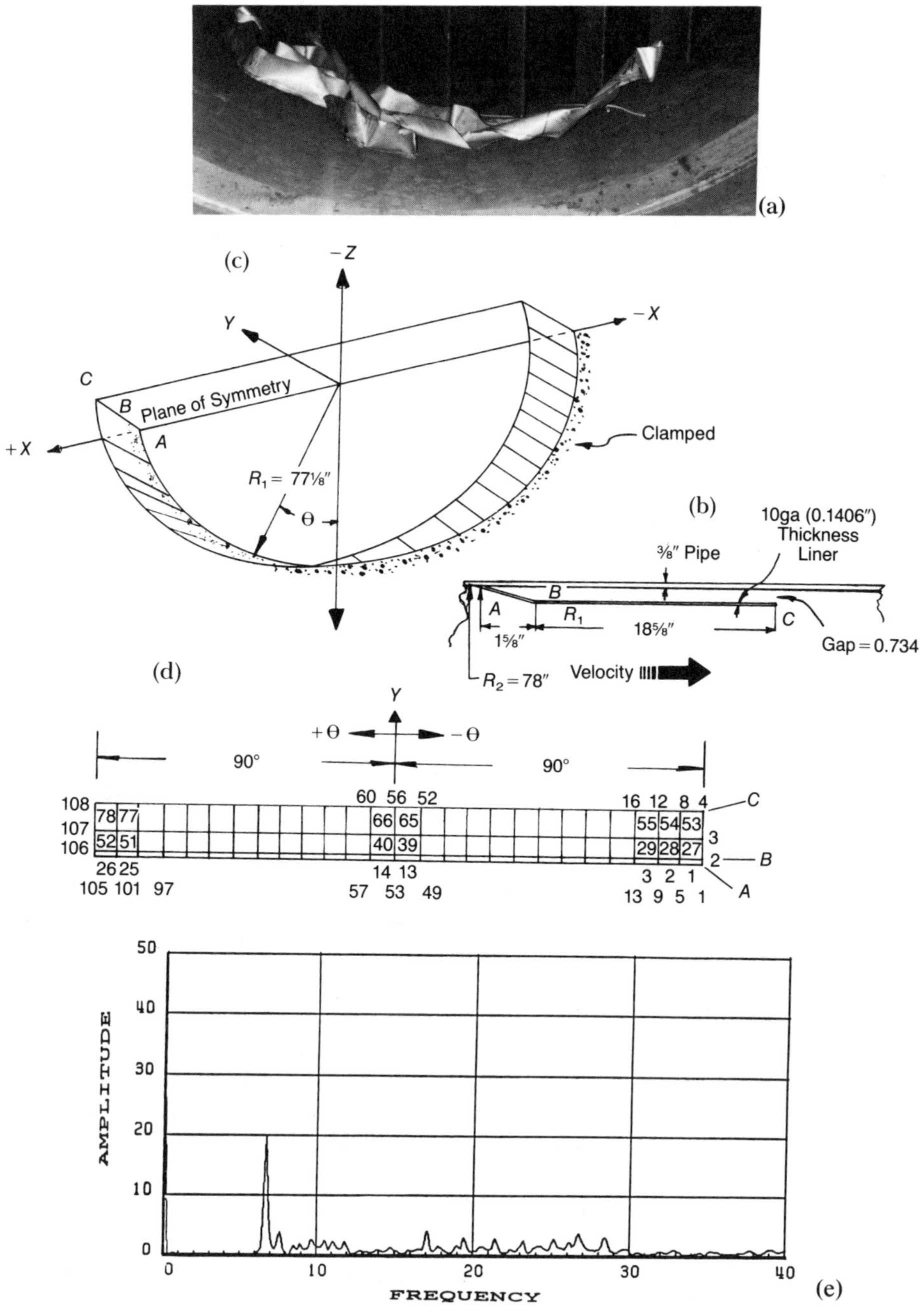

Figure 5-3. Pipeline failure. (a) Photograph of the failed liner. (b) Cross section of the pipe and liner. (c) Finite-element computer simulation of the liner. (d) Unfolded finite-element mesh. (Courtesy of Quest Engineering & Development Corp.) (e) Frequency versus amplitude response spectrum for the liner. (Courtesy of the U.S. Air Force)

first frequency *in vacuo* of 42.7 Hz was recorded for the same liner with the accelerometer placed at point C.

Figure 5-3(e) plots the frequency and amplitude of the response of the liner in the flow field that was recorded during an in situ experimental test. The presence of coupling effects (fluid environment and supports) and imperfections (including accelerometer instrumentation) lowered the frequency *in vacuo*. For higher frequencies, the amplitudes are considerably lower. Is the frequency of flow excitation critical at 7 Hz? And does fluid coupling reduce the first frequency of the liner *in vacuo* close to the excitation frequency? The answer is yes. The investigator concluded, however, that because of the complexity of specifying the performance of the liner on the basis of theoretically computed calculations of flow-induced vibration, the more desirable solution would be to "design out" the problem by placing liner stiffeners parallel to the flow field, which would raise the frequency and fatigue life of the liner beyond the critical range caused by flow-induced vibration.[3]

The cause of the failure in this case was inadequate specifications. The conclusions, however, reflect that the risk of failure can be reduced at a high level of confidence by revising the design to eliminate the need for (1) the stringent quality control procedures required by the original specifications for fabrication (e.g., tolerances and welding) to control the structural frequency of the liner, or (2) the modification of the flow field to control the vortex shedding that was exciting the liner vibration.

Rupture of a Gas Pipeline

Figure 5-4(a) is a photograph of a cement-coated pipeline that was ruptured by the cutter head of a dredge used to widen a riverbed. The walls of the pipeline were 0.375 of an inch thick (X = 42 as specified by the American Petroleum Institute (API)) and the inner diameter of the pipe was 12 inches. An examination of the ruptured section of the pipe indicated a classic axial fracture and the formation of one radial flap as the result of tearing. Because the pipeline was buried in the river bottom, the single-flap fracture shown in the figure did not progress to a two-flap or full 360° tear, which is more likely in most instances (Baum 1984; S. J. Brown 1985a; Nadai 1950; Shroder 1979).

On initial examination, nontechnical personnel assumed that the cutter tore the pipe flap, which was not the case. A fracture analysis of the pipeline indicated that the mechanism of failure was the propagation

[3]Chen (1985) and S. J. Brown (1985) recommend "designing out of the problem" rather than relying on accurate predictions about the flow-induced response of a system.

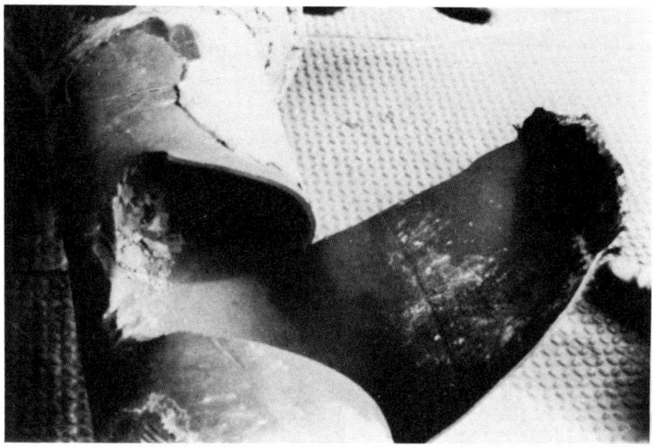

(a)

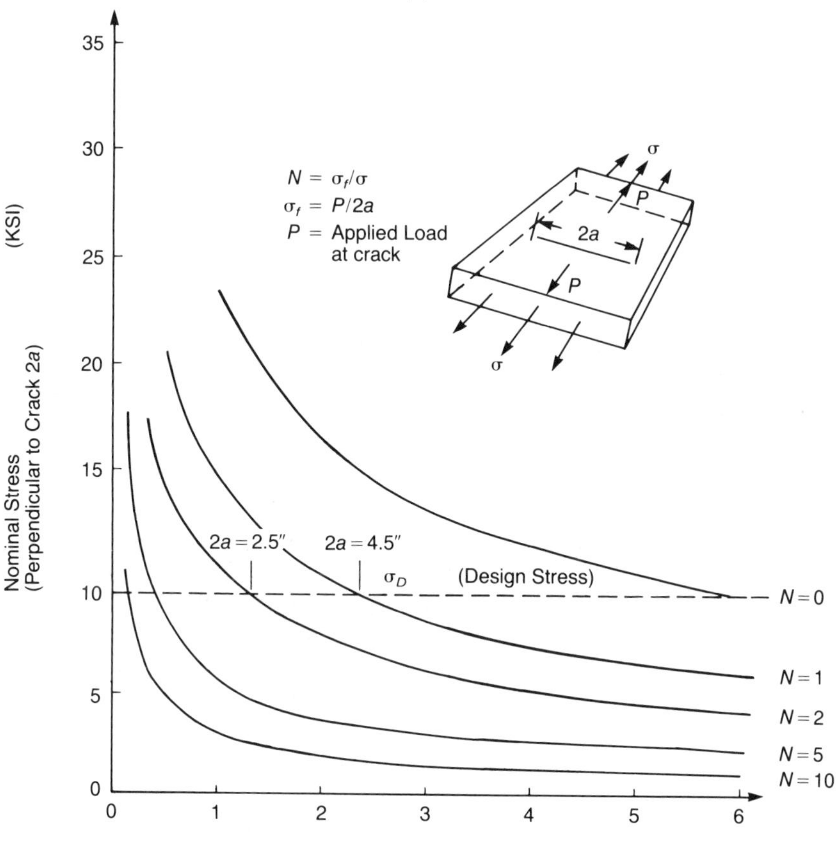

(b)

Figure 5-4. Pipe fracture. (a) Burst section of the 12-inch gas pipeline. (b) Nominal stress in the pipeline (with effects of varying impact-force coefficients) versus one-half critical crack size (a).

of cracks caused by (1) an axial groove or gouge caused by the cutter head, (2) the pressure within the pipeline, and (3) the impact of the cutter head.

The influences of these three effects are illustrated in Figure 5-4(b). The analysis of theoretical and full-scale burst tests indicated that the cutter head of the dredge initiated a fracture as a result of its dynamic impact on the pipe and therefore induced overload in a localized area. The pipeline ruptured in a classic manner. The following questions remain, however: Was the dredge in the proper location? Was the channel in the river properly marked? Was the pipeline placed at the proper depth? Should the pipeline have been designed to ensure that it could withstand the impact? The answers to these questions determine causality.

Failure of Flooring Under an Ore Crusher

Within a year after an ore-crushing plant was commissioned and operations began, large cracks appeared in the concrete flooring of the plant. The floor area was approximately 69.5 feet long by 54 feet wide and was supported by eight beams made of reinforced concrete: five beams (each approximately 40 inches deep) along its length and three transverse beams (each approximately 61 inches deep). Concrete support walls were located along the perimeter of the floor and at two locations along its length at the center. On the floor were two conveyor systems that received ore at 75 tons per load.

Figure 5-5(a) shows a schematic illustration of the truss work and surface area of the two parallel conveyor systems and the transverse concrete beams supporting the concrete floor. The specified nominal aggregate size of the load was approximately six tons (12,000 pounds), but could vary from 9645 to 18,289 pounds, depending on the density of the ore. The distance the ore fell was approximately 27.75 feet; thus, a significant amount of kinetic energy developed when the ore dropped into the conveyer system (see Fig. 5-5(a)). When the concrete floor began to deteriorate, a steel support system was installed immediately underneath it (see Fig. 5-5(b)). A review of the engineering calculations revealed that the design of the support flooring was based on a factor of approximately 16 times the static deflection.

To determine whether the system was properly designed to meet the owner's performance specifications, the engineering investigator needed to assess the stress caused by dropping 75 tons of ore per load at the specified rate. To accomplish this assessment, the investigator developed a finite-element computer simulation model representing the conveyor system on which the ore was dropped, the flooring on which the conveyor rested, and the support walls and columns to determine the parametric influence of stiffness on the structural system that absorbed the impact as well as the effective mass of the system. This technique was less expensive than evaluating an undamaged structure.

Figure 5-5(a) illustrates the finite-element idealization of the upper surface of the floor, and Figure 5-5(c) shows the finite-element representation of the supporting beams as viewed from underneath the floor during impact. Figure 5-5(d) indicates that the maximum dynamic deflection from the ore (dropped from $27\frac{3}{4}$ feet) can be represented as a factor K times the static deflection of the system versus a function of the ratio of the mass of the floor to the mass of the ore being dropped on it. (Static deflection means the displacement of the floor when the weight or mass rests on the floor.)

As Figure 5-5(d) indicates, other factors influence dynamic versus static deflection. For example, if the impact is perfectly elastic—that is, no dissipation of energy occurs during impact—the results will differ significantly from the results of an inelastic impact—one in which the ore is highly frangible and sticks to the floor mass on which it is dropped. Since the density of ore can vary, the weight of the maximum boulder specified by the owner to the designer also may vary. The stiffness of the target of the dynamic impact of the mass is influenced by where a hypothetical largest allowable boulder may strike the conveyor system. When the investigator in this case subjected small samples of ore to small-scale frangibility tests, the statistical results suggested that the impact was more likely to be inelastic than elastic.

Figure 5-5(c) illustrates the deformation caused by the impact of the ore predicted by the finite-element simulation coinciding with the fractures in the actual floor. The user's specifications spelled out the requirements of the design, but the designers were not formally trained engineers and were not registered as professional engineers. They estimated that the dynamic force of the mass would be about 16 times the static force load, whereas the actual load was between 100 and 280 times the static force load.

As a result of the finite-element (numerical), experimental, and theoretical studies performed during the evaluation of the structural adequacy of the flooring, the investigator determined that the failure was the result of the contractor's failure to design the plant to meet the specifications supplied by the owner's engineering firm, that the concrete selected to absorb the impact of the loads was inappropriate, and that dropping the ore from a distance of $27\frac{3}{4}$ feet created the problem of dissipating kinetic energy. The alternative to redesigning the ore off-loading system was to use dampers, distribute the load, and use a ductile material such as low-carbon structural steel to support the floor.

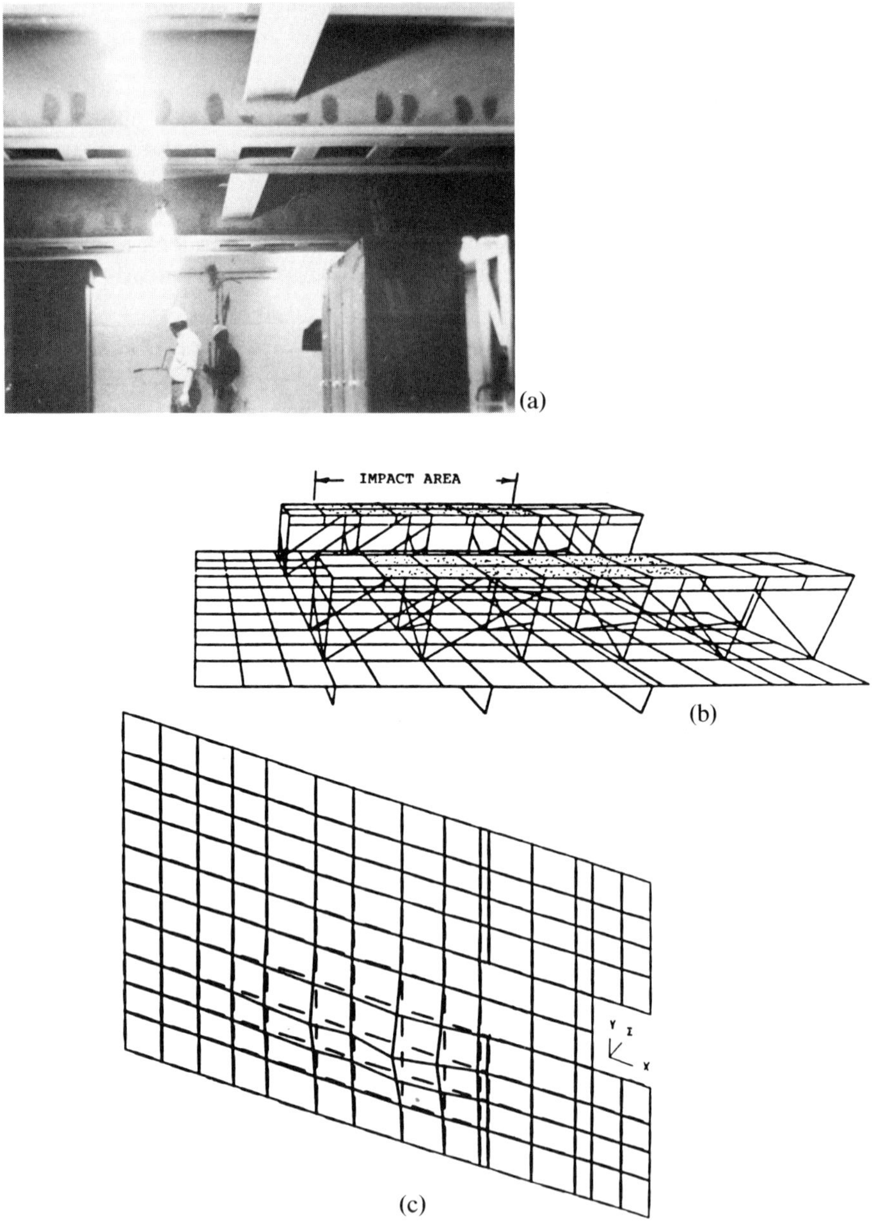

Figure 5-5. Ore crusher building failure. (a) Photograph of structural-steel floor beams installed under the concrete beams after the concrete beams failed. (b) Finite-element theoretical model of the ore-handling conveyors, floor, and support beams. (c) Finite-element deformation plot of the underside of the floor subjected to dynamic impact on one ore conveyor. (d) Mass ratio of ore (M_1) to effective mass of floor (M_2) versus dynamic-to-static deflection ratio (M_1g/k_B), where k_B is effective floor stiffness. Note that at a mass ratio of 10, k_B is a floor stiffened to 10 times the original design.

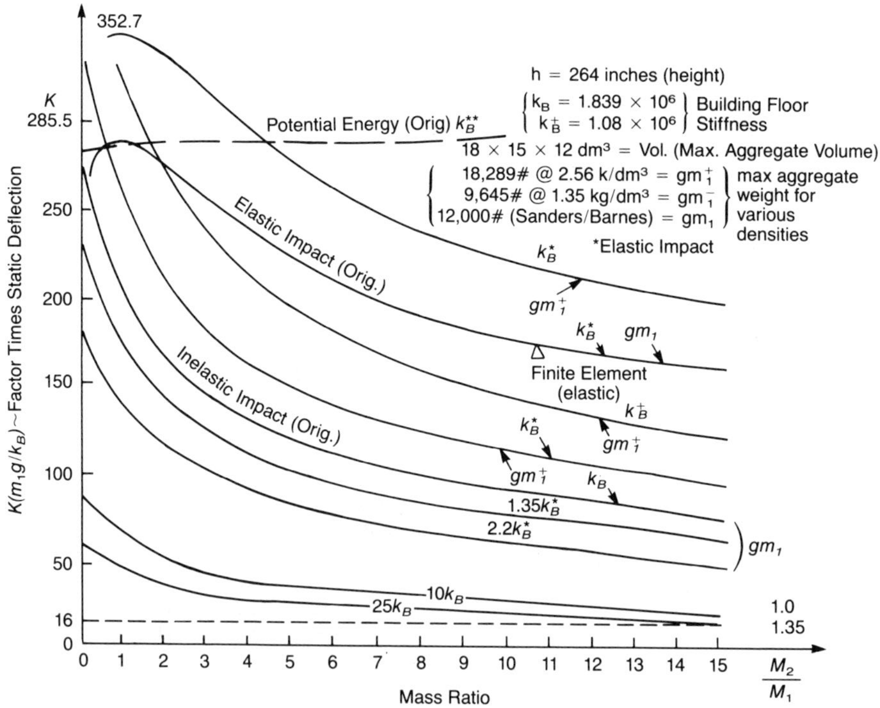

(d)

Separation of a Curtain Wall in a Building Complex

Figure 5-6(a) shows one wing of a three-wing, 10-story building complex. The architecture of the building, although elegant, is complex, and oblique angles of intersecting walls are the rule rather than the exception. Shortly after construction, the building was subjected to winds of about 50 miles per hour, which caused panels to sail and the curtain wall to separate from the building (see the bulge in the wall in the lower right-hand corner of the photograph).

A review of the architectural and structural drawings and the conditions existing at the time of the separation revealed that the causes of the wall separation from the superstructure and the panel "pop-out" were poor construction practices and inadequate structural drawings; the mode of failure was excessive deformation of the wall panels, and plastic and brittle failure of the fasteners by tensile and shear forces; the mechanism of failure was negative pressure induced by the wind vortexes and the aerodynamics of the building.

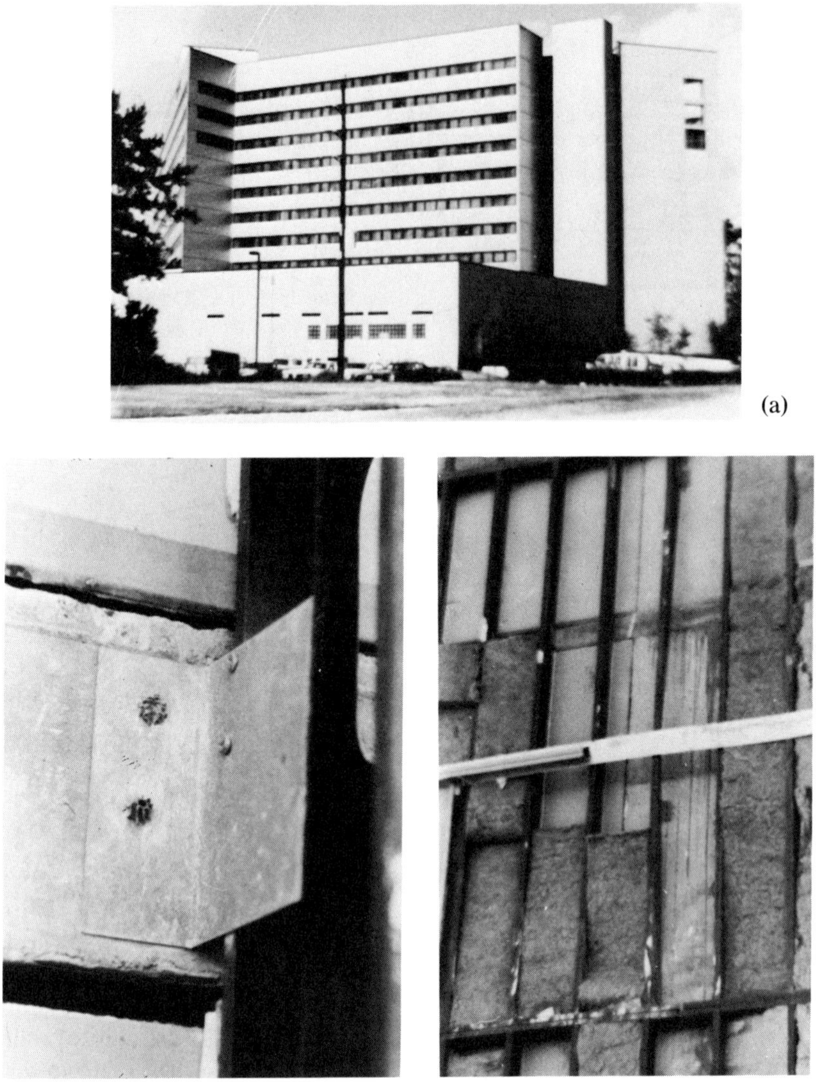

(a) (b) (c)

Figure 5-6. Building curtain wall failure. (a) Section of a building subjected to high wind loads. (Courtesy of Quest Engineering & Development Corp.) (b) Face-mounted clips attached to a vertical steel beam to resist negative wall pressure. (c) Damage to the curtain wall and separation of panels caused by vortex shedding.

Figure 5-6(b) illustrates the improper placement of structural clamps on the flange rather than on the web of the vertical beam. Excessively long unsupported spans of exterior composite (layered or sandwiched) paneling in addition to poor or no cross-bracing contributed to the pop-out shown in Figure 5-6(c). The manufacturer did not think a building designer would need information about the structural properties of the panels, and the contractor failed to follow standard construction practices for wind loads outlined in the Uniform Building Code.

A computer simulation of a section of the building using the finite-element method was used to evaluate a retrofit technique of reinforcing the wall with only minor changes to the interior and exterior of the building. Baseboard structural beams were devised to secure the exterior wall to the interior superstructure floor. In addition, exterior structural strips were added to match the existing window line and add cross-bracing and support to the excessively long vertical panels. In situ load tests were performed on existing exterior sections of walls that had the original support brackets. Load-deflection (flexibility tests) were performed on the sandwiched exterior panels to incorporate this "as-built" data into the finite-element model of the building wall along with all other structural components, including anchors.

The curtain wall separated because the contractor failed to follow the specifications provided by the architect/engineering (A/E) firm. The curtain wall contractor devised an inadequate face-mounted anchorage system to connect the curtain wall to the superstructure. However, the design provided by the original structural engineering firm retained by the architect-engineering firm was inadequate.

Overload of a Hot Tap System

When using any equipment, the owner or operator should be aware of the consequences of modifying the manufacturer's equipment, especially high-energy equipment. One example is the failure of a hot tap system caused by overloading.

Figure 5-7(a) illustrates that hot tap equipment can be used to install a bypass pipeline while still under operating pressure. In this instance, the 16-inch pipeline was pressurized to 600 psi and the equipment was being used to install the lock-ring plug into the flange section of the tee. The user-owner had placed an oversized valve into the adaptor (see Fig. 5-7(b)) to facilitate more rapid depressurization than was possible with the valve supplied by the manufacturer. During the operation, the replacement valve was opened before the lock-ring plug had been secured by the locking segments (see Fig. 5-7(c)). The result was a considerable differential of pressure across the lock-ring plug, to which the plug

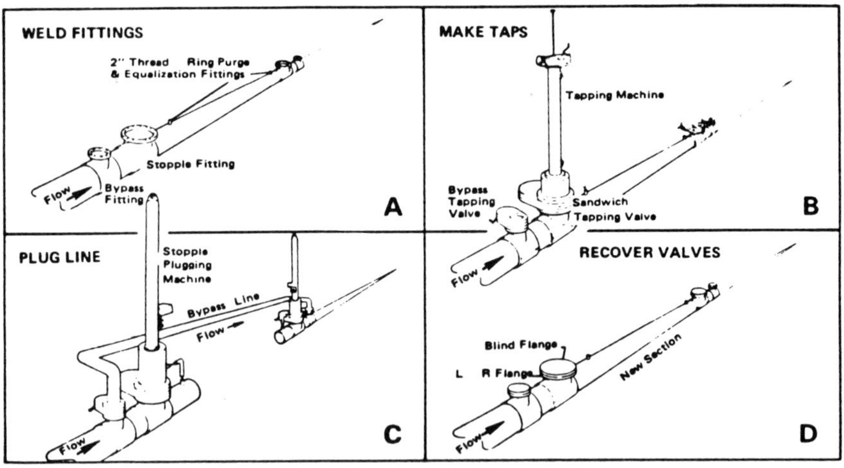

(a)

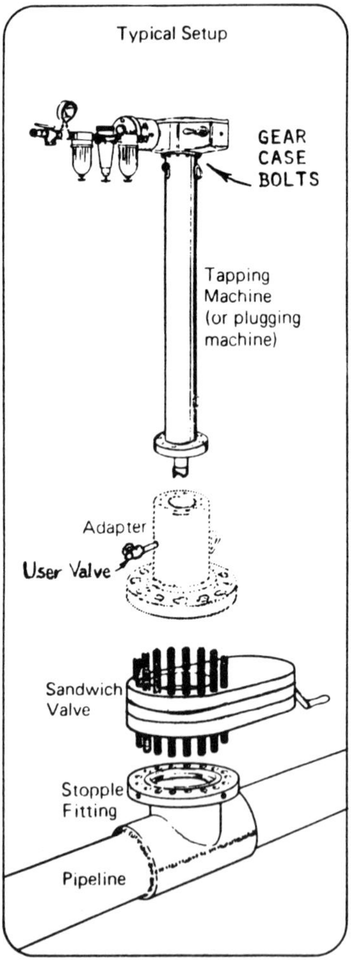

(b)

Figure 5-7. Hot tap separation. (a) Examples of how hot tapping equipment is typically used. (b) Arrangement of hot tapping equipment and the improper bleed valve installed by the user. (c) Arrangement of the lock-ring plug, valve body, and plug holder.

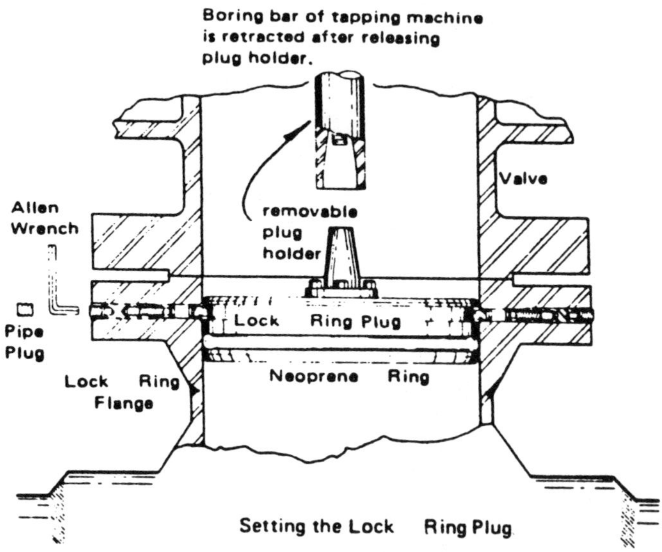

(c)

holder was still attached, and the subsequent exertion of force in a piston-like manner, which rapidly ejected the tapping machine and dynamically loaded the gear-box bolts. Because the operator was killed, interviews with eyewitnesses immediately after the accident were too emotionally charged to be of much value.

The investigation revealed that a high tensile loading rate fractured the gear-box bolt. (The rate at which the bolts were loaded is revealed in the amount of plastic deformation versus fracture.) An analytical simulation of the adapter depressurization indicated that the critical pressure differential was achieved within an eighth of a second after the oversized valve was opened. One can estimate analytically the kinetic energy available to drive an explosion or rupture; however, a physical measure of the minimum kinetic energy available can be determined by weighing the separated pieces, measuring their trajectory, and measuring the impact (S. J. Brown 1984, 1986a).

Although the user-owner of the equipment had years of experience and training and was aware of the instructions and warnings concerning its use, he modified the equipment, thoroughly understood the purpose of his modification, and operated the equipment improperly. In other words, he made a fatal error in judgment.

Failure of a Valve in a Power Station

An extremely important safety aspect of equipment is proper maintenance. Figure 5-8(a) is a photograph of a 900-psi gate-valve body and

(a)

(b)

(c)

Figure 5-8. Valve bolted assembly separation. (a) Photograph of a high-temperature steam valve body and bonnet. (b) Scanning electron microscopic view of nut threads showing heavy oxidation. (Courtesy of Pennsylvania State University) (c) Scanning electron microscopic view of graphite formation at the grain boundary in the nut. (Courtesy of Pennsylvania State University) (d) Cross section of stud and nut illustrating the creep bell mount of the nut. (Courtesy of the University of Florida)

(d)

bonnet from the control room of a power station. The valve was commissioned to operate at between 1000° and 1100°F. The valve was serviced for approximately 27 years before it failed, and all parts were essentially the same, with a few minor exceptions, throughout its life. The failure resulted in extensive damage, fire, and injury.

As a result of leakage around the valve gasket, the owner requested an in situ injection of a polymer to seal the valve gasket. The leakage may have indicated a failure of the original gasket or insufficient preloading of the flange bolts. During preparation for the process, but before the injection process was initiated, the valve separated. A section of the valve stud (see Fig. 5-8(d)) illustrates that a considerable amount of inelastic deformation (or creep) occurred in the nut and led to visible displacement of the threads at the washer end of the nut a length of one-half pitch. Examination with a scanning electron microscope revealed heavy oxidation of the nut threads (see Fig. 5-8(b)). Figure 5-8(c) provides evidence of void formation at the grain boundary in concert with graphite formation.

Studies such as those reported by Kenny and Patterson (1985) indicate that in bolted connections, a major portion of the load is normally carried by the first three threads adjacent to the washer or flange. As these threads creep and the nut spreads (a slow dynamic material deformation), the load is transferred farther up toward the top of the nut until the nut "bell-mouths" and only the upper threads carry the load transmitted between the stud and nut. Failure occurs when the shear-carrying capacity of the remaining threads is exceeded. If leakage does not lead to detection of the bell-mouthing during scheduled maintenance or inspection, the bolts are usually preloaded according to the manufacturer's design requirements. The chain reaction initiated when the load is transferred from the first failed nut-bolt threads to the remaining bolts leads to a rapid cascading effect ending in failure.

Damage to a Building by Subsurface Ground Shock

Finally, consider a not uncommon case that presents some aspects of causality that are not easy to define or compartmentalize. This type of case can be characterized as one in which the user does not take due care or exercise caution in relation to third parties, or bystanders.

During some geological studies performed in conjunction with a series of subsurface explosions with an energy equivalent as high as 30 pounds of TNT at a depth of approximately 100 feet, the foundations of some nearby buildings exhibited motion, subsidence, and cracking. According to the records kept while the 100-foot holes were drilled for the explosive charges, the first 100 feet of soil consisted of clay, with sand usually found at 70 feet or more below the surface and shells found between 70 and 90 feet below the surface.

Figure 5-9 illustrates the various waves propagated by an energy source. The types and intensity of the waves are influenced by the type of soil present, its homogeneity, and its layering or stratification. The figure provides a view of the relationship between the explosive charges and the two buildings that were damaged. It also shows a cross section of the surface of the ground, the composition of the soil, and the location of the blast relative to the two damaged buildings. The investigative engineer also included railroad beds, riverbeds, canals, roads, and pipelines in the drawing.

During the investigation, he noted that both buildings had suffered damage at the southeast face and that shear cracks in the walls indicated ground motion from a southeast direction. As shown in the figure, the *R* ground wave is the major tremor, and the *P* and *S* waves are the minor tremors: two-thirds of the total energy at the source of the explosion

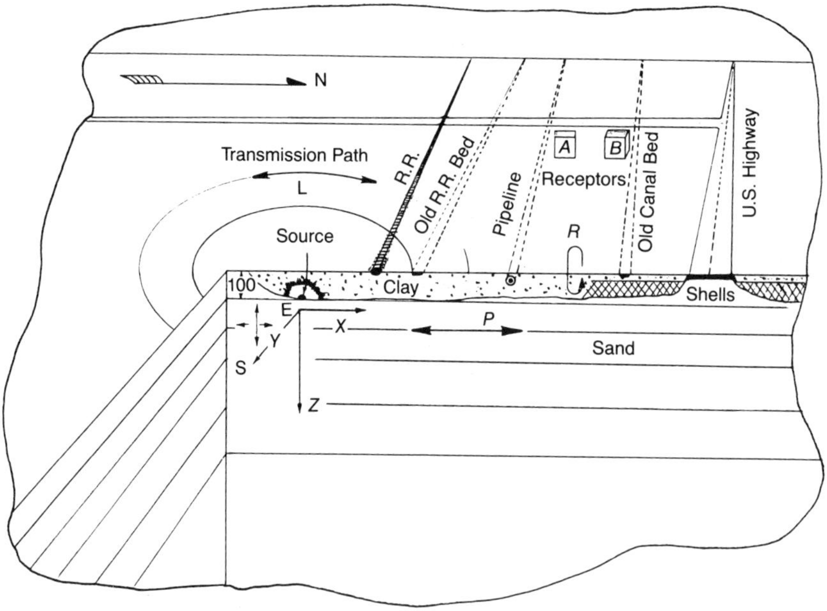

Figure 5-9. Illustration of the location of the subsurface detonation, the geologic characteristics of the terrain, and the location of the buildings and other structures. (Courtesy of Quest Engineering & Development Corp.)

went into *R* waves, which dissipated much more slowly with the distance than did the two less-energetic waves (see S. J. Brown 1986c; Richard, Hall, and Woods 1970; U.S. Department of Energy 1985). The motion of the *L* wave occurs in layered media. Waves transverse to the vertical plane usually arise from layering of various strata in soil. The presence of another layer of soil causes at least two different phenomena: (1) the appearance of the *L* wave, which causes transverse horizontal oscillations to occur and be sustained by repeated multiple reflections between the surface and the underlying layer (the *L* wave will not occur if the covering layer of soil is stiffer than the underlying layer), and (2) distortion of the propagation velocity of the *R* wave, which results in elliptical trajectories associated with the motion of these particles and changes the ratio of radial-to-vertical displacement on the surface of an upper layer when a stiff underlying layer is present. In the case described here, the underlying layer of sand tended to enhance the effect of ground motion over a longer distance than would be expected if the soil had been homogenous and sandbased, for example.

Initially, the service company claimed, after examining one building, that subsidence was the cause of the damage. The engineering firm that investigated the situation, however, surveyed all the buildings and demonstrated the relationship between the cause and the structural damage to all of them.

In this incident, the occupants of buildings A and B kept very good records relating to the commencement of subsurface detonation with the documentation of building motion and occurrence of wall movement and cracking. These data were even more meaningful in light of the fact that the owners and occupants of the two buildings were not the same and kept their own independent records of the dates of subsurface detonations and observable damage. This information correlated well with the physical evidence and with the records obtained from the organization performing the subsurface detonations. A degree of precision no greater than a tape and surveyor's level was required to measure the source of detonation relative to the buildings and other features such as roads, railroad tracks, and canals.

PREPARATION OF THE REPORT

Opinions

When an engineering investigation report is prepared, an opinion section is included for summarizing the cause of the failure and what information in the previous sections justifies the opinion. Mitigating circumstances may be cited. If physical evidence is lacking, the report may include a technical or legal opinion based on information from databases, case histories, eyewitnesses, and so forth.

In many instances, the section citing the cause of failure includes the opinion. For example, in the case history involving the ore crusher, the investigator's conclusion was that improper design was the reason why the flooring was inadequate to handle the impact loads. His opinion was that using reinforced concrete to support a floor subjected to repeated high-impact loads was a poor choice. In addition, he said that the arrangement of the processing equipment (conveyor-crusher) could have been designed to reduce the impulse loads. The engineer who designed the superstructure was, however, required to consider the specified loading conditions for the processing equipment.

In some jurisdictions, if several organizations are involved in the failure, the proportion of responsibility and liability assigned to each will depend on its contribution—or perceived contribution—to the cause of the accident. In the ore-crusher example, the owner's engineering firm had written *performance* specifications for the capacities of the plant and the equipment, not *design* specifications, and the general contractor had provided the specifications for the design and erection of the plant. Thus, the owner's engineering designer may have shared liability for the plant shutdown with the contractor's architect/engineer and the contractor. If the architect/engineer required the use of reinforced concrete, his responsibility would increase because concrete is a poor choice for the superstructure in the area affected by the failure. In fact, the architect/engineer was employed by the contractor, and none of the contractor's engineers had engineering degrees or were registered professionals.

In the example of the extension ladder with the faulty foot design cited in Chapter 3, manufacturing tolerances contributed to the hazardous situation because the feet seized on nonlevel ground. Test simulations showed that this effect was repeatable and that the ground at the site of the accident was indeed uneven. All the physical evidence pointed to the foot design as the cause of the failure and the injury; however, because no one observed the actual sequence of events, the expert's opinion, based on the data he gathered, was extremely important.

Recommendations

Having completed the investigative process of determining the mode and cause of the failure or accident, the engineering investigator must cite ways in which the incident could have been prevented. These recommendations can range from improving the design of the product or labeling the product more appropriately to revising the current codes and standards or using and maintaining the product more prudently. In summary, the Recommendations section of the investigator's report is a statement of the lessons learned in the accident investigation and what actions should be taken.

For example, in the ore-crusher case, the investigator suggested that the arrangement of the equipment could have been improved by distributing the load dimensionally and by using dampers or snubbers to slow down the rate at which the impact load is applied to the superstructure. With regard to future construction, he recommended that alternate processing layouts should be considered and that impact-resistant materials such as ductile steel should be used to support the flooring.

In the case of the hot tap equipment, the investigator recognized that no product can be completely foolproof and that an operator cannot be prevented from ignoring adequate labeling and documentation of the proper procedures for operating. In support of the manufacturer's desire to minimize the risk of operator error, however, he suggested possible improvements in the equipment through sequentially activated passive systems. For example, the bleed valve could be modified so that it could not be activated until the line pressure of the system was reduced or until the equipment was taken off the line pressure by closing its gate valve. On the other hand, engineers who conduct investigations for a defendant-manufacturer should exercise care, when discovering new or improved design features, not to provide what is called "20-20 hindsight data" for the plaintiff.

It is extremely difficult for a manufacturer to account for all the foolish actions or momentary lapses of judgment users are capable of. Consider Dentzer's (1984) description of a lawsuit in which the plaintiff sued a manufacturer of ceiling fans for injuries a child suffered when it was lifted into one of the company's fans. Ceiling fans were originally designed for use on ceilings that were 12 feet high or higher. Their current popularity means that they are now attached to ceilings as low as $7\frac{1}{2}$ feet. Should the manufacturer have limited the use of its fans to ceilings that users could not reach? If so, would placing a fan on a lower ceiling constitute negligence on the user's part?

PRODUCT LIABILITY LITIGATION

Discovery

If an accident leads to litigation, there will be a period before the trial called the discovery period, during which the opposing parties will have access to one another's technical material. When the discovery period ends, on a date set by the court, no additional technical work can be performed. The legal strategy during discovery will be influenced by (1) the perception of whether a settlement will be reached before trial, (2) the client's financial resources, and (3) the legal team's experience with product liability cases. Because the discovery period usually is a significant part of the legal strategy, the expert and the attorney need to work together to outline the objectives and assignments involved in the accident investigation and failure analysis.

During discovery, the expert may experience long periods of technical inactivity, punctuated by brief periods of intense activity. The process can be frustrating for experts who are inexperienced in litigation. However, the fact that their professional responsibilities are the same as those involved in a nonlitigious investigation can be reassuring. The only difference is that the technical expert is entitled to gain access to information gathered by the opposition. Only through technical strategies can data be kept from the opposition. Under no circumstances, however, should the investigator, manufacturer, or user turn over any files to anyone without the approval of the responsible attorney.

In one common strategy, data are "lost." This delaying tactic can be used to force the opposition to collect their own data or to prevent them from obtaining information that can be damaging, particularly when it demonstrates deficiencies in design, manufacturing, or use. Placing the opposition in the position of having to collect their own data can be particularly effective when their financial resources are limited, which is often the case for individuals and small companies or for law firms that work on a contingency basis. Another way to delay giving information to the opposition is to put off the failure analysis or the correlation of data until after the depositions have been given.

The Lawyer–Engineer Relationship

Clearly, a good working relationship between counsel and technical expert is necessary. Counsel should regard the technical expert as a resource from the beginning of the litigation, not as a tool to be used later in the trial to fill an evidentiary gap (Weinstein *et al.* 1970). The appropriate stance is to regard the engineering expert as an equal partner from the moment the two agree to undertake the lawsuit. The engineer's role is to explain the technology to the attorney. To maximize the skills of both, the relationship between them should be smooth and cohesive (Colangelo and Thornton 1981).

When several attorneys or technical experts are involved in an investigation or product liability case, appointing a coordinator, liaison, or manager for each team is a good practice. This is especially important if more than one law firm and engineering firm are involved. Directional responsibility for the legal work and authorization for the technical work should be decided, and put in writing, before the technical investigation begins. This is especially important in a personal injury suit, because law firms that handle such cases usually work on a contingency basis.

Engineering experts who conduct accident investigations and failure analyses should familiarize themselves with the experience of the attorney who handles the case, just as clients should research the qualifications, experience, and track record of law firms retained to handle product liability cases. Engineers must recognize when they are going beyond their own expertise, and thus may need the assistance of a consultant who is an expert in a particular discipline. Similarly, attorneys should recognize their technical and financial limitations. The failure to obtain the assistance of a legal consultant who has the necessary expertise in product liability and the financial resources to prepare an adequate technical case may jeopardize a client's case.

Consider a case involving a seriously injured worker in his twenties who fell while working at a construction site. Having engaged the safety snap in his safety belt, he began working, when suddenly the belt gave way and he fell about 50 feet. The accident investigation revealed that the manufacturer had failed to consider that left-handed users would try to secure the snap on the left-hand side of the safety belt (with the right-side snap already engaged). If this is tried, however, the buckle tongue and belt loops on the left side of the belt interfere with proper engagement of the snap—the snap feels completely engaged when it actually is not.

The plaintiff's law firm, having no experience in product liability, made the following mistakes. First, it had three of its attorneys communicate with the technical expert throughout the course of the investigation without assigning the role of coordinator to one of them. Second, it arranged for the technical expert to give his deposition without visiting the site of the accident or interviewing the victim and other eyewitnesses. Third, it waited until shortly before the end of discovery to ask the expert to develop his technical material.

The defense attorneys, recognizing that the opposition had failed to develop the case properly, simply avoided asking the plaintiff's expert any relevant technical questions and challenged the plaintiff's attorneys to take the case to trial. Lacking the confidence to try the product liability case and not wanting to invest the money required to develop the case, the plaintiff's law firm obtained a signed agreement to settle for $50,000—before the end of the discovery period and without reviewing the final technical evidence! Finally, the firm refused to pay the engineering firm for the expert's belt simulations and report, which the law firm had requested, because it had settled the case two days before the report was received. As Larry Johnson (1984 p. 62) stated in "The Ten Deadly Deposition Sins": "It is imperative to know when to use your expert witness."

The legal profession does provide a source of redress for clients with complaints against lawyers through policing organizations such as state bar associations. (Vance and Gould, 1985, proposed a similar source of redress for engineers if a client decides that a technical expert has not provided competent representation.)

The Importance of Engineering Credentials

Engineering qualifications can be an important issue during litigation, not only for the technical expert who performed the accident investigation and failure analysis, but for the engineer who designed the product involved in the incident. The sources of engineering credentials and the problems associated with them were discussed in Chapter 2. Here, the issue is discussed in the context of litigation.

The designer's qualifications can be important when deficient design is a causative factor in the product failure and when those deficiencies can only be the result of (1) lack of knowledge about the relevant engineering principles involved, (2) human error, or (3) oversight. However, the design engineer certainly cannot be held responsible for unforeseen conditions to which the product can be exposed. Clearly, having information about the design engineer's credentials helps the accident investigator understand why certain established engineering principles and practices were not observed in the design, manufacture, or use of the product involved in the accident.

Questioning the credentials of the opposition's expert witness also can help the technical expert understand how the adversary arrived at his or her conclusions. The following factors may also influence the adversary expert's testimony.

- Was the expert responsible for any engineering aspects of the failed product?
- Is the expert employed by one of the contesting parties, and thus feels obligated to say the "right thing"?
- Does the expert provide services to only a specific segment of industry, and thus be prejudiced against manufacturers or users?
- Does the expert refuse to provide services to personal injury litigants for fear of offending potential clients?
- Is the expert a professional expert witness?

The fact that the terms engineer and engineering have become a part of every conceivable job description or occupation regardless of education, registration, or experience accounts for the dilemma associated with the public's confidence in products that have been "engineered." The courts reflect the public's misunderstanding of the meaning of engineer and engineering (as illustrated by the Highlift case) and often exacerbate it.

CONCLUSION

If a common thread runs through accident investigations and failure analyses, it is that the cause of or responsibility for failure can be assigned with relatively high frequency to all six sources discussed in this chapter: design, materials, manufacturing, fabrication and processing, operating conditions, and maintenance and inspection.

In subsequent chapters, the eight steps involved in an accident investigation and failure analysis are discussed in more detail, principally in the context of failures involving structures, equipment, or components. Although economic losses related to product failure are not necessarily attributable to structural failures—for example, computer programs can cause economic losses when they malfunction or produce inaccurate results—we decided to emphasize structural failures because such failures are the greatest source of economic loss and personal injury and often precipitate chemical, electrical, and other failures.

6

Background Research: Documenting the Incident

When does an accident investigation begin? It should begin immediately following the incident when injury or economic loss is first identified. This is particularly true following a catastrophic event. It is also true that there are driving forces at an accident site to remove the incident or failure data and material in order to rescue personnel and to restore production, construction, and flow of traffic: in general, to return the site to normality as soon as possible. This is particularly true in vehicle accidents, building failures, and chemical-process hazardous releases. Frequently, two types (or more) of expertise are needed at an accident site: the experts needed to bring the accident site to safe conditions and the expert needed to perform the accident investigation and failure analysis. For example, government agents such as firefighters need to be cognizant of hazardous materials (Meidl 1970 and 1978) and industrial personnel such as plant or facility employees need to respond to such events (NCCHMS 1972; Shaver and Berkowitz 1984; CEP Manual 1987; Lees 1980).

There have been some efforts in various industries and professions to train personnel (who may be among the first to arrive at the scene of the incident) in some rudimentary scientific principles in order to document data pertinent to the incident. One such example[1] is *The Handbook for the Accident Reconstructionist*, by M. J. Lofgren (1983), which is directed to law-enforcement agencies involved in transportation accidents (particularly automotive or truck vehicles). The pros and cons of this parascientific approach to investigation of an accident is discussed later in this chapter.

Not all incidents that result in personal injury and/or economic loss are sudden and catastrophic, and those that are may be the result of an accumulation of incidences or may have been initiated by some incidences prior to failure and only precipitated by a more recent, less severe, or minor event.

INITIAL CONTACT WITH THE CLIENT

During the initial contact with the principals involved in the incident, the law firm representing one of the principals or the regulatory agency, the forensic engineering expert will ask questions to obtain a broad overview of the incident: What happened? Who was involved (i.e., who are the principal players)? Where did the incident take place? When? What information is available? Why, in the principal's opinion, did the incident occur? Has a lawsuit been filed (if appropriate)? If so, on what date will discovery end? The answers to these questions will help the forensic engineering expert or consulting firm determine whether a potential conflict of interest exists and whether the incident requires expertise that the firm cannot supply. If different expertise is required, the expert should recommend another consulting firm, if possible.

If the consulting firm agrees to conduct the accident investigation and failure analysis, it should stipulate how and when it wants payment (see chapter 12 for more on financial arrangements), then its investigator should ask to review the available data before he or she goes to the site of the incident. During this initial contact, the investigator tries to determine what equipment will be needed at the site. If the incident occurred some time ago, the investigator usually requests that eyewitnesses be available at the site.

---

[1]Other technical organizations associated with vehicular accident reconstruction information dissemination and individual membership are the American Association for Automotive Medicine (AAAM), the Society of Automotive Engineers (SAE), the Institute of Transportation Engineers (ITE), the Southwestern Association of Technical Accident Investigators (SATAI), the Traffic Institute—Northwestern (TI), the Society of Accident Reconstructionists (SAR), and the National Association of Professional Accident Reconstruction Specialists (NAPARS).

Before the actual on-site investigation begins, it may be necessary to get approval from the principals to go on the property and photodocument areas that may otherwise be restricted. If a lawsuit has been filed, these approvals are often handled by the law firm, which may have an attorney accompany the investigator to the site.

The forensic engineer should visit the site as soon as possible (1) if the source of the problem needs to be identified and remedied quickly, especially if production has been shut down, which is extremely costly; or (2) if litigation is involved, the date of the trial is approaching, and the end of the discovery period is at hand. Preparations for the on-site investigation can be relatively simple or extensive, depending on the circumstances. For example, the site may be in a foreign country or in a hostile environment such as a desert, a swamp, an offshore platform, or the Arctic.

A recent example of a situation where it is extremely difficult to perform an accident investigation immediately after the incident was reported by Suarez and Schupack (1988). In this incident, the L'Ambiance Plaza apartment building collapsed during construction in Bridgeport, Connecticut, and 26 workers were killed. As the authors reported:

> The site conditions immediately following the accident were chaotic and investigators were kept away from interfering with rescue operations. Since the main immediate task was to save as many lives as possible, no concern was given to any of the structural components that might have been in the way of the rescue operation. (p.4)

The National Bureau of Standards was commissioned by OSHA to investigate this building collapse. As a result of this incident, the Connecticut Legislature introduced legislation that would require a state-registered professional civil engineer to oversee all civil construction projects.

As was stated in Chapter 5, the order in which various aspects of the investigation are undertaken is dictated by the unique circumstances of the incident, the site, and the conditions surrounding the event. Thus, the forensic expert must exercise judgment concerning the sequence of the steps involved in the investigation. For example, the investigator may interview eyewitnesses first, then scan the site to obtain an overview of the incident, document the location and pieces of the failed product, and interview the eyewitnesses again in more detail. The investigator also may return to the site a second or third time or perform additional examinations of the failed product.

EYEWITNESS ACCOUNTS

Three types of witnesses are useful to the investigator: (1) those who observed the event; (2) those who were familiar with the operation and appearance of the product before the incident and, especially, immediately before the incident; and (3) those who arrived at the scene immediately after the incident and observed the condition of the physical evidence. Eyewitness accounts can be particularly useful if the investigator arrives at the site some time after the incident and after the physical evidence has been moved to a nearby storage area.

The forensic engineer must listen carefully and allow witnesses to tell their stories in their own way rather than attempt to guide them. Questions should evoke the witnesses' own interpretation of the events; however, the investigator should attempt to quantify what witnesses mean by nontechnical terms such as large, far, suddenly, and slowly.

If the incident occurred recently, the eyewitnesses's accounts may be colored by the trauma of the event; thus, the data they provide will carry less weight than the physical evidence (Abney 1986). In the aftermath of an incident involving a severe injury or a death, interrogating witnesses may be impossible or inappropriate. In such situations, the investigator should begin documenting the incident and interview eyewitnesses at a later date. In the interim, the eyewitnesses will have the opportunity to reconstruct the sequence of events with a minimum of emotional trauma. However, if the investigator needs to ensure that a system is safely shut down or to obtain vital information about the state of the equipment immediately before the event, what actions were taken after the event, and what actions should be taken, eyewitnesses will have to be interviewed. The information they can provide will be especially important if release of a hazardous substance must be avoided or minimized.

During the interviews, the investigator should take notes or tape record the witnesses's statements for later reference. If a witness illustrates the sequence of events or points out locations, parts, and so forth, the investigator should take photographs, videotapes, or both to document this information.

Consider the case of the death associated with the multifunction control pedal in the Highlift forklift truck. A number of people who had operated the forklift involved in the fatality said that the multifunction pedal tended to go out of adjustment and cause delays in shifting from forward into reverse. When the pedal functioned properly, operators tended to "punch" the pedal—that is, put the truck into reverse without coming to a full stop. A person who observed the accident reported that the operator suddenly looked surprised, then attempted to jump from the truck. Because the brakes were functioning, his action suggested that the directional pedal failed to reverse the direction of the truck and, in fact, accelerated it toward the edge of the loading dock. The information provided by this eyewitness correlated with data the investigator gathered using techniques discussed later in this chapter.

Finally, consider the following lawsuit involving Ms. Nicholich versus the government of a large city. While driving across a bridge on her way to work one cold morning, Ms. Nicholich's car suddenly went out of control on ice that had collected on the bridge. When her car came to rest, she got out and walked toward the right-hand walkway to avoid being hit by vehicles that were approaching the bridge and having similar difficulties. As she stepped onto the walkway and tried to put her hands on the railing, she slipped and fell over the railing, landing on a railroad car that had stopped under the bridge, then falling onto the tracks.

Ms. Nicholich's attorney asked the investigating engineer who was considering the study whether he was familiar with codes and regulatory statutes concerning bridges and whether the city was required to close the bridge under icy conditions or coat the surface to reduce or eliminate the hazard. The consultant said that he was familiar with these codes and statutes and that in order to proceed he would need to review the background data and would need a retainer or work authorization. He then asked the attorney to send him all relevant information collected by the law firm for review. The consultant also requested that Ms. Nicholich accompany him to the site if she was able to do so. The attorney said that his client had been seriously injured, but was now able to travel.

During the site visit, the consultant photographed and videotaped the site, took measurements, and, most important, interviewed Ms. Nicholich, who described the sequence of events leading up to the accident and indicated where different events had occurred (see Fig. 6-1). The investigator then correlated the data he had collected with weather reports, city records, and previous reports of accidents at the scene.

A key feature of the case was documentation that the bridge failed to comply with the standards of the American Association of State Highway and Transportation Officials. The walkway did not qualify as a true sidewalk according to any code or standard, although pedestrians used, and still use, it for that purpose; it did not even qualify as a curb. In addition, the handrail was not high enough to comply with the heights established for pedestrian handrails. Therefore, tall people such as Ms. Nicholich (who was almost six feet tall) were likely to fall over the railing if the walkway was slippery (see Fig. 6-2).

The consultant recommended to city officials that a barrier be placed above the walkway to prevent pedestrians from using it or that the height of the handrail be increased and that a safety strip be painted on the right shoulder of the bridge as a refuge for disabled vehicles. The

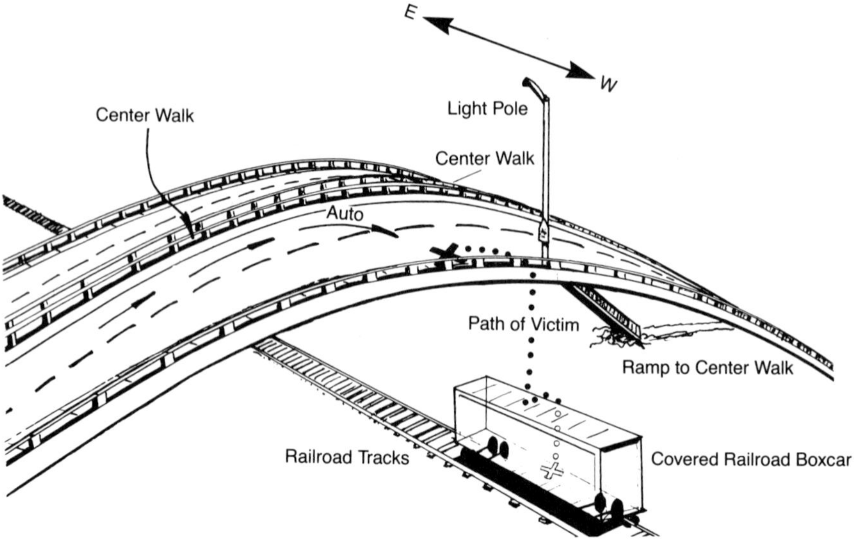

Figure 6-1. A drawing of the bridge where Ms. Nicholich was injured. (Courtesy of Quest Engineering & Development Corp.)

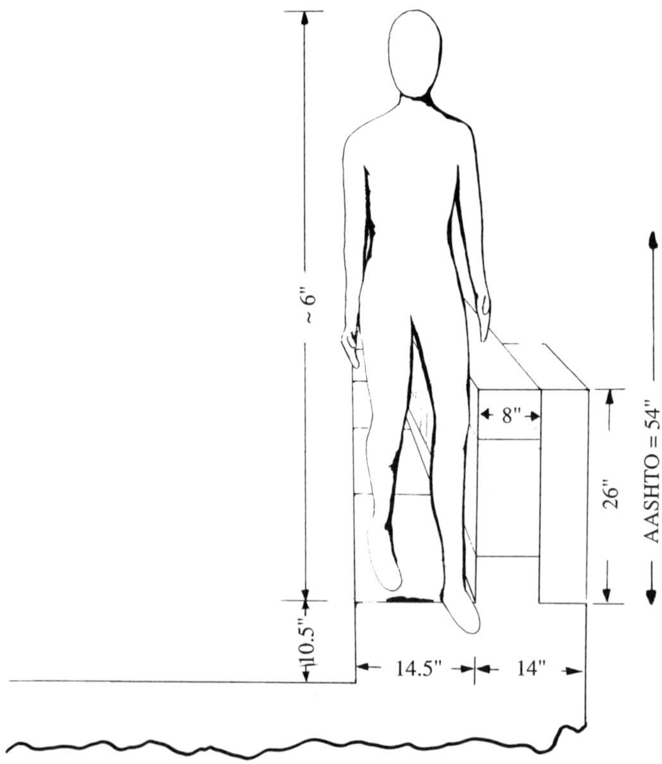

Figure 6-2. A drawing that illustrates the walkway and railing on the bridge. (Courtesy of Quest Engineering & Development Corp.)

city, however, never took any actions to reduce the risk of injury. Therefore, one must conclude that the city administration decided that the cost of retrofitting its bridges outweighed the risk of liability for injury or accidents.

MEASUREMENTS

When inspecting the site of an incident, an investigator should always bring along two basic measuring devices: a measuring tape and a micrometer. The need for additional portable instruments such as a surveyor's transit or level will be dictated by what kind of data are required and how accurate the measurements need to be. Pacing off distances and using the measuring tape are fundamental and inexpensive methods of documenting the distance between structures and the location of fragments that have been ejected during an explosion. Distances can be measured with sufficient accuracy with a measuring tape that is 100, 500, or 1000 feet long. A level or transit is particularly useful for measuring the displacement of foundations, soil, equipment, and buildings. With the micrometer, the investigator can measure inside and outside tolerances and the dimensions of manufactured parts.

Figure 6-3(a) shows the wreckage of a vessel that ruptured during a pressure test (Baker 1985). Figure 6-3(b) illustrates the layout of roads,

(a)

Figure 6-3. Rupture of vessel during a pneumatic test. (a) Photograph of ground zero after the rupture. (Courtesy of Quest Engineering & Development Corp.) (b) A map of the area showing the location of missile fragments (numbered items). (Courtesy of Sandia National Laboratory) (c) A missile map of the ruptured pressure vessel. The numbers correspond to those shown in (b). (Courtesy of Sandia National Laboratory)

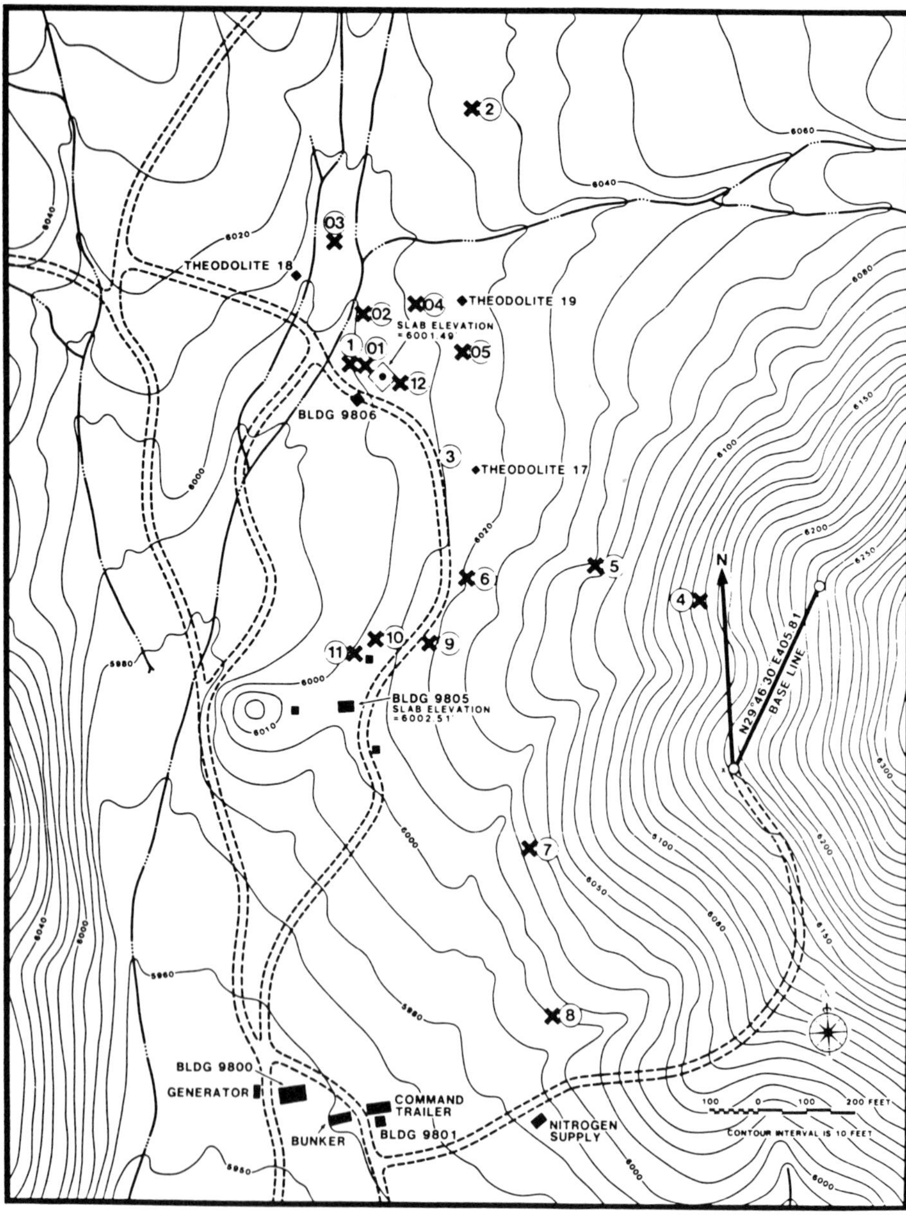

Figure 6-3b. *(continued)*

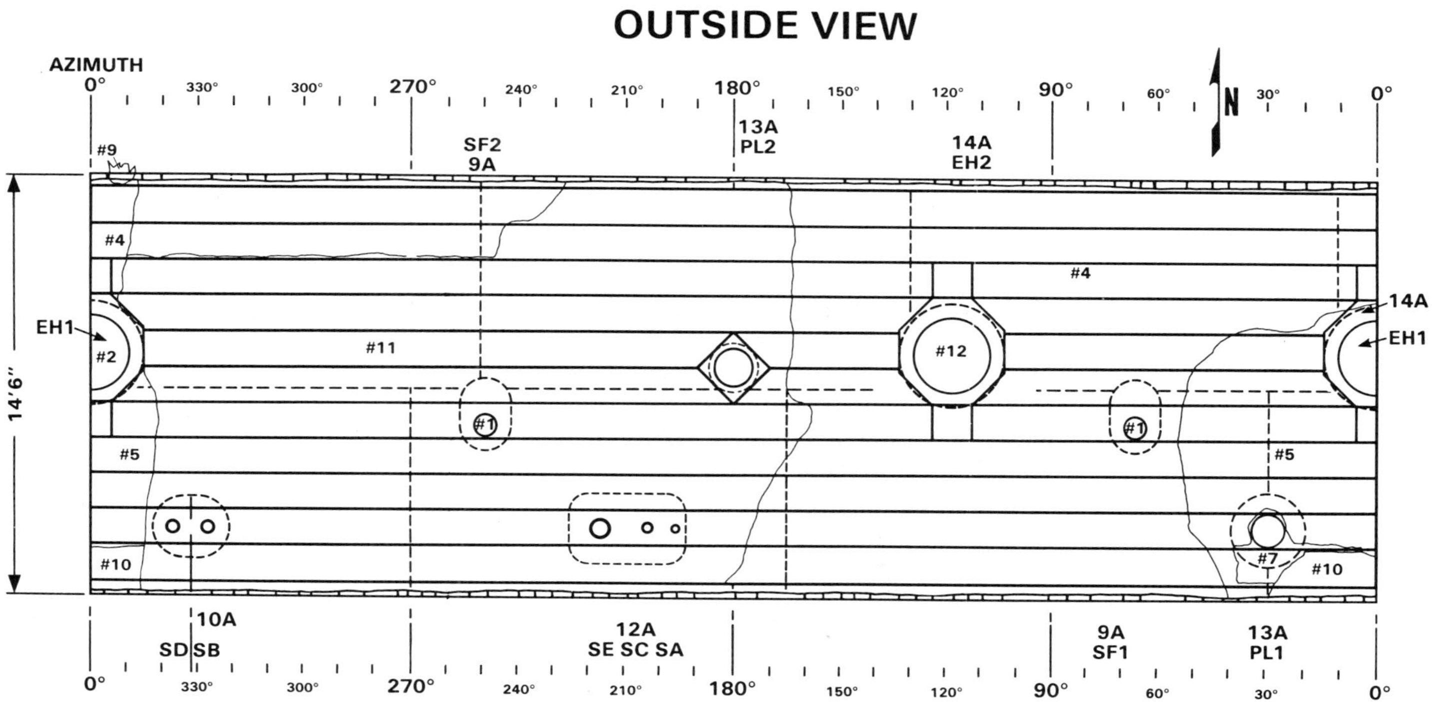

Figure 6.3c (*continued*)

buildings, and fragments; in this instance, satellite elevation readings as well as a transit were required to define the direction, elevations, and distances traveled by fragments from the source of the rupture (ground zero). Figure 6-3(c) contains a sketch of the fragments and their identification numbers. (Methods of mapping fragments and measuring their trajectories are discussed in Brown and Baker 1992, in press. Reviews of the literature on hazardous release, explosions, fragmentation, protection, and impact appear in Baker 1983; S. J. Brown 1985a; 1986c; Suarez and Shupack 1988; Zukas *et al.* 1982.)

VISUAL DOCUMENTATION

Because visual documentation of the scene of the incident usually provides information that will influence the testing of materials, simulations, and determinations of causality, its importance cannot be overemphasized. Furthermore, if litigation is involved, visual documentation is usually presented to the jury. For this reason, videotape is receiving more and more attention. It is a medium that judges, juries, attorneys, and experts are familiar with.

A significant number of engineers receive training in photography and darkroom techniques as part of their formal training, usually in metallurgy and materials testing. Thus, many investigators feel comfortable about taking their own photographs, a practice that can be extremely helpful to a client with a limited budget. In addition, many universities and professional development programs offer continuing education courses in the use of photography and videography in accident investigations, failure analyses, and preparation for trial. In April 1987, the American Bar Association's Section on Litigation sponsored a symposium on "The Use of Video Tape Evidence at Trial" (see Brown 1987b). Because visual documentation of the scene of the incident may play an important role in convincing a jury, the investigator must record the site with this in mind. This is especially important if the structures and equipment involved are unfamiliar to the audience: for example, a power station, a forge, the engine room of a ship, or a helicopter landing on an offshort platform. This is also true when economic losses were the result of a deficient design, lack of protection, or failure to warn rather than the result of a catastrophic failure. In other words, the investigator should not miss any opportunities to document visually what one would observe by visiting the site.

Before doing any visual documentation other than initial entry shots, the investigator should (1) review the information obtained from eyewitnesses, (2) make a grid map, pace off the site, and note the location of fixed structures, markers, and reference points as well as the structure and the equipment involved in the incident, (3) measure the reference points and the location of parts, (4) set up an indexing system for identifying objects that will appear in the photographs or video recordings, and (5) look for tags, serial numbers, model numbers, manufacturing dates, and other means of identifying the equipment.

Photographs

The section on measurements indicated that still photography is useful for recording spatial orientations of objects at an accident site. This type of photography offers imaging of the finest quality and "freezes" an instant. In addition, with a high-resolution lens, one can take closeup photographs of surface features that will help determine which features are related to the initial failure conditions and which features occurred afterward.

Most important, however, still photographs permit the use of reference points, usually in the form of a grid, which allow quantitative measurements of size, distance, location, orientation, and other dimensions. A grid is set up by establishing a baseline with a zero reference point, placing markers at regular intervals along the baseline, and taking a photograph at each interval. Whitmall and Playter (1987) suggested placing the markers at 20-foot intervals; however, the size of the interval will vary depending on the size of the photograph's field of view. Grids can be plotted using one of several techniques: graphic plotting using vanishing points, mathematical plotting, or computer plotting.

Grids also are useful for determining the dimensions of an accident site depicted in old photographs taken without a grid. In this procedure, called the reverse projection technique, a slide of the original photograph is compared with a new photograph containing grid lines. By intersecting the lateral, vertical, before, and after positions shown in the original photograph with the grid data in the new photograph, the dimensions of objects and areas depicted in the original photograph can be determined with a high degree of accuracy.

Figure 6-4 is a photograph of a collapsed dockside tower. To calculate the structural deformation, the investigator used specification and construction drawings, field measurements, photographs taken at the time of the collapse, and three-dimensional imaging software to calculate the structural deformation.

The usefulness of photographs is also illustrated by a failure analysis of a post-tension slab in Lufkin, Texas. A contractor retained a subcontractor to make the forms for the slab and another subcontractor to set the anchors and cables, pour and finish the slab, and tension the founda-

Figure 6-4. Photograph of a dockside ore-loading and off-loading tower that collapsed when hit by a ship. Courtesy QED.

tion cables. Figure 6-5 shows the cables and anchors used. When the slab was completed, the contractor proceeded with construction; shortly after, however, cracks appeared in the foundation. By the time the building was completed, numerous cracks were observed, and a local engineering firm and a "foundation repair" company were subsequently retained.

The contractor filed a suit for damages against the two subcontractors and the manufacturer of the anchor system, claiming that the foundation was constructed off-level, the tension cables were not properly secured, and the anchor blocks and cables permitted cable slippage. The contractor also claimed that the subcontractors should be responsible for the necessary work of lifting the foundation on one end and repairing the slope at the opposite end. He based his claims on conclusions reached by a consulting firm he had retained.

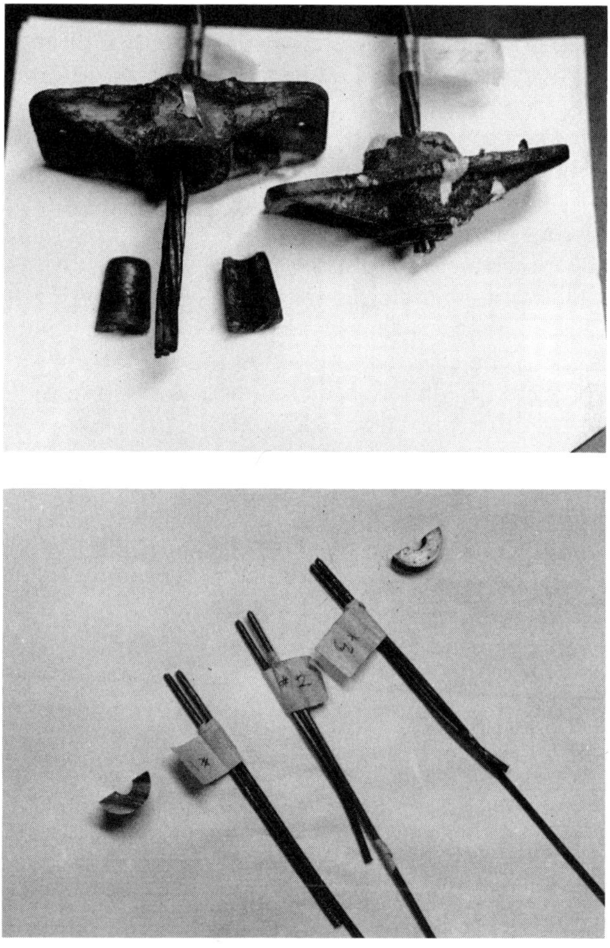

Figure 6-5. Photographs of the anchor blocks, wedges, and cables used in post-tensioning system.

An independent consulting firm retained by one subcontractor's insurance company obtained the tensioning records for the tendons in the initial installation, the records made by the contractor's foundation repair company, and photographs taken during the initial construction. The firm's forensic engineering investigator then went to the site to compare the original construction shown in photographs with the present condition of the foundation and to take new photographs and measurements. He also collected soil samples at various locations and depths around the foundation, and used a surveyor's level to record elevations over the surface of the foundation and along the building's exterior (Fig. 6-6(a)). The photographic data clearly indicated that when the foundation was poured, it had been level within generally accepted specifications. At the time of the current inspection, however, the elevation of the foundation varied by more than seven inches—topological data revealed that the foundation was settling significantly over the shallowest footings and lifting at the opposite end because of cable tension forces from the subsiding area. In other words, the foundation was behaving like a seesaw—its uplifting portion had the deepest footings, whereas its subsiding portion had the shallowest footings. Furthermore, analysis of the soil revealed a highly expansive clay that is found in many regions in Texas. The uplifting portion of the foundation was protected by a large concrete parking area slab (moist), and the subsiding portion was adjacent to numerous large oak trees (dry).

A metallurgical examination conducted by the insurance company's investigator revealed that the anchors had been properly heat-treated to an appropriate depth, and a hardness test conducted by the investigator and by an independent laboratory indicated that their strength and hardness were well within acceptable limits. Laboratory tensile loading tests of cable segments revealed that the strength of the cables exceeded the requirements for yield and tensile strength cited by the Post Tensioning Institute (1984, 1985, 1986).

Finally, data from the contractor's consultants and the foundation repair company about the magnitude of the tension load within the tendons before the foundation was repaired confirmed that, in fact, the proper preload of the cables was applied during the initial construction. The cables in which the preload was lower than the recommended amount (or was nonexistent), however, coincided only with the occurrence of foundation fracture (Fig. 6-6(a) and (b)). Therefore, the insurance company's consultant was able to reconstruct the events of the failure to show that subsidence of the foundation, followed by cracking, had caused the tendon load to decline in several discrete places. Ultimate responsibility for the subsidence was attributable to the contractor who prepared the site, selected the elevation for the foundation, specified the foundation drawings, and failed to consult a soils expert for guidance concerning how to prepare the foundation and soil properly. In this case, no catastrophic failure occurred and the timing of the investigation was not critical; however, a significant amount of money was required to repair the damage caused by the contractor's poor construction practices and planning. Photographs taken at the time the building was first constructed were used for comparison with photos taken after the foundation had subsided. This comparison provided proof that the structure had relative displacement subsequent to construction.

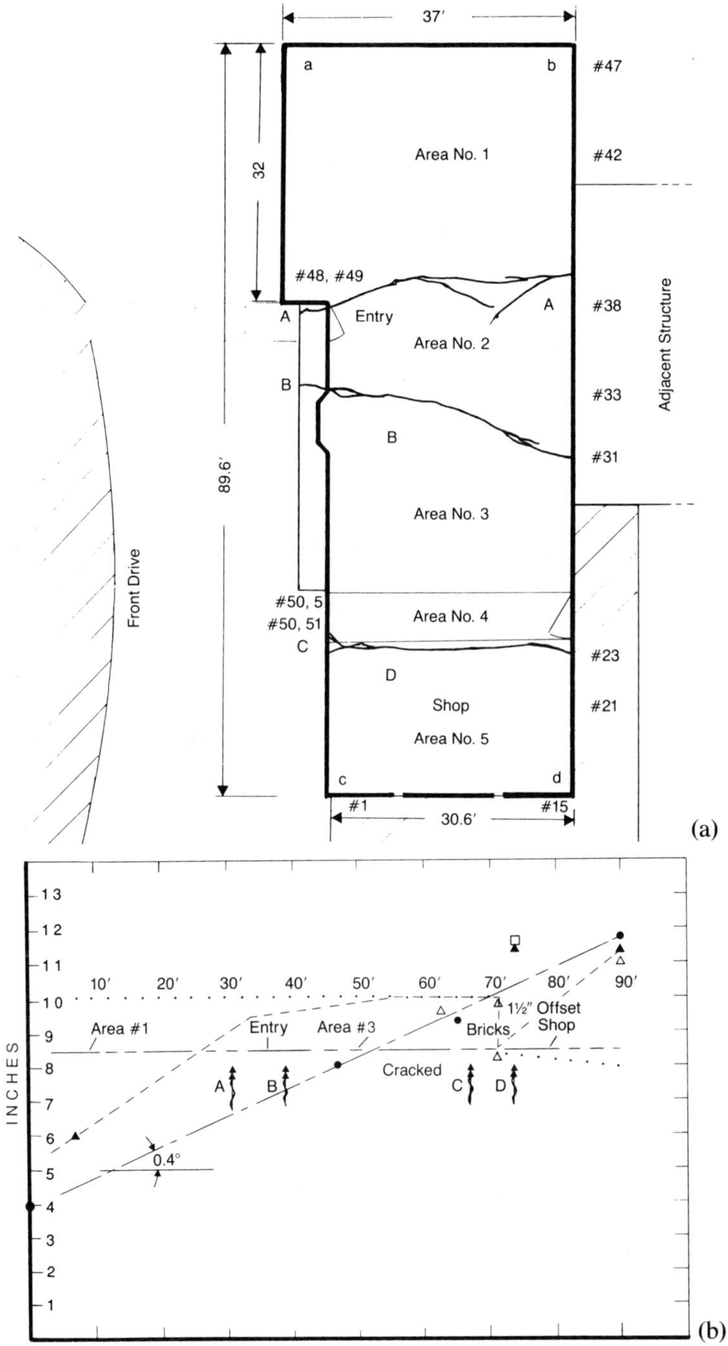

Figure 6-6. Post-tension slab fracture due to subsidence. (Courtesy of Quest Engineering & Development Corp.) (a) Plan view of the post-tension slab and cracks A, B, C, and D. (b) Floor and brick deflections versus the front elevation. △ --- present interior floor covering; ▲ = buildup by the owner; ● ——— — = present elevation of original brick exterior; —-- = approximate elevation of original brick exterior; . . . = approximate placement of original interior floor; ⤳ = visible cracks in the foundation.

Cinematography and Videotapes

Movie film is occasionally used during on-site investigations because literally thousands of photographs (each frame of the film) are available to identify something that was overlooked during still photography. Videotape is far more popular, however, because it can be reviewed instantly and is easy to edit.

One issue frequently discussed is hiring a professional photographer versus having the investigating engineer or scientist videotape his or her own data. The issues most frequently cited are: (1) the wreckage tells a story to the scientist in a global and micro sense, whereas the professional photographer is untrained in engineering failure analysis; (2) because he or she knows photography, a professional photographer can achieve the most desirable effect; (3) most scientists are trained in some aspects of photography; and (4) video tape recording is a relatively new medium. (The size of the equipment, portability, and cost have only recently made professional and semiprofessional services available; hence, there are not many engineering firms that have their own equipment, and only a few professional photography firms that offer professional video tape accident documentation services.)

Videorecording can be divided into two categories: scientific and artistic. Scientific videorecording involves documentation and data gathering, whereas artistic videorecording involves the demonstration of an event, usage, or application. In addition to recording eyewitness accounts and site data that may be useful when preparing for a failure analysis or a trial, videotaping can be used to document laboratory examinations and materials testing, simulations, reconstruction of the incident, causality, injuries, and repairs.

Both experimental and theoretical test data can be videotaped. Videotape can be used to document experimental tests on samples of material, scaled-model tests, full-sized tests at the site and in the laboratory, and destructive and nondestructive examinations. However, advanced training is usually required to set up the recording and testing systems and to interpret videotaped data.

Video recordings of dynamic events such as explosions; tire or vessel burst tests; vehicle, projectile, missile impact; flow-induced vibration; and seismic and shock soil-structure tests can be highly effective for obtaining and documenting data, as well as being illustrative to nontechnical interested parties (such as a client or jury).

Experimental tests include (1) parametric versus full functional simulations, (2) tests on scale models, (3) tests on prototype models, and (4) tests conducted at the site. Videotapes of controlled experimental studies can be used to enter data into engineering or scientific computer programs that interpret parameters such as motion, velocity, energy. In a manner similar to using photography and computer programs to determine the spatial relationships of objects, video images can be captured frame by frame on a digital computer, and key reference points can be correlated with the video image. Because the frame-by-frame data points are captured at known increments of time, this time-history—as well as data on structural properties of materials, such as density and elasticity modulus—can be used to predict what parameters are possible in a certain event. This technique is particularly useful for predicting people's ability to perform certain tasks (ergonomic studies).

Theoretical data that can be documented with videotape include classic methods, numerical methods, and parametric versus full simulation methods. Videotape in concert with computer-generated theoretical analyses is an excellent combination. Most structural and fluid computer-simulation (discussed in Chapter 10) programs have pre- and postprocessor graphic programs in addition to analytic capabilities. The preprocessor helps the engineer build the computer model, and the postprocessor helps interpret the output data. In both programs, the information is displayed on a video monitor and can be captured on videotape. Thus, computerized data such as the following can be recorded: (1) models, (2) simulations of the behavior of materials based on properties such as geometry, supports, loads (forces, displacements, temperatures, velocities), (3) displacements, temperatures, and velocities, and (4) graphic displays such as bar charts, pie charts, and curves showing temperature versus time, stress versus strain, deformation versus time at impact, and so on. Finally, videotape enables an investigator with expertise in computer simulations to summarize, interpret, and display data to viewers without having to bring along the computer.

A video tape recording can be used in the accident investigation and failure analysis for demonstration purposes in the following ways: (1) as a means of providing a reenactment based upon data and documentation studies; (2) to show personnel/equipment operation that is unfamiliar to the viewer; (3) to provide animation (cartoon), draft, and computer-constructed images for illustrations; and (4) computer-aided drafting image capturing.

Using videotape while reconstructing the incident and conducting the failure analysis is advantageous because it shows the relative positions of objects at the site of the incident, provides sound as well as motion, shows both closeup and panoramic views, presents sequences of events, and encapsulates ideas, particularly complex scientific concepts. For example, consider the case of *Marano vs. Ogden Marine* (1986), heard in the U.S. District Court in New Orleans, in which the main matter of contention between the two parties was the location of a power hoist in the engine room of a cargo ship. The defense counsel argued that one operator could, without difficulty, use the hoist, which was located on the fourth level of the engine room, to move equipment to the first level although the controls were located on the third level. However, the plaintiff's counsel showed that one person could not operate the hoist properly while trying to communicate with a person on the boiler room level (below the first level) through a maze of balconies, walkways, and ladders above the engine din. Verbal descriptions and drawings of the scene completely disoriented the jury. Not until the plaintiff's investigator introduced a videotape that took the jury on a tour of the engine room and showed them key locations that were referenced to a three-dimensional poster that provided a cutaway view of the engine room did the scene become clear. The judge and jury reacted positively to the videotape and the plaintiff won an equitable settlement.

Videotape also has disadvantages, however. For example, the resolution achieved is not always as good as that achieved in photographs. However, the introduction of high definition videotape recording (and recent developments) are producing higher resolution. In addition, because no unified system exists for indexing individual frames, one must refer to scenes during a trial. Text included in the videotape is difficult to read. And videotape is difficult to include in a file or technical report.

When preparing videotapes for use in a technical presentation, the following recommendations can be helpful (S. J. Brown 1987b):

- Keep the presentation short and simple.
- Do not make it too slick (be a teacher, not a salesperson).
- Narrate the presentation at the accident site or in the laboratory, if possible, but narrate the video in person during a trial: that is, make two versions, one with sound and one without sound.
- Number scenes when possible with graphic inserts so that they can be referred to at the trial.
- Use computer analysis and animation rather than cartoons, when possible, and explain the difference to the audience (see Chapter 10).
- Use the story to explain products, equipment, or structures that involve motion or complex spacial orientations.
- Use the presentation to focus attention.
- Use it to demonstrate simulations that cannot easily be performed in a court room or that require a number of iterations before one can demonstrate an incident successfully.

- Never forget Murphy's Law: If there is even an infinitesimal possibility that an important reenactment can go wrong, it will.

FIELD TESTS

The kinds of tests the investigator performs at the accident site will depend on what the interested parties agree is acceptable and necessary. They may not permit a nondestructive testing method such as the dye-penetrant test (see page 232) because they fear that the penetrant may alter crack surfaces. On some occasions, they will allow only optical documentation; on others, they may permit some destructive testing on failed components. The main concern is that such tests may adversely affect the outcome of later examinations and tests. Because even hardness tests leave an extremely small permanent indentation in the failed material, they may not be permitted.

Similar arguments often arise when an investigator suggests that a protective coating be applied to parts before they are stored because the coating may destroy evidence such as corrosion when it is removed. Large structures may have to be disassembled or be cut into pieces before removal to another location. If allowed to make such preparations, the investigator must take every step possible to ensure that no data are lost. If not, thorough visual documentation, measurements, nondestructive chemical testing, and other nondestructive testing should be performed.

The field testing should be arranged in a hierarchy beginning with techniques that do not contaminate the evidence and ending with those that do. Noncontaminating techniques do not affect the shape, geometry, or local surface chemistry of the specimen. Color photography and x-ray diffraction fall into this category, whereas ultrasonic inspection, which requires the use of a liquid couplant at the transducer, may alter the results of later analyses.

The investigator should keep in mind that the failure of a component may be uniquely related to the structure of which it was a part or to the environment. In addition, a component involved in an incident can be geometrically altered by an impact with another component after the failure. Furthermore, it may pick up foreign particles after the failure. Even the perspiration on the investigator's hands can complicate chemical analysis of the surface.

If corrosion or contamination is a concern, a variety of in situ chemical analyses are possible ranging from measurements of pH and magnetic probing to emission spectrography and energy dispersive spectroscopy. Corrosive materials or contaminants can be scraped into a clean vial or captured on a piece of cellulose acetate replicate tape. Either method allows later examination in the laboratory using scanning electron microscopy (see Chapter 8 for further discussion of electron microscopy), energy dispersive spectrography, or x-ray diffraction analysis.

The pH of a metal surface can be measured simply by wetting the surface with distilled water and touching the area with litmus paper. This method also enables the investigator to obtain qualitative measurements of environmental chemicals in crevices and in parts that cannot be disassembled easily. A high-intensity light source that produces both normal and oblique lighting is useful for optically examining the frac-

(a)

(b)

Figure 6-7. Cross-threaded large-diameter well casing casts. (Courtesy of Quest Engineering & Development Corp.) (a) Replicas of pipe threads. (b) Microscopic view of pipe threads.

ture site. Features of the fracture become visible because shadows accentuate slight variations in the height of the fracture surface.

Surface replication is a versatile technique because it provides a permanent record of the entire surface. The replica is generally made of some form of polymer; however, the most common form of surface replication is cellulose acetate film. An alternative is a plastically deformable columnar material; however, unlike cellulose acetate film, this material cannot remove corrosion deposits or heat. A number of resins, acrylics, latexes, clays, paraffins, and plasters are capable of preserving evidence (and each material has potentially adverse effects), particularly on large, frangible areas such as impact impressions in soil or large structural marks (Fig. 6-7(a)) that may be disrupted or lost when the equipment or structure is removed for storage. Replication techniques are preceded by photographs taken in situ (Fig. 6-7b). These techniques have permitted the representation of physical evidence associated with a valve system "Christmas tree" separation, thus allowing oil production operations to resume.

In summary, when conducting field tests, the investigator should avoid touching the fracture surfaces or trying to fit pieces together and instead do the following: (1) examine the interrelationship of the parts visually; (2) replicate the surface to determine the finish of the material and the presence of corrosion, degradation, and contamination; (3) examine the failed object to make a preliminary determination of the mode of failure; (4) document the evidence with photographs and videotape; (5) perform destructive tests only after obtaining the principals' permission; (6) weigh the pieces before they are removed, if necessary; and (7) remove or move pieces only after they have been labeled and visually documented and their position has been noted on a grid or map.

DATA FROM MONITORING DEVICES

As was mentioned in Chapter 5, the parameters of certain processing, power, transportation, and manufacturing equipment are monitored with recording devices. Typical examples of such devices are recorders in power or process facilities that measure and record data on pressure, temperature, revolutions per minute, flow rate, electrical current, chemical reactions, levels of radiation, deformation of components at various points, and so on. In addition, many facilities use video recorders and other equipment to record the activities of personnel and equipment. Obviously, the data obtained from these devices can be invaluable when attempting to reconstruct an incident and determine the mode, mechanism, and cause of the failure.

Some monitoring equipment is installed for purposes of prevention. Examples are lubricant sampling, acoustic and vibrational signature recording, seismic recording, emissions measurements, subsidence and precipitation documentation to name a few. Lubricant sampling in rotating equipment provides a record of particulate wear that indicates when maintenance is needed. The engineering investigator should ask about such devices before visiting the site as well as during the investigation. Instrumentation that lacks redundant systems and/or a recording scale that is not well below emergency levels (which have been identified during the design of the system) are indicators of deficiencies in the design of the facility, the specifications prepared by the owner-user for the architectural/engineering firm before the facility was constructed, or the operation of the facility, equipment, or product beyond its intended life.

Consider the case of a process facility in the Galveston Bay area that produced a highly reactive chemical in a relatively large reaction vessel. An overpressure activated the pressure-relief system consisting of a rupture disk and relief valve, and the blowdown line was ejected from the vessel. The ejected blowdown line behaved like a missile and punctured an adjacent vessel (see Brown 1984, Zukas 1980, 1981). An investigation of the charts printed by recording devices that monitored the internal pressure and temperature of the reactor vessel indicated that both the temperature and pressure had rapidly gone off the scale (see Fig. 6-8). Consequently, no one knew the magnitude of the overload. Some theoretical estimates were made, but the owner did not perform any significant tests to determine whether subsurface cracks had developed in the vessel or its attachments as a result of the overload.

The initiating mechanism of failure was a malfunctioning valve that allowed chemicals to be mixed in the reactor. The relief valve was not designed to keep the pressure below the maximum allowable working pressure (MAWP) specified in the ASME Boiler and Pressure Vessel Code or the Code of Federal Regulations. This precipitated the overpressure. Soon after the incident, a new blowdown line was installed and the system was put back into operation.

Three months later, another overpressure occurred, with catastrophic results. The vessel burst, and the shock waves and missiles caused extensive damage to the facility. A mixing of chemicals similar to the first event initiated this overpressure. In this instance, however, the failure of the two temperature-sensing thermocouples in the reaction vessel, not the inadvertent opening of a valve, was the mechanism that precipitated the overpressure. Because the thermocouples misled the operators about the temperature in the vessel, the operators initiated the mixing of chemicals. With only two sensing devices, determining which one was correct or in error was impossible. If both failed, the results

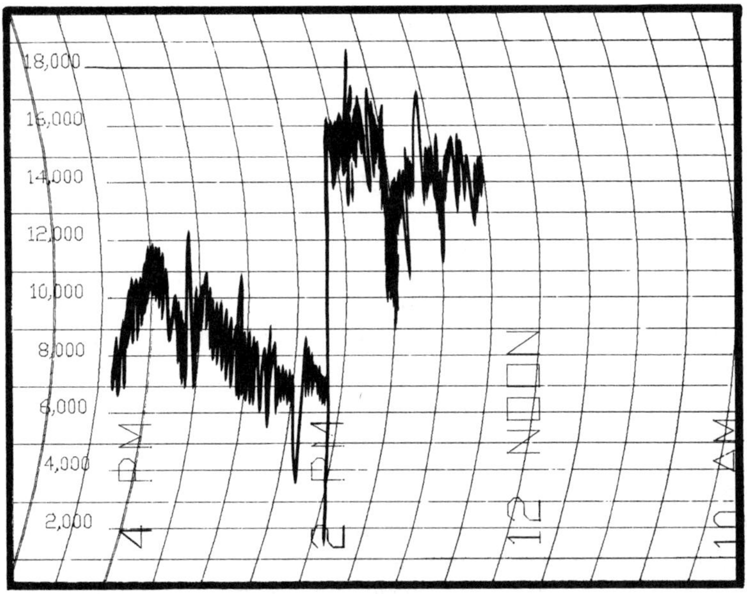

Figure 6-8. Illustration of the pressure strip chart recording from a pressure-vessel monitoring device.

would be catastrophic. Because the recording charts for temperature and pressure had gone off the scale again, no one could state with certainty what the temperature and pressure had been immediately before the incident until the investigator had examined the physical evidence and performed computer simulations. In the end, it was determined that the owner, designer, and manufacturer were negligent.

NOTE TAKING

Throughout the entire field investigation, the investigator needs to take accurate, copious notes. In addition to a trained eye, a pad and a pencil may be the second most important tools that an investigator brings to the site. The information recorded should include the following: (1) the name of everyone who was present at the site during the investigation; (2) the recording devices examined and the data obtained from them; (3) gridded measurements of the site; (4) sketches of all parts involved in the incident, including missile maps with key reference points; (5) a list of all visual documentation, such as the number of film reels or cartridges; (6) all field tests that were performed on labeled structures and parts and the results of those tests; (7) all photographs and all pertinent identifying information on equipment or parts; and (8) sketches of failed parts and fracture surfaces and observations about failure sites, contamination, degradation, and primary and secondary deformations.

PRELIMINARY DETERMINATION OF THE MODE AND CAUSE OF THE FAILURE

The investigator should be familiar with the modes of material failure (the mechanisms or mechanics of the process), the appearance of failure surfaces associated with different types of failure, and the principles of engineering mechanics, or structural analysis. The seven general modes of failure were summarized in Chapter 5, where we also pointed out that a structure can fail as a result of a combination of material, environment, temperature, geometry, time, and force-displacement. The modes of failure are discussed in Chapter 9 and the subcategories of the seven modes of failure are summarized in Box 9-1 on page 282.

Large *plastic* (ductile) failures are associated with structural distortions that preclude putting the pieces of the failed component together like a puzzle. Ductile failures are characterized by shiny, smooth surfaces with undulations, depending on the type of stress at the failure site.

Brittle fractures, on the other hand, are usually associated with small structural distortions in which the pieces of the failed component appear to fit together. The brittle fracture surfaces are characterized by fibrous (dull) features with chevronlike waves pointing toward the site of initiation.

Fatigue failures occur as a result of cyclic loads and crack growth in which ductile failure and fracture are associated with this failure process. The characteristic markings on surfaces that have been subjected to cycles of loading and progressive failure are referred to as beachmarks or tree-ring markings. (The general characteristics of fatigue failures are discussed in Chapters 8 and 9.)

Creep failures are associated with loads applied at relatively high temperatures over long periods. Visual evidence of creep deformation is usually not as clear as it is in brittle or fatigue failures; however, such deformations can be anticipated on the basis of operating conditions and observation of relatively large areas of permanent thinning or thickening. One can differentiate between plastic deformations caused by overload and those caused by creep strain by testing the material, examining its physical features, and performing experimental and theoretical simulations. An investigator should keep two concepts in mind, however: (1) plastic strain causes hardening of materials, and (2) high temperatures for prolonged periods remove the hardness—that is, "soften" the material. An obvious example of non-high-temperature creep can be seen in a window pane that has been in use for more than 100 years: it will be thicker at the bottom than it is at the top. Those who are unfamiliar with engineering mechanics may want to consult Appendix B or an introductory engineering text about the strength of materials (e.g., Timoshenko 1976).

Failures associated with erosion or corrosion are usually identified more easily than are those associated with other mechanisms. Erosion is characterized by a wearing away of the material surfaces by friction, such that the surfaces are rounded like sand dunes. Cavitation is a form of wearing away of material that is associated with impact that leaves the surface creviced, cratered, or pockmarked. Corrosion is associated with a chemical and electrical reaction or oxidation of the structure. Surface features will vary for material types but change in color is a typical symptom. However, failures that occur as a result of several mechanisms acting in concert—for example, material degradation, fatigue, and creep—may be more difficult to identify without a laboratory examination.

The foregoing is a brief discussion of the general characteristics of failure surfaces associated with the basic modes of material failure. In fact, identifying the mechanisms of failure on the basis of the surface features of the failed part requires considerable scientific expertise. Even ductile failure, which is usually considered to be a simple mechanism, takes the form of a variety of surface features that indicate the load or stress state (see Figs. 6-9 and 6-10).

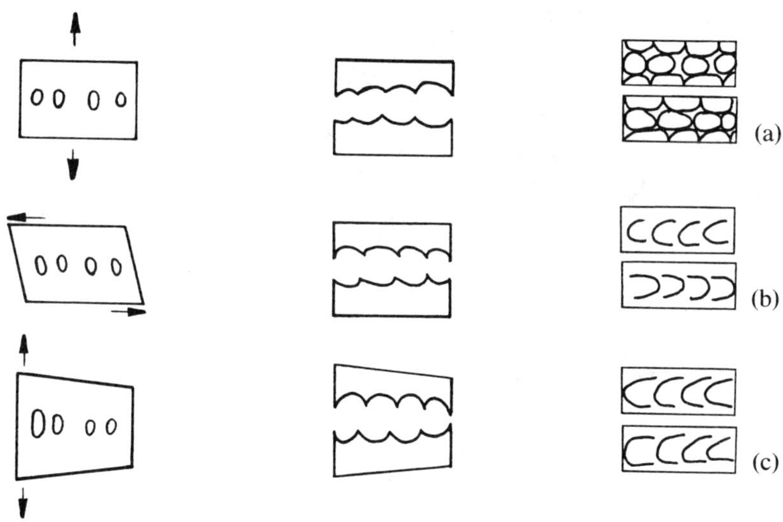

Figure 6-9. Formation of dimples by (a) tension, (b) shear, and (c) tearing. The variation in fracture morphology is shown at the right. (From LeMay 1978)

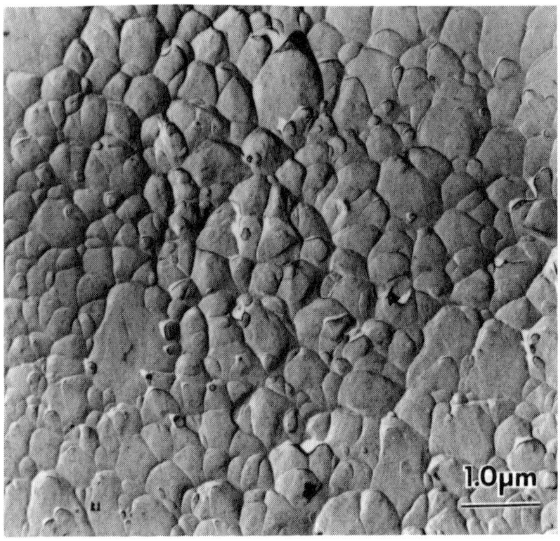

(a)

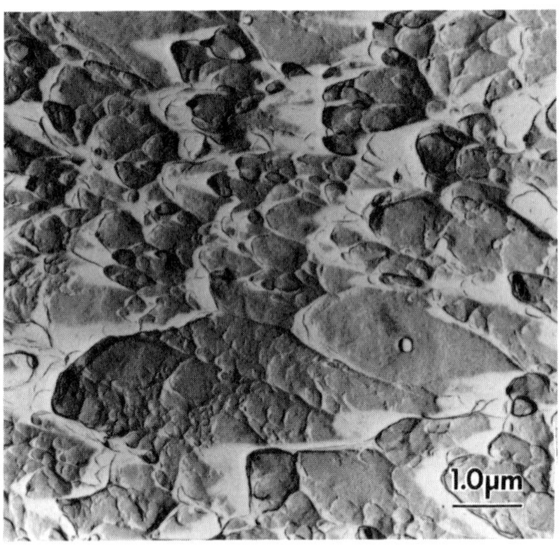

(b)

Figure 6-10. Photographs of quenched and tempered 4140 steel. (From LeMay 1978) (a) Equiaxed dimples on the fracture surface. Note the carbides at the point where local fracture was initiated (arrow) (see Fig. 6-9(a)). (b) Elongated dimples on the fracture surface (see Fig. 6-9(b) and (c)).

STORAGE OF THE PHYSICAL EVIDENCE

The physical evidence of the incident should not be removed from the accident site until the eyewitnesses have been interviewed (if possible), data recorded during the accident has been examined, measurements have been made, the site has been visually documented adequately, adequate field tests have been performed, a preliminary determination of the mode, mechanism and cause of the failure has been made, and all the information obtained during the site visit has been written down.

Clearly, the objective of removing the physical evidence from the site of the incident is to preserve it for later examination. Metal structures undergo oxidation during storage; the degree of oxidation will depend on the materials and the storage techniques. Wrapping a structure with noncontaminating or nonreactive material and placing it in an environmentally controlled area is often the best procedure unless the structure is large. Although coatings can either destroy or adversely affect physical evidence, this problem can be avoided by careful application and removal.

In the final analysis, failing to protect the structure or delaying implementation of the procedure can have devastating effects. Ultimately, decisions about moving the physical evidence and how to store it should be made only after a discussion with the principals involved in the incident and their representatives.

CONCLUSION

In Chapter 5, we noted that an accident investigation and failure analysis performed in-house is usually less formal than one performed in conjunction with a litigation. By formal, we mean that the requirements for collecting evidence that will be used in a litigation are more stringent. That is, records on the location and condition of the physical evidence must be maintained from the time of the initial investigation at the scene to the time that the trial is completed. At no time should the condition and location of the evidence be unaccounted for because the risk of alteration (real or inferred) can have a dramatic impact on the trial. Evidence may be disallowed if this becomes a point of contention.

For example, consider the instance in which an engineering consulting firm was retained to examine some high-pressure pumping equipment associated with an injury. The firm was initially retained to determine why the equipment failed unexpectedly. During the inspection, all the equipment was photographed in detail. A year later, the user-owner, who was contemplating a suit against the manufacturer, retained the consulting firm again. When the original investigator brought the equipment to his laboratory to run some tests, he noticed, on examining the original photographs, that some of the equipment in the laboratory was not the same as the original. He subsequently discovered that some of the equipment had been switched during repairs made after the original investigation. Thus, whoever has the authority—a principal, a regulatory agency, a law firm, or an engineering investigator—must make certain that the physical evidence is documented and safely stored.

Investigators must execute on-site investigation in as professional a manner as possible and bear in mind that, intentionally or unintentionally, interested parties may attempt to influence their conclusions. To a small manufacturer, user, or law firm, retaining an engineering consultant may seem prohibitively expensive. Initial use of such a consultant, however, is usually the most cost-effective approach when the potential for a product liability suit exists. The typical on-site investigation can be performed in one or two days and, rather than a formal report, may utilize a preliminary letter report indicating the mode, mechanism, or cause of failure. The interested parties can then decide whether they want the engineering consultant to proceed with additional aspects of an accident investigation and failure analysis, such as gathering data from sources such as manufacturer's specifications, codes and standards, current practices, case histories, and databases—the subject that will be discussed in Chapter 7. Tests and analyses discussed in Chapters 8, 9 and 10.

In fact, some engineering consulting firms report that once the preliminary determination of the mechanism of failure and causality has been made, they have been asked to stop the formal investigation and begin preparing the drawings, reports, and specifications necessary to get the failed system back into operation as soon as possible.

7

Background Research: Documentary Evidence

This chapter is devoted to documentary material that helps the forensic engineer investigating the accident assess the cause of the incident (who is responsible) and how the product involved might be improved. The sources of this information include (1) product specifications; (2) the user's maintenance and service histories; (3) applicable codes and standards; (4) current practices regarding the design, manufacture, and use of the product; (5) databases; and (6) case histories of product failures. The kinds of documentary material that investigators consult most often are product specifications, maintenance and service histories, and codes and standards. Readers should keep in mind that these three sources of information are interconnected—specifications typically cite codes and standards and mechanisms for implementing the gathering of service data.

PRODUCT SPECIFICATIONS

The primary reason for examining product specifications is to determine whether the design of the failed product complied with them. Therefore, a potentially complex written and implied relationship exists between the engineering designer, the manufacturer or fabricator, and the user. As was discussed in earlier chapters, specifications are significant to the investigator because they describe the requirements regarding function, performance, protection, and materials; tolerances regarding defects, dimensions, fitness, and testing; procedures for manufacture and operation; and requirements for inspection and maintenance of the equipment. These specifications may be described in manuals, drawings, labels, and citation codes and standards. Specifications are not prepared by the design engineer alone, but by the user and manufacturer as well.

Large institutional users such as public utilities, petrochemical companies, local governments, and federal organizations such as the Air Force prepare specifications with their own in-house technical organization or with an outside technical consulting firm. In cases where a large organizational owner or user is not equipped to construct and operate the plant, facility, equipment, or product, it will hire a consulting firm to act on its behalf in retaining an architectural or engineering firm or general contractor, who then prepares specifications and drawings to solicit bids from manufacturers of equipment that will be used to construct the equipment, system, or facility for the owner-user. In turn, each manufacturer may have suppliers for different parts or subassemblies. These factors are important for the investigator to consider. Thus, it is the engineering investigator's responsibility, together with the client (usually a law firm, manufacturer, or user), to sort through the interrelated responsibilities involved in preparing the specifications for the product involved in the failure.

Typically, specifications compiled by large owner-users cite the expected performance and function of the product and the voluntary codes and standards to which the product should conform (statutory codes and standards are automatically in effect). These owner specifications are also important to the investigator because they indicate what parameters of operation were defined for the architect or engineer. For example, if the processes to be carried out with the equipment are proprietary and classified, only the owner-user understands and can identify any unusual or dangerous events that might occur.

Specifications for the processes involved in producing the product are also important to consider because the processing procedures the designer specifies may not result in a product that performs as expected, or the quality control procedures may be insufficient. A perfectly good design can be rendered a dangerous one if the process chosen cannot be effectively implemented. As illustrated in Box 7-1, the designer must be knowledgeable about a wide range of processing techniques.

Engineering Design Reports

Engineering design reports provide information about the basis for sizing equipment and components to withstand specified loading condi-

BOX 7-1. Processes Used During the Manufacture of Products

Metal Casting and Molding Processes
- Sand casting
- Centrifugal casting
- Permanent mold casting
- Plaster-mold process
- Investment casting
- Plastic-molding
- Powder metallurgy

Welding and Joining
- Soldering and brazing
- Forge welding
- Gas welding
- Resistance welding
- Induction welding
- Arc welding
- Electron-beam welding
- Laser welding
- Friction welding
- Thermal welding
- Flow welding
- Cold welding
- Explosive welding

Metal Cutting
- Turning
- Drilling
- Milling
- Broaching
- Sawing
- Shearing and cutting-off
- Grinding
- Abrasion
- Finishing

Nontraditional Machining
- Abrasive flow
- Abrasive jet
- Hydrodynamic
- Low-stress grinding
- Thermally assisted
- Total form
- Ultrasonic and rotary ultrasonic
- Water jet
- Electrochemical deburring, discharge grinding, grinding, honing, machining, polishing, sharpening, turning
- Electro-stream
- Shaped-tubed-electrolytic
- Electron-beam
- Electrical-discharge grinding, sawing, and wire cutting
- Electrical-grinding
- Laser-beam machining and torching
- Plasma beam
- Chemical
- Electro-polishing
- Photochemical
- Thermochemical

Metal Forming
- Hot and cold working surface treatments

Surface Cleaning
- Chemical: vapor bath, spray, brush, dip, blast
- Mechanical: blast, tumble, vibrate, brush, belt grind, polish, buff, burnish
- Electrochemical: reverse plate

Surface Coatings
- Organic coatings
- Paints, varnishes, enamels
- Lacquers
- Vitreous metalizing
- Metalizing
- Hot dipping
- Electroplating

Chemical Conversion Surface Treatments
- Anodizing
- Chromate coatings
- Phosphate coatings
- Oxide coatings

tions: that is, the mechanism of failure (or the failure-mode parameters) discussed in Chapter 5 concerning the behavior of materials and the modes of failure. Failures associated with improper design are usually the result of the design engineer's lack of familiarity with the complex nature of the loading mechanisms involved, the failure to have the design evaluated by a qualified professional engineer, or the owner-user's failure to specify the loading conditions accurately to the designer and manufacturer. For example, in the investigation of the failed flooring under the ore crusher discussed in Chapter 5, engineering reports indicated that the designer had underestimated the impact loads by a factor of 5 to 10 (he was not a graduate professional engineer). In the case of the overpressure in the chemical reactor vessel in Galveston, Texas, the owner's engineering design firm was unable to design a properly sized venting system because the owner had failed to inform the firm about the type of proprietary chemical reaction to which the vessel would be subjected. Figure 7-1 shows the failure of a header containing relief valves that failed during preoperational testing of a nuclear power system in Florida. In this instance, the designer failed to account for the thrust caused by opening the relief valve. Figure 7-2 compares the original and revised designs.

Some products are never evaluated by an engineer; thus, no engineering reports on them exist. This can be the result of several factors: (1) lax state and federal requirements for engineering certification of

Figure 7-1. General view of a south "A" steam header showing fracture openings 2 and 3 of valve attachments 1402 and 1403.

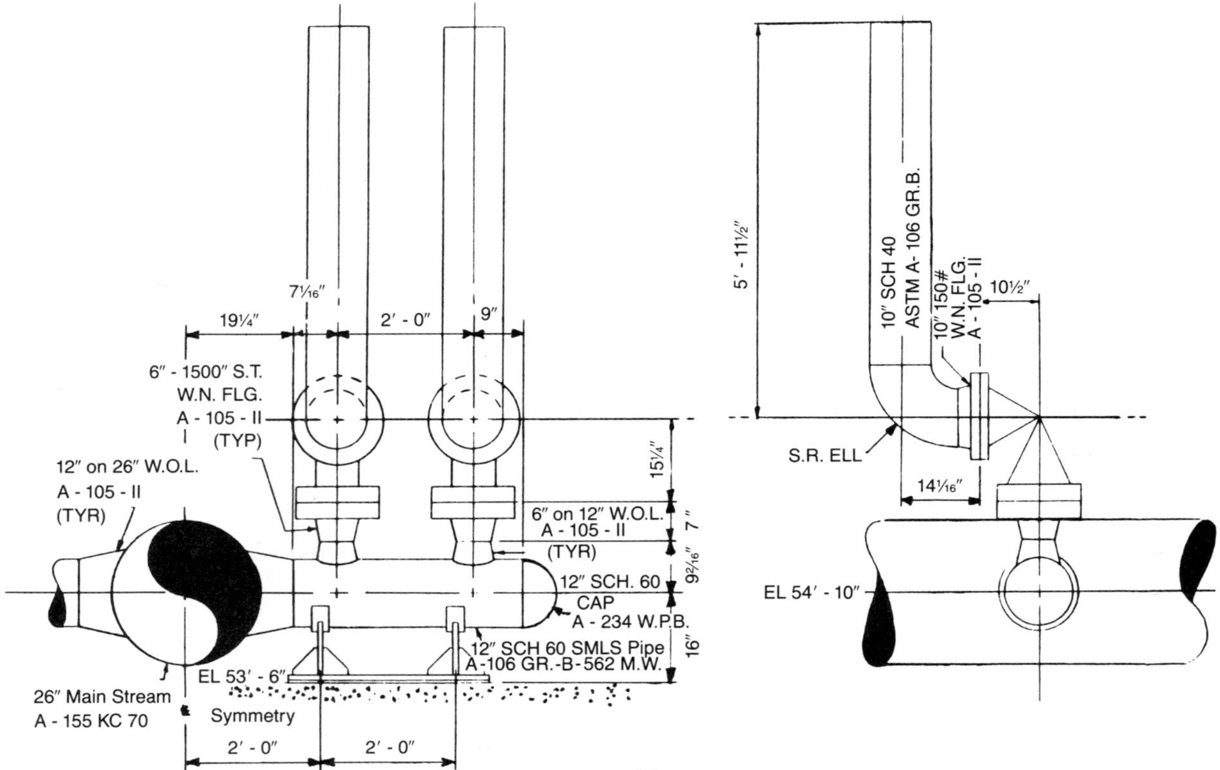

(a)

Figure 7-2. Failure of relief valve as a result of improper design. (a) Original design of mainsteam piping unit No. 3 steam header "A." (b) Revised design of the mainsteam safety relief valve assembly.

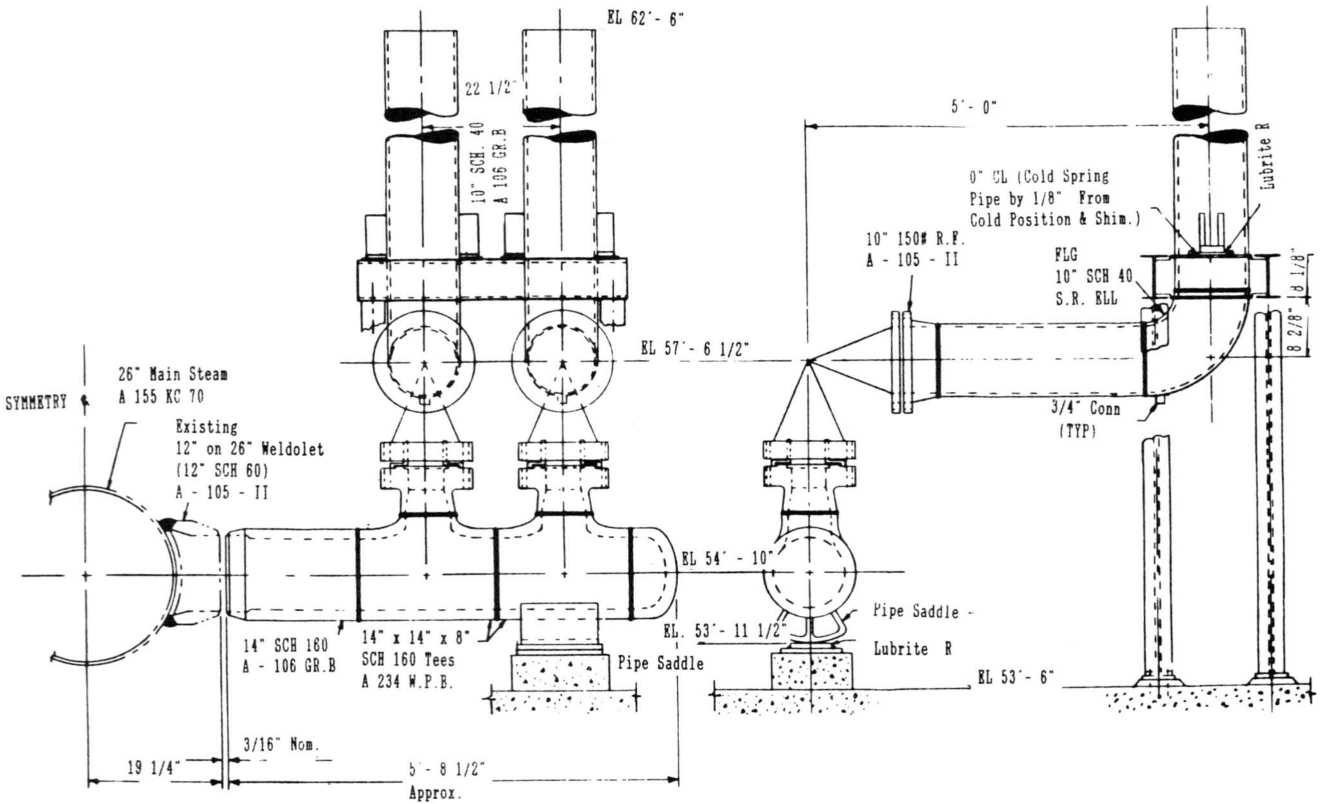

(b)

Figure 7-2. *(continued)*

new designs and retrofit or revised designs; (2) a belief on the manufacturer's part that a failure is so unlikely that an engineering evaluation is unnecessary; (3) a lack of awareness on the manufacturer's part concerning the need to have the product certified by a qualified engineer; or (4) the absence of requirements in statutory or voluntary performance codes and standards for certification or recertification by a qualified professional.

Paradoxically, a well-designed product produced by a conscientious, competent manufacturer may fail structurally precisely because of its initial success. Such a product may function exceptionally well for years in the application for which it was originally designed. Eventually, however, it may be used for applications that were not outlined in the original specification, or it may be modified. Either situation can extend the limits (or envelope) of knowledge about the failure mode parameters of the product. Unless engineering analysis and testing "design" and "redesign" reports keep pace with these expanding applications or modifications, a failure and possibly litigation may result. The temptation to extrapolate from the known to the unknown or to make guesstimates about the adequacy of a product for a new application is ever present. The collapse of the Quebec Bridge on the St. Lawrence River in 1907 is a good example (Tarkov 1986) (see Fig. 7-3). Unfortunately, the advanced technologies of today are often not as forgiving of errors in judgment

Figure 7-3. Nineteen thousand tons of steel lie twisted and smashed on the south bank of the St. Lawrence River after the collapse of the Quebec Bridge. (National Museum of American History; Courtesy of the Smithsonian Institute.)

as were those of the early industrial revolution—the consequences of nuclear, biological, or chemical catastrophes are frequently severe.

Recertification reports are required to provide an engineering designer justification for and approval to use a product for a specified life beyond its original specified life of use before being removed from service. If a product has performed well in the past and if previous test results have shown that the product would function well in other than the originally intended applications, some managements tend to frown on performing tests for new applications because of the cost involved, and view the designer's concerns about the possibility of failure as overcautiousness. Such views are likely to be revealed in the quality and character of engineering design reports. Some believe this problem could be reduced by introducing statutory requirements with respect to redesign and to recertification by a registered professional engineer.

Finally, the following questions can serve as a guide to the forensic engineer investigating the accident during an examination of engineering design reports.

1. Were force-displacement magnitudes, rates, frequency, and location adequately described (e.g., ground motion or foundation motion, impact, flow-induced vibration, impulse, thrust, thermal interaction, blast waves, gravitational or magnetic body forces)?
2. Were the effects of time considered? Was a design life specified for the product? Will the product fail because of fatigue, cracks, crack growth, creep, or creep fatigue in 5, 10, or 30 years? How was the product to be removed from service, and how could it be recertified?
3. Were the geometric dimensions selected to reduce the risk of various modes of failure and to minimize the risks with respect to manufacture and use?
4. Were the effects of temperature (thermal stress, displacement, influence on material properties) adequately considered?
5. Were environmental effects such as erosion, corrosion, degradation of material properties accounted for?
6. Did the design account for the characteristics of materials, such as elasticity, plasticity, viscoelasticity, and memory?
7. In concert with consideration of failure-mode parameters, were human factors, performance, and methods of protection specified?

Drawings

Drawings are usually prepared to indicate the conceptual design, specifications for suppliers of materials, manufacturers and fabricators, and for transportation and installation. Thus, they are a source of information on dimensions and tolerances, inspections, and manufacturing details such as welding, machining, and plating, assembly and disassembly, lubrication, and surface finish. Figure 7-4(a) illustrates the retrofit design to correct the inadequate anchorage that caused the curtain wall failure discussed in Chapter 5. Note in Figure 7-4(a) the original angle of clip anchorage (item 9) that was specified in the original design by the architectural firm. Figure 5-6(b) (p. 144) shows the clips used by the original curtain wall contractor, who may have anticipated that the original design would fail because of fracturing of the concrete. His field correction was inadequate, however, because it did not have sufficient pull-out strength to resist negative wind loads from vortex shedding. The absence of horizontal interior bracing further weakened the original curtain wall because the vertical beams were not braced. This was corrected in the subsequent retrofit design by incorporating an exterior structural brace that was concealed by a decorative strip (see Fig. 7-4(a), items 2 and 3, and Fig. 7-4(b)) and extending the length of the clip (see Figure 7-4(a), item 9).

Drawings are usually referred to as "conceptual" when the purchaser supplies them without fixed dimensions or tolerances, and the vendor's or manufacturer's engineering designers fill in these final specification on the basis of their design. A "shop" or "as-built" drawing is one prepared by the fabricator. A "build-to-print" drawing refers to one supplied to the fabricator or manufacturer with the requirement that it is stipulated as the only working drawing; usually, no variations are permitted without approval. Revision numbers and dates indicating when changes were made should correlate with specifications cited in the drawings. Changes made in the field or shop without revision of the drawings denotes poor quality control and are a source of errors leading to product failures. (Refer to the American National Standards Institute, document Y14.5, for a discussion about engineering drawings and related documentation.)

In hard copy, drawings vary in size from A to J designations, and since the early 1970s are commonly stored as computer files by a computer-aided drafting software. Examples of software programs for general microcomputer application are AUTOCAD, CADKEY, Anvil, VersaCAD, CAD VANCE, and In-a-Vision.

Manuals

Manuals can be an invaluable source of information for the investigator. Typically supplied by manufacturers, manuals can be divided into two categories: 1) service or repair and 2) operation and maintenance. These manuals are available for almost all types of products—from small in-

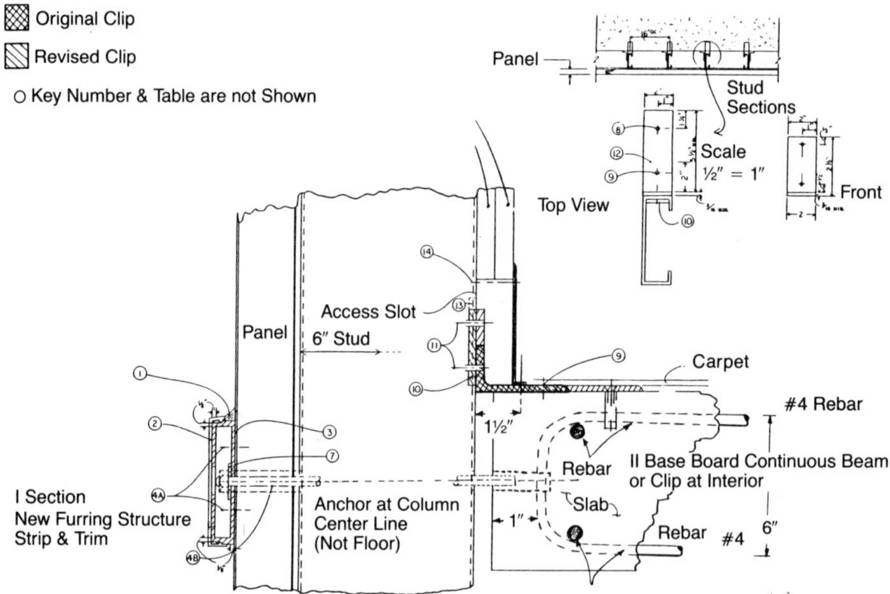

(a)

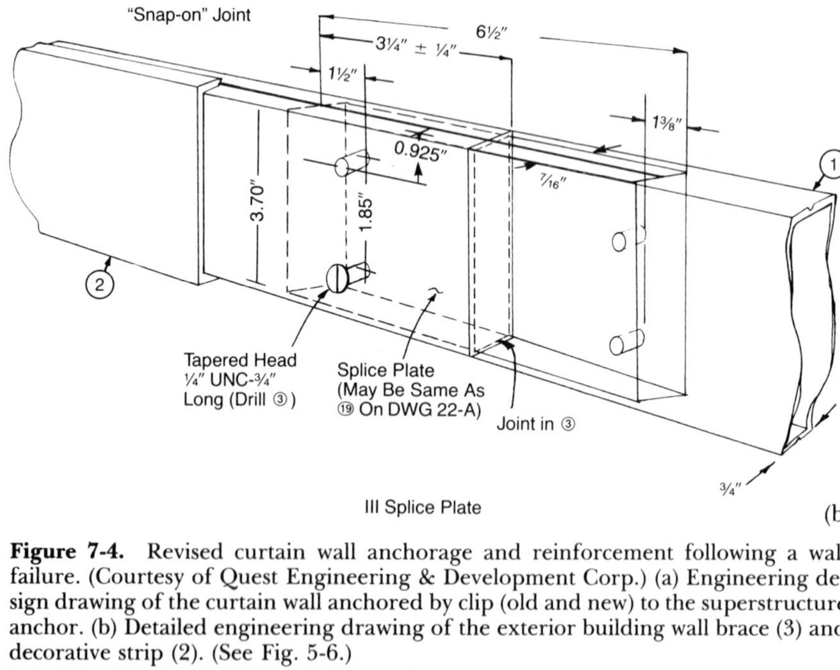

(b)

Figure 7-4. Revised curtain wall anchorage and reinforcement following a wall failure. (Courtesy of Quest Engineering & Development Corp.) (a) Engineering design drawing of the curtain wall anchored by clip (old and new) to the superstructure anchor. (b) Detailed engineering drawing of the exterior building wall brace (3) and decorative strip (2). (See Fig. 5-6.)

dustrial equipment and consumer products to large processing, civil, maritime, and military equipment.

Many failures are associated with poorly prepared manufacturers' manuals or with the user's failure to follow the procedures outlined in a manual. In the hierarchy of methods of communicating how a product is to be operated and serviced properly, written instructions are regarded as the least effective means of instruction, whereas hands-on experience combined with instruction is the most effective. For example, in order to reduce gymnastic injuries, a number of safety manuals have been written to augment instruction: The NSGF Safety Manual, The AACA Cheerleading Manual, and the US Diving Safety Manual. Figure 7-5 illustrates the relationship between verbal, visual, and participatory experiences with respect to learning. This illustration demonstrates that training programs are significantly more effective than reading a manual. Pictorial warnings or instructions are more effective than reading. Audiotapes, photographs or drawings, and audio-video tapes are listed in increasing order of effectiveness. Because manuals and other documentation are often lost when products change hands, the source of user and service manuals should be indicated on the product. Instructions and illustrations that are embossed on the product can be a helpful supplement or alternative to a manual.

Warnings

During an accident investigation, the investigator must determine whether warnings were present on equipment associated with the failure. If warnings were originally issued with the equipment, did the user remove them at some point before the incident?

The investigator—and the designer, manufacturer, and user—should be aware of the mandatory and nonmandatory guidelines for warnings. According to the legal requirements regarding the failure to warn, the plaintiff must prove that (1) a duty to warn existed, (2) no warning was present or the warning was inadequate, and (3) the injury was caused by the lack or inadequacy of the warning. As Baldwin (1986) pointed out, the manufacturer or engineering designer of the product cannot discharge total responsibility to the user by placing a warning on the product. The duty of the designer, manufacturer, or both is supported by *Uloth vs. Cititak Corp.* (1970), which states that one cannot escape responsibility for negligence simply by providing a warning. Three important factors for determining whether a duty to warn exists are (1) the probability that an accident will happen when the product is put to a foreseeable use, (2) the seriousness of the injury that is likely to result, and (3) the feasibility of issuing a warning (Noell 23 SWLJ 256).

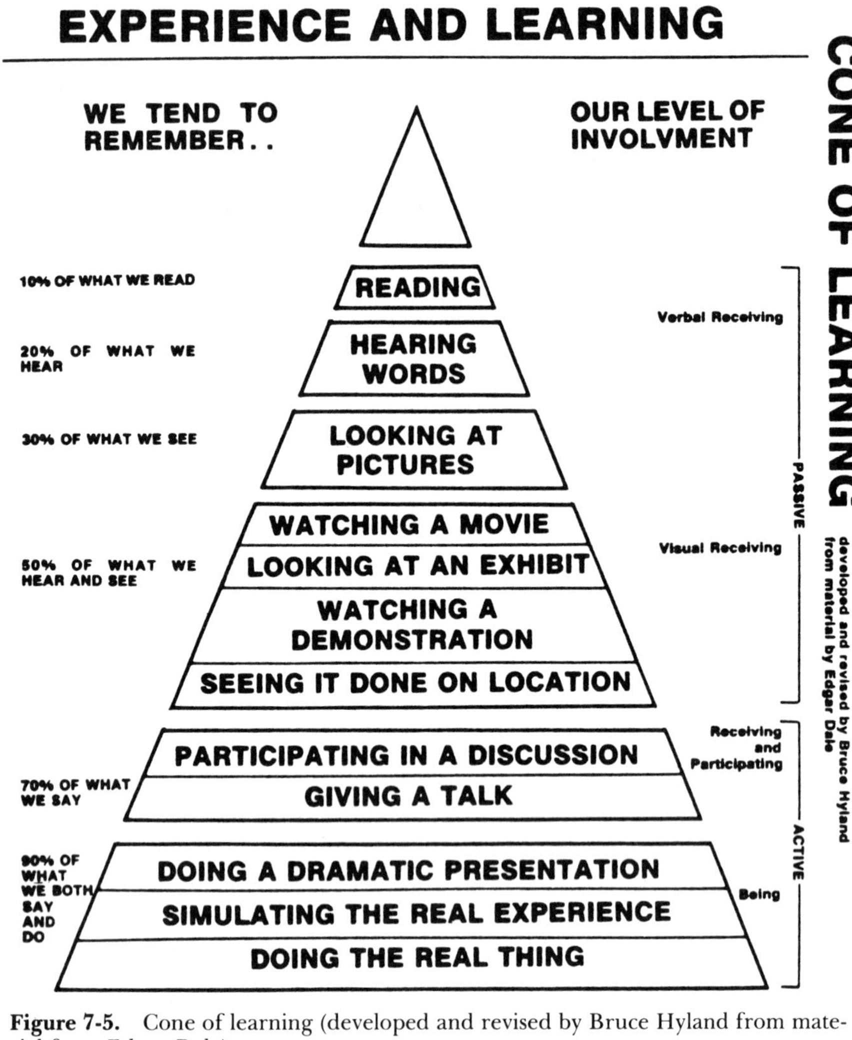

Figure 7-5. Cone of learning (developed and revised by Bruce Hyland from material from Edgar Dale).

The legal requirements for the adequacy of warnings were defined in *Muncy vs. Magnolia Chemical Co.* (437 SW 2 15).

1. The warning must be in a form that could reasonably be expected to catch the attention of a reasonable person under the circumstances in which the product is used.
2. The content of the warning must be comprehensible to a reasonable person and must convey the extent of the danger.

3. The warning must be conveyed with a degree of intensity that would cause a reasonable person to exercise caution commensurate with a potential danger.

Hall (1985) described the following factors that investigators and designers should consider regarding warnings: Was the warning in a specific color? Were the colors used appropriate? Was the warning sign of adequate size? Were pictographs such as standard warning symbols used? Was the warning sign or label located where it would attract the user's attention? Was it placed in a logical position—that is, where the danger would be related to the product? Was the sign adequate for a person who does not speak English or is illiterate? Weinstein *et al.*'s (1978) outline of the steps to consider when designing a reasonably safe product also apply to assessing the need for warnings. In addition, the National Safety Council (1978) has published a checklist that is useful when writing instructions and warnings (see Box 7-2).

In summary, warnings must be available when and where needed, attract immediate attention, control or modify the user's or consumer's potential behavior to avoid injury, and explain how to avoid the danger. (Additional discussions about warnings can be found in Coskuntuna and Clmauro 1980; Cunitz 1980; Ferry 1985; Fowler 1980; McGuinness 1977; Woodson 1981.)

Although instructions or warnings placed on equipment can be extremely helpful, they should not be used if a design feature can eliminate the need for them. Designers should always strive to design products that can be maintained, whenever possible, with readily available tools and that incorporate passive safety features. Consider, for example, the frequent injuries associated with the use of radial arm saws. Two common causes are the lack of an antikickback feature in some models and the failure to state in the instructions that the saw should be pulled toward the user when performing a cross cut. For users who are familiar with hand-held circular saws, pushing the radial arm saw through the work piece (cutting at the nose end) may seem logical unless the instructions state otherwise. In the final analysis, however, no product can be made foolproof.

The warnings aspect of engineering design—considered with function, performance, human factors, and protection—is complex because manuals are lost and some products do not lend themselves readily to obvious warnings and instructions. For example, some products are too small or are constructed of a material on which an embossed warning will wear away, fall off, or become soiled. For instance, quick-release couplers that are used to attach high-pressure hoses or flexible tubes are often too small to hold an embossed warning. Some design configura-

BOX 7-2. Safe Product Design: Preliminary Steps and Warnings

A. Weinstein's (1978) steps to consider in designing reasonably safe products and assessing the need for warnings.

1. Determine the scope of the product or environment uses.
2. Identify the environment the product will be used in.
3. Describe the user population.
4. Determine all possible hazards.
5. Determine the probability of certain hazards.
6. Determine the nature or the seriousness of the possible personal injuries.
7. Postulate alternative design features, including warning instructions, that will mitigate or eliminate the hazards.
8. Determine whether other hazards are introduced by the alternative design, warning, instructions.
9. Investigate similar products or environments.
10. Determine whether the alternative design, warnings, or instructions affect the usefulness of the product.
11. Determine the cost of the alternative design, warnings, instructions.
12. Determine whether the warning or instructions will be followed by the user, that is, test the warning and/or instructions.

B. The NSC checklist for writing instructions and warnings (see also the Michigan Mutual Shop Manual, Sept./Oct. 1978).

1. If the product can harm someone with normal use, are there sufficient warnings and cautions on the product and the instructions? Are the warnings and cautions complete, written clearly, and well presented?
2. If the product can harm someone through misuse, treat this misuse as normal and use warnings and cautions for this type of action.
3. Is there a warning or caution on the product referring the user to the instructions before beginning to operate the product?
4. Are there harmful chemicals contained within the product? Are there markings on the product warning the user of the danger? If it's necessary to induce vomiting, does the literature reflect current knowledge that salt should never be used for this purpose?
5. If the product is heavy and must be lifted, are the weight and lifting weight displayed on the product?

BOX 7-2. *(continued)*

6. Are weights and limits clearly displayed on the product? Are there warnings and cautions against exceeding the ratings?
7. Did someone else who is not familiar with the product give instructions for understanding the use of it? Did they operate the product for test observations? Are the results reflected in the product operation?
8. Are there warnings to the user for maintenance and the possible consequences from the lack of maintenance?
9. Are there warnings in the instruction brochure that the customer should train personnel on safety procedures, proper use, or possible misuse?
10. Is a caution in the instruction literature to the customer that he or she should make sure that personnel (users) have basic knowledge about their job and the potential danger of their work environment? Also that instructions are not intended as a substitute for training in or experience of the product's proper use?
11. Warnings and cautions of potential hazards should be given to get attention. Try to answer the "why."

C. The warning handbook's list of circumstances that imply the duty to include warnings.

1. A machine with a real or potential hazard involved in its operation.
2. A consumer product that has inherent dangers and/or can become hazardous under specific types of uses or misuses.
3. Man-made physical arrangements that have dangerous features (such as a stair system without handrails).
4. Natural physical arrangements that are inherently dangerous (e.g., swimming hole or cliff).
5. Physical settings that have been made temporarily or permanently dangerous by man (e.g., an oil spill).
6. Physical settings that have been made dangerous by the forces of nature (e.g., snow or sleet).

tions, however, present a hazard of unexpected disengagement and injury. In this instance, no real economic or functional deterrents prevent the design of the coupler from having a passive protective feature in which a positive engagement interlock, such as a breach lock, thwarts accidental disengagement by the prudent user. In the ideal passive design, the line pressure should not be activated unless the interlock is engaged. Another remedy is to use a protective system such as a restraint cable (Higgins 1988).

Even simple tools such as the common carpentry hammer can cause injuries if misused (Colangelo and Thornton 1981). Its intended use is to drive or pull carpentry nails, but if used on extremely hard objects, fragments from the face of the hammer can injure the user's eyes. This example raises the issue of the user's responsibilities. To meet these responsibilities, the user needs to do the following.

- Read the operating instructions and warning labels.
- Ask a qualified person (preferably the manufacturer's representative) how to operate and shut down the product safely.
- Do not operate the product until instructions are available.
- Always use the proper tools and equipment.
- Never modify the product without the manufacturer's written approval.
- Never operate the product when your faculties are impaired by illness, fatigue, or medications, or when you are distracted.
- Do not use the product for a purpose that was not intended.
- Use common sense: for example, wear safety goggles if fragments can become missiles, protect yourself from burns, wear protective headgear, or use a safety belt.
- Know the risks associated with the product before buying it.
- Have an emergency plan for coping with hazards associated with the product: for example, have a fire extinguisher or neutralizing chemicals nearby when using the product.

In other words, ignorance of the risks associated with a product is no more of a defense for the user than it is for the manufacturer if the information is reasonably easy to obtain.

Finally, all products have a "design life." One can compute when a product will fail because of normal use on the basis of some confidence level and probability of failure. Some products are designed to fail in a short time; others are designed for long-term use. Many products, however, lack warnings or instructions about design life that would help the user determine the normal life expectancy or fitness of the product above or below this norm. Some two-piece interlocking truck rims fall into this category. Although the results of a separation can be catastrophic, no mechanism is usually provided to help the user determine when wear, corrosion, erosion, or fatigue has caused excessive degradation at the interlocking ridges.

SERVICE AND MAINTENANCE HISTORY

The owner-user's records on service and maintenance of the failed equipment may help the investigator determine whether the user-owner was negligent with regard to operating, inspecting, or maintaining the equipment, or whether the specifications were deficient.

Figure 7-6 is a flow chart showing the activities and actions to consider when examining maintenance and operating procedures. Note the relationship between the maintenance history and the availability of the system. Records of predictive maintenance contain plots or digital data reflecting either periodic or continuous monitoring of parameters such as temperature, flow, and pressure, as discussed in the section on users' monitoring devices. Predictive maintenance of rotating equipment consists of analyses of oil and engine performance. Preventive maintenance consists of scheduled maintenance programs. Corrective maintenance addresses deficiencies in the design of the product. Such service histories exist in a variety of formats. Written records may be the only histories available for older systems; computerized data are more likely to be available on newer ones. Many organizations have incorporated their maintenance and operating data into completely computerized and readily accessible information retrieval systems.

The service history can provide significant information about the failure as well as the satisfactory performance of equipment, structures,

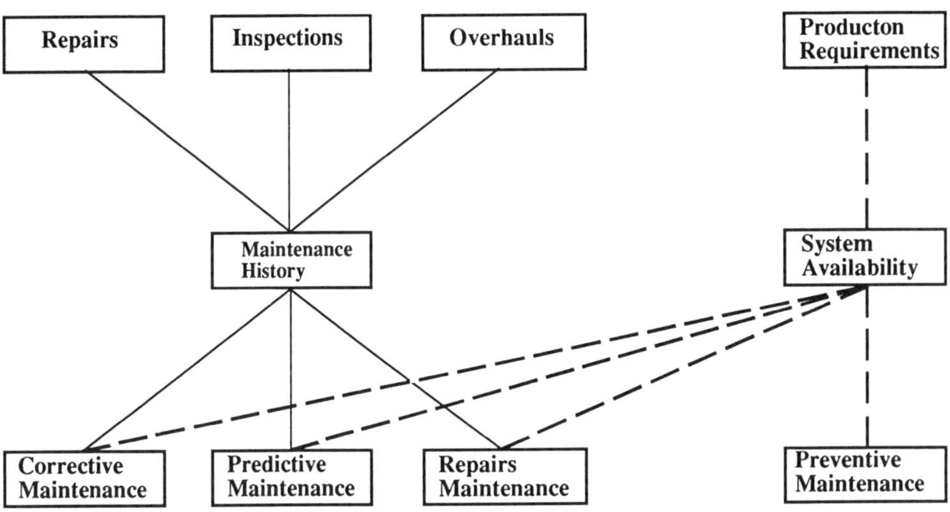

Figure 7-6. Flow chart for maintenance operating procedures. (From Higgins *Maintenance Engineering Handbook,* 4 ed., © 1988, McGraw-Hill.)

or systems. All well-run commercial organizations that use equipment, structures, and components have integrated a systematic maintenance program into their overall operations. As Higgins (1988) pointed out, maintenance is an essential part of performance no matter what product is involved. In large organizations, the maintenance department is most likely to have records on maintenance, inspection, lubrication, generation and distribution, and alterations of existing equipment and on purchases of new equipment. Maintenance records can show not only when repairs were made but also who performed them and what parts and materials they used. Dow's Fire and Explosion Index (1987) provides an example of a processing plant checklist of potential hazards for the plant maintenance and service. Maintenance records also can reveal the failure to perform the proper servicing procedures. For example, consider the case of the failed chemical reactor in Galveston. The company's service records revealed that after the first incident, the company had examined the equipment visually and had replaced the blow-down line; it had not conducted any nondestructive or destructive tests to determine whether the first overpressure had damaged the reactor vessel.

A final incident illustrating the importance of good maintenance records involved the explosion of 38,000 gallons of liquified natural gas in a railroad tank car. The operator at a tank farm was almost finished filling the car when he decided to vent the tank through the vent line to the flare stack. The venting process caused an electrostatic charge in the tank and combustion (not detonation) occurred. As pressure built up in the tank, the relief valve opened and released a 50-foot "pencil-like" flame that was followed by a breach in the tank car caused by overpressure. A secondary explosion then ignited the liquid expelled from the tank. How did an oxidizing agent get into the tank car?

The tank car had been repaired shortly before the incident in a nearby shop and had been pneumatically tested for leaks after the repair. The car was scheduled to go to another facility, and coincidentally markings indicating that it had been used for the same type of liquified natural gas that had been used at the tank farm had not been removed after it had been repaired. Through a number of paper and physical transfers of the tank car from the maintenance and repair company to the tank farm, the error was not detected.

CODES AND STANDARDS

Reviewing codes and standards that apply to a product involved in an incident allows an investigator to determine whether a manufacturer or fabricator has produced a product that complies with its own standards.

As we stated in earlier chapters, voluntary codes and standards are typically self-imposed "consensus" documents written by competitors in language that both protects and promotes their industry.

Codes and standards address four aspects of a product: (1) function (uniformity), (2) performance (e.g., fatigue, fracture, creep), (3) protection (e.g., shields, chemical neutralization), and (4) warnings (which can be combined with protection). Thus, the codes and standards can differ significantly, depending on the unique characteristics of the product, and can require varying degrees of sophistication regarding the application of fundamental engineering principles. The character and style of the code and standard for a product also can be significantly affected by the developmental history of the product. According to the *Guide to Writing ASME Codes and Standards* (ASME 1986), a code is a standard that has been adopted by governmental bodies, whereas a standard is a set of technical definitions and guidelines developed so that products are manufactured uniformly and are safe. According to the guide, a standard should be suitable for repetitive use, enforceable, definite, realistic, authoritative, complete, clear, and consistent and should not be too broad in scope. Useful guidelines for the forensic engineer to review codes and standards during an accident investigation and failure analysis are listed in Box 7-3.

BOX 7-3. Guidelines for Reviewing Codes and Standards that Affect the Failed Product

1. The product should be approved and certified or stamped by organizations such as the National Board of Boiler & Pressure Vessel Inspectors (NBBI) or Underwriters Laboratory (UL).
2. The product should comply with specified and statutory codes and standards. (Remember that these are minimum criteria and in some instances no code exists at all.)
3. Compliance with codes and standards does not ensure a properly designed, manufactured, or used product. Typically, there are no formal professional (P.E. registration) nor technical educational requirements of the members of the code-writing committee.
4. Codes and standards are no substitute for qualified engineers who apply accepted scientific and engineering principles. These principles are not legislated or approved by committees; they are intrinsic to the universe.
5. Applicable codes and standards may conflict or overlap. Therefore, be sure to check the applicable version of the code that was in effect when the product was manufactured.

BOX 7-3. *(continued)*

6. The language or intent of the code and standard may be unclear. If so, try to determine whether an official interpretation has been rendered. If an interpretation cannot be found, seek one from the committee that wrote the code and standard.
7. The absence of an applicable code and standard for a product or the absence of a section that addresses a particular feature must *not* be interpreted as meaning that the product does not violate or fail to comply with any code or standard.
8. Performance criteria for a product are preferred over codes and standards that rely on design formats. The design approach encourages the use of codes and standards as a handbook to educate rather as a checklist for the qualified engineering designer—its preferred use.
9. Codes and standards usually require shielding when the product is high-energy equipment that poses severe consequences if failure occurs. If shielding is required, determine whether it and other protections were removed. If it is not required, do not assume that the equipment was safe to operate without protection.
10. Although codes and standards provide a service that a highly industrialized society cannot do without, some standards are ineffectual and a few fail to address hazards that have existed for years. These codes and standards are sometimes called "motherhood" codes and standards because, in essence, they say that the product "shall be (designed and manufactured) good."
11. Codes and standards are often influenced by factors other than technical correctness and safety. For instance, representatives of organizations often want to protect their industry, are motivated by fear that a governmental agency will write a standard if they do not, or do not want codes and standards that will adversely affect their competitiveness.
12. Codes and standards that are design rather than performance oriented usually stifle creative design concepts and can be hazardous if they promote use without stating that the user must understand the limits of the product's application. Furthermore, they may not apply to new product configurations or to all configurations. They are typically derived from assumed boundary conditions, geometry, loading, material properties, and principles.
13. When two or more products are combined into a system, no individual component product can increase the probability of the failure of the system greater than the codes and standards that apply to the safer or lower probability of failure system.

The case of the exploding chemical reactor in Galveston provides an interesting issue regarding codes and standards. The relief valve on the reactor vessel was designed according to the ASME code published in 1977, as the owner of the equipment had requested. The Code of Federal Regulations, however, mandated that design of relief valves must comply with the ASME code published in 1968. The two versions were

(a)

(b)

Figure 7-7. Tire explosions as a result of tire-rim mismatch. (Courtesy of Quest Engineering & Development Corp.) (a) Comparison of a 16-inch tube tire rim and a 16.5-inch tubeless rim. (b) A 16-inch tube rim with a 16.5-inch tubeless tire mounting.

essentially the same, except that some paragraphs had been added in the later version concerning the use of multiple valves. During hearings held by the Department of Labor, the attorneys spent a considerable amount of effort and time dissecting the language of the applicable sections. The meaning of the new sections may have been clear to a professional engineer, but they had a different meaning to the attorneys. In other words, the clarity or lack of clarity was extremely important in this case.

Codes and standards can often prevent catastrophic failures by requiring a passive means of preventing the combination of incompatible components. A classic example of hazards caused by such combinations is the injuries associated with tires placed on rims of the wrong size. The diameters of rims and tires usually increase in one-inch increments (e.g., 14, 15, and 16 inches) and cannot be combined easily. However, when tubeless tires with intermediate diameters such as 14.5, 15.5, or 16.5 inches were introduced, a number of serious injuries occurred because of mismatched tires and rims—especially a 16-inch tube tire, which can be mounted relatively easily on a 16.5-inch rim for a tubeless tire (see Fig. 7-7). In some instances, the tire bead will engage the sidewall of the rim when inflated; more often, however, the tire ruptures because the tire bead fails to engage, either because it is defective or is broken during the mounting process (*Yearbook of the Tire and Rim Association* 1986). A statutory code or standard that set tire and rim sizes might eliminate the problem, but no such code or standard exists.

CURRENT PRACTICES

In the absence of codes and standards concerning the design, manufacture, or use of a product, an investigator should review current practices with respect to the product under investigation to determine whether a principal acted in a reasonable manner. Are new proprietary methods or techniques for the design, manufacture, or use of a product unacceptable if they do not conform with general industry practice or with codes and standards? The answer is no. In such cases, however, the appropriate organization (designer, manufacturer, or user) must conduct enough tests and analyses to verify the safety of the product (which will be presented as evidence for a petition to the voluntary and/or statutory code organization that affects the product—if it exists—for approval of variation or exception).

Because technology is constantly changing, it is incumbent upon designers, manufacturers, and users to keep abreast of new techniques that are routinely available, such as nondestructive and destructive methods of testing materials, other methods of analysis, and information gathering. Ignorance of these methods is no excuse; designers, manufacturers, and users can be held responsible for a product failure caused by the failure to use widely available technology or preventive safety procedures.

For a system to function safely and effectively, all parts must operate in a measurable manner. ReVelle (1986) outlined methods of analyzing systems that are used routinely in industry and, in addition to the analysis of failure modes and mechanisms, listed several other types of analysis that can be useful during an accident investigation.

- Gross-hazards analysis, which is performed early in a design stage and considers overall system performance instead of individual components.
- Classification of hazards identifies the types of hazards disclosed in gross-hazards analysis and displays them according to potential severity (remedial to catastrophic-type failures).
- Failure-mode and failure-mechanism (mechanics) analysis (which was discussed in Chapter 5 and will be discussed further in Chapters 8 and 9). Refer to Figures 10-31a, b, c.
- Hazard-criticality ranking analysis ranks hazards from the most critical to the least critical.
- Reliability-risk analysis utilizes statistical, database, and real-time data to assess confidence levels and probability of failure or success.
- Fault-tree analysis, which outlines the possible sequences of events leading up to an incident (illustrative fault-tree analyses are shown in Fig. 7-8(a) and (b)).
- Energy-transfer or energy-release analysis, which determines the interchange of energy that occurs during a catastrophic event.
- Catastrophic analysis, which identifies the modes of failure that would create a catastrophic accident.
- Systems–subsystems analysis, which reveals the interfaces between systems.
- Maintenance hazards analysis, which evaluates the performance of a system from a maintenance standpoint and addresses whether the maintenance procedures produced new hazards.
- Human-error analysis, which defines the skills required to operate and maintain a system, considers the question of whether the failure was initiated by human error, and if so, how the error affected the system.

- Transportation-hazards analysis determines the hazards to shippers or innocent bystanders.
- Consequence analysis assesses the consequences of a system failure with respect to personnel injury, property damage, facilities loss, litigation (real cost), and public relations (indirect losses).
- Hazardous-release-protection analysis considers the design of protective systems to reduce energetic (blast, missiles, ground (foundation) motion) and degenerative (radioactive, chemical, biological, thermal) hazards to acceptable levels (motivated by risks, reliability, and consequence analysis). (See Figure 10-31c, which is taken from ASME-SC6000 Hazardous Release Protection.)

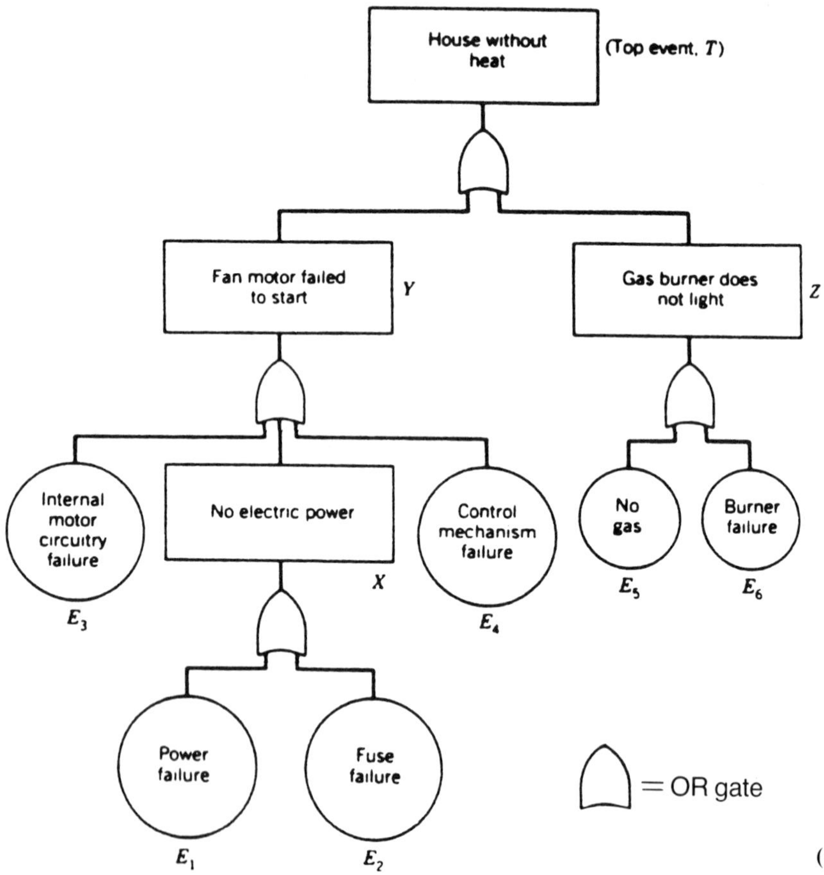

(a)

Figure 7-8. Examples of fault-tree analysis. (a) A simplified fault tree of a house without heat (from Dhillon "Fault-Tree Analysis", ch. 20. *Mechanical Engineers Handbook,* J. Kutz, 1986, John Wiley & Sons.) (b) Fault tree for a temperature feedback loop (Lapp & Powers 1977).

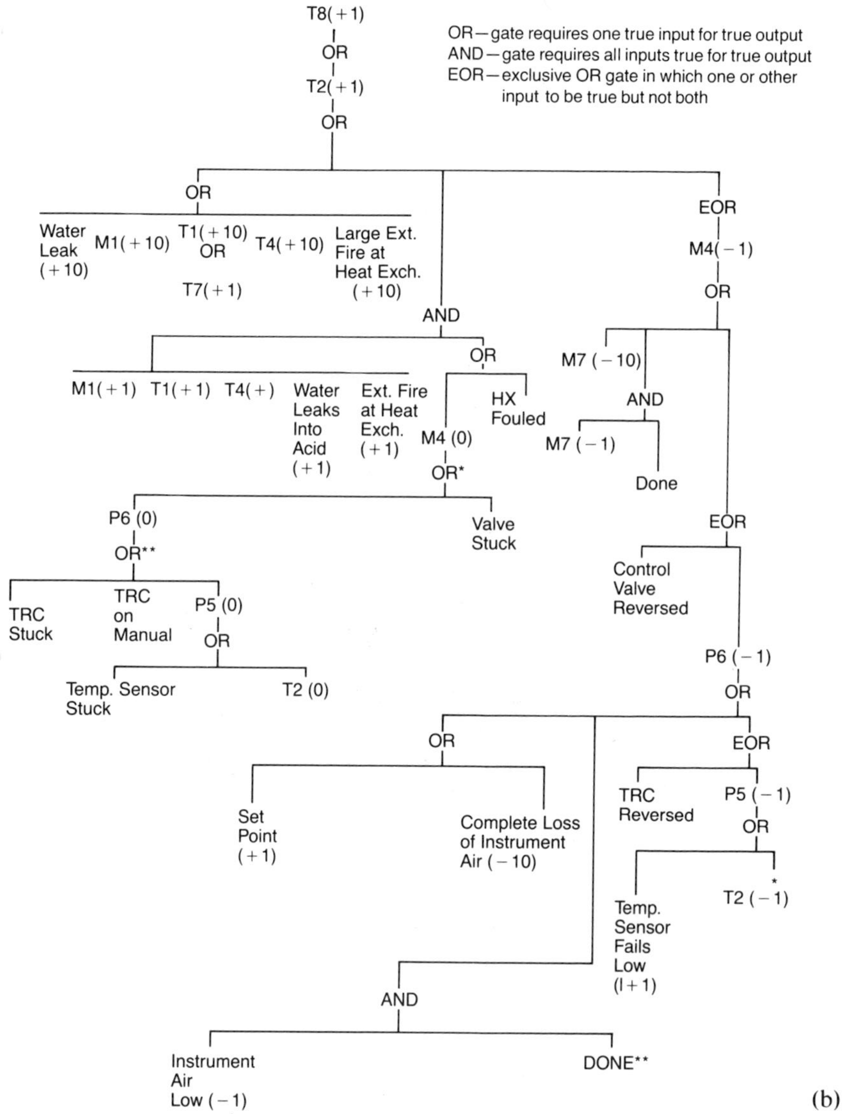

Figure 7-8. *(continued)*

Figure 7-9 illustrates the interrelationship between various analyses discussed by ReVelle (1986) (see Bennet, Feates, Wilder 1982; Green 1982; D. B. Brown 1976). Statistical analyses are excellent decision-making tools for assessing risks associated with design, manufacturing, operation, maintenance, and inspection (e.g., see Burgoyne 1985). Examples of these curves are shown in Figure 7-10. For example, in the case of the oil derrick that collapsed, the scientist/investigator did not have to exam-

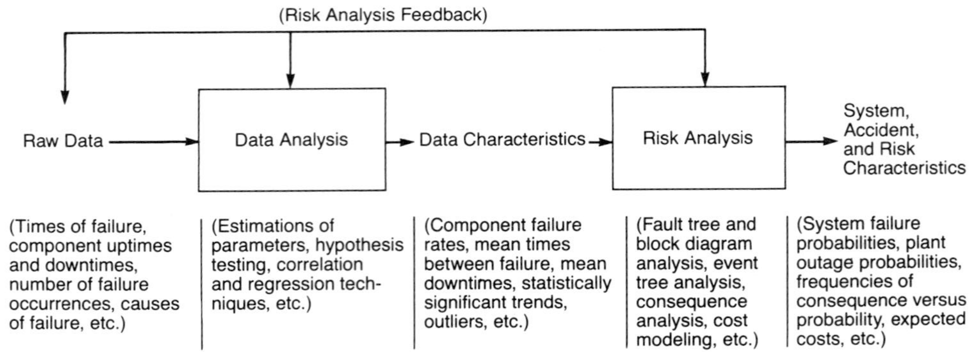

Figure 7-9. Flow of information data analysis and risk analysis. (From Vesely; see Ganghadharan, 1987)

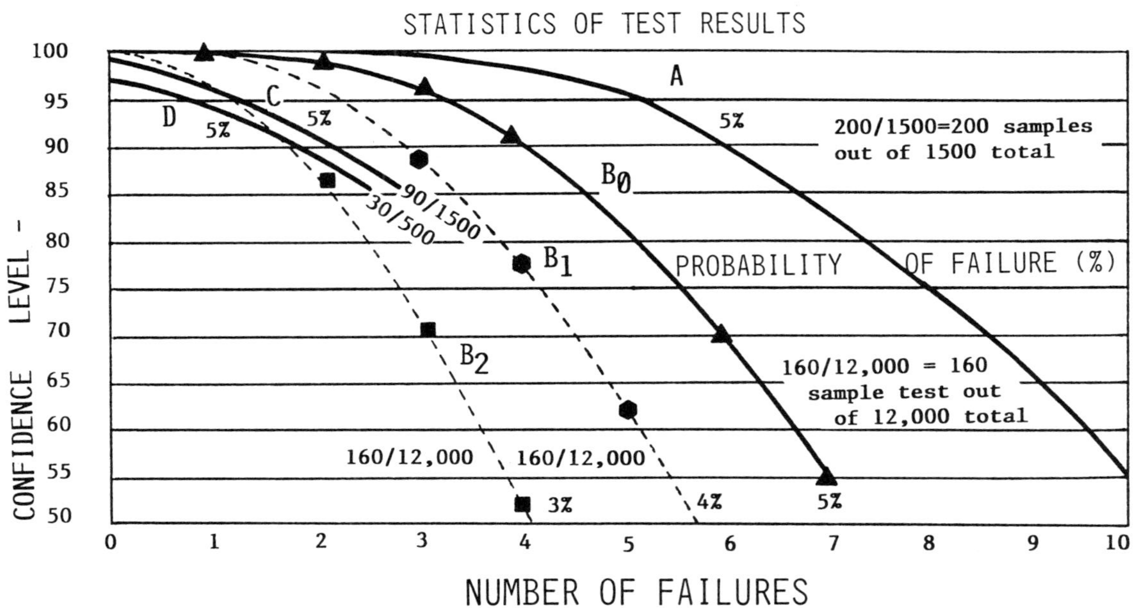

Figure 7-10. Examples of the use of statistical analysis to ensure quality control during fabrication. (Courtesy of Quest Engineering & Development Corp.) (a) Plots for probability of failure (percent) percentage confidence level versus the number of failures (NF). (b_1) Frequency of natural events versus early fatalities. (Vesley 1977). (b_2) Frequency of man-casual events versus early fatalities. (Vesley 1977). (c) Confidence level versus the probability of failure for number of good samples (NGS's) for inspection of product (10 samples out of 500). △NGS = 0; □NGS = 2; ■NGS = 4; ▲NGS = 1; ◪NGS = 3; ⬡NGS = 5. (d) Confidence level versus the probability of failure for NGS's for inspection of product (20 samples out of 500). △NGS = 0; □NGS = 2; ⬡NGS = 4, NGS = 8; ▲NGS = 3; ⬢NGS = 5.

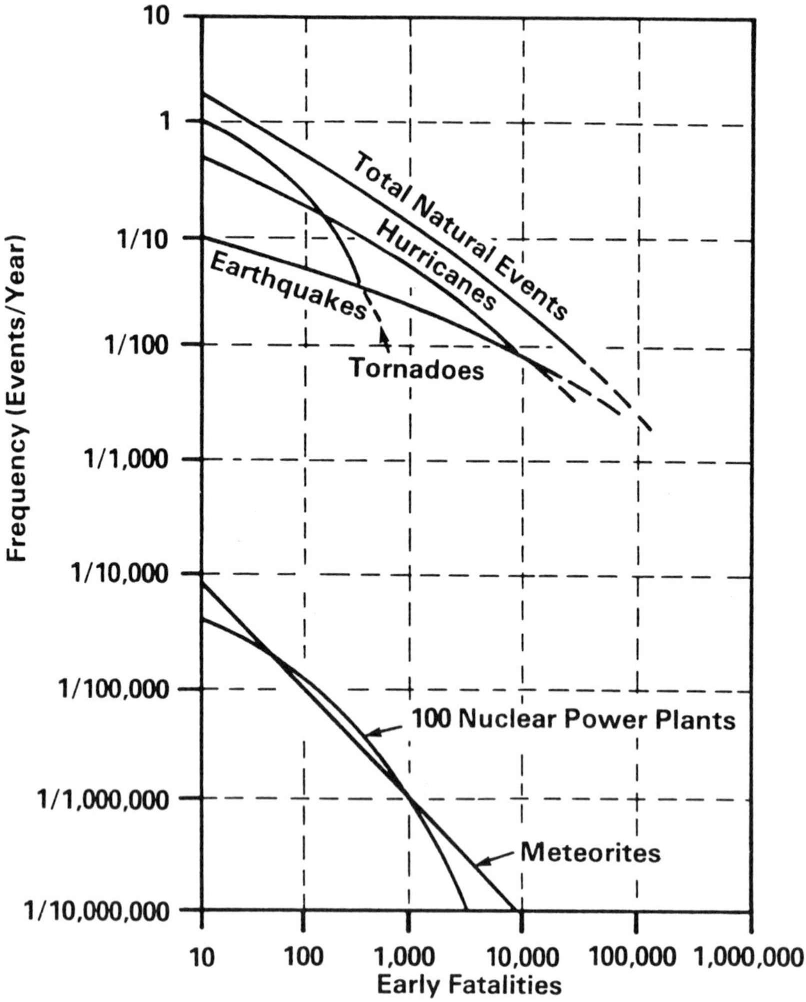

Figure 7-10. *(continued)* (b₁)

ine all 400 or 500 derricks that might have been affected by defective material in order to assess the magnitude of the potential problem. Instead, one could develop curves for the probability of failure and levels of confidence based on a random sampling of the population of rigs by "heats" or batches of steel. This method provides means by which a database can be used in a statistical analysis for predicting the probability an event will occur.

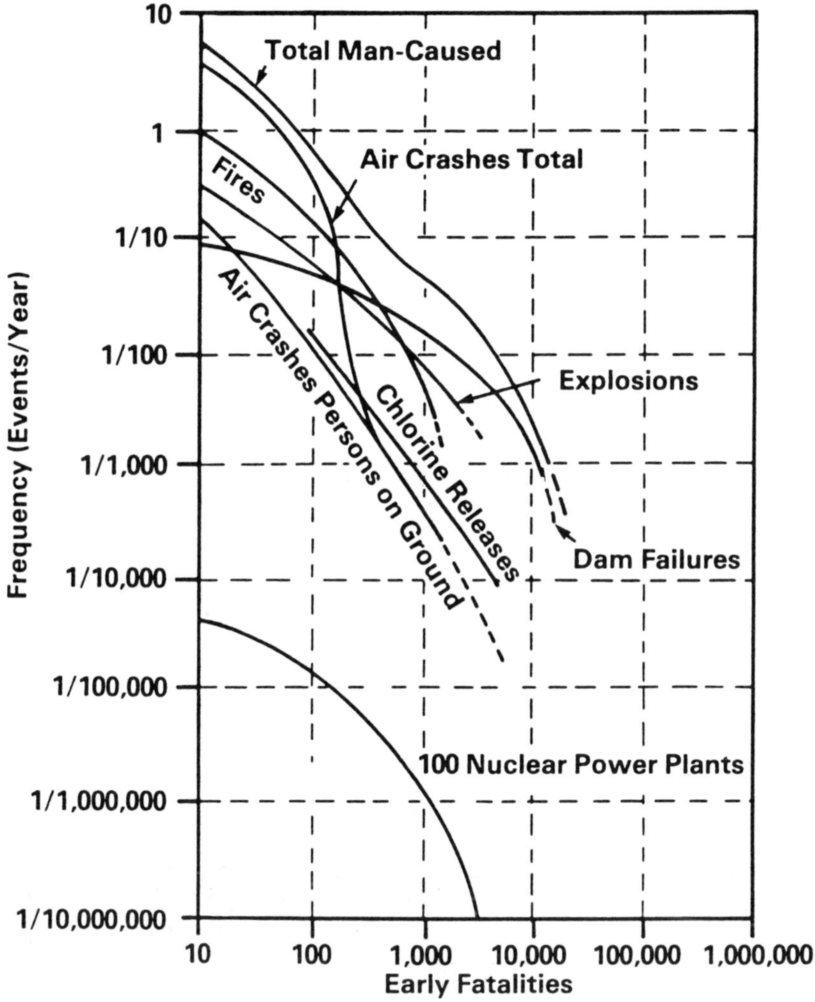

Figure 7-10. *(continued)* (b$_2$)

The deterministic approach to analysis is a theoretical method of predicting specific responses to a given set of conditions. Many industrial organizations and governmental agencies have adopted different deterministic formulas for predicting how structures will respond to a blast, an earthquake, vibration, temperature fatigue, fracture, and so forth—in other words, to predict the modes and mechanisms of failure.

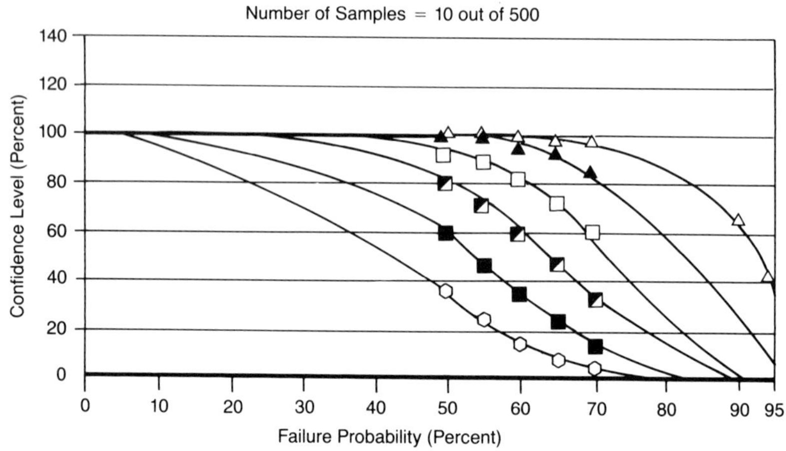

(c)

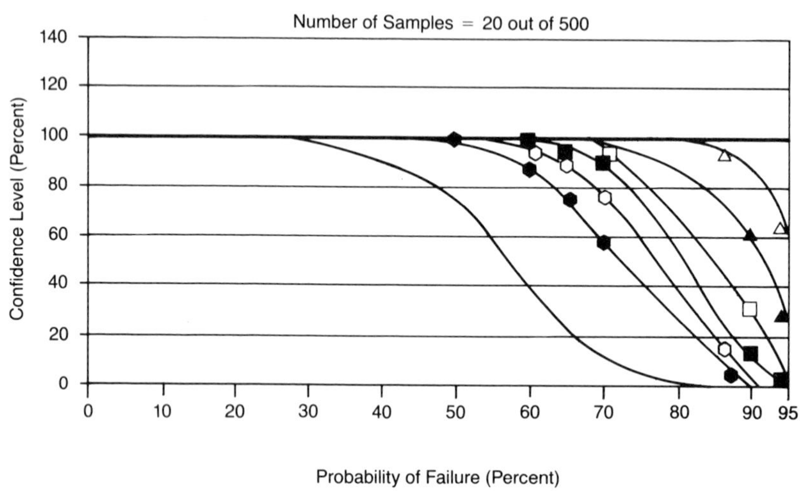

(d)

Figure 7-10. *(continued)*

Some of these methods are simple, others are complex, and either type can produce accurate or completely inaccurate results. (For a discussion about the advantages and disadvantages of probabilistic and deterministic approaches to analyzing hazards, see, for example, Apostolakis *et al.* 1989; D. B. Brown 1976; S. J. Brown 1986c; Burgoyne 1985; Fong, 1978; Fussell, Powers, and Bennetts, 1974, Gangadharan 1976; Gangadharan and Brown 1977; Green 1982; Powers and Thompkins 1974; ReVelle 1986; Schilling 1982; Sundararajan 1985; Tobias and Trindad 1988.)

Product design practices in various industries are often available in textbooks and handbooks: for example, practices concerning the design of pressure vessels (Bednar 1986; Jawad and Farr 1984; Megyesy 1983), metallic bellows and expansion joints (S. J. Brown and Reimus 1984; Jetter, Brown, and Pamidi 1981), and valves (Lyons 1982). These sources usually discuss the basic scientific and engineering principles that are important for designers, manufacturers, users, and especially maintenance organizations in specific industries. Some provide such information about specific products as well as practical information about theoretical, experimental, and empirical data relevant to the product. (Deterministic methods of analysis will be discussed in more detail in Chapter 10.)

Investigators who need background information on a failure may be pleasantly surprised to discover that many consultants, companies, and other organizations use the latest probabilistic and deterministic methodologies to reduce failures, injuries, and the risk of litigation. They also may be surprised to hear that a significant number rely primarily on older analytic tools and that some even resist applying deterministic technologies that were developed 200 years ago. In any case, investigators must be aware that such methods exist and can be used effectively to corroborate whether a principal behaved in a responsible or irresponsible manner.

DATABASES

When high-quality data on cases of failure are combined with population and performance data for products, invaluable survey data are available, not only about specific incidences of failure, but about problems associated with the design and use of products as well (S. J. Brown 1976). The engineer who investigated the incident involving the collapsed oil derrick used databases to assess the impact of using defective steel to construct derricks. The development of the CIMOS system (Cast Iron Maintenance Optimization System) of the Gas Research Institute

(1988) consists of an historical database on cast-iron piping systems, probability models for predicting performance, and optimization decision models.

How does one identify an industrywide problem? Databases containing histories of product operation and failure are typically collected by federal, state, and local agencies; manufacturers; users such as the military, corporations, utilities, and profit and nonprofit consumer organizations; industry associations; insurance companies; technical societies; publishers; and research organizations.

For designers, manufacturers, and users alike, access to formal records and databases are a means of "remembering" cyclic failures, identifying high-risk products, and remedying problems before catastrophic failures occur (Brown 1976). No uniform national program exists, however, in the United States for collecting and documenting statistical data on product performance and failures. And many private organizations are reluctant to share their in-house databases for fear of exposing themselves to the risk of litigation. One concern is that a thorough documentation of the mode, mechanism, and cause of failure cannot be performed satisfactorily and still disassociate the relationship between the principal organization and the failure. For an early example of efforts to create a database on case histories of explosions, see Assheton (1930). Kolar (1985) provided an interesting insight into relatively recent difficulties associated with collecting failure data from organizations within the power industry. Stress-corrosion cracking appeared to be the mechanism of failure in a number of reports of turbine disk failures, beginning in the USSR in 1959. In 1968 a similar failure occurred at the Connecticut Yankee nuclear power station, followed by a catastrophic failure in England a year later. In 1974 two more failures were reported: one at the Rancho Seaco Power Station; the other at the Arkansas Nuclear One. In 1977 another failure occurred at the Oak Creek plant. This potential for catastrophic failure prompted the Electric Power Research Institute to undertake a study to address the concerns of American turbine designers, manufacturers, and owners. Interestingly, between 1950 and 1980, the average capacity of electric-power generators increased from about 100 megawatts to 1300 megawatts (with rotors weighing as much as 247 metric tons constructed from a 500 metric-ton steel ingot). The yield strength of the steel was shown to be the combinant factor governing the rate of crack growth. Thus, as steels with higher yield strengths (e.g., 165 ksi) were used, the rate of crack growth increased 5 to 10 times faster than the rate for steel with lower yield strengths (e.g., 105 ksi).

Haueter (see Ganghadharan & Brown, 1977) discussed some of the issues with respect to establishing a national database for the electric utility industry that was undertaken in a study by the Electric Power Research Institute, the Edison Electric Institute (EEI), and the American National Standards Institute. As stated in Ganghadharan and Brown (1977), Frankle and Dapkenus discussed the establishment of a failure analysis reporting and dissemination system that was organized to take advantage of power plant operation and enhance workability in future nuclear power plants.

As illustrated by the story just cited, the databases of other countries can be extremely helpful to designers, manufacturers, and users in assessing high-risk products. This is particularly true in recent decades, which have witnessed the growth of multinational corporations and the wide dissemination of products and technology. The use of international databases is not new. A report by the U.S. Atomic Energy Commission published in 1974 referred to databases from the Federal Republic of Germany (Kellerman 1966; Kellerman and Ciepel 1968), the United Kingdom (Gene 1968), the American Boiler Manufacturers Association, and the EEI–TVA. Smith and Warwick (1974) discussed the use of conventional power plant data in the United Kingdom to assess potential problems in nuclear power plants, and Dorey and Gachot (1978) discussed data on the reliability of pumps derived from operating experiences in France. Studies such as these direct attention to the fact that cracks or defects account for 89 percent of failures; 35 percent of these defects occur during manufacturing, and 15 percent are the result of fatigue, initiating at cracks, defects, or welds in predictable places on the equipment.

The development and use of databases in the power industry illustrates that databases are a useful tool for identifying the mode and mechanism of failure and the cause of failure. But what about the need to improve methods of prediction? Can the need for better analytical techniques be identified with the use of databases? Yes. State-of-the-art reviews and surveys are often undertaken by organizations and industries (see, for example, S. J. Brown's survey (1980) for the U.S. Department of Energy).

Because of the special hazards associated with nuclear energy, the nuclear power industry relies heavily on databases and on cooperative efforts between organizations. This is also true of the chemical processing and aerospace industries. As S. J. Brown (1980) noted, fundamental technological research and development would not only benefit the nuclear power industry but many other industries and product applications as well. Industry-oriented associations, societies, and research institutes would provide an invaluable forum for identifying the need for databases and how to interpret them. All too often, however, these organizations fail to communicate with one another about common problems that could be solved more cost effectively through unified interindustry programs. In some instances, this lack of communication is intentional: for example, technical societies want to protect the research they sponsor from outside influences; others are solicitous about input or feedback from other technical programs and organizations. Thus, organizations formed to coordinate interindustry sharing of information often become ineffectual, de facto representatives of a single industry, or both.

This raises an interesting question. Can an industry or company be held liable for or be deemed negligent about a problem it is aware of but cannot remedy because it lacks the resources to do so on its own? Consider the following example (S. J. Brown 1982, 1985d). In the mid-1980s, a survey of theoretical and experimental efforts to predict the effects of viscosity and inertia on fluid structure dynamic coupling revealed that in the 1800s, the rotating equipment industry provided major funding for investigations on the effects of viscosity, particularly in relation to journal bearings. By the mid-twentieth century, the industry was concerned with the effects of inertia on journal bearings. By the twentieth century, the power and processing industries that performed the nineteenth century investigative work into predicting fluid structural responses associated with the effects of inertia, quite independent of the rotating equipment industry, began inyestigating the effects of viscosity. The two industries seemed to be unaware of each other's efforts to fund fundamental studies to develop design formulas to predict component of product behavior during operation.

Another example is a survey funded by the National Aeronautical and Space Administration (S. J. Brown 1986a), the purpose of which was to identify industries and government agencies that had developed technical literature, databases, codes and standards, and practices that were complementary or overlapping and to determine what investigative research was necessary concerning blast, missile generation, and terminal ballistics. The number of diverse organizations performing interrelated research was significant. Some of the industries surveyed had contributed to the handling of liquid and solid missiles, both military and nonmilitary; ultra-high-pressure systems; nuclear power; chemical processing, transportation, and storage; ordnance, turbine, and rotating equipment; and missiles associated with natural phenomena such as tornados.

Many departments and bureaus of the federal government have large databases relating to their areas of responsibility. For example, the National Highway and Traffic Administration of the U.S. Department of Transportation provides failure data associated with motor vehicles. The U.S. Department of Labor, through its Bureau of Labor Statistics, obtains statistics on labor and personnel through (1) an annual survey of injuries and illnesses related to the workplace, which are available on diskettes and in bulletins; (2) complementary data systems on injuries and illnesses by state from Worker's Compensation records, which are available from the National Technical Information Service (NTIS); and (3) records on work-related injuries sorted by type of injury and by incidences, which appear in a bulletin published by the NTIS. The NTIS acts as a clearinghouse for data collection sponsored by the federal government.

An impressive amount of data on fatalities caused by overturned forklifts has been collected by the Division of Labor Statistics and Research, the California Department of Industrial Relations, the Industrial Trucking Association, and the Society of Automotive Engineers, (SAE) (see Table 7-1). An interesting facet of this information is that manufacturers of automotive and farm equipment were studying the need for protective cages, safety restraints, and better operator controls long before some forklift manufacturers began to conclude that corrective measures needed to be taken on their equipment.

In some instances, accident investigators will find that no data are available on certain types of failures or injuries associated with certain types of products because none have been collected by manufacturer, consumer, insurance, or governmental organizations. In other instances, the data that are available do not address the specific type of incident under investigation. In these cases, investigators have two options: conduct a survey of their own or develop an in-house database via questionnaires and interviews or via reviews of the literature using on-line databases. For example, after a preliminary evaluation of a consumer product, a consulting firm identified three design deficiencies that would cause personal injury, but was unable to find any data on the product. Thus, the firm developed an in-house database on injuries associated with the product by mailing a questionnaire to distributors and asking them what users complained of most often. The distributors identified the same three design deficiencies as the major problems associated with the product.

Although computerized databases have existed for several decades, on-line databases are relatively new. These databases can be classified as reference databases and source databases. *Reference databases* consist of bibliographic and referral information, whereas *source databases* contain complete numeric, textual-numeric, textual, or software files of the original source information. The user obtains data by connecting his or her computer with the remote computer via a modem. (For a listing of on-line databases, consult the *Directory of On-Line Databases,* published by Cuadra Associates, Inc., Santa Monica, California.)

TABLE 7-1. Work Injuries and Fatalities Involving Forklifts (California, 1951 to 1969)

Year	*Fatalities*	*Nonfatal Disabling Injuries*
1951	2	521
1952	4	550
1953	5	562
1954	6	500
1955	7	733
1956	8	781
1957	4	798
1958	4	842
1959	7	1092
1960	9	1042
1961	5	1106
1962	3	1355
1963	13	1587
1964	12	1856
1965	12	1967
1966	15	2405
1967	14	2190
1968	6	2486
1969	10	2795

CASE HISTORIES

Case histories can be an invaluable tool for designers, manufacturers, and users as well as accident investigators. One should remember, however, that rarely are two incidents identical in all respects. Thus, investigators should exercise caution and not be unduly influenced by technical or legal case histories.

Technical case histories can be found in a variety of media such as journals, magazines, industrial and governmental publications, proceedings of symposia, books, and newsletters. Sources of technical and legal cases include governmental archives, libraries, industrial and professional organizations, and user associations. Technical societies usually organize their information according to scientific discipline (e.g., vibration, creep, flow-induced vibration), product or component (e.g., valves, bridges, pumps), or *industry* (e.g., trailer truck, power generator, chemical process).

After examining the data on failure or injuries associated with the product of interest, the investigator's next step is to explore whether any organization has identified incidents of such failures or injuries and whether anyone has sponsored research to address the problem. Examples of such organizations are the Design Institute for Emergency Relief

Systems, the Program on Emergency Relief Systems sponsored by the Center for Chemical Process Safety via the Chemical Process Industry, and the Pressure Vessel Research Committee of the Welding Research Council. If litigation is involved, the investigator also should determine whether either of the contesting parties is a member of these organizations. Keep in mind that data on one type of industry or application can often be found in a database or the technical literature for another industry.

The need for design engineers, manufacturers, and users to keep abreast of the literature makes economic sense. On-line databases have made the job much easier. Engineering organizations keep up with the technical literature as a part of their service, and other organizations, such as the Architecture and Engineering Performance Information Center, the National Fire Protection Association, and the U.S. Department of Interior's Bureau of Mines, provide forensic engineering databases. In addition, publications such as *The Engineering Index* and *Applied Mechanics Review,* published by the ASME, provide access to the international literature in the engineering sciences.

Databases such as LEXIS and WESTLAW provide references to litigation involving specific manufacturers and products. Once the law firm and technical consulting firm involved with a law suit are identified through these databases, they can be contacted regarding the extensiveness of their legal and technical files on the case and whether all or only some files are available. This means of gaining access to information can be particularly helpful to a small law firm that is working on a contingency basis. It also can be a valuable resource for the technical organization, because it could identify aspects of the current incident that might have been overlooked. The main concern, however, is that if a law firm acquires such information before hiring the technical consulting firm, it may attempt to find a firm that will tailor its investigation to that of a previous case, especially if the case was successful.

CONCLUSION

In this chapter we have focused on the types and sources of documentation. Discussions of the closely related topics of the impact of discovery upon the investigation and the credentials of the investigating engineer can be found elsewhere in this book. Discovery is covered in Chapter 5, and there is a lengthy look at the engineer expert in Chapter 2; the latter is also covered, though more briefly, in Chapters 5 and 12.

Not only is proper documentation necessary in product design, when its relevance and content are determined, but, as we will emphasize in Chapter 12, it is extremely important in helping the investigating forensic engineer prepare a competent case. If he or she does not properly document the failure that led to the accident, or cite sources, such as the codes and standards that may have been violated, to help establish responsibility for that product failure, then that engineer will not fare well either in court or with his or her client.

But documentation is only one of several tools, and should be used in conjunction with other tools, such as the tests used to measure the behavior and condition of materials. These tests are discussed in the following chapter.

8

Measuring Material Behavior and Condition by Testing

This chapter will acquaint readers with a number of testing methods that are commonly used by engineers to assess the characteristics of and defects in materials. To put the topic in perspective, one (especially the nonengineer) should keep three points in mind.

First, in the context of an accident investigation and failure analysis, the physical evidence is tested for three primary purposes: (1) to determine the mode of failure through laboratory examination of the fracture surfaces, (2) to obtain information required to reconstruct the incident, and (3) to obtain data that will help determine the cause of the incident. Were the properties of the failed material different from those originally specified? Did the mill provide material that was defective because of its composition or because of inclusions and cracks? Or did the fabricator introduce defects during welding or heat treatment, for example? Or was the product exposed to environmental conditions during operation that affected its characteristics? Second, the testing methods described are typically used in design and quality control when producing the product. And third, in the engineering and scientific community, the following elements are intimately related: (1) recognition or identification of failures through the examination of failed parts, (2) physical tests used to correlate the composition of the material, its constitutive behavior (how it responds to stress, strain, temperature, load, time, and state), and the results of microscopic examination, (3) analytical theories used to predict its structural behavior, and (4) experimental techniques used to investigate how a product responds when subjected to a variety of loading conditions and environments.

In Chapter 5, we said that the testing techniques used to assess the characteristics of materials and locate defects are usually divided into two categories: *nondestructive* tests, which cause little damage to the object being examined, and *destructive* tests, which cause major damage to or may even destroy the object. Box 8-1 lists some illustrative applications of both types of tests that are commonly used (Box 11-1 for applications when examining power plant components). Many of the methods listed will be described in this chapter.

Techniques for testing materials evolved because design engineers needed a standard against which to measure the performance of materials used in their designs. A multitude of materials are available, ranging from metallic alloys, composite materials, elastomers, polymers, and exotic materials such as pressure-bonded ceramics and single-crystal components. The various classes of materials exhibit properties that can be vastly different and yet, under certain conditions, they may behave in a similar manner. For example, at extremely low temperatures, some alloys shatter but show little or no deformation (ductility), whereas at room temperature they may show a considerable amount of ductility. Small changes in the composition of other alloys such as stainless-steel alloy can significantly affect their behavior. For example, stainless steel that contains nickel and chromium may perform well under fatigue conditions, whereas stainless steel containing these elements in slightly different proportions may perform poorly under fatigue conditions but be highly resistant to corrosion. The American Society of Testing and Materials (ASTM) cites a long list of materials with different compositions that can be used as recipes during testing and quality control procedures through the processing cycle.

One important aspect of quality control the design engineer and failure analyst must keep in mind is that the properties of materials can be affected by heat treatment, the forming process (e.g., extrusion, forging, or rolling), and the fabrication process, especially welding. Unlike the alloys that are produced in the mill, the quality of welds is a function not only of the welding process and the weld rod but of the welder's qualifications as well. An additional and extremely important set of facts about welds is that they are usually found in the areas of structural components that receive the most stress, they are the most common sources of discontinuities and defects, and they can affect the properties of the base metal in the heat-affected zone (HAZ) to which the weld material is being added. For these reasons, the weld material is as important as the welding procedures, the welder's qualifications, and inspec-

BOX 8-1. Applications of Nondestructive and Destructive Testing

Nondestructive Testing

1. Macroscopic examination of fracture surfaces to identify the following:

   Changes in color or texture produced by corrosion, oxidation, or heat treatment

   Distinguishing surface features such as fatigue shear marks, fracture chevrons, ductile shear lips, gross plasticity, voids and inclusions, and secondary cracks

   The direction in which the failure was propagated

   The origin of the fracture

2. Examination to identify surface and subsurface defects using methods such as the following:

Electromagnetic inspection	Interferometry
Magnetic-particle inspection	Strain gauges
Liquid-penetrant inspection	Photoelastic coatings
Ultrasonic inspection	x-ray defraction
Radiography	Thermography
Stress coats	Radiation tomography
Infrared	Acoustic tomography

3. Macroscopic and microscopic measurements of hardness
4. Chemical analyses with spectrographic methods, spot tests, x-ray defraction or fluorescence, and electron beam microprobes
5. Fractography using transmission electron microscopy or scanning electron microscopy

Destructive Testing

1. Metallographic methods, both macroscopic and microscopic (grain size, microhardness, structure, cleanliness)
2. Mechanical methods, including tensile, impact, fracture, toughness, hardness, fatigue, creep, friction, special
3. Corrosion testing
4. Wet chemical analysis
5. Testing of physical properties such as electrical conductivity, thermal conductivity and emissivity, radioactivity, and biological characteristics

tion—the sources of standards outlined by the American Welding Society and the American Society of Mechanical Engineers.

Can designers expect the properties of the materials they use in their designs to be identical from batch to batch? Certainly not. The intent of the ASTM's standards, procedures, limits, and ranges were established to serve as a guide concerning the chemical constituents of alloys and reasonable limits concerning mechanical behaviors such as yield strength, tensile strength, and conductivity.

Design engineers want materials with well-defined behavior and homogeneous properties. In reality, however, the mill processing of materials can produce nonhomogeneous characteristics, either intentionally or unintentionally. For example, in the rectangular extruded-steel sections of the tube shown in Figure 3-3 (p. 75), the yield point, the tensile strength, and, in particular, the Charpy impact properties in the axial direction of the tube, differed considerably from those in the thickness direction. Laminations resembling layers in a cake appeared through the thickness of the tube.

A design engineer can mathematically model the directional properties of structural materials (S. J. Brown 1983a). This, however, can make the engineering analysis of structural performance more complex and less certain than if the material is a homogeneous continuum. When contemplating a design concept and the function of the proposed product, access to tables and charts in engineering handbooks published by organizations such as the ASTM, SAE, ASME, and the American Society of Metals (ASM) regarding properties such as relative yield and tensile strength and fatigue, creep, and fracture properties of various materials provides the designer with a rational approach to the selection of materials. Of course, other factors such as cost, availability, and whether the material is manufacturable play a role in which materials the designer will select for the final design.

In a mathematical context, the performance of materials refers to the constitutive equations that define the relationship between the rates of deformation and stress in a material. These relationships include field variables, material constants, material functions, and material functionals, such as elastic modulus, Poisson's ratio, thermal conductivity, hardenability, fracture toughness, and activation energy. (These functionals are described briefly here and in more detail in Appendix B.) These constitutive equations and parameters are necessary to build mathematical models and perform theoretical or computer simulations to predict how a product will perform during some prescribed loading condition. For example, a design engineer may want to determine how much deformation will occur and what levels of stress and strain will be reached during operation of a product.

The preceding discussion raises another question: How can the design engineer predict that product failure will occur based on mathematical models that are solved theoretically or numerically by computer simulations using constants (functions and functionals), constitutive relationships (stress–strain), and field equations (equilibrium, momentum, continuity, compatibility)? The answer is that in performance tests such as those discussed in this chapter, structural integrity is measured over a range of failure-mode parameters such as the following:

- Yield strength—the performance stress criteria listed in most design codes; and plastic strains—path-dependent loading that usually requires proof of secondary stress shakedown, recertification, or both
- Tensile strength—the value of stress in a material at which fracture occurs with a tensile testing machine
- Fatigue-life—alternating stress (or strain) versus the number of cycles required for failure
- Time-dependent strain or stress to rupture, which ranges from slow rates of strain (creep) to high rates of strain (impact and shock waves)
- The relationship between stress-fracture toughness, crack size, crack growth, and type of material
- Environmental effects—the effect of the environment in which the product material functions on the above properties

Design engineers have historically correlated levels of stress (in combination with material constants) to predict material failure and thus structural failure when attempting to use mathematical models to predict, theoretically and numerically, the behavior of structural components by computer. (The mechanics of material failure are discussed briefly in Appendix B. See also Dieter 1961; D'Isa 1968; Kurtz 1986; Malvern 1969; Marin 1968; Seeley and Smith 1963; Shigley 1977; and textbooks on applied mechanics.)

NONDESTRUCTIVE TESTING

The typical nondestructive test is oriented toward finding cracks in the material and determining the shape and orientation of these flaws. When a crack grows to a critical size, catastrophic failure of a structure or part is the result. Some nondestructive tests, however, provide information on the performance properties of materials, such as correlations between their strength and constitutive behavior.

Liquid-Penetrant Method

The liquid-penetrant method is simple, inexpensive, and easy to use in the field, and it has a variety of applications. A liquid dye with wetting characteristics is spread over a surface and penetrates into discontinuities such as cracks and flaws. The liquid is usually a bright color and may contain fluorescent particles that reveal discontinuities under ultraviolet light. This method, which is limited to the detection of surface flaws that are 1/10,000 of an inch or larger, is used to inspect ferrous, nonferrous, plastic, and ceramic materials. It can be used on objects with irregular shapes and on surfaces with temperatures as high as 1500°F.

Interpretation of liquid-penetrant test results depends to a great extent on the analyst's experience and the type of developer used to draw the liquid from the discontinuity. Liquids used in this method should have low viscosity and high surface tension. (Codes and specifications for penetrant testing can be found in ASTM-E-165; the ASME's Boiler and Pressure Vessel Code: Section IX. Welding Qualifications; and American Society for Non-Destructive Testing (ASNT), SNT-TC-1A guide, Supplement D.)

Magnetic-Particle Method

The magnetic-particle method is used to locate flaws on the surface and under the surface of a ferromagnetic material. The material being tested must be magnetized so that a magnetic field that can detect discontinuities is created. The magnetic fields set up at cracks are called leakage fields, the strength of which determine the number of magnetic particles that will form, and thus indicate the location and magnitude of cracks or discontinuities.

The magnetic-particle method has three limitations: the materials must be ferrous (i.e., conductive), penetration decreases with depth (the best results are obtained on the surface), and the surface must be clean to achieve good contact. Although elaborate electronic equipment is required, the method is inexpensive and detects surface flaws quickly. (For the specifications and standards regarding this method, consult the ASME's Boiler and Pressure Vessel Code, Sections III, V, and VII; and the SNT-TC-1A guide issued by the American Society for Non-Destructive Testing.)

Eddy-Current Method

The eddy-current technique is used to inspect any material that conducts electricity, including semiconductors and many composite materials, although it is used primarily for inspecting metals. The method is based on the principle that eddy currents are produced in a conductive material whenever the currents change. The currents are detected in a nearby pickup coil.

Eddy currents are affected by changes in the electrical conductivity, geometry, and magnetic permeability of the conductor (the object to be inspected). The method is useful for inspecting cylinders such as tubes, vessels, and rods as well as sheets and plates. By moving the pickup coil over the conductor, one can measure such properties as size, shape, location of flaws, heat treatment, heat-affected zone, and cladding interface. The driving coil is usually placed as close to the conductor as possible to maximize the induced currents. Because the eddy currents induced in the conductor are proportional to the rate of change of the current in the driving coil, the frequency of the eddy current must be high enough to produce a good signal, which is usually as high as or higher than an audio frequency.

Rapidly changing currents in the driving coil produce large currents on the surface of the conductor, but penetrate the surface to only a negligible extent. Thus, there will usually be an optimal operating frequency for any given property. This fact, combined with the probability that a variety of properties will be present in the conductor at the same time, can prohibit one from measuring the defects one wants to measure. Large coils produce eddy currents over a large region, and the result is an average of the conductor's properties in that region. Thus, a small coil must be used to detect localized changes such as small flaws and defects.

Because a single-frequency measurement of eddy current can give only two pieces of information, multiple-property techniques have been developed to overcome the problem of having a number of properties present such as the geometry of the conductor, coil-to-conductor distance, electrical conductivity, magnetic permeability, thickness, cladding thickness, and defect size. Three fixed frequencies are considered a practical minimum because they allow one to calculate six independent pieces of information. Computers have made it possible to process the information from multifrequency, multiproperty tests and to evaluate separate variables in real time accurately.

In addition, analytical techniques are available that allow one to calculate eddy-current signals that will be produced by a given configuration of coils and conductors at a given frequency. Analytical solutions are usually classified as classic, numerical, and dimensional analyses to develop scaling of the data. When these analytical techniques are combined with the processing of the field information, they provide an accurate nondestructive technique that can be used remotely. For this rea-

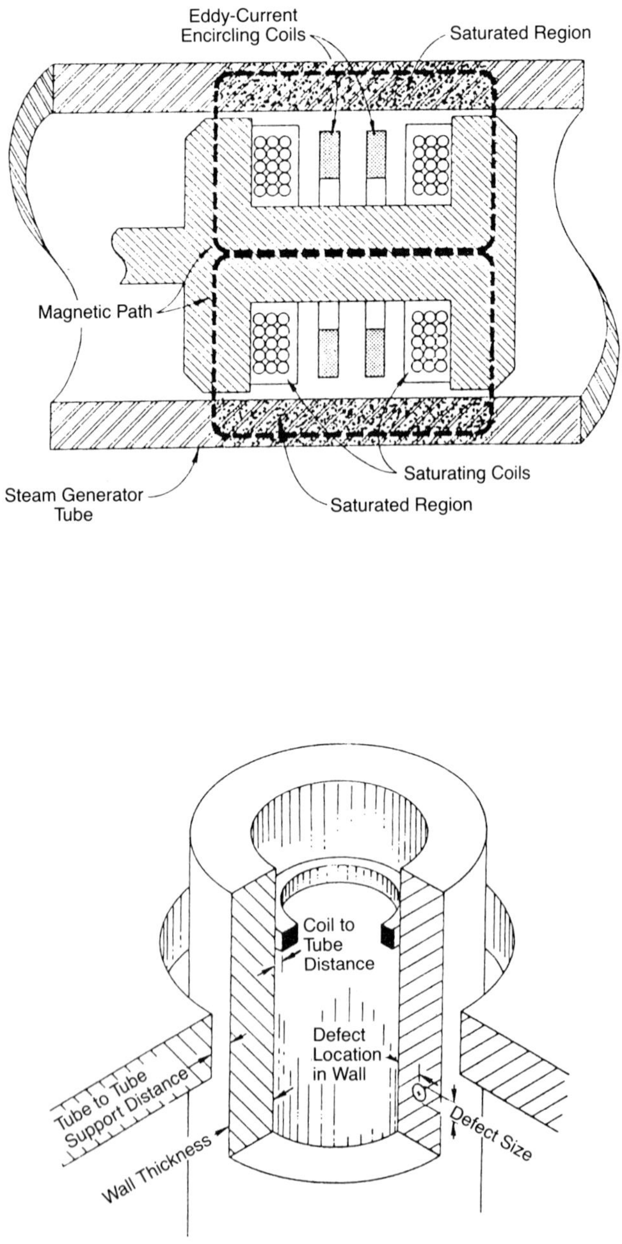

Figure 8-1. Use of the eddy-current method of inspection in steam generator tubes. (From Deeds and Dodd 1985)

son, the eddy-current technique has also been used in situ to examine heat-exchanger tubes that have been shut down temporarily for inspection (see Fig. 8-1).

Radiographic Method

Pressure vessels manufactured in accordance with the ASME Boiler and Pressure Vessel Code are inspected periodically with nondestructive procedures. Radiographic testing is one basic inspection method that is widely used in the aerospace, processing, power, and marine industries. Either x-rays or radioisotopes can be used as the source of radiation. The basic elements of this method are shown in Figure 8-2.

The basic principle of this method is differential absorption of penetrating radiation. When an object is bombarded with a beam of x-rays, any part of the radiation that is not absorbed by the object exposes a sheet of film. The film then shows the object as an image or "shadow picture." The intensity of the unabsorbed radiation depends on the thickness of the object. Because the method can detect flaws that are below the surface, it is used to inspect castings and weldments. Both ferrous and nonferrous alloys can be radiographed; however, this method is expensive and its practicality is limited to cracks that are

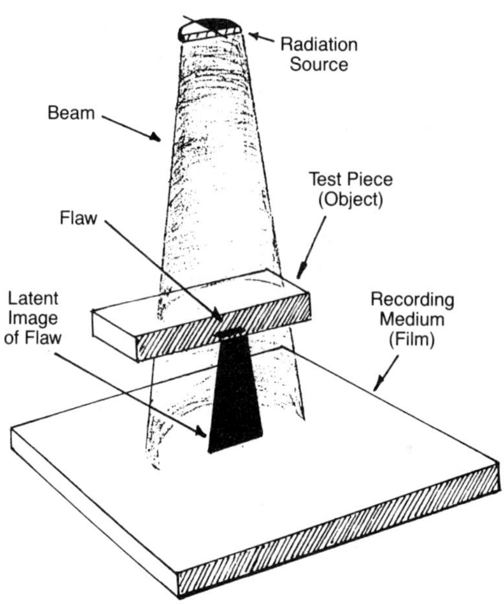

Figure 8-2. Schematic diagram showing the basic elements of a radiographic testing system.

parallel to the radiation beam. Also, minute cracks can be hard to detect unless exposure is closely controlled to achieve optimum contrast between the shadowed areas and unless the film is developed properly.

Because radiographic techniques and interpretation of the results are closely associated, skill is required to direct the inspection, determine the optimal period of exposure, and develop the film. Therefore, unless the inspector is highly qualified, the results can be seriously misinterpreted and artifacts can be confused with actual flaws.

Ultrasonic Method

In the ultrasonic method of inspection, a beam of high-frequency sound waves penetrates the material in a manner similar to sonar. The waves traveling through the material are reflected when they encounter a discontinuity or a surface. The reflected beam is then analyzed to determine the type of internal or surface flaw or crack that is present or the thickness of a part and hence extent of corrosion (see Fig. 8-3).

Ultrasonic testing methods have not received wide acceptance because a common reference or standard is unavailable. For example, a variation in grain size in a material can alter the reflected beam in a manner that may cause errors in interpreting the signal, or, in the case of a large object with a rough surface or a contoured shape, the pattern of the reflected echo is likely to be misinterpreted.

Although recent developments in ultrasonic equipment have improved the reliability of this inspection technique, interpretation of results is still far from being well understood. In quality control inspections, however, ultrasonic techniques have been used with great success.

Acoustic Emission Method

When a structure undergoes deformation or fracture, acoustic emissions (elastic energy) are released that are associated with crack growth in the discontinuities. Using an acoustic pickup device, one can literally hear the material cracking as loads are applied.

The mechanisms for the acoustic emissions in metals are nucleation and propagation of cracks, fracture of inclusions, realignment or growth of the magnetic domain, phase transformations, moving dislocations, and sliding of grain boundaries. The energy that is detected as a result of the acoustic emissions can occur *anywhere* within or on the surface of the object, and it occurs when the internal stresses are redistributed. Because defects affected by loads are sources of acoustic emissions, they are likely to be significant indicators of a loss of structural integrity. Interestingly, the acoustic emissions associated with loading up to a cer-

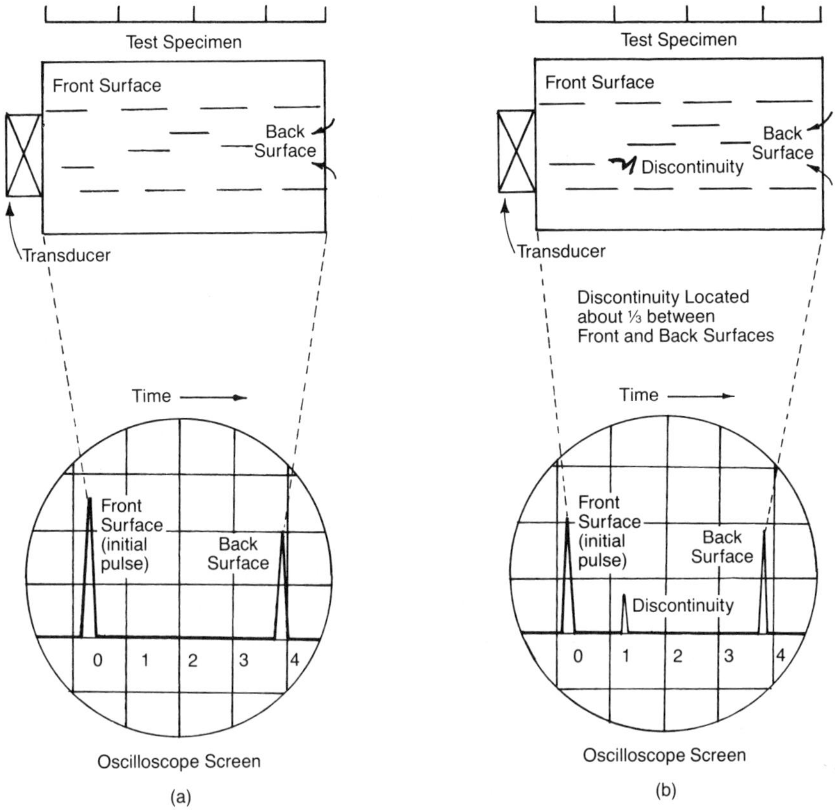

Figure 8-3. A typical ultrasonic normal-beam pulse-echo test with oscilloscope display. (a) No discontinuity. (b) Discontinuity.

tain level of stress usually will not occur if the structure is unloaded, then loaded again, unless the levels of stress associated with the first emission are exceeded.

The technique has proved to be a reliable way of pinpointing subsurface defects. Because of this ability to locate flaws, acoustic emission technology is viewed as a complement to other nondestructive methods such as ultrasonic testing and radiography. The test locates the flaws, and radiographic or ultrasonic techniques then define the extent of the flaws. Spanner (1985) outlined the following applications of acoustic emission testing: (1) the characterization of mechanical properties, (2) preservice testing of materials, (3) in-service testing or requalification, (4) on-line monitoring, (5) in-process weld monitoring, (6) mechanical signature analysis, (7) detection and location of leaks, and (8) geological applications. Although acoustic emissions locate flaws successfully, the relationship between the acoustic signals and levels of stress, size of flaws, rates of crack growth, and the like is not well defined.

Tomography

Tomography is a relatively new technique that offers an increased capability for nondestructive evaluation of defects, dimensional parameters, and variations in the composition of materials. The process is also referred to as computerized tomography (CT), which traces its origins to the computerized axial tomography (CAT scan) developed in the 1970s for medical use. Because three-dimensional imaging of the internal features of a structural component are possible, the dimensions and spacial orientation of defects can be determined accurately. One limitation of the three-dimensional CT approach is the time required for the computer to generate the images. An alternative is laminography, which generates a planer image and thus requires fewer computations and generates an image more rapidly.

Currently, two types of CT systems are being developed: 1) radiography (both film and real time) fluoroscopic and 2) ultrasonic. Figure 8-4(a) compares the images achieved with radiographic CT and radiography; note that the tomograph provides a complete picture of the defect in the slice plane, whereas the radiograph only shows a shadowlike projection of the defect (Colwell 1981). Figure 8-4(b) illustrates a CT scanning system, and Figure 8-4(c) illustrates the source–object–detector system of an industrial CT system (Frank, Kohutek, and Hopkins 1983). Figure 8-4(d) illustrates the cathode-ray-tube display of a computer-generated radiography scanned image. With radiography computer tomography (RCT), dimensional measurements can be recorded with an accuracy on the order of 0.001 of an inch, the accuracy of density analysis ranges from 1 percent to a fraction of a percent, and objects as large as 60 by 72 inches can be scanned. Implementation of CT systems requires three subsystems: the scanning system, a computer equipped with data-processing and -storage software, and graphic display hardware and software.[1]

Computerized x-ray tomography uses the methods discussed in the section on the radiographic method, and computerized ultrasonic

[1]Engle (1987) and Ellenger, Buchele, and Hopkins (1988) provide excellent illustrations of CT applications. For overviews of the research, development, and application of these systems underway at various facilities, see Eberhard 1982; Onoe, Yamada, and Liu 1983; Reimers and Goebbels 1983; Schlapper *et al.* 1981; Segal, Notea, and Segal 1982; Tonner and Tosello 1986; Tsao 1983.

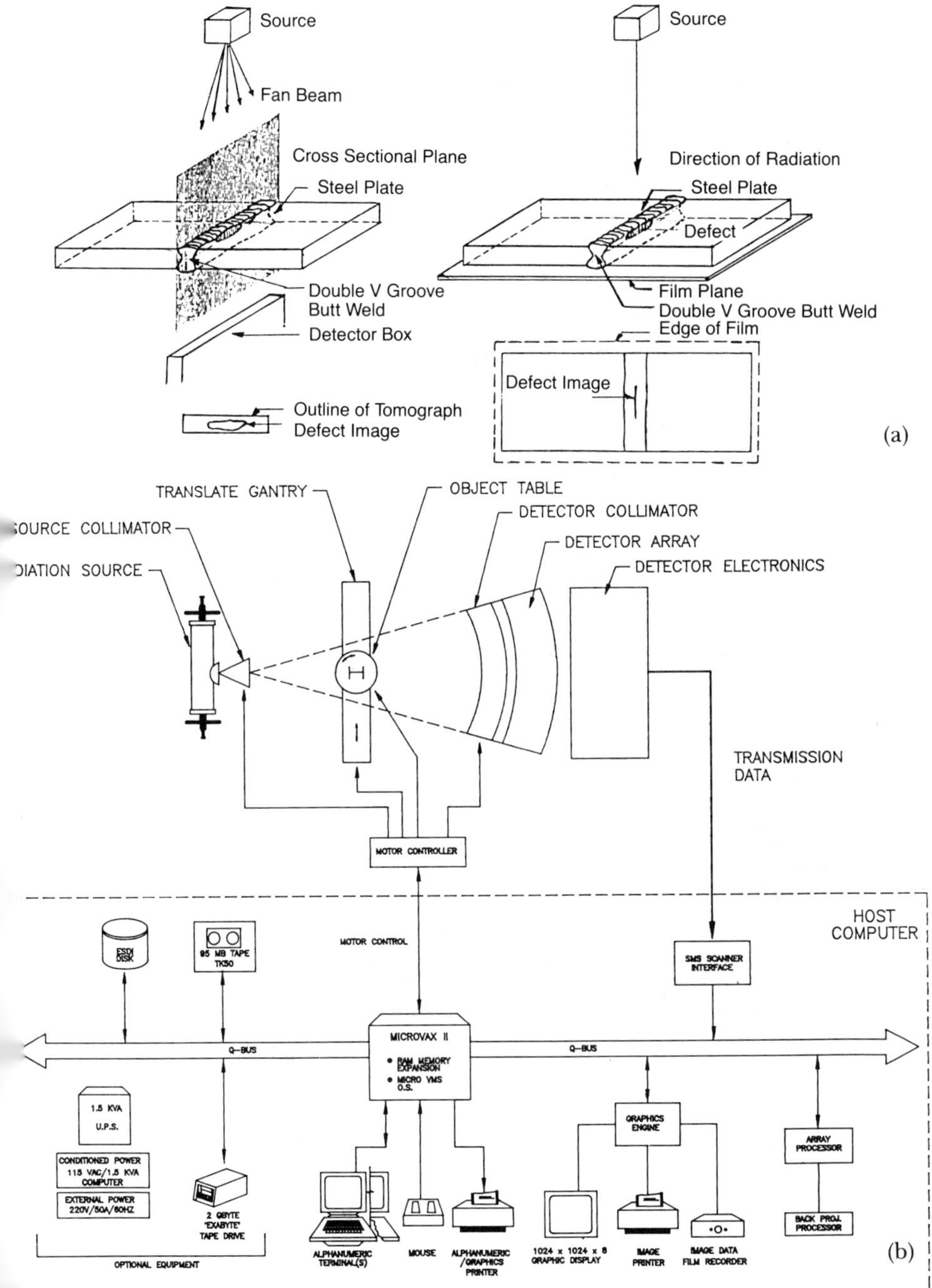

Figure 8-4. Computerized radiation tomography. (a) A comparison of the computerized tomography and radiography imaging processes. (From Colwell 1981) (b) The SMS tomographic scanning system. (Courtesy of SMS) (c) An illustration of the SMS tomographic apparatus. (Courtesy of SMS) (d) A photograph of a cathode-ray-tube–computer display of a thermographic image. (Courtesy of Scientific Measurements Systems (SMS))

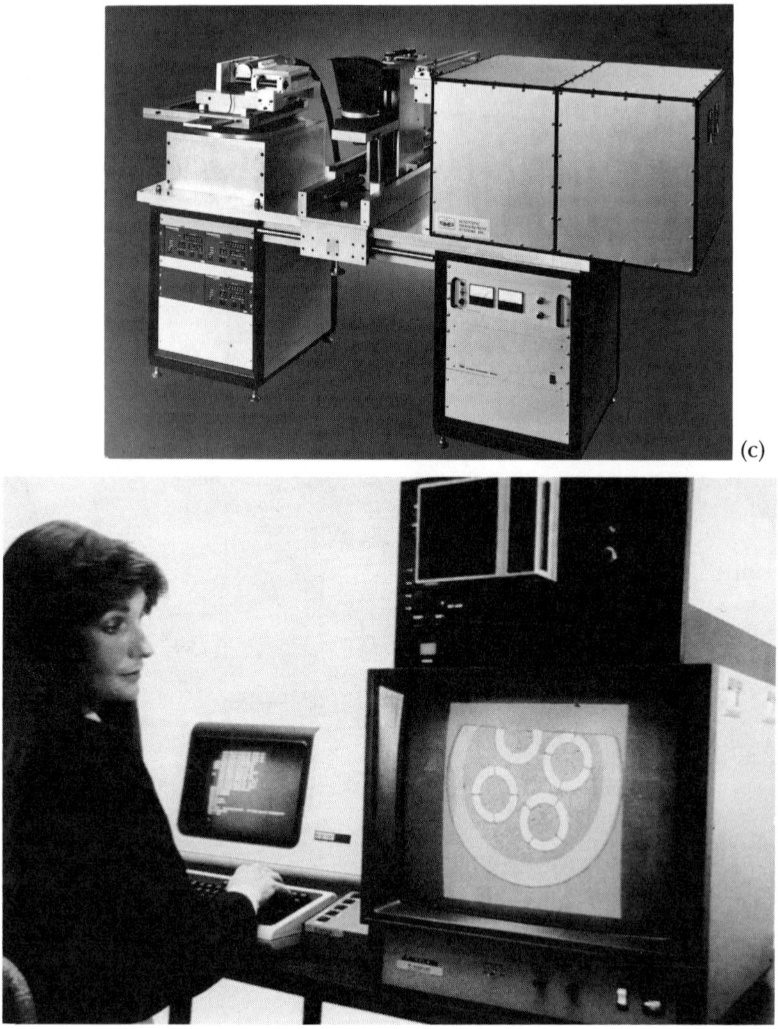

(c)

(d)

Figure 8-4. *(continued)*

tomography uses the methods discussed in the section on the ultrasonic method.[2]

---

[2]For a discussion about models or algorithms for dimension information through CT images, see Segal, Notea, and Segal (1982). For a discussion of the development of computer axial tomography for industrial applications, see Reimers and Goebbels (1983). Tonner and Tosello (1986) have demonstrated the use of CT for locating and sizing cavities in valve castings. With regard to ultrasonic tomography, Eberhard (1982) discusses research on two-dimensional image reconstruction being conducted at the General Electric Research and Development Center. For a discussion of transmission- and reflection-type ultrasonic CT, see Onoe, Yamanda, and Liu (1983).

Hardness Testing

Hardness testing is used to evaluate heat treatments and is easily performed in the field. It has also been used to evaluate the resistance (tensile strength) of materials to deformation. The measurements taken with these tests have been related to the yield stress and ductility of materials. The objective of this test is to determine whether the hardness or tensile strength prescribed in the specifications have changed as a result of operating conditions such as load and temperature or as a result of processing and fabrication. Hardness testing is generally regarded as nondestructive unless a microhardness test, which requires a specimen from the product, is performed.

Several mechanical properties have been defined as measures of hardness: indentation, rebound, wear, machinability, and scratch hardness. Relative numbers have been assigned as measures of hardness, which are usually associated with tensile strength. The most commonly used method of testing hardness is the indentation hardness test, consisting of forcing an indenter of defined size and shape into the surface. Examples of indenters are the spherical-shaped for the Brinell test-number, the square-based pyramidal for the Rockwell test-number, and the rhombus-based pyramidal Tukon-Knoop for the test-number. The depth of the indentation at a prescribed load is the measure-of-hardness number. Some instruments use a dynamic indentation mechanism (i.e., a traveling weight with an indenter attached that impacts the part). The portable equipment required ranges from the size of a pen to the size of a fixed unit.

The measurements made by these indentors are expressed in terms of scales that have been calibrated against the mean diameter of the indentor and the surface area of the indented material (see Table 8-1).

TABLE 8-1. Hardness Conversion Table

Brinell Number	*Vickers Number*	*Rockwell Numbers* "C"	*Rockwell Numbers* "B"	*Scleroscope Number*	*Tensile Strength 1000 psi*
780	1150	70	—	106	384
745	1050	68	—	100	368
712	960	66	—	95	352
682	885	64	—	91	337
653	820	62	—	87	324
627	765	60	—	84	311
601	717	58	—	81	298
578	675	57	—	78	287
555	633	55	120	75	276
534	598	53	119	72	266

TABLE 8-1. *(continued)*

Brinell Number	*Vickers Number*	*Rockwell Numbers*		*Sclero-scope Number*	*Tensile Strength 1000 psi*
		"C"	*"B"*		
514	567	52	119	70	256
495	540	50	117	67	247
477	515	49	117	65	238
461	494	47	116	63	229
429	454	45	115	59	212
415	437	44	114	57	204
401	420	42	113	55	196
388	404	41	112	54	189
363	375	38	110	51	176
341	350	36	109	48	165
321	327	34	108	45	155
302	305	32	107	43	146
293	296	31	106	42	142
285	287	30	105	40	138
277	279	29	104	39	134
262	263	26	103	37	128
248	248	24	102	36	122
241	241	23	100	35	119
235	235	22	99	34	116
229	229	21	98	33	113
223	223	20	97	32	110
217	217	18	96	31	107
207	207	16	95	30	101
202	202	15	94	30	99
197	197	13	93	29	97
192	192	12	92	28	95
183	183	9	90	27	91
174	174	7	88	26	87
166	166	4	86	25	83
159	159	2	84	24	80
153	153	—	82	23	76
146	146	—	80	22	74
140	140	—	78	21	71
134	134	—	76	21	68
131	131	—	74	20	66
126	126	—	72	—	64
121	121	—	70	—	62
116	116	—	68	—	60
112	112	—	66	—	58
107	107	—	64	—	55
105	105	—	62	—	54
101	101	—	60	—	52
99	99	—	59	—	51
97	97	—	57	—	50
95	95	—	56	—	49

The ASTM provides a comprehensive discussion on hardness testing.

Infrared Stress Analysis

The infrared stress analyzer is an optical instrument that records changes in temperature on the surface of a structure as the structure is loaded mechanically and converts this information into stress contours. The equipment usually consists of an optical sensor, which views the area to be evaluated; a portable computer; a video monitor; and a printer (Fig. 8-5). Because the equipment is portable, it can be placed some distance away from the test structure. This test is based on the principle that an object subjected to tensile or compressive stresses undergoes a reversible conversion between mechanical and thermal forms of energy, provided that adiabatic conditions are maintained. The relationship between the reversible temperature change and the change in the sum of the principal stresses is linear and is independent of the loading frequency. Once the equipment is calibrated with either a strain

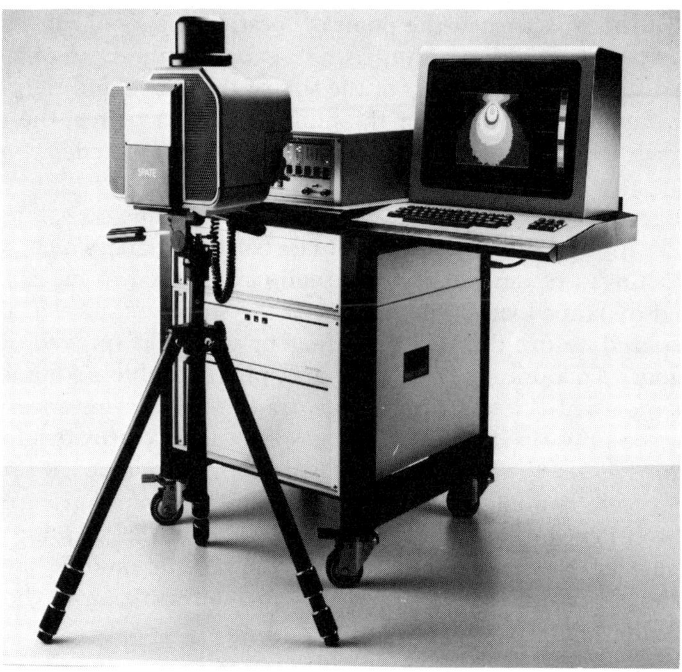

Figure 8-5. A photograph of thermography apparatus (Spate 8000 Stress Analyzer, Spate). (Courtesy of Ometron, Inc.)

gauge or a theoretical correlation, sensitivities on the order of 0.001 K equivalent to 145 psi in steel or 58 psi in aluminum can be achieved. While the optical system scans the structure, the computer collects the data and produces a stress contour image on the monitor. Depending on the software enhancements included with the system, the data can be converted into line- and three-dimensional data plots.

Surface Deformation and Interferometry

A number of nondestructive techniques are available to measure surface deformation of and stress on a component. Because infrared techniques are not appropriate for use on some objects, strain gauges, photoelastic (birefringent) coatings, interferometry, stress coatings, and x-ray defraction have been developed for certain applications (Buck and Buck 1961).

Strain-Gauge Technique. This technique consists of a small gauge or rosette, whose electrical conductivity or resistance changes linearly as a function of its extension or deformation, being connected to a calibrated electrical-resistance recording device (essentially a wheatstone bridge) (see Fig. 8-6). To record deformation data, the gauge is bonded to the surface of the structure at the point or location that is of interest. The degree of accuracy with which the strain-gauge method can obtain concentrated stresses is a function of the size of the gauge and the capacity of the equipment. The greater the number of data points, the greater the number of gauges and instrumentation channels needed.

Birefringent-coating Technique. In this technique, a photoelastic coating is applied directly to the actual structure from which surface-stress readings are desired. The method can be used to measure both static and dynamic loads. The surface of the structure must be cleaned and prepared before the clear birefringent coating is sprayed onto the component. An optical polariscope and photographic equipment are used to record the stress contours or trajectories. In instances where dynamic loadings are present, a strobascope usually provides the light source to "freeze" the action at some spatial or chronological reference point to obtain a photographic record. The stress contours are manifested as either dark and light bands or as bands of varying colors, depending on the type of light source. Photoelastic models have been used to assess the distribution of stress and deformation in the actual part being tested.

Interferometry. Interferometry utilizes the optical rings or interference rings produced when rays from a deformed object interfere with

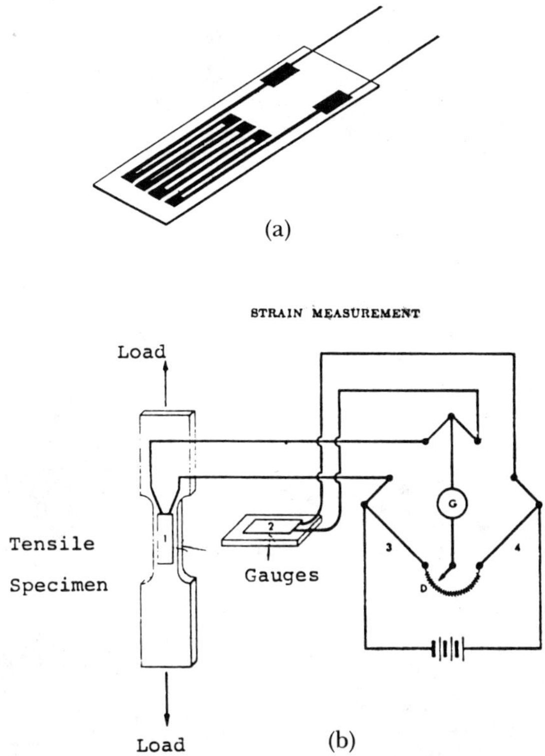

Figure 8-6. An illustration of strain gauges. (a) A strain gauge before bonding to a component. (b) A strain-gauge measurement system applied to a tensile specimen.

the light rays from an undeformed object. This can be accomplished in one of two ways: holographic interferometry or Moiré fringe analysis.

In holographic interferometry, light rays stored in the recording medium can be made to interfere with light rays from the real object, thus generating a set of dimensional differences between the actual object and the recreated object (see Fig. 8-7). This method is a precise way of determining how an object responds when minute stresses are applied. One powerful feature is that information is obtained over the entire surface of the test object rather than at a specific point. The surface can be three-dimensional, and no contact with the actual object is required. The method can be applied to static, dynamic, and thermal strains.

In the Moiré fringe analysis, grid lines must be etched, marked, embossed, or photographically imposed on the test object. During the

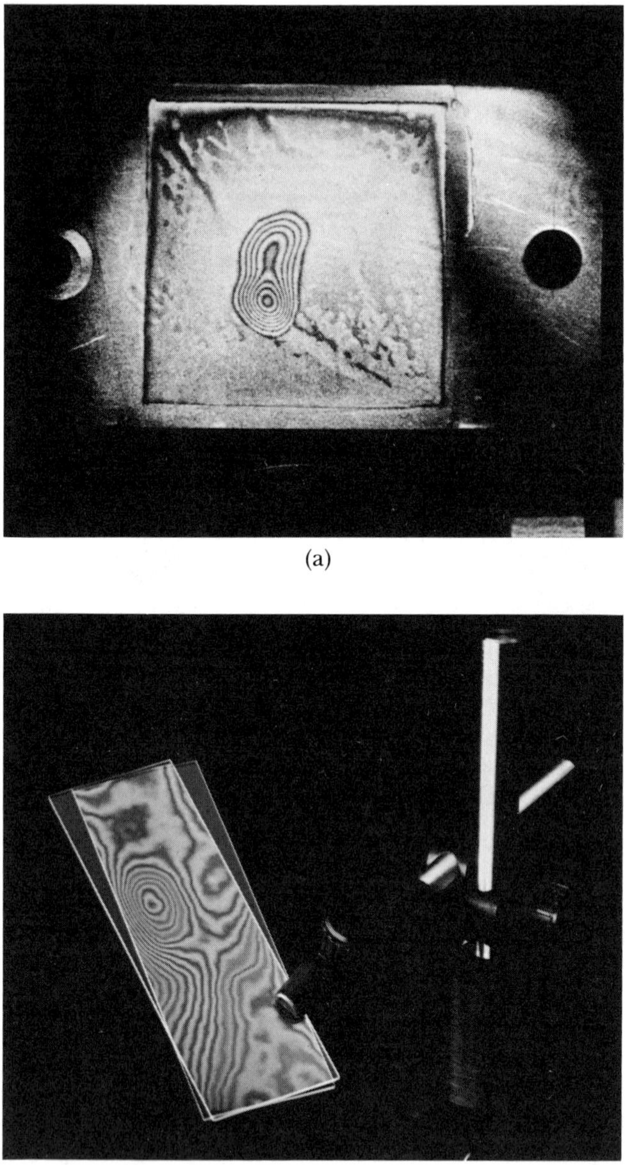

(a)

(b)

Figure 8-7. An example of holographic interferometry. (a) The void in the glue bond of this clutch plate facing is easily spotted from the pattern of contour lines. (b) Concentric, nearly circularly colored bands known as Newton's rings are formed by the interference of light reflected in the nonuniform air gap between two glass plates. (Courtesy of Newport) loading process, the deformation of these grid marks relative to a master grid plate sets up a pattern of interference that is recorded optically.

Stress-coating Technique. The stress-coating technique involves spraying a lacquer or lacquerlike paint on the surfaces of the test object. During loading, areas undergoing a high degree of stress can be identified, and the directions of the principal strains and the approximate magnitudes of tensile and compressive strains can be determined. This method of assessing the characteristics of the stress and deformation the object undergoes when subjected to loading is suitable for use before another nondestructive method is used for more detailed analysis.

X-Ray Defraction. X-ray-defraction is the only method available for direct measurements of surface residual stresses in crystalline materials. Stresses are determined by measuring the angle at which the stressed crystal defracts an x-ray beam. This method is very useful as a measure of fatigue damage. A technique referred to as "line broading" provides a measure of plastic deformation.

DESTRUCTIVE TESTING

Destructive testing is usually done after nondestructive testing has been completed, or at least after the areas that will not require nondestructive testing have been identified. These tests are usually conducted to identify the constitutive properties of structural parts (e.g., elastic, plastic, viscoelastic [creep], fatigue, dynamic, impact, and corrosion characteristics) in the context of the history of the loads and environment to which the object was subjected.

Tensile Testing

The main purpose of tensile testing during a failure analysis is to determine the mechanical properties of the failed material. The investigator's analysis and interpretation of the results will determine the mode and cause of the failure. Because most designs are based on the mechanical properties of a material, such as its yield strength or ultimate strength and its microstructure, any significant changes in the strength values could lead to structural failure (see Fig. 8-8).

The tensile test furnishes adequate data about a material regarding its yield, ultimate and fracture strength, ductility (total and uniform elongation and reduction in area) or lack of ductility, especially if brittle failure was encountered (see Fig. 8-8 and Appendix B). The lack of

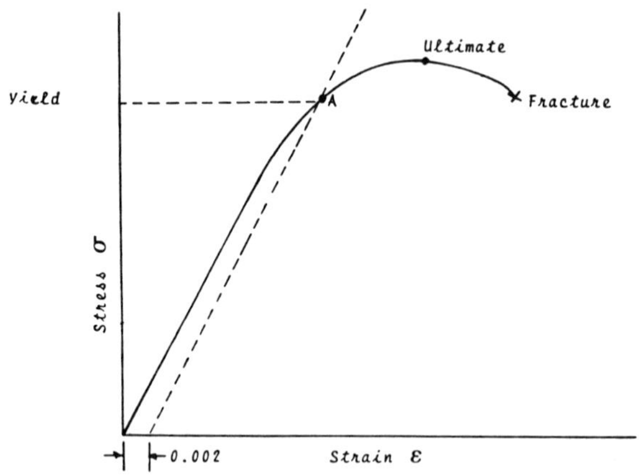

Figure 8-8. Engineering stress–strain diagram showing the location of the yield stress using the offset method of 0.2 percent strain.

ductility is obtained by measuring fracture ductility. In addition, anisotropic properties can be determined from a transverse tensile test.

Changes, if any, in the mechanical properties of the failed material, taken alone, do not conclusively pinpoint the cause of the failure. The data should be used in conjunction with other failure analyses because macroscopic properties may indicate changes but certainly not causes.

Figure 8-9(b) illustrates tensile-testing equipment with a tensile specimen in the cross beads. Two types of loading are available in tensile-testing equipment: stress-controlled loading and strain-controlled loading. Some equipment allows the analyst to control the path and rate of loading and the presence of environmental factors such as temperature and corrosive chemicals. (The effects of strain rates are discussed later, in the section entitled "Dynamic Impact Testing." See Myer 1986 and Appendix B for reviews of the basic engineering concepts regarding the behavior of materials.)

Fractography (Microscopic Testing)

Microscopic analysis is essential to understanding the basic mechanism of the failure process. It is a penetrating journey into the behavior of a material. This journey requires equipment that can reveal changes in structure that cannot be seen by the naked eye or detected with macroscopic testing methods. It is essentially a detailed examination of the fracture surface that reveals features and changes in the internal struc-

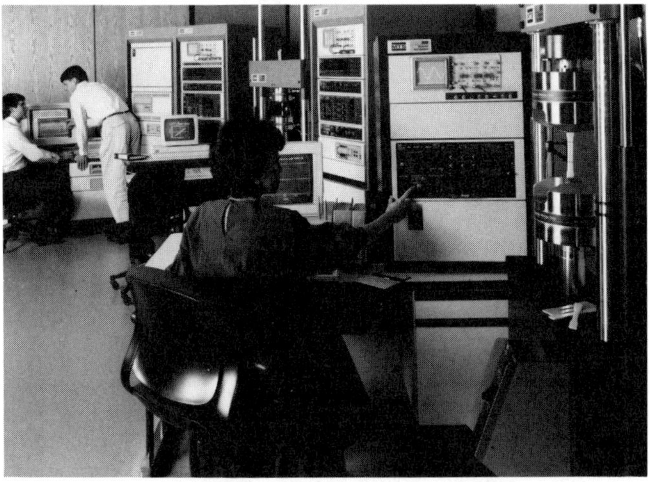

(a)

(b)

Figure 8-9. Illustration of a computer-controlled tension-testing (a) strain/load controlled tension test. (b) Tensile specimen with extensiometer. (c) Environmental controlled tensile test. (Courtesy of MTS Systems Corp.)

Figure 8-9. *(continued)* (c)

ture of the material caused by service conditions. The microscopic investigation and analysis is an essential part of fractography—the investigation of failed surfaces (Boyer 1975; Dorn 1961). Fractography requires one or all of the following pieces of equipment, depending on the depth of the analysis.

Light Microscope. Because they are simple and easy to use, light microscopes are available in most facilities. Because the depth of magnification is low, a deep-field photographic microscope with better magnification was developed that can focus on a selected spot on the fracture surface. Although the degree of magnification can reach 1500×, magnification of 1000× or less is more useful for interpreting surface features. A photograph of the spot is produced that shows all the features of the magnified surface. If the surface is etched, a detailed view of grain sizes and boundaries can be observed. Therefore, the instrument can serve as a scanning device to reveal surface features in a minimum amount of time, and it does not require skilled personnel. The photographs can show the origin of the crack, environmental effects such as oxidation, the type of fracture, the path and pattern of crack growth, and void formation.

Transmission Electron Microscope. The transmission electron microscope (TEM) has greater depth of resolution and magnification than does the light microscope, and trained personnel are required to perform the microscopy and analyze the results. Basically, the equipment consists of an electron "gun" that ejects electrons in a stream from a heated filament into a chamber evacuated to a pressure of 10E−4 torr or less. One or more condenser lenses enable the operator to focus a reduced image of the electron source on the specimen; the image can be adjusted to an area with a diameter of only 2 microns. TEM has a range of magnification of 210 to 300,000×.

The TEM requires extremely thin specimens or replicas of the fracture surface. The most common type of replica is made by pouring a plastic solution over the fractured surface. When the solution dries, it forms a film containing a replica of the fracture surface. When removed from the object, the replica is shadowed with a heavy metal and coated with carbon. The plastic is then dissolved, leaving a thin carbon-film replica of the fracture surface that is ready for TEM analysis. (This technique of producing replicas is described in detail in most metallurgy texts.)

The great advantage of replicas is that they allow the analysis of objects that are too large to be transported or inserted under a microscope. The replica of the fracture surface is made at the site, then transported to the laboratory for analysis without losing any of its features. The technique also has several disadvantages. The diameter of the replica that can be placed in the TEM is extremely small (3 mm).

The disadvantage of the technique is that features that are not accurate representations of the fracture surface may be present because of improper preparation or cleaning of the fracture surface or because of holes in the plastic coating that occur during stripping ("pull off of the replica"). Another common problem arises when the plastic coating scrapes against the fracture surface during removal, producing marks that may resemble striation marks observed in the fatigue process.

Scanning Electron Microscope. The scanning electron microscope (SEM), a relatively new instrument, combines electron-optical, vacuum, and electronic control devices that direct a beam of electrons on a specific site on the surface of an object. The SEM is a powerful tool for total characterization of a material; its magnification ranges from 5 to 240,000×, and it has a useful fractographic power to about 30,000×. Its resolution varies from 25 to 100 angstroms, and its depth of field is 300 to 500 times greater than that of the optical light microscope. Specimens can be tilted as much as 45 degrees and still be in focus (see Fig. 5-8b, c, p. 150). The basic component is an electron gun that emits electrons from a heated filament and accelerates them in a constant stream down a column to form a beam, which travels through a series of electromagnetic lenses and hits the specimen located at the lower end of the column. The size of the spot determines the resolution of the SEM; the smaller the spot, the higher the resolution. Because specimens must fit into the chamber of the SEM, the specimen must be cut into sections. The device can be used for direct examination of a specimen (no replicas are needed); therefore, it provides a realistic three-dimensional image of the topographic features on the surface of the specimen. An experienced operator is required to interpret these features.

When selecting TEM or SEM for fractography, the operator must keep the following four features in mind: magnification, resolution, area of coverage, and imaging. Although both microscopes provide high magnification, the SEM offers a better range of magnification. Whereas the SEM can resolve to about 70 to 100 angstroms, the TEM has a better range at 25 to 50 angstroms, and therefore reveals much finer details of the fracture surface. The SEM can analyze the fracture surface area on a larger scale than the TEM and also reveals the edges of a crack and its contours. Although the SEM provides a three-dimensional image and the TEM does not, it can only handle specimens of limited size, and thus may disguise some important features of the fracture surface. Therefore, the TEM replica may be more advantageous, since the integrity of the fracture surface is preserved.

Fatigue Tests

Cyclic loading can cause a component to fail even if the load is lower than the yield stress of the component. Failure in this case is called *fatigue failure*. Thus, fatigue tests are sometimes necessary to determine the life of samples taken from a failed component. Life is defined as the range of stress or strain relationships versus the number of loading cycles before failure occurs. The samples or specimens are subjected to cyclic loading until failure occurs; the fracture is then examined visually and microscopically to assess the fracture mechanism that may have led to the actual failure.

To achieve a statistically meaningful result, a number of tests are required for each data point (alternating cycles of stress). A widely used high-speed apparatus for testing fatigue is the R. R. Moore rotating beam machine, which loads specimens $3\frac{7}{16}$ inches long by 0.30 inch in diameter in pure bending by means of weights, and thus constant load is applied.

Tensile, torsion, and beam-bending fatigue tests are performed on equipment such as that shown in Figure 8-9. Multiaxial fatigue testing is performed on specialized specimens with standard equipment that has been modified or with equipment designed specifically for this type of testing.

Some materials—for example, composite materials such as epoxy–fiber glass—do not lend themselves readily to the tensile-testing type of machine (heads, grips, nor specimen). For example, composite materials are more easily tested in the beam-bending-type apparatus (four point).

Fatigue depends on a number of variables that can act collectively or independently. Fatigue failure can be defined as a process resulting from repetitive loading—that is, it must be a cyclic event. The mode of failure is influenced by the stress state, rate of strain, temperature, frequency of cycling, surface condition, corrosion, the size of the component, the type of loading (axial, torsional, bending, or multiaxial, mean stress, or localized plasticity), and of course the possible hardening or softening characteristics of the material resulting from the cyclic conditions (see Fig. 8-10).

The process of fatigue failure requires the initiation and propagation of a crack. These two stages of crack growth occur in two types of fatigue tests: high cycle (HCF) and low cycle (LCF). The basic information obtained from these tests is the life of the material under a known cyclic loading condition (see, for example, Fig. 8-11). The crack growth order is important in the two tests. In high-cycle tests, the propagation

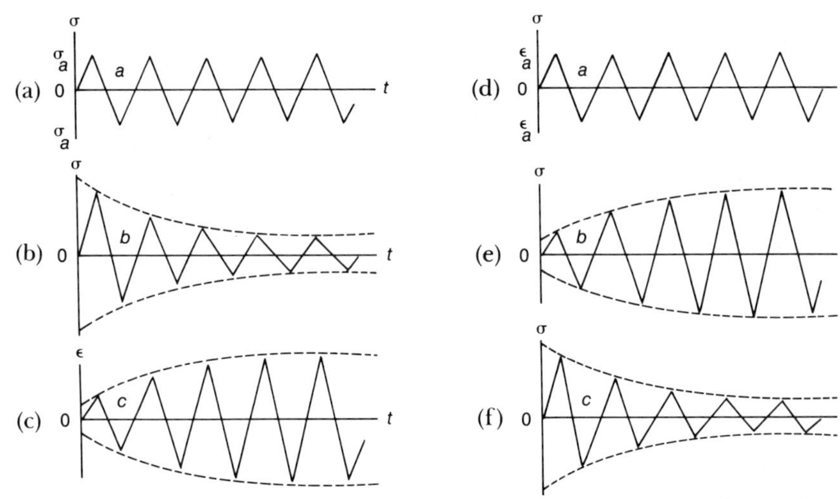

Figure 8-10. Low-cycle fatigue data showing the range of strain versus the cycles to failure: (a) controlled stress cycling; (b) strain hardening followed by strain softening; (c) cycle softening followed by strain hardening; (d) controlled strain cycling; (e) cycle hardening; and (f) cycle softening.

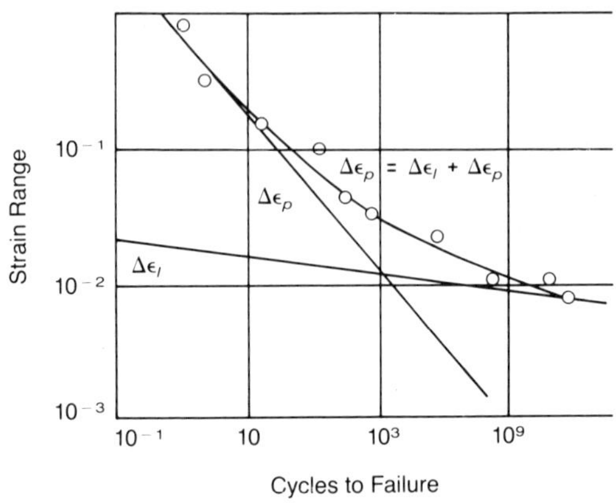

Figure 8-11. Low-cycle fatigue data showing the component of strain.

dominates the life; however, in fatigue studies, the two stages of crack growth are not defined for design analysis. Instead, the common design tool is the "S-N" diagram: stress versus number of cycles to failure (refer to Appendix B for further discussion and illustrations). If the process of crack propagation is the issue, then the analysis can be fully explored according to the concepts of fracture mechanics. (See the section below entitled "Fracture Mechanics and Fracture Toughness Tests" and Appendix B).

Failure caused by fatigue has a basic macroscopic feature that is often regarded as a classic identification mark—progressive markings called beachmarks (see Fig. 8-12) that indicate the progressive growth of the crack as the number of loading cycles increases. These marks serve as a life measurement in the same sense that the age of a tree can be measured by counting the number of rings it has accumulated.

Comprehensive laboratory testing is the major source of fatigue data. Although these tests are not sufficient to predict the life of a structural component accurately, they serve as an approximate guide. For this reason, codes and standards such as those outlined in the ASME Boiler and Pressure Vessel Code incorporate safety factors. These modified fatigue data can be used during design with a high degree of confidence.

One requirement for the growth of fatigue cracks is the presence of a tensile stress that acts as a driving force. Therefore, the majority of existing data is generated under uniaxial tensile cyclic loading. Few laboratories have the capacity to conduct tests under multiaxial loading systems despite the fact that multiaxial fatigue is an extremely important issue in design analysis.

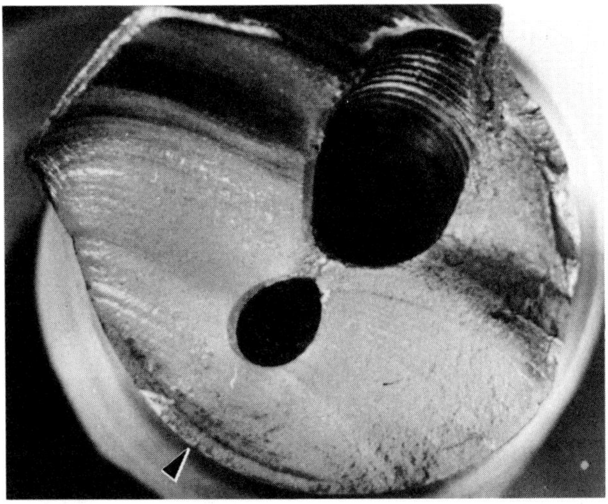

Figure 8-12. Beachmarks formed as a fatigue crack grew from the outside of a crankshaft. The arrow indicates the origin of the fracture.

Although macroscopic examination of a fatigue failure reveals beachmarks, microscopic examination generally reveals striations that can be related to the number of loading cycles to which the component has been subjected. In some cases, striations are absent although fatigue was the mode of failure.

Fatigue cracking is normally initiated in a highly stressed region where one or more cracks are nucleated at persistent slipbands or at grain boundaries. Once the crack starts moving—the result of opening and closing at the crack tip because of the cyclic effect—striations begin to form parallel to the front of the crack, which continues to grow until catastrophic fracture occurs. The direction of growth and the rate of propagation are controlled by the magnitude of the localized stress and the mechanical properties of the material. If the load is alternating, a crack will open under one-half of the cycle and close in the compressive mode under the other half, producing a rubbing and shinning effect that causes pronounced beachmarks. If the load is fluctuating, the rubbing effect may not be pronounced and the beachmarks may be less visible. The effects of loading observed on the fracture surface can be the result of axial, bending, pressure, or combination loading. The characteristic of this type of loading is a normal tensile stress that, in the case of low-stress–high-cycle fatigue, produces a flat face with a fracture surface of fine-grained and lightly polished areas near the starting point of the crack. As the crack grows, the surface texture becomes more fibrous and rough because of the increased intensity of the stress. In the case of low-cycle–high-stress fatigue, the fracture surface has a rough, fibrous texture similar to that associated with plane-stress loading. In the case of cyclic tensile stress, the crack takes a path that is normal in relation to the direction of maximum stress; however, in a cyclic torsional loading system, cracks will develop in a longitudinal or transverse shear direction, depending on the ductility of the material. In laboratory testing on a shaft, cracks have been observed to begin in a shear direction, then become normal or tensile-mode oriented at 45 degrees with respect to the longitudinal axis of the shaft. In other cases, cracks remain in a shear mode. This behavior is attributed to the ductility of the material when strain hardening or softening controls the mode of failure.

If the cyclic loading is alternating, two sets of cracks will appear normal to each other at an orientation of 45 degrees with respect to the axis of the specimen. If the cyclic loading is fluctuating, then one crack will develop normal to the direction of tensile stress and continue to grow in one direction. If the load is monitored on a strip-chart recorder, a one-sided rather than symmetrical drop in the loading magnitude will begin. Because the presence of localized plasticity is a basic requirement for a crack to grow, the microscopic structure of the material dictates the amount of plasticity available for crack growth. If the degree of plasticity is small or nonexistent, then the mode of failure will change from ductile to brittle. For example, grain size—a discontinuity that causes a concentration of stress—influences the mode of failure. A shear failure that changed to normal failure at high temperatures because of the grain size has been observed in a nickel-based super alloy.

Fatigue life is also influenced by the elements used in the alloy. Strain or work hardening is another variable that may influence the mode of fatigue failure. If the work hardening does not soften with stress cycling, a small plastic deformation will form at the crack tip, reducing the rate of crack growth and the fatigue life. Other factors such as the condition of the surface, heat treatment, and overloading greatly affect fatigue life.

In a case study on catastrophic failure of a high-strength-steel pressure vessel, specimens were taken from different sections of the failed vessel for quasi-static tensile fatigue testing. The performance properties of the quasi-static material were within those required for acceptable design and did not reveal any variations in processing that could have contributed to the failure. LCF tests also were conducted under strain cycling to failure on several specimens taken from different sections, and metallurgical examinations were conducted on all cycled specimens. The investigator observed that the fracture surface of some samples showed evidence of carbon precipitation, whereas no precipitation was observed on others. A comparison of fatigue life data from the two sets of samples revealed that the life of the samples containing carbon precipitation was reduced. Thus, the microanalysis pointed to the manufacturer's heat-treatment process. Most metals and alloys used in high-stress components undergo heat treatment to improve their properties, and improper temperature control can cause the heat to be applied at uneven rates, which inevitably causes imperfections that will adversely affect the safe operation of the component. Further inquiries confirmed that this was the problem in the case just cited.

Creep and Stress-Rupture Test

Failure also may occur as a result of applying a constant load over a long period at high temperatures (e.g., T over 800°F.). Figure 5-8(d) illustrates the failure of a bolted connection because of "creep" deformation of the nut. Tensile testing machines such as the one shown in Figure 8-9 are used to perform constant-load (creep) tests and constant-strain (relaxation) tests (see Appendix B).

Creep is a type of deformation that accumulates over time. It differs from the immediate deformation that occurs when a load is imposed on a component. The temperature needed to produce creep under a steady load is about half the melting temperature (0.5 Tm) of the material. The creep time can vary from a year to many years, depending on the load and the temperature.

The analysis of creep is based on data obtained from tests conducted over a period of a few weeks to many weeks (see Table 8-2), but certainly not over the design life of most components exposed to this type of failure. Because long-term testing is not a common practice, designers must estimate creep life using procedures that provide approximate data.

Creep deformation in metals occurs in three stages: transient or primary, steady or secondary, and tertiary (see Fig. 8-13). These stages are time dependent and not instantaneous; each one has a distinguishing characteristic that reflects the response of the material to the service load and temperature. The microstructure of the material plays a significant role in the development of the three stages.

Transient Stage. During the transient or primary stage of creep, plastic strain increases rapidly over time, followed by a decrease. This behavior is attributed to the realignment of crystallographic imperfections in the material, resulting in strain hardening. Microscopically, strain hardening is a process of resistance to dislocation or movement caused by

TABLE 8-2. High Temperature Strength Values for Several Wrought Alloys

Alloy (Balance of Composition Is Essentially Iron)	*Temperature (°F)*	*Short-time Tensile Strength (lb/in.²)*	*1,000-hr Rupture Strength (lb/in.²)*	*Creep Strength (lb/in.² 0.1% per 1,000 hr)*	*Creep Strength (0.01% per 1,000 hr)*
Carbon steel	800	50,000	—	20,000	14,000
0.15% C	1,000	30,000	11,700	4,000	2,700
	1,200	15,000	2,000	600	300
Carbon–molybdenum	800	60,000	—	30,000	21,000
steel 0.15% C					
0.55% Mo	1,000	45,000	25,000	10,800	6,700
	1,200	27,000	4,400	2,000	700
18-8 stainless steel	1,000	55,000	35,000	18,000	11,000
18% Cr 8% Ni	1,200	42,000	14,000	7,000	4,000
	1,500	21,000	3,500	2,000	1,300
Timken alloy (hot rolled)	1,200	86,000	35,000	18,000	10,000
16% Cr 25% Ni 6% Mo	1,350	60,000	20,000	13,000	7,000
0.1% C 0.15% N_2	1,500	41,000	10,000	6,000	3,000
N-155 (low carbon)					
20% Cr 20% Ni	1,200	91,000	39,000	20,000	17,000
20% Co 3% Mo	1,350	65,000	22,000	15,000	11,000
2% W 1% Cb	1,500	56,000	13,000	9,000	6,000
0.12% C 0.12% N_2					
S-816					
44% Co 20% Cr	1,200	110,000	52,000	28,000	14,000
20% Ni 4% W	1,350	95,000	28,000	20,000	12,000
4% Mo 4% Cb	1,500	70,000	17,000	10,000	7,000
0.36% C	1,600	50,000	10,000	6,000	4,000
Inconel "X"					
(solution heat	1,200	120,000	65,000	62,000	54,000
treated and aged)	1,350	92,000	40,000	39,000	28,000
73% Ni 15% Cr 0.9% Cb	1,500	52,000	18,000	17,000	12,000

TABLE 8-2. *(continued)*

Alloy (Balance of Composition Is Essentially Iron)	*Temperature (°F)*	*Short-time Tensile Strength (lb/in.²)*	*1,000-hr Rupture Strength (lb/in.²)*	*Creep Strength (lb/in.² 0.1% per 1,000 hr)*	*Creep Strength (0.01% per 1,000 hr)*
2.5% Ti 0.9% Al 7% Fe	1,600	34,000	6,000	8,000	7,000
Cupronickel					
68.9% Cu 30.0% Ni	750	40,000	—	18,800	9,100
0.6% Mn 0.5% Fe					
Ti–^{6}Al–^{4}V titanium alloy					
6% Al 4% V	800	98,000	75,000	48,000*	—
bal. Ti	1,000	79,000	—	10,000*	—
2024 aluminum alloy (naturally aged)	200	62,000	59,000	—	—
4.5% Cu 1.5% Mg	400	26,000	19,000	21,000*	—
0.6% Mn bal. Al	600	7,000	2,700	3,600*	—
HK31A magnesium alloy	400	23,000	—	5,400	—
(artificially aged)	500	20,000	—	3,100	—
3.0% Th 0.7% Zr	600	18,000	—	900	—

*One percent per 100 hours.

Taken from F.A. D'Isa, *Mechanics of Metals*, Addison-Wesley, Reading, MA, 1968.

temperature and loads. Because dislocation is responsible for the deformation, the substructural changes in this stage caused by loading and temperature slow down and form a line of resistance to the dislocation movement. Thus, primary creep is retarded with time. Because this behavior is commonly observed in metals, engineering/mathematical expressions have been developed to relate this type of creep strain to time and strain hardening.

Steady Stage. The main feature of steady or secondary creep is the linear relation between creep strain and time, the result of a balance between strain or work hardening and softening processes. Thus, a

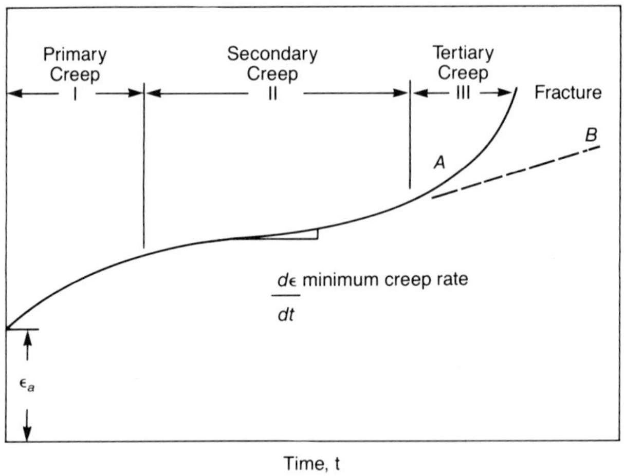

(a)

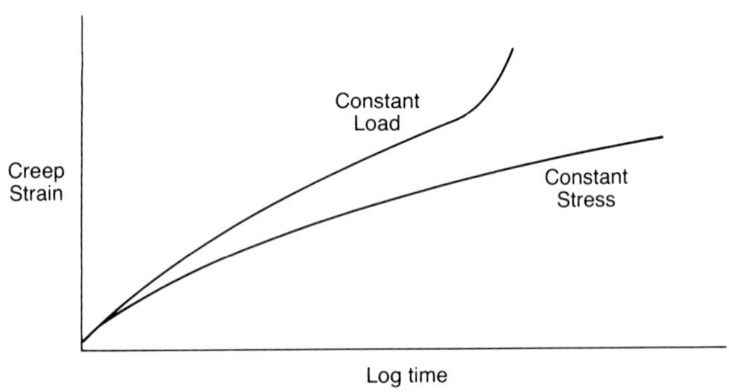

(b)

Figure 8-13. Creep behavior of materials. (a) The three stages of creep. (b) Types of creep loading tests.

steady state is reached during which a minimum amount of creep takes place. The steady state indicates that the material is stable, with no changes in substructure. A number of engineering/mathematical expressions also have been proposed to relate the rate of creep to applied stress and temperature.

It is generally recognized that with low stresses and with temperatures near the melting point (Tm) the rate of steady-state creep varies linearly, whereas with high stresses and with temperatures above 0.5 Tm, the rate varies nonlinearly. Dorn (1957, 1961) and Sherby, Orr, and Dorn (1954) suggested that for T/Tm greater than 0.5, the creep strain depends on an activation energy process H, which was found to be constant in the steady stage over a wide range of temperatures. The activation energy can be calculated with the following equation:

$$\Delta H = R\left(\ln\frac{\mathring{e}_1}{\mathring{e}_2}\right)\left(\frac{T_1T_2}{T_1 - T_2}\right)$$

where $\dot{e}_1$ and $\dot{e}_2$ are rates of creep at T_1 and T_2 and R = universal gas constants.

Tertiary Stage. The tertiary stage of creep ends in catastrophic failure. The creep strain is a nonlinear function of time, and the time at failure is known as the rupture time. Tertiary creep begins at the end of the minimum creep rate segment of the curve and features unstable metallurgical changes in the material that increase the deformation process. The tertiary stage is also the period in which softening overtakes hardening, resulting in rapid microstructural changes such as microvoid formation and intercrystalline fracture. In addition, the strain-hardened grains may recrystallize in such a way that the balance between softening and hardening observed in the second stage is destroyed.

Tertiary creep is reported in terms of three factors: stress, temperature, and rupture time. The rupture time defines the life of the material at a given stress and temperature. The rupture time can be 100 seconds in the case of a rocket engine nozzle, 100 hours in the case of a turbine disk in the engine of a military aircraft, or 12 or 13 years in the case of a component designed to have a long life. In any case, the data are derived from stress versus rupture time or creep resistance, which refers to the magnitude of stress required to produce certain creep strain in about 11 to 23 years. Obviously, an experiment that lasts 11 to 23 years is impractical, if not impossible. To overcome this problem, one must rely on short-term data that can be extrapolated to the long term. The most widely used are the Larson–Miller and Manson–Haferd parameters.

The Larson–Miller parameter is expressed as $P(\sigma) = T(C + \log tr)$, where P is the Larson–Miller extrapolation parameter, which depends only on stress (σ); C is a constant with a value of 20 for most materials; T is the absolute temperature; and tr is the time to rupture in hours.

The Manson–Haferd parameter is the result of plotting log tr versus temperature T at a given stress. This plot shows a series of stress curves that converge at a point with coordinates labeled T_a and t_a. Because the plot is a log-log plot, the parameter is expressed as

$$P(\sigma) = \frac{(T - T_a)}{(\log tr - \log t_a)}$$

where T_a and t_a are the convergence points of the curves. The difference in the second approach is that constant C is replaced by T_a and t_a. In either case, utmost care must be exercised when interpreting the results because the extrapolation process is not linear.

Fracture Mechanics and Fracture Toughness Tests

When analyzing fatigue data, one is concerned with the number of cycles to failure at a given cyclic load and with other environmental variables, not the failure process. If the failure process must be considered, however, one must look at the origin of the crack and its pattern of growth. This process includes three stages: crack initiation, crack propagation, and the failure mode that resulted from the first two stages (see Fig. 8-14). Therefore, to analyze the failure process, one must rely on the concepts of fracture mechanics, which considers crack size (a), material toughness (K_c, K_{Ic}, K_{Id}), and levels of tensile stress (σ) needed to bring about a brittle fracture. Factors such as loading rate, concentration of stress, residual stress, and temperature affect the three factors (see Figs. 3-2, 3-4, 5-3, 7-1).

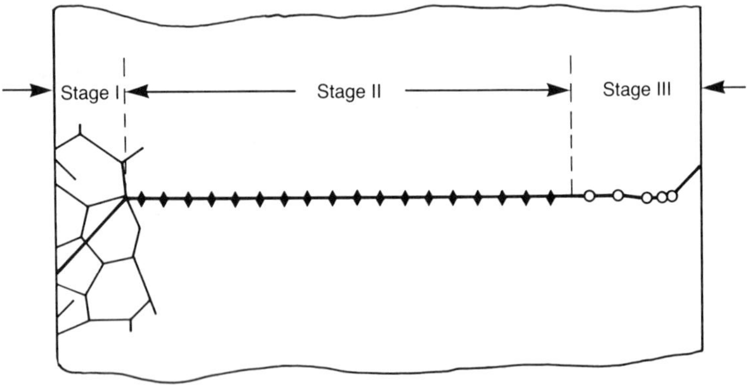

Figure 8-14. The three stages of crack propagation in fatigue. Stage I is along a slipband, Stage II growth is oriented normal to the applied stress, and Stage III represents the final fracture, with formation of voids and a final shear lip.

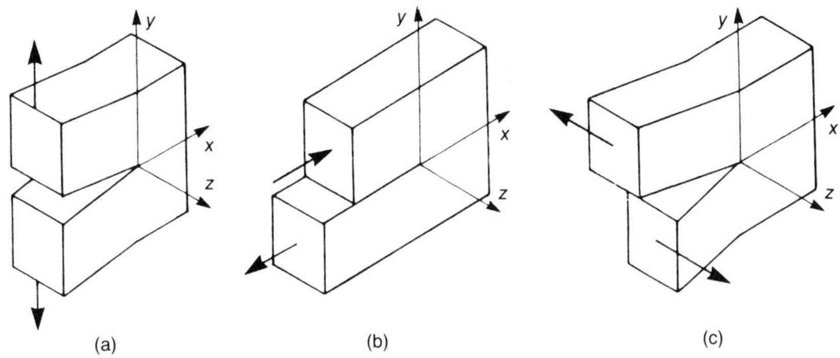

Figure 8-15. The three loading modes for a crack: (a) Mode I, the opening mode; (b) Mode II, the edge-sliding mode; and (c) Mode III, the anti-plane-strain or tearing mode.

The mode of a fracture failure can be one or a combination of three types of loading: opening mode (Mode I), sliding or shear mode (Mode II), and tearing mode (Mode III) (see Fig. 8-15). Fracture concepts were developed to deal with each type (see Rolfe and Barsom 1977).

Like fatigue analysis and testing, where data are generated for different materials and are presented in terms of S-N curves, fracture mechanics data are also developed in terms of the rate of crack growth versus a stress-intensity range known as K. K is related to a cyclic stress amplitude and a known initial crack length. The rate of fatigue crack propagation per cycle, expressed as da/dN, is then plotted versus K, which results in an S-shaped curve consisting of three different parts with different slopes (Fig. 8.16). Each part is related to the crack-growth phenomenon. For example, a deviation at the upper part of the curve indicates that the crack is reaching a critical size that leads to total failure.

At that point, da/dN becomes infinite, and the stress-intensity factor K becomes critical and takes on the value K_{Ic}—a critical value for Mode I failure under the plane-strain condition that is determined with a number of fracture toughness tests. The term *toughness* describes the ability of a material to absorb energy by deforming plastically before it fractures. For example, in a slow tension-loading test of a specimen taken to failure, the area under the stress–strain diagram represents the toughness of the material being tested (see Figs. 8-8 and 8-9).

Material toughness represented by K, however, is a measure of the ability of the material to carry a load or deform plastically in the presence of a notch. The value of K can be obtained under static or dynamic loading. In the first case, K can be expressed as K_c or K_{Ic}. The term K_c

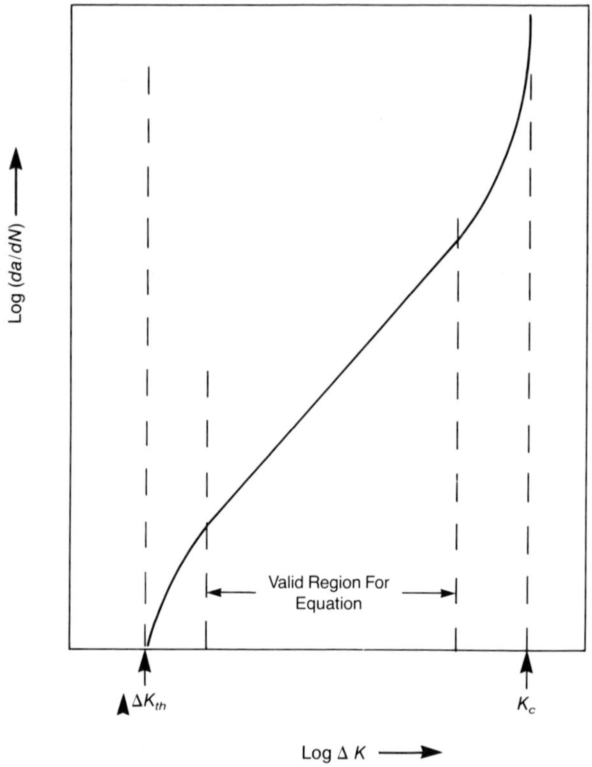

Figure 8-16. The fatigue-crack growth curve for a long crack, which illustrates the valid region of growth according to the following equation: $da/dN = C(K)^m$, where a = crack length, N = number of load cycles, K = the range of stress intensity, and C and m are material constants.

describes the critical value of the stress-intensity factor under the plane-stress condition. Range K depends on the thickness and the geometry of the specimen and the size of the crack, whereas K_{Ic} represents the critical value under the plane-strain condition (small crack-tip plastic deformation) and becomes independent of thickness. Therefore, K_{Ic} can be regarded as a property similar to the yield strength assessed in mechanical testing.

K_{Ic} tests. The ASTM E-24 established two acceptable tests for determining K_{Ic} values: the compact-tension specimen test and the notch-bend specimen test (see Figs. 8-17 and 8-18). The ASTM established strict standardized rules for both tests.

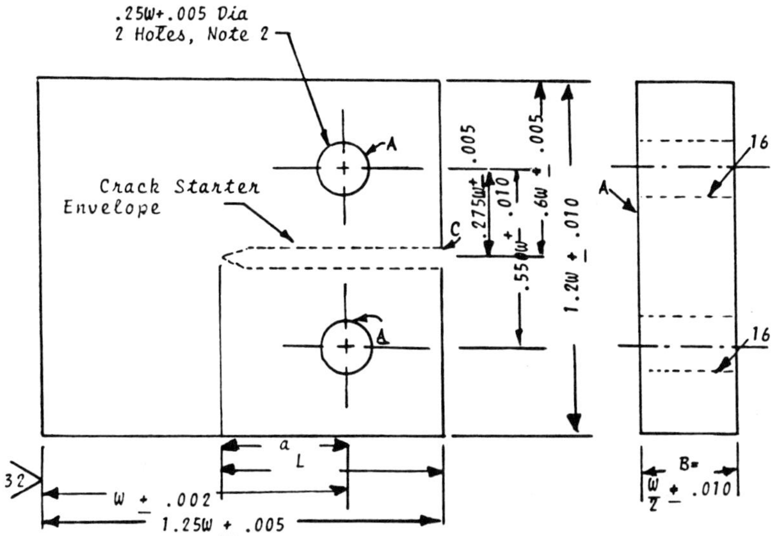

Figure 8-17. Illustration of compact-tension specimen used to test for plane-strain fracture toughness (ASTM E399-72).

Thickness is an important parameter that determines whether the plane-strain condition has been met. The initial machine crack-length in either type must be 0.45 × width (W) so that the crack can be extended to 0.5 W. The machined crack will act as a notch and crack starter that can be extended by fatigue. The fatigue-crack extension requirement is necessary to eliminate the effect of the geometry of the notch. (This

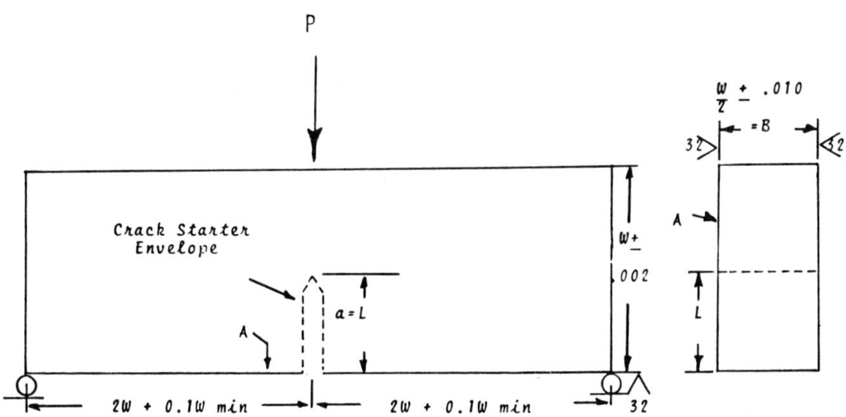

Figure 8-18. Typical bend specimen used in plane-strain fracture toughness tests (ASTM E399-72).

testing procedure is described in the ASTM Annual Standards, Part 10 [E399-74]).

The size and dimensions of the specimen are critical for establishing an acceptable K_{Ic} test. The dimensions must meet the following criteria:

$$a = \text{crack depth} \geq 2.5\left(\frac{K_{Ic}}{\sigma ys}\right)^2$$

$$B = \text{specimen thickness} \geq 2.5\left(\frac{K_{Ic}}{\sigma ys}\right)^2$$

$$W = \text{specimen depth} \geq 5.0\left(\frac{K_{Ic}}{\sigma ys}\right)^2$$

where σys = yield stress. The ASTM standard describes how to select and prepare the test, fatigue-crack the notched specimen, and measure the load deformation so that a critical load value can be defined. The critical load P value is called P_Q or P_5 (see Fig. 8-19). The two values can be equal or different. In either case, a procedure to determine the P value is established by drawing a line, the slope of which is 5 percent less than the original slope of a load deformation diagram. Once the critical P value has been determined, it is related to a stress intensity factor

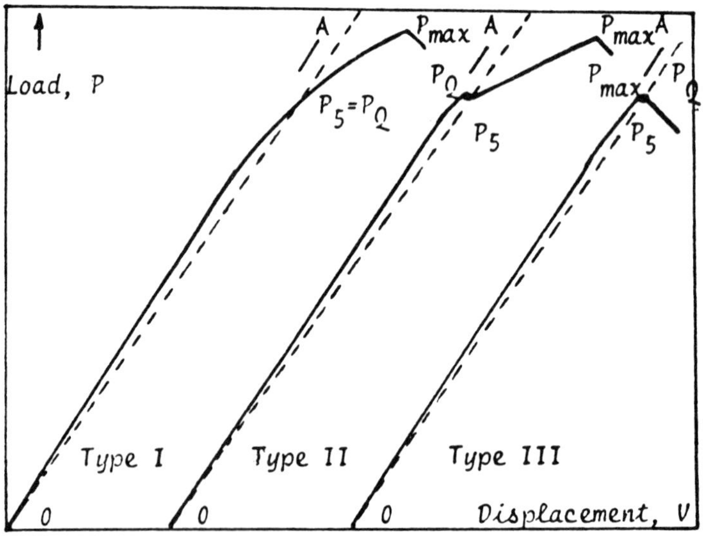

Figure 8-19. Schematic diagram showing load-displacement curves used to measure crack-opening displacement (see Table 8-3). A 5 percent secant was used to calculate K_Q.

TABLE 8-3. Comparison of Various Fracture Criteria

Name	*General Use*	*Level of Performance*	*Test Specimen Used*	*Applications*
K_{Ic} or K_{Id}	All materials	Plane strain	K_{Ic} standard (ASTM E-399) or K_{Id} nonstandard	Linear-elastic behavior Quantitative stress-flaw size calculations Aerospace applications Very high-strength materials Low-temperature applications
15-ft-lb CVN impact	Low and medium strength (30–140 ksi) structural steels	Upper plane strain to lower elastic–plastic	ASTM standard (E-23); Charpy V-notch impact	Ship hull steels Bridge steels Applications where moderate level of toughness is required
Loading-rate shift	Structural steels σys = 30–140 ksi that exhibit a load-rate shift	Elastic–plastic	CVN impact (ASTM E-23) or K_{Ic} slow-bend (ASTM E-399)	AASHTO bridge steel toughness specifications; structures that are loaded at slow or intermediate loading rates
Fracture analysis diagram	All materials exhibiting transition-temperature behavior	All levels; plane strain, elastic–plastic, fully plastic	Nil-ductility transition NDT (ASTM E-208); dynamic tear- DT (Mil Standard 1601)	Transition-temperature region Applications where intermediate to high levels (elastic–plastic) of toughness are required; defines NDT, FTE, and FTP criteria
Through-thickness yielding	Structural steels	Elastic–plastic	CVN or K_{Ic}	Accounts for desired increase in toughness with increasing yield strength and plate thickness, Large thin plates with edge or through thickness cracks

TABLE 8-3. *(continued)*

Name	*General Use*	*Level of Performance*	*Test Specimen Used*	*Applications*
Leak-before break	Primarily pressure vessels to ensure "leaking" before unstable crack growth	Elastic–plastic	K_{Ic} or K_c (nonstandard plane stress) *R*-curve	Pressure vessels Design stress-flaw size calculations possible as a function of vessel thickness
Ratio-analysis diagram	All materials primarily used as a comparison of materials rather than as a criterion	All levels, but primarily fully plastic	K_{Ic} dynamic tear, or CVN (upper shelf values)	Limited to upper shelf behavior for very high-strength materials that do not exhibit a transition behavior; this may be plane-strain behavior
The following three methods of analyzing the fracture toughness of materials are still in the research stage and for purposes of comparison are included in this table.				
Crack opening	All structural materials, but primarily low- to medium-strength steels	Elastic–plastic to fully plastic	Crack-opening displacement bend specimen; British standard—similar to slow-bend K_{Ic} test specimen (E-399)	Widely used in United Kingdom and Japan; Alaskan pipeline
R-curve	Primarily high-strength sheet materials	Elastic–plastic	Oversize compact tension specimen	Primarily used in studies to measure resistance to stable crack extension; plane-stress behavior
J-integral	All structural materials; general extension of K_{Ic} into elastic–plastic behavior	Elastic–plastic (directly related to K_{Ic} in elastic region)	Similar to K_{Ic} test specimens	Research extension of linear-elastic K_{Ic} behavior; quantitative stress-flaw size calculations possible

Taken from Rolf and Barsom, *Fracture and Fatigue Control in Structures*, Prentice-Hall, Englewood Cliffs, NJ., 1977.

known as Kq. Factor Kq becomes equal to K_{Ic} if it was calculated according to an established mathematical relation.

In addition to the recommended tests for K_{Ic}, additional testing methods and empirical correlations are available to estimate K_{Ic} for most structural materials. Data obtained from these tests are correlated with K_{Ic}. These additional tests, which are described below, are called dynamic because they involve impact loading.

Charpy V-Notch Test. Materials used in structural components may exhibit ductile- or brittle-type failure, depending on the state of stress, strain rate, and temperature. Therefore, the transition from brittle to ductile occurs at a temperature called the *transition temperature.* The most widely used test to determine the transition temperature is the Charpy V-notch (CVN) test, a standardized impact test that correlates its results with K_{Ic} values. (This is described further in Appendix B.) The transition temperature is defined as the temperature at which the fracture surface shows a 50 percent shear fracture, or as the temperature below which the fracture is brittle and is dominated by cleavage and above which it is ductile and dominated by microvoid coalescence.

The major cause of brittle fracture is the inability of the material to resist propagation because the stress relaxation that normally takes place at the crack tip is absent because of localized plastic deformation. Therefore, if the time is insufficient for relaxation to take place, a crack will spread rapidly through the material and lead to brittle fracture. Because the rate of stress relaxation is temperature dependent, low temperatures inhibit relaxation and tend to promote rapid propagation of cracks. This observation has related the values of K_{Ic} to brittle fracture because the lower the K_{Ic} value, the more brittle the behavior.

The standard CVN impact-fracture-toughness test specimen is shown in Figure 8-18. The impact machine known as Charpy has a maximum striking force of 220-ft-lb. The pendulum strikes the specimen at the center of the unnotched side and breaks it in a single blow (see Fig. 8-20). The energy of the impact absorbed by the specimen is measured by the difference between the height of the fall and rise of the pendulum.

CVN tests are conducted at different temperatures. The data are plotted in terms of the energy absorbed (ft-lb) versus the test temperature (°F). A smooth curve is fitted to the results, and a transition temperature is found that separates ductile from brittle fracture (see Fig. 8-21). Barsom and Rolfe (1977) found a good correlation between CVN values and K_{Ic} and stated that K_{Ic} values and CVN values in the region of the temperature transition can be correlated if (1) the test results for slow-bend K_{Ic} specimens are related to the test results for slow-bend fatigue-

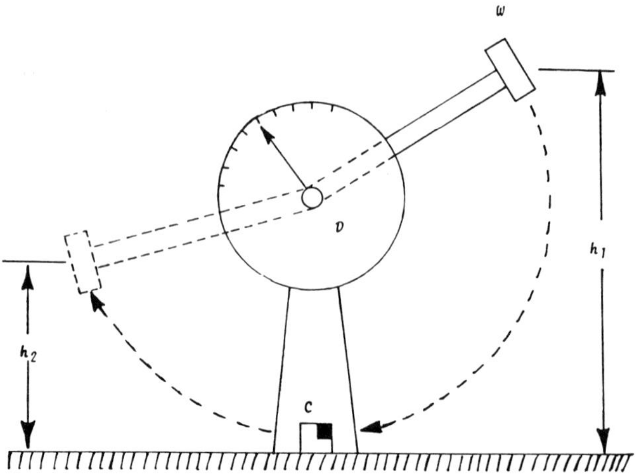

Figure 8-20. Impact test showing weight (W) and its impact on specimen (C). The test is used to measure the energy absorbed by the specimen.

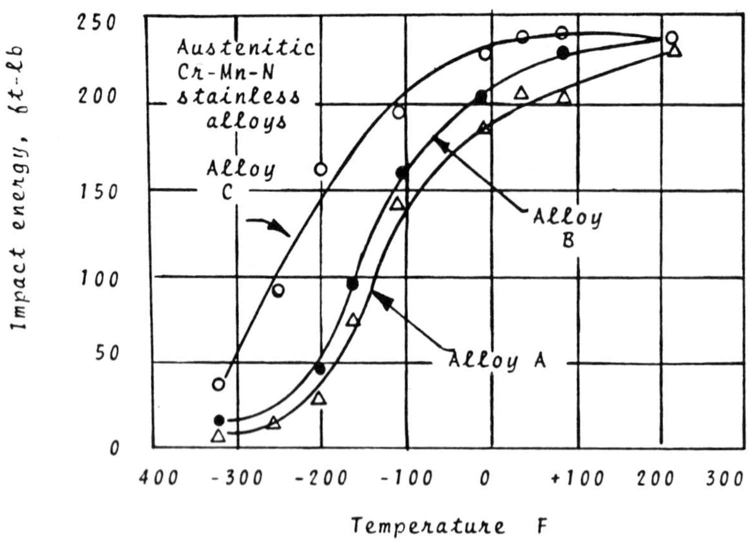

Figure 8-21. Transition-temperature range of three alloys determined by Charpy V-notch tests. A = austenitic; B = Cr–Mn–N; C = stainless alloys.

cracked CVN specimens, and (2) the test results for dynamic K_{Ic} specimens are related to the test results for dynamic fatigue-cracked CVN impact specimens. Therefore, the correspondence between K_{Ic} and CVN energy-absorption values obtained at a particular test temperature and at the same rate of strain can be expressed as

$$\frac{K_{Ic}^2}{E} = A\ (\text{CVN})$$

where E = Young's modulus and A = a constant. The constant A takes into account the effect of size as well as notch acuity. In the transition-temperature region, one can expect conservative estimates of K_{Ic} values (see Rolfe and Barsom 1977 for other restrictions).

Drop-weight Test. The objective of the drop-weight test also is to determine the transition of fracture from brittle to ductile. The transition is measured at a temperature where the ductility of the material becomes nil. (The effect of temperature on brittle fracture is shown graphically in Fig. 8-22.) Therefore, at the transition temperature, the fracture mode changes; this temperature is called the *nil-ductility transition temperature* (NDT). The NDT specimen is standardized according to ASTM-E-208 and is used primarily for ferritic steel materials. The specimen used in this test is $\frac{1}{2} \times 14 \times \frac{3}{4}$ inches and has a bead weld at its center that acts as a crack starter. The specimen is then subjected to three-point bending by a 60-pound falling weight that strikes the center. The impact causes a sharp semicircular surface crack to form at the center of the bead weld. Because the deformation results could be ductile, a mechanical stopper is placed under the center of the specimen to limit the amount of deformation. A number of tests are repeated at different temperatures until a NDT temperature at which nil ductility is observed is defined—that is, the specimen breaks in a brittle manner before reaching the mechanical stopper.

Test results from this type of test correlate with the dynamic yield stress of the material and the dynamic stress-intensity factor K_{Id}. Irwin *et al.* (1967), Shoemaker and Rolfe (1969), and Pellini (1971) proposed relations that have similar forms but differ slightly in the proportionality constant. The best estimated relationship is expressed as

$$K_{Id} = C\ (\sqrt{\text{in.}})\ \sigma_{yd},$$

where C has a range of 0.05 to 0.76; K_{Id} is the dynamic critical stress-intensity factor for the plane-strain condition at NDT temperature, ksi $\sqrt{\text{in.}}$, and σ_{yd} is the dynamic yield strength at the NDT temperature,

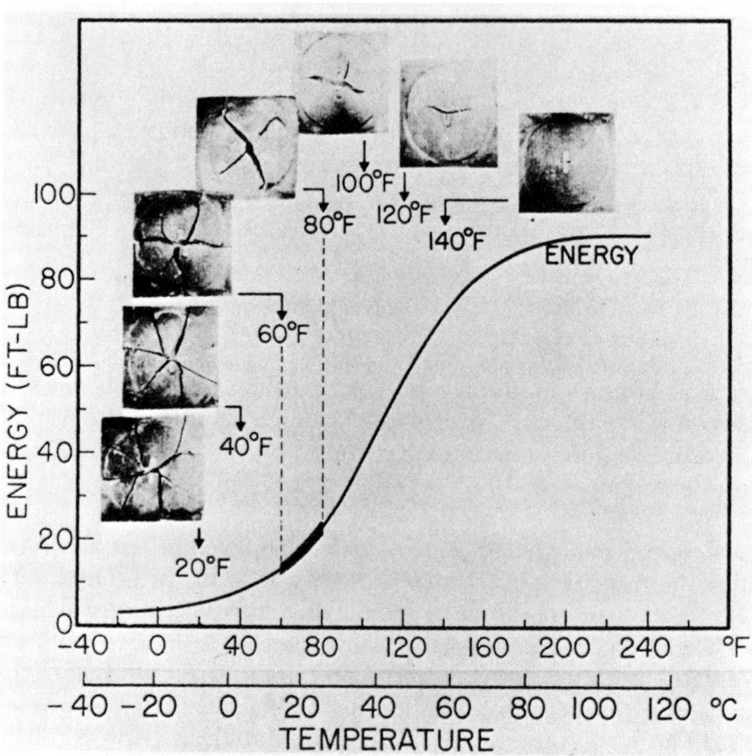

(a)

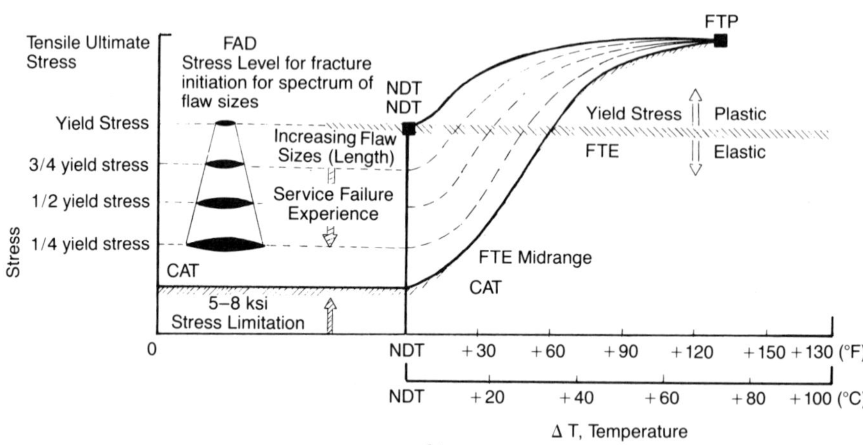

(b)

Figure 8-22. (a) Typical correlation of explosion crack-starter test performance with C_v transition curves for ship fracture steels. Flat break (NDT) fractures are obtained at temperatures below the 10 ft-lb transition temperature. Arrest characteristics are developed in the range indicated by the solid band, which spans the 10–20 ft-lb transition-temperature range. Full ductility is attained at what approaches shelf temperatures. (b) Fracture analysis diagram. (From Pellini 1971) ksi. The σ_{yd} also can be related to σ_{ys} (static yield stress at 0.2 percent offset limit) by

$$\sigma_{yd} = (\sigma_{ys} + 20 \text{ to } 30 \text{ ksi})$$

Notch Slow-bend Tests. The slow-bend tests evolved as a result of concerns about the effect of the high strain rate encountered in the dynamic impact tests and concerns about the applicability of those test results when applied to components that usually do not encounter the high strain-rate loading encountered in most service conditions. The test is similar to the CVN test, except that the specimen is loaded in a three-point bending where a static load is applied at a low rate of strain similar to that of a standard tension test. A load-deflection diagram is recorded, and the area under the diagram is a measure of the total energy to fracture. The strength of the material is calculated from the breaking load, and the recorded deflection measures the relative ductility.

Test results conducted on low-strength steel show that if the material is sensitive to the rate of strain, the transition from brittle to ductile will begin at a lower temperature than it does in impact CVN tests. The CVN impact shows an NDT at about 30°F, whereas the slow-bend test showed an NDT at about −150°F.

Lehigh Slow-bend Test. The Lehigh slow-bend test is another widely used slow-bend test. The specimen is 3 × 12 × $\frac{3}{4}$ inches and has a notch across its entire width. It also may have a longitudinal weld bead. Similar to the notch slow-bend previously described, the Lehigh specimen is subjected to a slow bending load until fracture occurs. The ductility transition temperature is defined as the point at which lateral contraction of 1 percent is obtained. This is accomplished by measuring the lateral contraction just below the notch at the time of failure at specimen temperature. This process is repeated at different temperatures until the 1 percent criterion is met. The definition of transition used in other types of tests also is applied; that is, at the transition temperature, the fracture appearance is 50 percent of the shear fracture.

Dynamic Tear Test. The dynamic tear test evolved as a result of the need to evaluate the fracture toughness of materials that are ductile and cannot achieve the plane-strain condition. The test was developed at the Naval Research Laboratory, where Pellini (1971) obtained a correlation between dynamic tear and K_{Ic}. The test is similar to the CVN test, except that the specimen has a sharper pressed notch. Two specimens can be used for testing: One is small (7″ × 1$\frac{5}{8}$″ × $\frac{5}{8}$″); the other is larger (17″ × 4$\frac{3}{4}$″ × 1″). Both have a sharp crack tip induced by pressing with a knife edge, and both are designed to produce the maximum strain condition (see ASTM, Part iv, 1975). The crack in the small specimen is 1.13 inches deep, whereas the crack in the larger one is 1.75 inches deep. Both specimens are subjected to three-point bending and are broken by a load that, similar to the CVN test, strikes on the side opposite the notch.

The complete transition curve obtained from dynamic tear tests is similar to the one obtained from CVN tests. Pellini (1971) proposed correlating the results of dynamic tear tests with K_{Ic} using ratio-analysis diagrams (RADs), which are plots of dynamic tear energy (ft-lb) versus yield strength (σys-ksi) and K_{Ic} values. The RADs show three regions of deformation: plastic, elastic–plastic, and plane-strain. The division of these regions is based on the ratio of $K_{Ic}/\sigma ys$ that relates the critical depth of the crack and the critical thickness of the specimen. RADs are used primarily to compare materials by placing them in either the ductile regime or in the fracture mechanics regime (see Fig. 8-23).

Chemical Analysis

In some failure analyses, particularly those performed as part of a product liability case, a routine analysis of the material is important to deter-

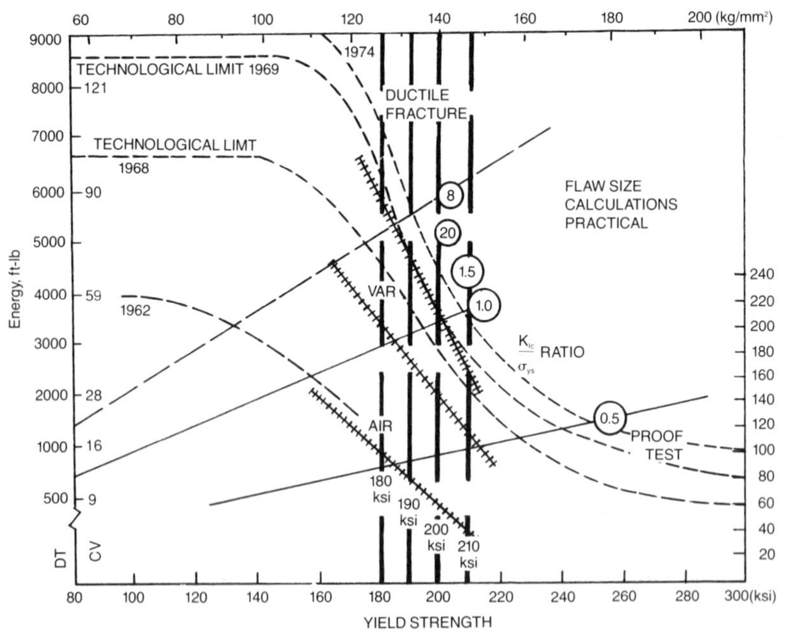

Figure 8-23. Ratio-Analysis Diagram that defines the effects of metallurgical and mechanical parameters that classify ductile and brittle materials. (From Pellini 1973)

mine whether its initial chemical composition met specifications or whether the composition changed as a result of environmental effects. In the case of creep failure of the bolted closure nuts illustrated in Figure 5-8(d), the investigator determined that some nuts did not contain the 9H stamp mark verifying specific chemical components and material properties. In addition, a number of destructive and nondestructive tests revealed that the tensile properties (yield and ultimate stress) of the nuts had declined considerably after long-term exposure to high temperatures. Chemical analysis of the nuts addressed the following questions: Were inappropriate chemicals or alloys used in the manufacture of the unstamped nuts: that is, was quality control at fault? Or were the unstamped nuts essentially the same as the stamped nuts? In other words, analyzing the chemical composition of a material may play an important role not only in identifying the mode of failure but who is responsible for the failure. Improperly specified material would be attributed to the designer's negligence, and improper material supplied by the mill would be the supplier's responsibility. On the other hand, chemical analysis of failed components often reveals that the user has subjected the parts to environments that are more severe than those specified by the manufacturer. This is especially true in instances where corrosion and stress corrosion of steels can have significant negative effects. For example, hydrogen and nitrogen can cause strain aging and quench aging, and hydrogen can induce brittleness.

Techniques for identifying the chemical composition of materials include emission spectrography, spectrophotometry, atomic absorption spectroscopy, x-ray defraction, x-ray fluorescence spectrography, infrared and ultraviolet spectroscopy, electron-microprobe analysis, and wet chemical analysis. Combustion methods are generally used to determine concentrations of carbon, sulfur, nitrogen, hydrogen, and oxygen. The presence and concentration of the anions Cl^-, NO_3, and S^- are usually determined by wet chemical analysis, and crystalline compounds can be identified via x-ray defraction.

Atomic absorption spectroscopy is useful for identifying minor trace elements, and x-ray fluorescent spectrography can be used to analyze a variety of crystal and amorphous solids, liquids, and gases. Infrared and ultraviolet spectroscopy are used to analyze inorganic materials. Dispersive and nondispersive x-ray spectrography are used in conjunction with SEM to permit simultaneous viewing of the specimen during chemical analysis at the surface.

The electron-microprobe analyzer can identify the chemical composition of surface areas as small as 1 micron in diameter. At the other end of the spectrum, spot tests that involve small samples taken in the field or in the laboratory are dissolved in acid or some other solution. The dissolved material is then combined with reagents that indicate the presence of specific substances.

Dynamic Testing (Rate of Strain)

Rates of strain of material on the order of 10^{-6} to 10^{-5} per second and slower are a primary concern in cases of creep. At higher rates of 10^{-4} to 10^{-3} per second, quasi-static data are obtained. In the case of the hot tapping system discussed in Chapter 5, the system failed (fracturing the assembly bolt) because the user subjected it to rapid overload. Thus, different types of testing equipment are required to analyze the failed bolt than have been discussed so far. Some of the considerations involved in dynamic testing are shown in Figure 8-24 (Lindholm 1971). A constant strain-rate test can be performed at strain rates as high as 10^4 per second. Figure 8-25(a) illustrates the stress–strain curves at various rates of strain for mild steel at room temperature, and Figure 8-25(b) illustrates the influence of rate of strain on the tensile properties of mild

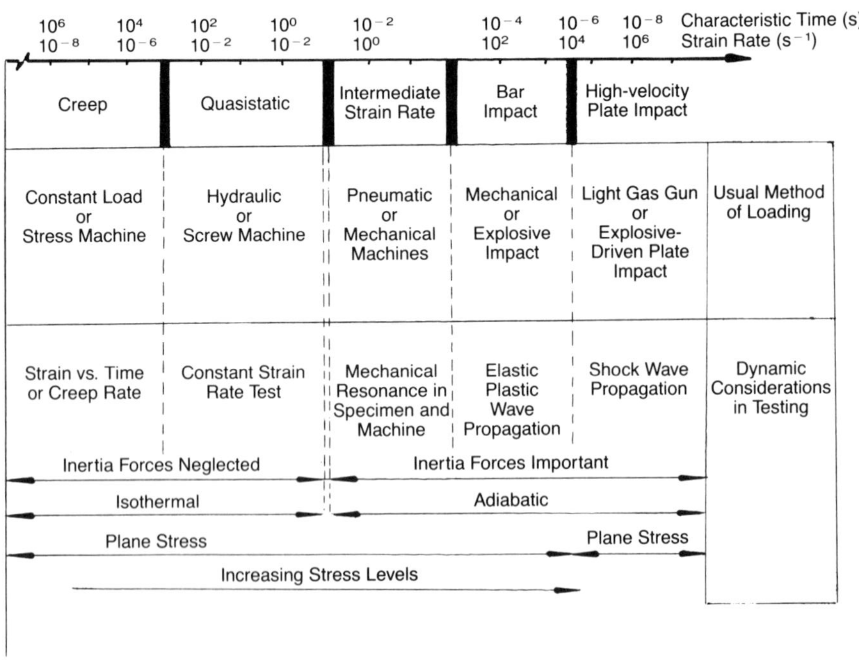

Figure 8-24. Rate of strain on a material versus material regime or characterization usual methods of loading, and dynamic consideration in testing. (From Zukas *Impact Dynamics*, © 1982, John Wiley and Sons.)

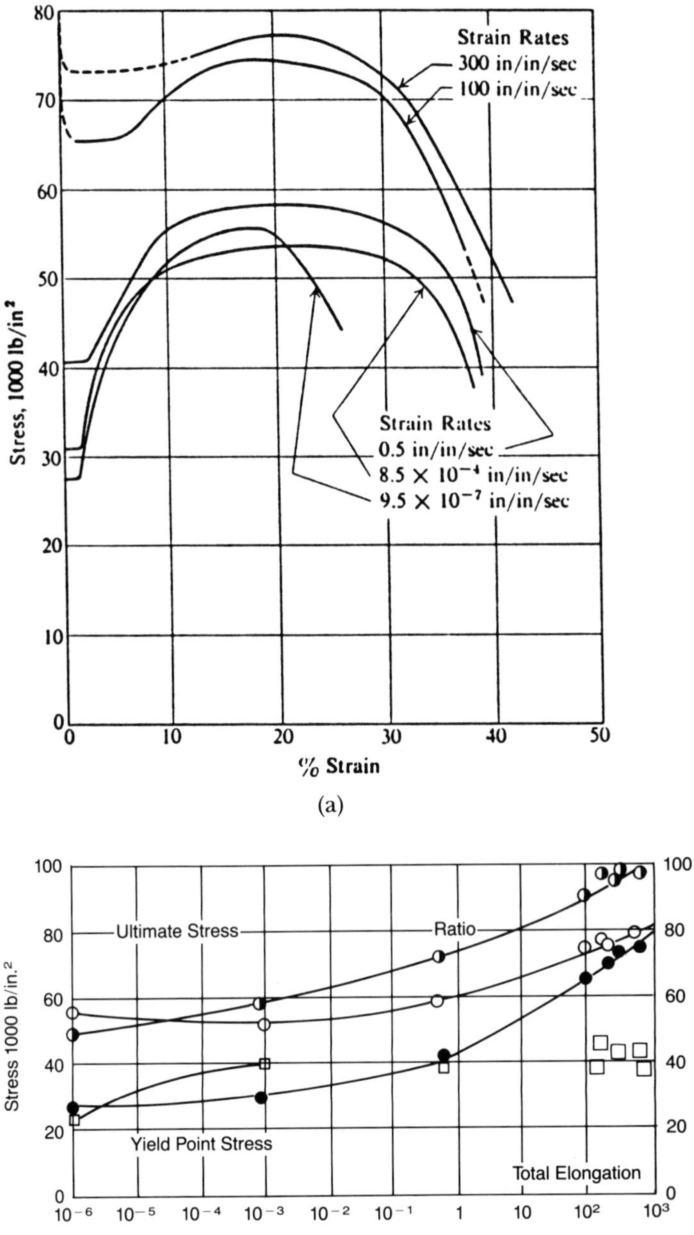

Figure 8-25. Rates of strain. (a) Stress–strain curves of mild steel at room temperature for various rates of strain. (b) Influence of rate of strain on tensile properties of mild steel at room temperature. (From Manjione 1944)

steel at room temperature. Note that the ultimate stress and the yield point increase as the rate of strain increases.

Strain rates of materials on the order of 10^3 per second or more are generally considered to be in the range of high strain rates. At high rates of strain, inertia, rate of crack propagation, thermodynamic considerations, and the transition from isothermal to adiabatic conditions become important. (See Zukas *et al.* 1982 for excellent discussions of impact dynamics and testing techniques used at high rates of strain.)

The principal dynamic testing methods are the uniaxial methods discussed earlier and the well-known and widely used Split–Hopkinson bar test (see Fig. 8-26). In the Split–Hopkinson bar test, the striker bar is accelerated by mechanical or explosive energy. The area affected and its mechanical behavior determine the shape of the reflected and transmitted pulses after impact. The strain time of the traveling pulses is usually determined by strain-gauge stations on the bar that permit one to determine the strain-time history at the end of the bar. This apparatus has been adapted for use in tensile testing, and its use has been extended to other tests such as Charpy impact tests and dynamic three-point tests.

In addition, other tests have been developed to determine the behavior of materials at high rates of strain: for example, the Taylor cylinder, which determines the high yield stress of dynamic material; the

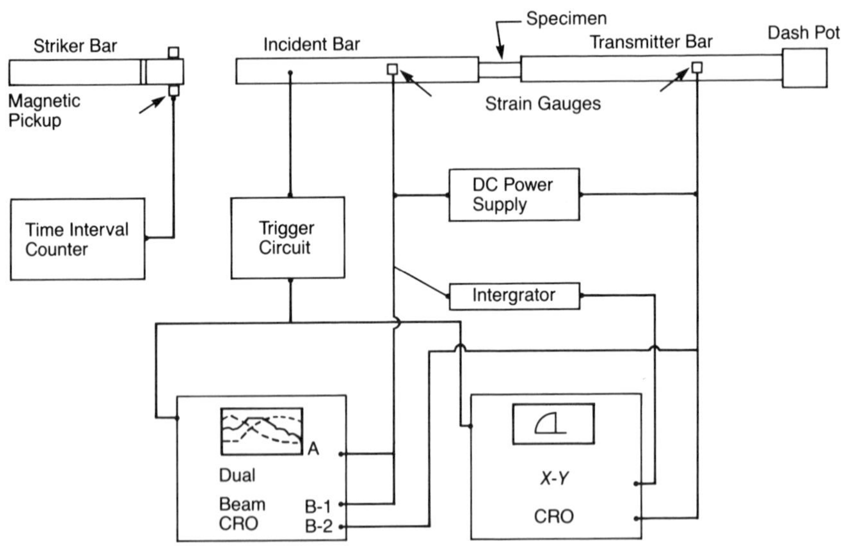

Figure 8-26. Schematic of apparatus and instrumentation for the Split–Hopkinson bar test. (Nicholas, 1982)

expanding ring or cylinder test developed by Clark and Duwez (1950); the dynamic shear test developed by Dowling *et al.* (1970); the drop-forge test; and the dynamic bend test (which can distort the load-time data because of inertia for impact velocities greater than 1.5 meters per second). Nicholas (1982) pointed out that the person doing the testing must know the first mode of failure when using the Charpy impact test because, unlike isotropic materials, anisotropic materials (e.g., composites and some extruded materials) will fail as a result of interlaminar shear.

Because the features of a fracture usually indicate the mode as well as the rate of failure, strain-rate testing in combination with fractography is a valuable tool for assessing a failure in which the rate of failure may be a crucial factor when reconstructing the accident (see, for example, the case involving hot tapping equipment discussed in Chapter 5). Many engineering structures and equipment are subjected to impact velocity during both normal operation and accidents. Examples are protection barriers, shields, and containments for rotating equipment, pressure vessels, pipes, valves, energetic chemicals, vehicles ranging from airplanes to trains, and construction equipment (Brown and Baker; S. J. Brown 1984, 1985a, 1986a, 1987a).

CONCLUSION

The methods of testing materials discussed in this chapter are the same as those used during the design phase to assess the performance of materials and during the processing phase to assure quality control. Some of them have broad applications, whereas others have only limited applications. Some are expensive, and others are relatively inexpensive. We have refrained from discussing cost, however, because techniques that were formerly expensive and difficult to perform have become more reliable, inexpensive, and available because of rapidly changing technology. In addition, a number of new techniques that were not discussed show promise.

Finally, a considerable number of experimental techniques, such as accelerometer, laser, maser, and fiber optics, are used to characterize how structures respond to various loads. These methods are usually oriented toward describing the *global* response of a component, system, or structure rather than its *local* characteristics, which were the subject of this chapter. That is, they record how a product deforms; conducts, both thermally and electrically; and redistributes forces as a function of time during actual or simulated loadings. Test simulations, which are performed on scaled models, prototype, or actual products to measure and predict global responses of the product, will be discussed in Chapter 10. (Excellent technical overviews of local and global testing methods are provided in Society of Experimental Mechanics (SEM) 1988; ASM 1986 (particularly Vol. 8: Mechanical Testing, and Vol. 10: Material Characterization); ASTN Handbook 1988; and specialized ASTM publications treating topics such as creep fracturing, fatigue, and components.)

9

Modes of Failure

In previous chapters, we indicated that engineering components and systems can fail during service for many reasons and that the root cause (responsibility) can be assigned to one or more of the following:

- Inadequacies in the design
- Selection of materials that are inappropriate for the particular operating conditions
- Improper processing or heat treatment of materials
- Defects in the structure of materials that are sometimes introduced during fabrication
- Unforeseen loading conditions, including those arising from changes in the use of the system, and
- Environmentally induced effects resulting from inadequate maintenance and inspection or from time-dependent fracture processes that occur to a greater, and often far more localized, extent than predicted

Again, we want to caution that the cause of failure must not be confused with the mode of failure nor the mechanism of failure.

In Chapter 5, we said that the structural failure of a product is a function of six failure-mode parameters (also referred to as mechanisms of failure) to varying degrees: (1) force displacement or load deflection, (2) time, (3) geometry or dimensions, (4) temperature, (5) chemical or physical interaction with the environment, and (6) the properties or characteristics of the material. If one considers all the possible combinations of these parameters and their subsets, more than 100 modes of failure are possible. In practice, however, the modes of failure are classified according to a predominant effect (usually material), which usually involves one or possibly two failure-mode parameters. These classifications are dictated primarily by engineering analysis considerations—that is, mathematical predictive methods. Box 9-1 provides an illustration of an expanded list of modes of failures that considers a number of combinations of material behavior with other failure-mode parameters.

Good designers attempt to limit the operation of a product to regimes that are most reliably assessed with respect to possible modes of

BOX 9-1. Modes of Failure*

- Yielding or inelastic deformation
- Excessive elastic deformation
- Ductile rupture
- Brittle fracture
- Fatigue
  - High-cycle fatigue
  - Low-cycle fatigue
  - Thermal fatigue
  - Surface fatigue
  - Impact fatigue
  - Fretting fatigue
  - Corrosion fatigue
- Wear
  - Adhesive wear
  - Abrasive wear
  - Corrosion wear
  - Impact wear
  - Fretting wear
- Brinelling
- Impact
  - Impact fracture
  - Impact deformation
  - Impact fretting
- Fretting corrosion
- Corrosion
  - Direct chemical attack
  - Galvanic corrosion
  - Crevice corrosion
  - Pitting corrosion
  - Erosion–corrosion
  - Cavitation corrosion
  - Hydrogen damage
  - Biological corrosion
  - Stress corrosion
- Creep
- Thermal relaxation
- Stress rupture
- Thermal shock
- Galling and seizure
- Spalling
- Radiation damage
- Buckling
- Creep buckling
- Combined creep and fatigue

*From Collins (1981).

failure. Consider the criteria for evaluating failure modes outlined in the Design by Analysis section (III and VIII—Division 2) of the Boiler and Pressure Vessel Code published by the American Society of Mechanical Engineers (ASME):

1. Excessive elastic deformation, including elastic stability
2. Excessive plastic deformation
3. Brittle fracture
4. Stress rupture or creep deformation
5. Plastic instability–incremental collapse
6. High-strain–low-cycle fatigue
7. Stress corrosion
8. Corrosion fatigue

As we pointed out in Chapter 5, failure is associated not only with material fracture or failure but also with the effects of other failure-mode parameters that render a product unusable or unsafe. Two examples of these types of failure modes are excessive deformation and instability (which are associated with failure-mode parameter 3, Table 5-1—excessive changes in the dimensions or **geometry** of the product). Excessive elastic deformation is usually associated with slender geometries, with excessive compressive loads, or both, but it includes all forms of excessive loading in the elastic range. Excessive plastic deformation is associated with a loading history that causes unacceptable permanent distortions resulting from plastic strains.

The methods of measuring the behavior of materials (see Chapter 8) often do not involve intrinsic values, but instead provide relative measures of what is taking place in the material at a local or microscopic level. In recent decades, the relationship between the methodology of testing, prediction, and the microscopic behavior of materials has become closer. This relationship will be discussed in Chapter 10 and, to a limited extent, in this chapter, particularly with respect to fracture mechanics. The focus of this chapter is a discussion of the modes of failure in a metallurgical context.

Although the purpose of structural design codes is to provide a safe design, these codes cannot foresee all factors and events that may arise when a product is in service. Thus, the design engineer must take these factors and events into account at an early stage to minimize the risk that a product will fail while in service. The adequacy of designers' concern about possible modes of failure has increased in recent years for several reasons. First, designs have become more sophisticated as a result of the widespread use of computers and more sophisticated methods of stress analysis. Second, materials tend to be subjected to increasingly severe service conditions, some of which are beyond the range of current experience: for example, this is true of cryogenic containment equipment, high-temperature petrochemical plants, pipelines operating in cold climates, aircraft-engine components, and nuclear-reactor cores. Third, the number and size of containment facilities for hazardous substances has increased; these substances range from radioactive liquids and gases to highly explosive gases such as hydrogen. Finally, many new materials have been developed for which no background of experience is available; examples are composite materials used in aircraft components, polymer containment vessels, and adhesively bonded structural joints. Thus, an appreciation on the designer's and fabricator's part of the possible modes of failure that can and occasionally do occur has become increasingly important.

In the following discussion about the various modes of failure, it is important to remember that one mode of failure may be apparent on the surface of a product, whereas another failure mode may have been responsible for the origin of the failure. For example, a small preexisting defect may have grown by fatigue or some other mechanism over time to the point where loading finally initiated a final fast fracture. Therefore, in reviewing the modes of failure, we will also discuss the mechanisms of crack initiation.

FAILURES ASSOCIATED WITH OVERLOAD

The types of failures associated with overload that most engineers are likely to be familiar with are those resulting from tensile and impact testing of laboratory specimens, in which the entire failure surface is rapidly formed. In most service situations, the complete fracture does not occur rapidly from overloading; in most cases, an overload failure occurs in the remaining material after a crack has developed over time, after repeated loading, or as a result of a preexisting defect. Because the *final* failure is almost always attributable to overloading, it is important to anticipate such failures and produce a design that will reduce or eliminate their occurrence. As will be discussed below, fractures associated with overload can be ductile or brittle in nature.

Definitions of Ductile and Brittle Fractures

Ductile fractures are characterized by extensive plastic deformation during separation of the two halves of the fracture surface, whereas brittle fractures involve only localized deformation, which gives the fracture surfaces the appearance of brittle separation. In an ideal brittle material, separation of atoms takes place along crystallographic planes with no plastic deformation, and the amount of energy dissipated during fracture is only that required to separate the planes of the atoms—that is, the amount needed to create the new surfaces. In practice, some plastic deformation occurs on a localized scale even in the most brittle materials. Thus, the energy required to fracture them is significantly greater than that predicted for an ideal material. However, classifying fractures as ductile or brittle on the basis of their appearance and the *relative* extent of plastic deformation is useful.

Ductile fractures can give rise to surface features ranging from dimples, produced when pores form and link up, to relatively flat surfaces, initiated in cavities produced at grain boundaries during creep failure at high temperature. In general, however, the term *ductile* is used to indicate rapid tensile or shear fracture involving extensive plastic deformation. The surface morphology of fast "brittle" fractures is usually distinctive and easily recognized. With the naked eye, one often can observe chevron marks that point back to the origin of the fracture; with a microscope, cleavage steps and river patterns may be observed.

Most brittle fractures occur at relatively low temperatures, and they are promoted by triaxial tensile stresses, which reduce the level of shear stresses required for plastic flow. Normally, brittle fractures tend to be initiated at defects or in regions where stress is concentrated in localized areas such as notches. For example, brittle fractures have been observed in storage tanks (LeMay 1982) and in pipelines that failed in service (LeMay, Justice, and Jamieson 1984).

The Mechanisms of Ductile and Brittle Fracture

Ductile fractures involve plastic flow, the extent of which can vary greatly. In situations involving limited plastic flow, the crack propagates in a plane normal to the applied tensile stress and, macroscopically, may appear to be brittle. The localized plasticity at the crack tip, however, precludes the formation of cleavage facets and steps.

Although pure ductile metals deformed in tension may neck down to a point, more commonly ductile fractures occur by means of the nucleation of voids, frequently at second-phase particles. Thus, the mechanism of ductile fracture is generally associated with the subsequent growth and coalescence of these particles. The fracture surfaces that result contain dimples that often show evidence that the fracture was initiated in this manner. Figure 9-1 shows a good example of a ductile rupture; inclusions are clearly evident within the voids.

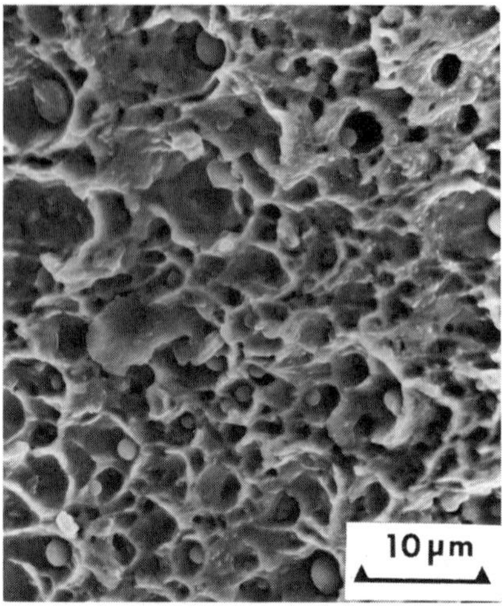

Figure 9-1. Ductile fracture in the welded region of a structural steel as seen in the scanning electron microscope. Note the inclusions within the voids, from which the voids nucleated.

Because of its crystallographic nature, cleavage leads to a fracture surface consisting of flat, shiny facets. The crack changes its orientation slightly when crossing a grain boundary because of the different grain orientations and because the crack propagates on specific families of planes (e.g., on $\{100\}$ in α-iron). Many small parallel cracks with cleavage steps between them may form as the crack crosses a tilt or a high-angle grain boundary. These parallel cracks gradually become joined, with coalescence of the fine steps, to form a larger one. The result is the characteristic river pattern of cleavage fractures shown in Figure 9-2 (LeMay 1981). River patterns are produced even when some degree of plastic deformation occurs adjacent to the crack tip, as in the fracture of ferritic steel at relatively low temperatures. In addition, ductile tear ridges are often formed between cleavage steps, which again increases the quantity of plastic flow required to fracture the material.

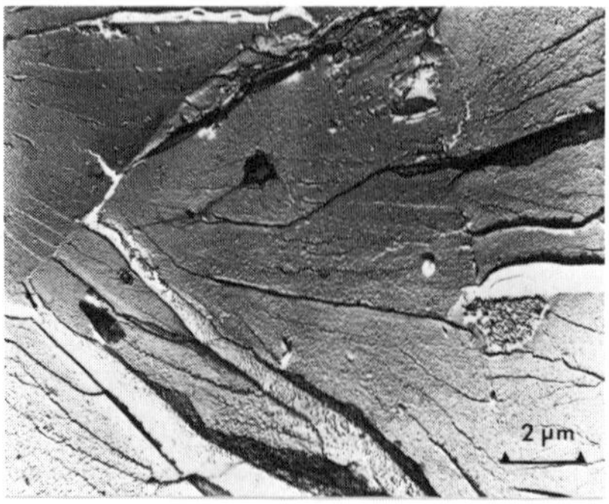

Figure 9-2. A characteristic river pattern on a cleavage fracture surface of a mild steel bolt. The specimen is a two-stage plastic-carbon replica viewed in the transmission electron microscope.

Various detailed mechanisms for cleavage fracture have been proposed, and these depend on initial movements of dislocations that are blocked to produce a high local tensile stress. The ease of initiating cleavage cracks declines with decreasing grain size because of the limited dislocation pileup that is then possible. The fact that initiation of brittle fracture becomes more difficult with finer grain sizes is exploited by using fine-grained steels for pipelines and pressure vessels that will not be subjected to high temperatures. Initiation of cleavage cracks also is facilitated by the presence of second-phase particles such as carbides (E. Smith 1966). To guard against unexpected brittle fracture, both the size and quantity of such particles should be minimized.

Figure 9-3 shows an example of quasi-cleavage in the weldment of a structural steel. In this case, brittle fracture was promoted by a high degree of triaxiality and the rapidity of the overload, which allowed little time for plastic flow.

Fast Fracture from a Preexisting Defect or Crack

Griffith (1920) was the first to recognize the relationship between the work required to cause the growth of a crack from a defect and the energy of the newly created surfaces. He suggested that, for an ideally brittle solid, the reduction in elastic strain energy as the crack grows under the influence of a stress field must be at least equal to the energy of the new surfaces created. Orowan (1949) and Irwin (1948) subsequently modified this equality to account for the work required because

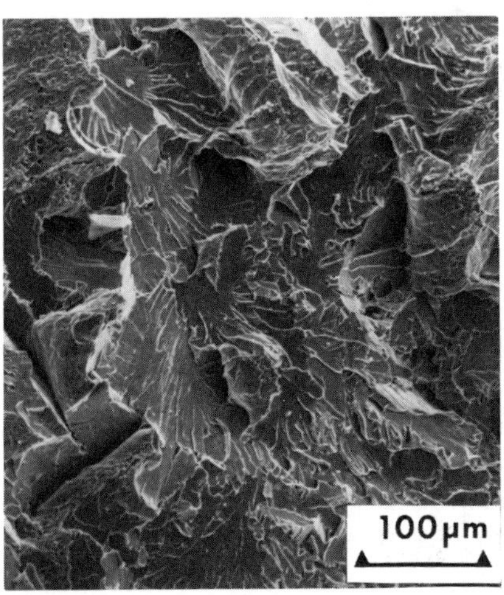

Figure 9-3. Quasi-cleavage in a structural steel viewed in a scanning electron microscope.

of the local plastic deformation that occurs in the region around the crack tip. Nevertheless, Griffith's criterion was the fundamental idea that in due course gave rise to the present-day concepts of fracture mechanics.

Griffith's criterion for the stress to cause failure was

$$\sigma_F \geq \left(\frac{2E\gamma}{\pi a}\right)^{1/2}, \qquad \text{plane stress} \tag{1}$$

or

$$\sigma_F \geq \left[\frac{2E\gamma}{\pi(1 - \nu^2)a}\right]^{1/2}, \qquad \text{plane strain} \tag{2}$$

where σ_F = fracture stress, E = elastic modulus, γ surface energy/unit area, a = half-crack length (for an internal crack of length 2 a), and ν = Poisson's ratio.

Orowan's modification of Equation 1 has the following form:

$$\sigma_F \geq \left[\frac{E(2\gamma + \gamma_p)}{\pi a}\right]^{1/2} \tag{3}$$

where γ_p represents the energy expended in plastic work per unit area of crack surface. Because $\gamma_p \gg \gamma$, Equation 3 can be rewritten as

$$\sigma_F \geq \left(\frac{E\gamma_p}{\pi a}\right)^{1/2} \tag{4}$$

Irwin postulated that a critical amount of energy was required to create a crack with additional surface area and that this quantity could be measured for a specific material in a fracture test. Thus, by comparing the quantity of energy released when a given crack extended under a specified level of stress with the known critical energy requirement, one could predict whether crack growth would occur.

Correlation with Fracture Mechanics

Today, Irwin's (1957) linear-elastic-fracture-mechanics (LEFM) approach is commonly used in design and in failure analysis to evaluate the load that causes final failure. As we showed in Figure 8-16, a crack can be initiated by three basic modes of loading. Because Mode I is by far the most important in engineering applications, it will be the primary focus of this discussion. Mode II and Mode III fractures are discussed elsewhere (LeMay 1981).

In the LEFM approach, a linear elastic distribution of stress and strain fields is assumed to exist around the crack tip, and one can demonstrate that for a given remotely applied stress, σ, the local values of stress and strain depend only on the angle and distance from the crack tip. The magnitude of the stress field at the crack tip depends on the stress intensity factor, K_I, which is defined as $K_I = \sigma\ (\alpha\pi a)^{\frac{1}{2}}$, where α is a geometric factor that equals 1 for a crack of length $2a$ in an infinite plate. As long as K_I does not exceed a critical value, the crack will not propagate in an unstable manner because the energy balance is unfavorable. This critical value is a constant for a particular material under plane-strain conditions (e.g., in a thick section containing a small defect); this value is termed K_{Ic}: the plane-strain fracture toughness.

Figure 9-4 indicates that the critical value of the stress intensity factor will vary, depending on the thickness of the specimen. For thinner specimens, the fracture mode tends to be Mode III; only when the specimen is thick does the plasticity (produced under plane-stress conditions) have a negligible effect, and thus the K_{Ic} fracture criterion fully applies.

The plastic zone at the crack tip increases the work required to cause fracture and the toughness of the material; however, this can readily be accounted for when evaluating K, provided the material is not too ductile and the plastic zone is not too large. In any event, K_{Ic} represents a lower bound for fracture that is safer from the design standpoint.

Tables and formulas are available today to evaluate K_I (and K_{II} and K_{III}, if required) for a wide variety of crack, component, and loading

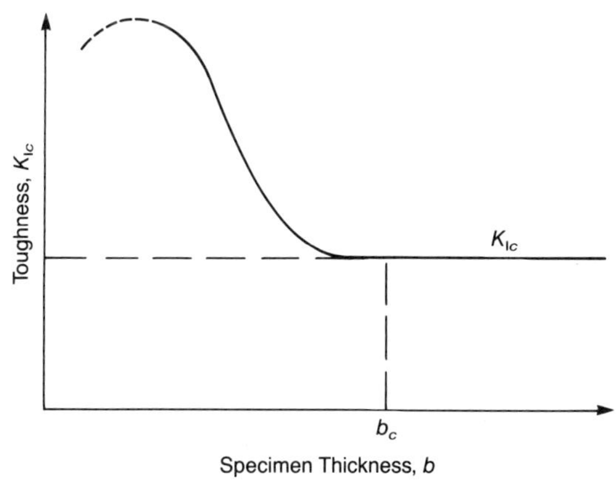

Figure 9-4. Variation of the critical value of the stress intensity factor with the thickness of the specimen.

geometries (Tada, Paris, and Irwin 1976; Rooke and Cartwright 1976). Therefore, K_I can be compared with the known toughness of the material at either the design stage or in later analyses, and the critical level of stress required to cause unstable growth of a given defect can be established. In addition, methods of extending the analysis to more ductile materials with greater degrees of plasticity are now available. One of these methods is covered in a standard of the American Society of Testing and Materials (ASTM 1981).

In most engineering components, the most important potential mode of fracture is plane strain (Mode I) arising from a small defect in a restrained thick section. Shear fracture is important in certain structures such as pipelines: for example, running shear fractures, rather than running brittle fractures, are sometimes observed in pipelines, the fracture mode being dependent on the toughness of the steel (LeMay, Justice, and Jamieson 1984).

Using the concepts of fracture mechanics, one no longer needs to assume (unrealistically) that materials are free of defects. The fracture resistance and safety of engineering components can be assessed, and specifications can be written to limit the size of defects quantitatively on the basis of observed defects and the limits of available nondestructive testing techniques. Similarly, the service life of a structure can be monitored throughout using nondestructive techniques to examine the critically loaded areas. Thus, cracks can be allowed to grow until the value of K approaches K_c, at which time repairs should be undertaken or the component should be replaced. This practice is followed today for components as diverse as aircraft structures, turbine rotors, pressure vessels, and pipelines.

Finally, it should be noted that fracture toughness declines at extremely high (impact) rates of loading and that low temperatures cause a loss of toughness in many structural materials. This emphasizes the need to select materials with properties that are appropriate for the conditions under which a product will operate.

FAILURES ASSOCIATED WITH FATIGUE

Fatigue failure involving the initiation and growth of cracks under repeated loading is probably the most common mechanism responsible for structural failure. In many cases, a diagnosis of fatigue fracture is possible almost immediately if the surface has distinctive "clam-shell" or "beach" marks (see Fig. 8-12), which are produced when the crack either stops growing temporarily because the equipment is shut down or the load level fluctuates markedly. Such markings also indicate the origin of the fracture.

Similarly, the presence of microscopic striations, produced as the crack advances during each load cycle to maximum tension, is an almost certain indicator of fatigue (see Fig. 9-5). In some components, however, distinguishing such features on the fracture surfaces can be difficult, although fatigue was the operative mechanism, either because an environmental factor such as corrosion has damaged the fracture surface or because elevated temperatures have led to progressive ductile rupture during each loading cycle.

Fatigue and Initiation of Cracks

According to the classic view, a fatigue fracture is produced in three stages: crack initiation, crack propagation, and final fracture. In many structural components, the first stage may be ignored at the design phase because preexisting defects such as inclusions or weld defects are already present. The designer can then consider the load cycles required to propagate the initial crack to a length that will cause final, rapid fracture (see the previous section, "Failures Associated with Overload").

The initiation of fatigue cracks in the absence of preexisting defects occurs by means of local slip, leading to the formation of persistent slipbands and extrusions and intrusions on the surface of the material (Klesnil and Lukáš 1980). Although such processes in pure metals have been well documented in the laboratory, it is still impossible to make accurate predictions about the number of cycles required to initiate a microcrack of what may be considered the initial size for subsequent crack-growth-rate relations to be applied.

The initial direction of crack growth from a free surface is along a slipband (Stage I); the direction then changes to one that is roughly

Figure 9-5. Fatigue striations on the fracture surface of AISI 4140 steel (Fe, 0.4% C, 1% Mn, 1% Cr, 0.2% Mo) in the quenched and tempered condition. The specimen is a two-stage plastic–carbon replica in the transmission electron microscope.

normal to the tensile axis (Stage II) in a uniaxially loaded situation (see Fig. 8-14). Most structural components are loaded in a multiaxial manner, however, and recent work has shown how the directions of Stage I and Stage II crack growth can be related to different loading situations (M. W. Brown and Miller 1973; Miller 1985).

Miller (1985) has recently argued, convincingly, that the matter of crack initiation in engineering structures can be ignored because surface discontinuities effectively initiate extremely small cracks from the time the first load is applied. The analysis then becomes one of predicting the extent to which these extremely small cracks must grow before they begin to propagate as a macrocrack or else become nonpropagating. The concept of the *fatigue limit* becomes one of the level of stress below which cracks stop propagating and thus do not lead to failure.

Propagation of Cracks

For an engineering structure, especially one containing welds, there are inevitably sufficiently large defects present such that the analysis of fatigue-crack growth may require only the consideration of long cracks. For the growth of such cracks to continue, the peak stress level during cyclic loads must exceed some limit. The criterion is that K_{max}—the maximum value of the stress intensity factor—must exceed the threshold K_{th} for crack growth to continue. The growth of such cracks occurs because of the local plasticity generated at the crack tip during each load cycle. When the cracks are small—such as surface discontinuities at slip bands, surface roughness steps, and the like—their growth depends on the magnitude of the local plastic strain field produced when each load is applied. Figure 9-6 shows a schematic plot of the growth characteristics of short and long cracks and indicates the behavior of a nonpropagating crack. During Stage II, a long fatigue crack propagates with a small forward movement during each load cycle. During a large part of this growth, the following relation applies (Paris and Erdogan 1963):

$$\frac{da}{dN} = C\,(\Delta K)^m \tag{5}$$

where a = crack length, N = number of load cycles, ΔK = the range of stress intensity factor, and C and m = material constants. Refer to Figure 8-16 for an illustration of the growth curve for a long fatigue crack. Throughout this period of growth, surface steps or fatigue striations are commonly observed on the fracture surface at right angles to the direction of growth, and the growth rate can later be determined by measuring the spaces between these striations with a scanning electron micro-

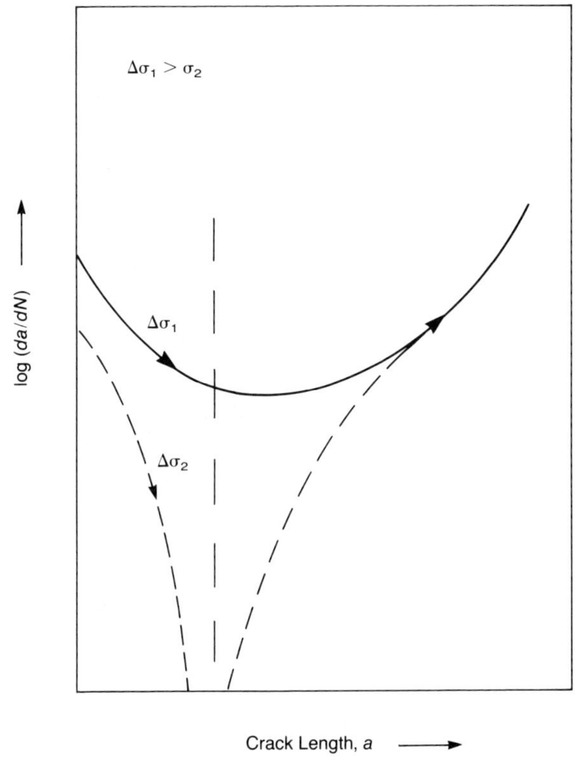

Figure 9-6. A schematic plot of the growth characteristics of short and long cracks. For the stress range $\Delta\sigma$, the crack grows from the short-crack regime to the long-crack regime after initial deceleration. For $\Delta\sigma_2$, the short crack becomes nonpropagating because ΔK never reaches the threshold condition, ΔK_{th}.

scope either directly or using a replica or with transmission electron microscopy and a surface replica.

Figure 9-7 illustrates the regimes for the growth of long and short cracks. The incline of the sloping line is $-\frac{1}{2}$ on the log-log scale because $K\alpha(a)^{\frac{1}{2}}$. At some lower limiting value, the surface roughness becomes unimportant as a starter of small fatigue-crack growth. This limit appears to be on the order of 2 μm (Miller 1985). Again, note that the fatigue limit shown in the figure designates the level of stress below which small cracks will begin to grow but become nonpropagating. The limit does not represent a level of stress below which cracks will not be initiated.

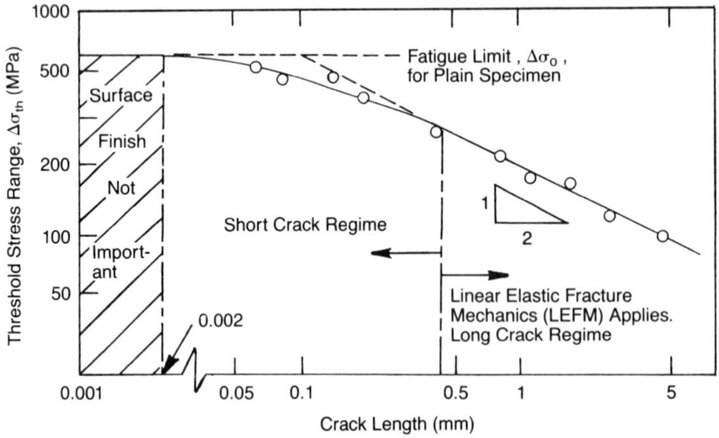

Figure 9-7. The regimes for the growth of long and short cracks as a function of crack length. The superimposed experimental data shown for threshold stress range are from Kitagawa and Takahashi (1976).

High-Cycle and Low-Cycle Fatigue

The only real distinction between high-cycle fatigue (HCF) and low-cycle fatigue (LCF) relates to the number of load cycles required for failure. In low-cycle fatigue, failure normally occurs after fewer than 50,000 cycles, but may occur after only a few hundred cycles. Instead of being load or stress driven, low-cycle fatigue is generally strain controlled; the damage often occurs at some local discontinuity such as a nozzle, perforation, or notch in an otherwise elastically loaded structure. Like high-cycle fatigue, low-cycle fatigue is characterized by striations on the fracture surface. However, because of the larger stresses involved and the extent of the crack growth per cycle, together with the presence of high mean stresses, the fracture surface may have the appearance of void coalescence, with the crack front moving forward in a series of ductile-rupture steps. The distinction between high- and low-cycle fatigue is more of a concern in design because, in the case of low-cycle fatigue, the Coffin–Manson relationships between the number of stress cycles or reversals to failure and the amplitude of the cyclic plastic strain can be used to predict life (Coffin 1954; Manson 1953). Refer to Appendix B for an illustration.

ENVIRONMENTALLY ASSISTED FRACTURES

Environmentally assisted fracture is a large and complex subject. In a number of cases, a combination of stress and environmental factors leads to fracture. These factors include stress corrosion cracking, corrosion fatigue, and fretting and fractures involving elevated temperatures, hydrogen, or irradiation. The first three factors are discussed below; the others will be dealt with in later sections.

Stress Corrosion Cracking

Stress corrosion cracking is an important mode of failure in many components used in the chemical and petrochemical industries as well as in areas such as marine applications. This mode is a serious problem in pipelines. It occurs externally through the interaction between the metal with the soil, in some cases because the protective coating holidays or breaks down, and it occurs internally because a corrosive chemical such as hydrogen sulfide is present. The incidence of stress corrosion cracking is minimized by avoiding crevices where high concentrations of aggressive ions can build up and local stresses are likely to be high.

Although stress corrosion cracking can occur by several mechanisms, a simple model will serve as the basis for discussion (see Fig. 9-8).

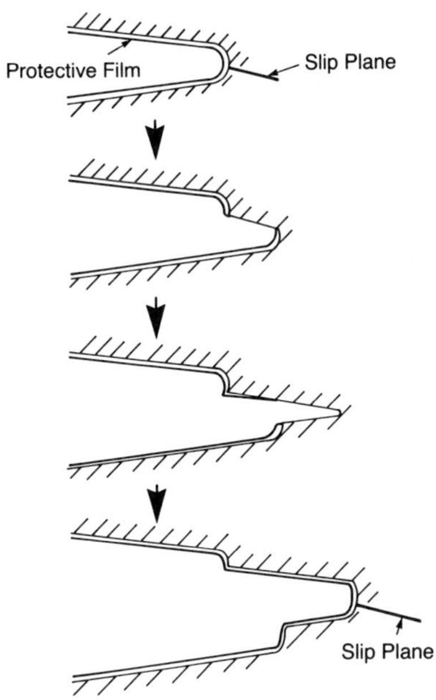

Figure 9-8. A schematic illustration of the mechanism of stress corrosion cracking. The crack is propagated by rupture of the protective film on plastic deformation occurring at the crack tip, subsequent corrosion, repassivation, and buildup of stress to repeat the cycle.

When the protective film or passivated layer on a base metal ruptures, corrosion occurs at the tip of the crack as a consequence of slip that takes place at the tip. The corroding metal becomes passivated or otherwise protected in its turn, and the process is repeated, with the crack progressing in a stepwise manner. The path of the crack can be intergranular or transgranular, depending on the material and the environmental conditions, whereas the fracture is characterized by little ductility, and an oxide coating is often seen on the crack surfaces. Figure 9-9 shows a typical cross section through a region of stress corrosion cracking. The morphology of the fracture surface can vary widely and may include striations as well as linear features running parallel to the direction of crack propagation (LeMay 1978).

In some combinations of metal and environment, an additional mechanism of stress corrosion cracking is believed to involve the evolution of hydrogen at the crack tip. Hydrogen may be released during corrosion in an aqueous medium, or it may arise from dissociation of water vapor on newly exposed metal at the tip. In either case, the hydrogen may diffuse into the metal ahead of the crack, causing local embrittlement. It should be noted that the environmental conditions at the tip of the crack are more important than those of the bulk fluid (which is located away from the structure; e.g., seawater versus seawater in the crack of a propeller) because the pH and solute concentration at the crack tip may differ greatly from the bulk fluid.

The importance of the fracture toughness, K_{Ic}, of a material was explained earlier. In the absence of a synergistic effect between stress and environment, as in stress corrosion cracking, failure under static loading will not occur when the combination of stress and defect size gives a rise to an initial value of a stress-intensity factor, K_{Ii}, that is much lower than K_{Ic}. In a corrosive environment, however, subcritical crack growth may take place until the value of K approaches K_{Ic}, at which time

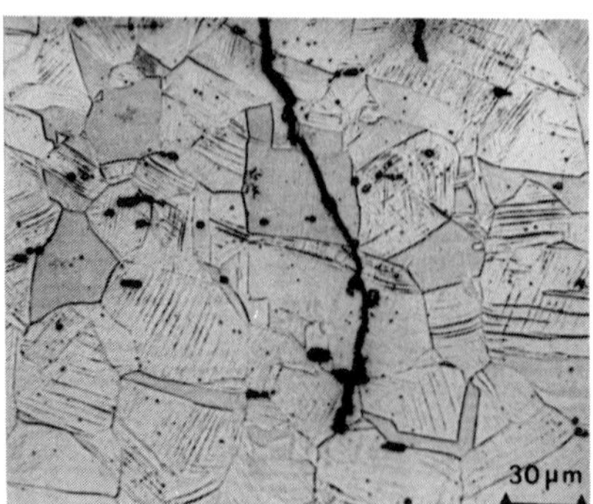

Figure 9-9. Stress corrosion cracking in an AISI 316 austenitic stainless steel (Fe, 0.08% C, 17% Cr, 12% Ni). Etched in diluted aqua regia.

catastrophic failure occurs. A threshold value of K called K_{Iscc}, below which crack growth does not occur, appears to exist. But this value is not unique for a particular combination of material and environment; it is dependent on the test time and the geometry of the specimen. Although some values of K_{Iscc} are available as a guide, using them to ensure the safety of a design and to avoid crack growth caused by stress corrosion cracking in most combinations of material and environment where such cracking is a factor is extremely difficult. Therefore, periodic monitoring of such situations is important to ensure that existing cracks have not reached a length that is likely to result in failure or leakage (critical length).

Corrosion Fatigue

A corrosive environment can greatly reduce the resistance of a material to cyclic loading. The corrosive fluid may assist the rapid initiation and growth of short cracks and significantly reduce the time required to reach the long-crack phase. The fluid may attack the freshly exposed metal at the crack tip upon loading, or water vapor may dissociate at the crack tip, causing penetration of hydrogen and embrittlement ahead of the crack, which again increases the growth rate.

When materials are tested in a corrosive environment, they usually have no definite fatigue limit, as is the case when they are tested in air. As we noted earlier, the current definition of a fatigue limit is the level of cyclic stress that is insufficient to maintain crack growth. In the presence of a corrosive medium, this condition is apparently not met, and the cracks continue to grow.

Because fracture surfaces caused by corrosion fatigue are commonly devoid of striations and also may exhibit brittle facets, distinguishing them from fractures caused by stress corrosion cracking can be difficult. When attempting to determine the nature of the failure, the service conditions can be an important guide. It also is difficult to predict the rates of crack growth without conducting tests on the specific combination of material and environment. The matter is further complicated by the fact that the rate of crack growth depends on loading frequency and waveform (LeMay 1981).

Fretting

Fretting involves relative oscillatory motion between contacting surfaces under the influence of vibration of a cyclic stress. The amplitude of motion is on the order of 25 μm or less. Surface debris consisting of small metallic platelets covered in oxide is produced; eventually, the layer may consist entirely of oxide. When cyclic stressing of one of the contacting surfaces gives rise to the fretting action, fatigue cracks can be produced. Fretting can affect couplings, mountings for machinery, and other surfaces that are bolted or clamped together. Tightly fitting pins and rivets also may have enough relative movement to initiate fretting and fatigue in the component in which they are set. The susceptibility to fretting can be reduced by applying a good film of lubricant and by applying a substance such as a water-displacing compound to inhibit moisture or other corrosive elements from entering the areas in contact. (See Waterhouse [1979] for a detailed discussion of the fretting phenomenon.)

FAILURES ASSOCIATED WITH CORROSION AND EROSION

The concern here is with the mechanisms of corrosion that may lead directly to failure. Situations in which corrosion acts in conjunction with other factors to promote failure will be discussed separately. Erosion–corrosion will be covered later in this section, and hydrogen-assisted cracking will be covered in a later section. General corrosion will not discussed further because the attack is, by definition, relatively uniform and easily recognized.

Corrosion

Pitting corrosion, or localized attack, eventually leading to perforation occurs when one area of a metal surface becomes anodic to the remaining areas or when localized changes in the concentration of the corrosive material cause the attack to accelerate. The depth of the pits is easily observed in a failed component, but determining their depth by visual inspection in a component in service usually requires ultrasonic means. Because of their local stress-concentrating effect, pits can provide the nucleus for stress corrosion cracking or corrosion fatigue.

Crevice corrosion can occur at a joint between metallic surfaces, between a metal and a nonmetal, or under a deposit on a metallic surface. All of these can lead to a concentration cell and consequent local attack. In heat exchangers, for example, the junction between tube and tube plate is a favored place for crevice corrosion. Other examples include breaks in mill scale, under deposits in boilers if correct cleaning procedures are not followed, and at breaks or holidays in the linings or protective coatings of vessels used to process or store chemicals.

Selective leaching involving the dissolution of one element in an alloy by corrosion can affect a number of alloys: for example, alpha brass, from which zinc tends to be lost in the presence of water or salt solutions, and alloys containing iron and chromium, which can become depleted in high-temperature oxidizing environments. Sectioning and metallographic examination will reveal the porous structure of the depleted material.

Intragranular attack involving preferential dissolution of grain-boundary phases or adjacent zones can occur in several alloys: for example, austenitic stainless steels in a sensitized condition and some nickel-based alloys in extremely hot water containing low concentrations of chloride or other ions. Again, metallographic sectioning will reveal the nature of the problem.

Other mechanisms of corrosion that are matters of concern in an examination of failures involve the formation of different types of electrochemical cells leading to localized attack. Examples are galvanic cells, where dissimilar metals are in contact in the presence of an electrolyte and the anodic metal is exposed to accelerated attack (see Table 9-1); differential concentration cells, which are commonly found on metals buried in soils having different degrees of aeration or moisture content; and differential temperature cells, in which the metal varies in temperature between the anode and the cathode.

Erosion

Erosion occurs for two principal reasons: (1) wear caused by the direct impact of particulate matter, and (2) cavitation caused by contact with liquids, which involves the formation and collapse of bubbles within the liquid. Collapse of the bubbles produces local stresses on the surface attributable to the shock waves produced during collapse (Preece 1979). Impellers, pumps, nozzles, and sharp discontinuities in the flow path of a fluid can produce cavitation and subsequent erosion (see Fig. 10-11).

Face-centered cubic metals fail as a result of cavitation erosion either by means of void growth and coalescence or by ductile rupture (Preece 1979). Body-centered cubic iron can fail by brittle fracture or by ductile tearing (the mechanism depends on the local strain rate). Generally, a scanning electron microscope is required to study the morphology of failures associated with cavitation.

Many industries, but particularly mining and other earth-moving industries, are concerned about erosion by solid particles. The hardness of a material is a critical requirement for resistance to erosion by abrading particles. To significantly improve the resistance of a material to abrasion, its hardness should be at least half that of the abrasive medium (Richardson 1968). Also, because resistance to the initiation and propagation of cracks is required for good resistance to abrasion, the toughness of a material becomes important beyond a certain level of hardness.

TABLE 9-1. Galvanic Series of Several Commercial Metals and Alloys in Seawater

↑	Platinum
	Gold
Noble or	Graphite
cathodic	Titanium
(protected	Silver
end)	[Chlorimet 3 (62 Ni, 18 Cr, 18 Mo) Hastelloy C (62 Ni, 17 Cr, 15 Mo)]
	[18-8 Mo stainless steel (passive) 18-8 stainless steel (passive) Chromium stainless steel 11–30% Cr (passive)]
	[Inconel (passive)(80 Ni, 13 Cr, 7 Fe) Nickel (passive)]
	Silver solder
	[Monel (70 Ni, 30 Cu) Cupronickels (60-90 Cu, 40-10 Ni) Bronzes (Cu-Sn) Copper Brasses (Cu–Zn)]
	[Chlorimet 2 (66 Ni, 32 Mo, 1 Fe) Hastelloy B (60 Ni, 30 Mo, 6 Fe, 1 Mn)]
	[Inconel (active) Nickel (active)]
	Tin
	Lead
	Lead–tin solders
	[18-8 Mo stainless steel (active) 18-8 stainless steel (active)]
	Ni-Resist (high Ni cast iron)
	Chromium stainless steel, 13% Cr (active)
	[Cast iron Steel or iron]
Active or	2024 aluminum (4.5 Cu, 1.5 Mg, 0.6 Mn)
anodic	Cadmium
(corroded	Commercially pure aluminum (1100)
end)	Zinc
↓	Magnesium and magnesium alloys

Erosion–Corrosion

When general or uniform corrosion takes place, a protective layer of corrosion product often builds up on the surface and reduces the rate of attack. As the velocity of the corrosive fluid passing over the surface increases to the turbulent regime, however, the protective film may be stripped off in localized areas, thus increasing the rate of corrosion in these areas.

Figure 9-10 schematically illustrates the effect of stream velocity on the erosion–corrosion rate of copper–nickel alloys in oxygenated solutions containing chloride. In Zone A the supply of oxygen is poor, but a protective oxide film is formed, the flow is laminar, and the shear stresses are not great enough to disrupt the oxide. In Zone B the flow is partially turbulent, but the oxide remains protective. If the supply of oxygen increases, however, the cathodic reaction takes place at the surface and a decreased concentration polarization results. In Zone C the turbulence increases and the oxide is removed in some areas by the large shear stresses. These newly exposed areas are severely attacked because of the galvanic activity between them and the oxide, which results in pitting. If the kinetics of repassivation are favorable, however, the pits eventually heal and other areas are attacked. In Zone D the removal of oxide is more general, galvanic effects are reduced, and both anodic and cathodic reactions proceed rapidly at the surface. The degree of turbulence and the shear stresses are high, and repassivation is difficult. Finally, in Zone E the removal of oxide is complete, repassivation is impos-

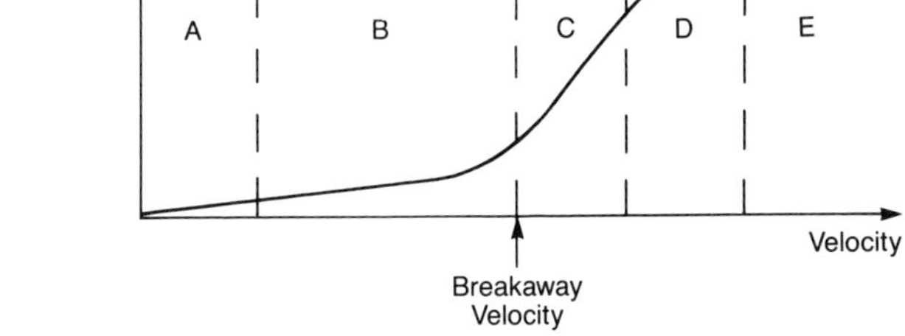

Figure 9-10. The effect of stream velocity on the erosion–corrosion rate of copper–nickel alloys. The different regions are identified in the text.

sible, and both the anodic and cathodic reactions proceed even more rapidly.

A metal surface that has suffered erosion–corrosion has a distinctive sculpted or carved appearance and the object weighs considerably less than it did originally. Although erosion–corrosion is observed in tubes and other components of liquid-carrying installations where the velocity of the liquid is high, the problem is much more severe in slurry-handling pipelines and vessels, where the solid particles scour off the protective film more easily.

FAILURES ASSOCIATED WITH HIGH TEMPERATURES

Creep Deformation

Components that operate at high temperatures are subject to time-dependent deformation, or creep, under stress. The temperature at which creep becomes important is roughly 0.4 to 0.5 times the melting point of the material on the absolute scale (Tm), but this value varies for different alloys. If the deformation becomes excessive, the component may fail without fracturing. Thus, it must be removed before the risk of rupture arises.

The safe life of many components continues throughout the steady-state or secondary stage of creep (see Chapter 8), and beyond to a limited degree. In the secondary stage, a minimum amount of creep occurs, indicating that the material is stable and that no changes in its substructure are taking place. Rarely, however, is much tertiary-stage deformation allowable because of the uncertainty of strain measurements in service and the difficulty of assessing the amount of safe life remaining. Although distinct bulging may be observed occasionally on overheated superheater tubes, for example, intentionally allowing this situation to continue would pose unacceptable risks. Many groups are making a serious effort to assess damage associated with creep and estimate the remaining safe life in steam generators, petrochemical plants, and other large units that operate at high temperatures. (For a review of this phenomenon, see LeMay and da Silveira 1984.)

Creep Fracture

The tertiary stage of creep (accelerating deformation) is associated with the nucleation and growth of voids on grain boundaries. The growth of voids is promoted by stress and temperature. Therefore, components that have a relatively long service life at high temperatures tend to develop intercrystalline cracks, whereas those operating at lower temperatures but subjected to high degrees of stress tend to display transcrystalline cracks and subsequent transcrystalline fractures. Generally, components in which creep is a significant factor at high temperatures undergo intercrystalline cracking.

Effects of Environment on Creep

The gaseous environment in which creep takes place can affect the creep behavior and the rupture life. The strength of a material may be increased or decreased over that found for tests in air or in a vacuum, but of particular concern are environments that may cause severe loss of creep strength. This can occur in superalloys when exposed to hydrogen or oxygen under specific conditions (Woodford 1980). Also, in some pressure components operating at high temperatures in catalytic cracking units, for example, severe carburization can result, which greatly reduces ductility and promotes the formation of voids. Predicting creep behavior and design lives for components that operate under adverse conditions is difficult unless pilot tests are conducted; in addition, systematic checks of the components during their operating life are needed to assess damage.

Effects of Precipitation and Metallurgical Factors

One possible effect of long-term creep behavior after an extended period at high temperatures is precipitation on grain boundaries. In the past, such effects have led to premature failures in austenitic stainless-steel superheater tubes. This particular problem no longer exists, but one must watch for similar effects when newer alloys are put into service. Because weld regions are particularly susceptible to such effects, the need for a sufficient amount of delta ferrite in austenitic welds relates directly to such potential problems. Thus, when examining creep failures, it is important to determine whether precipitates are apparent on the fracture surfaces and, if so, what role they may have played in the fracturing process.

FAILURES ASSOCIATED WITH IRRADIATION

Embrittlement

Nuclear-reactor components are at risk for radiation damage and consequent *embrittlement*. In general, for significant embrittlement to occur, the neutron fluence must be large, as it is in the reactor core. The chemistry of steels used in pressure vessels and reactor cores is strictly controlled to obviate potential embrittlement and loss of toughness (ASTM 1977; U.S. Nuclear Regulatory Commission 1975). Embrittlement of components is of greater concern under fast reactor conditions; however, evaluating fracture toughness after extended service and in the event of unplanned operating incidents is an appropriate measure. Surveillance programs are currently in place to undertake this evaluation (Steele 1984).

Radiation Damage

Radiation damage is associated with the production of vacancy–interstitial pairs and thermal spikes from the passage of fast neutrons and from transmutation damage caused by the absorption of thermal neutrons. Passage of fast neutrons may be a direct cause of embrittlement, whereas absorption of thermal neutrons may release alpha particles or protons that cause the formation of a displacement spike and the production of damaging helium gas within the lattice. Again, the result is an embrittled structure. Elevated temperatures can provide simultaneous annealing and damage removal in particular cases.

Under fast-flux conditions such as those in fast breeder reactors, voids can occur through the formation of vacancies and interstitials that migrate to sinks. The interstitials migrate more rapidly to dislocations, and the excess vacancies agglomerate as voids. In stainless steels subjected to temperatures above 650°C. and the irradiation conditions found in thermal reactors, bubbles of helium gas are precipitated throughout the material, but preferentially on grain boundaries (LeMay 1981). The effect of these reactions is swelling and high-temperature embrittlement.

Another effect of radiation damage is an increased rate of creep at elevated temperatures. This, together with the formation of embrittling precipitates, has been observed in pressure tubes made of a zirconium alloys that are used in the reactor core.

HYDROGEN-RELATED FAILURES

Despite extensive studies over the years, there is still much that is not understood concerning the effects of hydrogen on the deformation behavior and fracture of pure metals and engineering alloys. Apart from the role that hydrogen plays in stress corrosion cracking and corrosion fatigue, questions remain concerning the mechanism of hydrogen embrittlement and its assistance in initiating cracks. In addition, there are many unanswered questions about how to transport and store gaseous hydrogen safely, particularly if this gas becomes more widely used as a fuel in some countries, which seems likely. Thus, hydrogen-related failures and ways of preventing them are topics that merit serious attention.

Hydrogen-Assisted Cracking

The fact that hydrogen can reduce the fracture resistance of steels and affect their plastic deformation is well known. In their decohesion model, Troiano (1960) and Oriani (1972) envisaged the accumulation of hydrogen as weakening the atomic bonds, and Beachem's model (1972) suggested that hydrogen aids the deformation process. Beachem noted that various fracture modes may operate in the presence of hydrogen, including intergranular fracture, quasi-cleavage, and microvoid coalescence. Thus, not only embrittlement but ductile fracture could result as well. Lynch (1979), who further developed Beachem's model, proposed that hydrogen atoms (and, in a similar manner, liquid-metal ions) adsorbed at a surface facilitate slip through nucleation of dislocation movement. As Oriani (1981) noted, although both models are consistent with observations, neither one has been proved.

Even brief exposure to hydrogen helps to initiate cracks. Hydrogen-assisted cracking can occur in many steel components, such as pipelines in the presence of carbonate solutions or hydrogen sulfide, both of which contribute to reactions when hydrogen is released. The fractures produced in this way are normally transgranular, depending on the microstructure and the level of stress present. Although cleavage facets are often seen on such fracture surfaces, Lynch (1979) and others noted that small dimples are present on apparently cleaved areas in specimens fractured in the presence of hydrogen. As we mentioned earlier, stress corrosion cracking and corrosion fatigue in aqueous solutions are believed to involve hydrogen, but the exact mechanism is still in doubt.

The surfaces of most fractures caused by hydrogen are clean and free of corrosion products. In some cases, however, the hydrogen solution may contaminate and react with the fracture surfaces after the hydrogen-assisted crack has developed.

Hydrogen Embrittlement

The fact that embrittlement and premature failure occur in steel containment vessels has long been known (Smialowski 1962). Avoiding these problems requires adequate precautions during design, selection of materials, fabrication, and inspection. One method of avoiding such problems, albeit an indirect one, is to use carbon-fiber-reinforced plastics in composite vessels. Some investigations of this have been conducted, particularly in the fabrication of liner mantles (Tronsberg 1982). The traditional method of fabricating light-weight, high-strength pressure vessels involves cold forming, which increases the susceptibility of the metal to hydrogen embrittlement (Clugston, Weertman, and Shewmon (1983). Therefore, either this method should be avoided or the material should subsequently be annealed to eliminate the effects of cold working.

The potential failure modes for storage or containment vessels are not limited to embrittlement and consequent low fracture toughness. Subcritical crack growth, delayed fracture, and reduced resistance to fatigue (Silva, Pascual, and de Miranda 1984) are all promoted by exposure to hydrogen.

Petrochemical Industry Problems

Many reports of hydrogen-related failures relate to petrochemical industry applications. The problems caused by the presence of hydrogen sulfide in pipelines have already been mentioned, but hydrogen-assisted cracking can occur preferentially in sour gas pipelines and in oil pipelines carrying high-sulfyr crude (LeMay, Justice, and Jamieson 1984). Cracks are initiated by hydrogen buildup in regions containing nonmetallic inclusions, and one often sees stepwise cracking along aligned inclusions. The problem is mitigated by controlling the shape of inclusions when making the steel. The first incident of such stepwise cracking was reported when crude oil leaked from an underwater pipeline in the Persian Gulf in 1974. Leaking girth welds in pipelines are often associated with the use of the cellulosic coatings on electrodes and with hydrogen pickup in a susceptible heat-affected zone (Justice 1983). Figure 9-11 shows schematic examples of blistering and cracking in a steel plate caused by hydrogen damage.

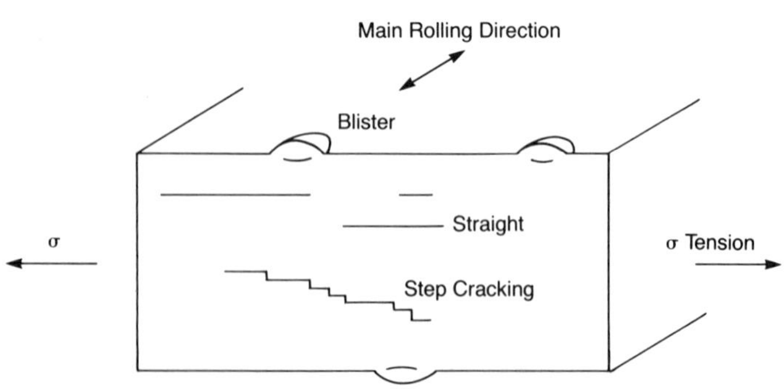

Figure 9-11. Schematic view of features found in hydrogen-embrittled steel.

Another major concern in the petrochemical industry is the possibility of hydrogen damage in high-temperature, high-pressure hydrogenation reactor vessels. Because most of the hydrogen currently produced comes from petroleum and natural-gas feed stock, the prevention of fractures is extremely important.

Watanabe (1984) discussed the failure modes and the methodology for preventing fracture in hydrogenation reactors. The $2\frac{1}{4}$ chromium, 1 molybdenum steel commonly used for such vessels is resistant to the formation of methane within it because of the stable carbide formers chromium and molybdenum. Hydrogen-assisted cracking can occur

BOX 9-2. Mechanisms of Failure and Countermeasures in Hydrocracking Reactors*

Problem	*Mechanism*	*Countermeasures*
Initiation of cracks	Fabrication defects Low-cycle fatigue cracking from stress raisers	Improve quality control during fabrication Eliminate or reduce stress raisers Reduce levels of stress
Subcritical growth of cracks	Hydrogen-assisted crack growth	Maintain safe levels of tensile strength Maintain safe levels of hydrogen Reduce the stress-intensity factor Reduce temper embrittlement
Fracture	Brittle fracture from defects	Improve toughness by suppressing temper embrittlement and increase the temperature of the metal Monitor dangerous defects by in-service inspection

*From Watanabe (1984).

upon cooling, however, because the hydrogen has entered the steel at high temperatures (~846°F). The mechanisms of failure in vessels that have suffered hydrogen-assisted cracking and the various countermeasures are summarized in Box 9-2.

The problem of temper embrittlement that arises during extended service is an especially important one. The risk of fracture can be reduced in older steels by warm start-up and shutdown. Newer steels, including vacuum carbon deoxidized ones, however, have a much reduced shift in their ductile–brittle transition curves due to temper embrittlement (Watanabe 1984). This reduces the risks of hydrogen-assisted fracture, but well-planned in-service nondestructive testing is required to maintain safe operation over the life of the plant.

FAILURES OF NONMETALLIC MATERIALS

Modes of Failure in Polymers

Failure modes and fractography have been studied to a much lesser extent for polymers than for metals. As the use of polymers in engineering applications has increased, however, a substantial amount of information in this area has accumulated. Thermoplastics are the subject of the majority of published work because of their widespread application.

All the fracture modes described for metals also apply to polymers; however, *crazes* form in many polymers, particularly in those with an amorphous structure. A craze represents a preliminary stage in the formation of a crack (Frank 1981); it is an area of fine cracks and voids bridged by extended molecules. Brittle fracture surfaces of amorphous plastics may display Wallner lines—fine parallel cleavage steps that intersect each other—similar to those seen on the fracture surfaces of glass and other brittle solids. These lines are caused by the interaction of a propagating crack front and elastic stress waves in the material (Boyer, American Society for Metals 1975).

At high loading rates, as in impact loading, polymeric materials usually behave in a brittle manner; the first cavities exceed the local critical strain for fracture, and the crack progresses rapidly because it is not arrested by plastic deformation. At more normal rates of deformation, such as that occurring in components that are overloaded in service, the microcracks that develop act as notches, plastic deformation takes place in front of the cracks, and shearing deformation occurs. Additional deformation leads to slip, necking, and rupture. At low loading rates, creep deformation takes place, with material diffusing through the microcracks. Extended regions gradually weaken and break (Frank 1981). Lüder's lines also are visible on the surfaces of many failed polymeric components.

In some polymers such as amorphous polyvinyl chloride (PVC) and polyethylene, a shiny mirrorlike region can be observed at the origin of the fracture. This phenomenon occurs in polyethylene under slow (creep) conditions and consists of many shallow, uniform hollows.

One important difference between the fracture surfaces on polymers and metals is observed in studies of fatigue fractures. In metals, one load cycle produces one striation on the fracture surface; in polymers, this is not always the case. In PVC the fatigue-fracture surface contains a series of crack arrest lines corresponding to the advance of the crack in sudden increments, each of which occurs after several hundred loading cycles (Hertzberg, Skibo, and Manson 1979).

Similar features have been observed on the fatigue-fracture surfaces of a number of other polymers such as polycarbonate, polystyrene, and polymethyl methacrylate (PMMA), when tested under conditions of low ΔK. At higher ΔK values, fatigue striations are observed (Hertzberg, Skibo, and Manson 1979). The reason for this intermittent crack growth is a craze that develops ahead of the crack during cyclic loading; the crack then breaks through the craze, and the process is repeated at intervals of hundreds or even thousands of cycles. Consequently, it is difficult to predict the service values of ΔK that the failed polymeric component experienced by observing the striations on the fracture surface. Good data relating ΔK to surface features and the rate of crack growth are required for the specific polymeric material before such an analysis can be made.

Failure of Composite Materials

The increasing application of fiber-reinforced composite materials means that designers must appreciate the load-bearing capacity, limitations, and failure modes of these materials. (For a discussion about the load-carrying capacity and the mechanics involved in load transfer between the generally tough matrix and the generally strong, more brittle fibers, see Hull 1981; LeMay 1981; Piggott 1980.) Figure 10-14d provides an illustration of a fiber-reinforced specimen.

The strength of fiber-reinforced composites is normally highly directional because fibers are arranged in the direction in which loading will occur. Thus, in an aligned fiber composite, the strength may be extremely high in the direction of fiber orientation and extremely low at right angles to the fibers. In an aligned-fiber composite loaded in the direction of the fibers, tensile failure results in fracture and pull out of the fibers (see Figure 9-12), whereas loading at right angles to the fibers results in matrix failure and exposed fibers along the fracture surface.

Shear failure is an important potential mode of failure when deep-sectioned beams fabricated from aligned-fiber composites are loaded in

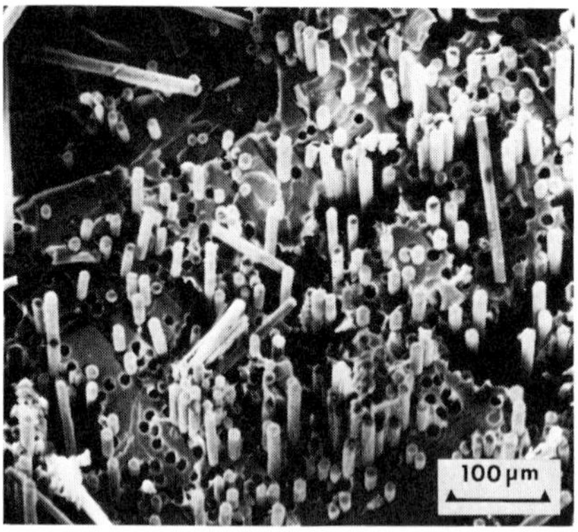

Figure 9-12. Fracture of a carbon–fiber polyester matrix composite as seen in the scanning electron microscope. (Courtesy of Professor K. K. Chawla, Instituto Militar de Engenharia, Rio de Janeiro, Brazil)

bending. The mode of failure that must be considered is shearing along the fiber–matrix bond and, if the strength of the fiber–matrix bond is great, in the matrix itself.

Compressive failures commonly occur on composite beams during bending. Failure occurs first on the side under compression, essentially because the fibers buckle, and debonding occurs as the matrix acts to support them against buckling.

The fatigue properties of fiber composites are generally excellent. Usually, the fatigue strength and ultimate strength of these composites differ only slightly. The exceptions are polymers reinforced by glass fibers, which have a high margin between fatigue strength and ultimate strength, giving rise to good fail-safe properties (Bellavita 1978). The mechanisms of damage that accumulate under fatigue are complex and include matrix cracking, fiber–matrix debonding, and fiber fracture. The initial damage commonly involves a large volume of material; severe local damage takes place at a later stage and leads directly to final fracture. Refer to Figure 10-14 for an illustration.

A variety of nondestructive testing techniques can be used to monitor and evaluate accumulated damage in fiber-reinforced composites. These include C-scan ultrasonic methods, penetrant-enhanced x-ray radiography, and acoustic emission (Sendeckyj 1983).

Failure of Ceramic Materials

Ceramic materials have long been used as construction materials in the building industry. Because of their brittle nature and low tensile strength, the stresses are extremely low or strictly compressive in nature. In addition, because these materials are extremely stable at high temperatures, they are used as refractory solids for furnace construction and linings. Furthermore, unlike superalloys and refractory metals, ceramics maintain their strength and are highly resistant to creep at temperatures as high as 1000°C. Thus, they are being used in many new high-performance engineering applications in which materials may be stressed to a considerable extent. Examples include blades and rotors for gas turbines and turbochargers operating at high temperatures, and cutting tools, wire-drawing dies, and bearings that must be highly resistant to wear.

Ceramic materials fail in a brittle manner, with fracture resulting from a preexisting crack or defect. Because their tensile strength is much lower than their compressive strength, they have a low resistance to bending. The multiaxial loads that brittle ceramic materials can be designed to carry without failing have been summarized by LeMay (1981).

The distribution of small defects within a brittle solid is statistical in nature. Therefore, failure strength in the presence of large cracks can be predicted using linear-elastic-fracture-mechanics (LEFM) procedures; however, the LEFM approach cannot be used to describe the size effect found in smooth bar tests nor the role of stress concentrations and blunt notches in the failure. Pankow and Finnie (1980) have used statistical methods to determine the distribution of flaws and the strength distribution function in ceramic materials, and further refinements of such techniques are likely. (One interesting aspect of ceramic rotors is that, if they fail, they break into many small particles, which eliminates the need for a heavy and expensive containment housing.)

Failures of Reeinforced Concrete

Steel reinforcing bars have been used in concrete structures for many years to enable the structures to carry tensile loads. Tensioning of the reinforcing bars significantly increases the tensile load-carrying capacity. Structures fabricated of prestressed concrete are now used in a variety of situations where the tensile loads are large and where failure might result in serious consequences. Examples are civil structures such as bridges, commercial buildings, reactor containment vessels, tanks for cryogenic storage of liquified gases, and offshore platforms. Thus, a review of the possible modes of failure in such materials is appropriate.

Conventional reinforced concrete may develop cracks and splitting if the reinforcements are corroded by carbon dioxide from the air in the pores of the concrete reacting with the calcium carbonate in the concrete or when the chloride content of the concrete is high. The product of corrosion on the steel surface causes high tensile stresses in the concrete because of the expanded volume, and the low tensile strength of the concrete results in cracking.

In prestressed concrete structures, the steel rods must be kept free of corrosion to preserve the integrity of the structure. If they fracture or are reduced sufficiently in section to preclude carrying the tensile load, one concrete member or the entire structure may collapse. Unbonded tendons, which lie within unfilled channels in the concrete members, are protected from corrosion by a coat of grease and subsequent wrapping with paper or sheathing with plastic. Clearly, resistance to corrosive attack depends on a well-applied coating of grease and the avoidance of damage during fabrication of the structure. Bonded tendons that lie in channels filled with a grouting material (and are not coated with grease) are protected against corrosion by a passive layer that forms on the steel as a result of the reaction between the metal and liquid with a relatively high pH that is present in the pores of the cement. If this passive layer is incompletely formed or is destroyed by some external factor, corrosion of the tendons will occur.

The failures reported for prestressed concrete structures have been related to stress corrosion and hydrogen embrittlement as well as to direct corrosion. Some of these failures, such as the collapse of the roof of the Berlin Congress Hall in 1980 (Isecke and Stichel 1984), have been spectacular. However, in the small percentage of structures that have failed, the root causes have been inadequate attention to fabrication detail or improper construction. The failures cannot be blamed on the choice of steel or any inherent lack of resistance to corrosion or stress corrosion cracking. In the case of the Berlin Congress Hall, the corrosion and stress corrosion cracking observed were a result of improper fabrication and finishing. It is also worth noting that brittle fractures have not been encountered in reinforcing steel tendons in cryogenic vessels at temperatures as low as −392°F, because these tendons are ductile enough to ensure structural integrity (Elices, Planas, and Corres 1984).

CONCLUSION

In this review of the modes of failure that may be relevant when examining a failed structure, we have emphasized the more fundamental aspects of the various types of failure. We hope, however, that we have provided a reasonably comprehensive and up-to-date overview that will be interesting to undergraduate and graduate students in engineering, professional engineers, practicing attorneys, and law students. The information also should be of special interest to engineers and engineering designers who seek to minimize the risks of failure. (More detailed information about the modes of failure can be found in the following sources:

American Society of Metals 1986 (see vol. 9, Metallography and Microstructure; vol. 13, Corrosion); McCall and French 1978; Rider *et al.* 1975; Thielsch 1965.)

As the costs of shutdowns escalate and the responsibility for safe operation of plants and components falls increasingly on the shoulders of designers and of engineers in charge, knowledge about the potential modes of failure becomes of greater and greater importance. In the chapter that follows, we will examine the theoretical and experimental methods of predicting the response of structural components to various mechanisms affecting their loading (the mechanism of failure). Once the response of behavior of a product can be predicted, then the risk of various modes of failure can be reduced or avoided.

10

Simulations: Experimental, Theoretical, and Ergonomic Analyses

The design engineer's chief asset is the ability to measure the performance of a product by a reproducible simulation. In fact, one can say that this ability is what defines a design engineer. As was stated in Chapter 5, theoretical or experimental analyses or a combination of the two are the means by which engineers predict the performance of an unproved product design. The purpose of both types of simulations is to address the modes and mechanisms of failure and, in the case of forensic engineers, causality. In other words, both design and forensic engineers must consider the following in theoretical and experimental analyses, all in the context of predicting a failure mechanism: (1) specifications for materials, (2) the effects of defects, (3) manufacturing and processing, (4) operation and use, and (5) maintenance and inspection.

Theoretical and experimental simulations perform the same functions to the engineer/scientist designer and forensic investigator (as a part of an accident investigation and failure analysis), namely, (1) quantitative data acquisition, and (2) demonstration. Strictly speaking, most theoretical and experimental tests are performed to gather quantitative data; however, there is a distinction. A demonstration test differs from a quantitative data acquisition test in that a demonstration test may be performed not only to collect quantitative data but also qualitative data. For example, to observe if there are any effects present that were not anticipated, and hence were not monitored with respect to quantitative data acquisition. By quantitative data acquisition we mean the numerical-value measurement (direct or indirect) of one or more of the failure-mode parameters (such as temperature) for purposes of correlation to some acceptable *performance criteria* (such as maximum allowable temperature). An example of performance criteria is presented in Chapter 10. Demonstration-type simulations help define the tolerances for the fabrication of the product (acceptable dimensions, material properties, shape, defect size, etc.) and for the operation of the product (limits on load, duration of use, temperature, operational procedures, maintenance, etc.). In general, most people are familiar with demonstration tests associated with publicized prototype aircraft, automotive vehicles, chemical process facilities (fossil and nuclear pressurized water reactor, breeder, and fusion), industrial equipment, and consumer equipment. The series of demonstration tests by Wilbur and Orville Wright from 1900 to 1905 consisted of a series of parametric and prototype scaled and full-sized aeronautical structures and parts that were tests in situ and in a wind tunnel, and are discussed in an interesting article (Crouch 1987).

Global product performance criteria can be stated in terms of failure-mode parameters (the variables that define the mechanism of loading the product—the mechanism of failure): (1) force or displacement, (2) time, (3) geometric characteristics, (4) temperature, (5) environment, and (6) material state and behavior. But all materials have their own characteristic behaviors and modes of failure that are a function of the failure-mode parameters (the mechanisms of failure) reviewed in Appendix B. Methods of quantifying this behavior experimentally are discussed in Chapter 8.

Hence, theoretical and experimental simulations or products with respect to structural performance are investigated to determine, for example, whether the stress or strain state within the component exceeds the material's ability to perform safely (i.e., it will not fail by one of the modes of failure discussed in Chapter 9). But not all product failures are a result of material failure (material performance). And products or systems that do fail, injure or damage not only *directly* but also *indirectly* by means of improper ergonomics or substances contained within the system, parts from the system, or secondary hazards. Hence, a need to assess *function, protection,* and *warnings* can be a motive to perform theoretical and experimental simulations.

Hazards associated with product or system failure can be categorized as (S. J. Brown 1986a): (1) kinetic energy (force–motion)—blast/explosion, missile or product impact, foundation motion; or (2) degenerative (personnel/structural degradation)—temperature/heat, chemical, biological, radioactive, electrical.

The focus of theoretical and experimental simulations of hazardous release is to predict the consequences of failure and to devise methods of reducing the severity of or containing the hazard. Clearly, a nuclear power plant requires a containment structure because the consequences of a failure are far more costly, as the Soviets discovered during the release at the Chernobyl nuclear facility. The protective system itself must meet performance criteria to ensure that it can contain the kinetic energy or degenerative condition to which it might be subjected. Green (1982) presents a discussion of existing techniques to assess risks that have evolved in various industries; D. B. Brown (1976) provides a text on analytic modeling for safety decision making in the framework of systems engineering and its application to safety (as is briefly discussed in Chapter 7); Brown and Baker (1992) present a design, analysis, and applications overview of hazards associated with vessel, pipe, and rotating equipment failure in various industries; Bennett *et al.* (1982), Scott (1979), and the EPA (1972) provide discussions of in situ detection systems and measurement equipment in a context of prevention, protection, and cleanup of hazardous material spills; Sax (1979) provides hazardous analysis information for 15,000 industrial materials, and Sax and Lewis (1986) provide a guide to 700 materials in an abbreviated format.

This chapter provides an overview of the concepts of experimental and theoretical simulations and their application. These methods, which are the culmination of thousands of years of investigation, are thoroughly discussed in the engineering and scientific texts used in university programs. Thus, the overview provided here will be cursory indeed. Because failure of materials is the most frequent cause of product failures and injuries, the primary emphasis will be the modes of failure. For the nontechnical reader, the examples will provide a "feel" for the engineering concepts that provide a basis for predicting product performance and how these concepts are related to the behavior and modes of failure of materials discussed in Chapters 8 and 9.

Theoretical analyses are an economical means of assessing how a product will respond to changes in variables such as thickness, rate of loading, contour, defect size, temperature, and corrosion—all of which can affect the safety of a product. For example, performance criteria for structural components typically set the maximum allowable amount of stress or strain that is permissible for a given alloy. Some alloys can undergo far greater loads or temperatures than others before they fail. Such effects can be simulated theoretically.

The engineer can isolate phenomenological effects such as constitutive material behavior by limiting a theoretical analysis to the prediction of the response of a structural product that is composed of linear material behavior. This type of theoretical analysis is considerably less expensive and less susceptible to error than a representation of the material beyond yield or a nonlinear prediction. Sometimes such simplified (or parametric) methods are relatively accurate in predicting the response of a material and sometimes they are not. But they permit the engineer to try out different materials, geometries, and environments and the effects of time, temperature, and force/displacement before a product is built, or to assess in retrospect what properties a material should have had.

The advantage of parametric, scaled, prototype, and in situ experimental tests is that the response of an actual product to parameters such as stress, strain, and temperature can be determined. Although such tests can be expensive, the data they provide are often correlated with the results of theoretical rests to establish a confidence level in the theory, to readjust the theory, and, on many occasions, to assess the safety of a product by deliberately overloading it to discover the conditions that will cause the product to fail.

Finally, statistical analysis and probability theory/methods are an important analytical tool used with theoretical and experimental predictive methods for design, manufacture, and use. As we discussed in Chapter 7, statistical methods are used to assess data on failures contained in databases. Statistics and probability theory permit engineers to predict that an event will occur or that a certain condition will exist with some level of confidence based on the size of the sample. As Tagart (1985) pointed out, one cannot make predictions with a high confidence level based on an inadequate sample.

THEORETICAL METHODS OF ANALYSIS

Theoretical methods of analysis are classified as classic (formulas), numerical (numerical output for numerical input), and parametric versus full simulation (that is, some effects are excluded versus all effects are included). Theoretical predictive methods can be classified into five categories: (1) closed-form analysis, (2) series-expansion methods, (3) numerical integration, (4) finite-difference, and (5) finite-element methods—all of which can be implemented in a computer.

Although predictive solutions involving structures and fluids represent the greatest application, the theoretical methods of prediction used in other disciplines of engineering can be roughly categorized as these deterministic methods. Whereas closed-form and series-expansion methods can yield formulas, the other three methods cannot. The three numerical methods became practical only with the development of the digital computer.

The five methods of predicting product performance satisfy some or all of the following laws of matter and energy, which can be expressed as or "called" the governing differential equation (Chu *et al.* 1983; Frederick and Chang 1965; Malvern 1969; Sears and Zemansky 1975):

- Conservation of mass, or continuity
- Conservation of momentum (motion and equilibrium)
- Conservation of energy (the first law of thermodynamics or energy balance, which can include electrical and other forms of energy)
- Constitutive equations (entropy and the second law of thermodynamics, caloric equations of state)

(The first three laws are discussed in Appendix B, "Fundamental Physical Laws," pp. 505–550; the fourth law is discussed in the first two major sections of the appendix.) Chapter 8 illustrated how constitutive stress–strain relationships (properties) are determined for structural materials. Chu *et al.* (1983, chap. 19) provided an overview of constitutive relationships of interest to structural engineers, and Appendix B correlates basic constitutive formulas with the material test methods.

Boundary conditions can be described as known or prescribed conditions imposed upon a product, structure, or system during an actual or simulated event that is to be predicted (e.g., such as failure-mode parameter, values of initial or fixed force, displacement, velocity, and temperature that do not need to be predicted). The following predictive methods that are discussed in the following sections satisfy the first three laws of matter and energy either explicitly or implicitly and either exactly or approximately. A theoretical predictive solution defines a type of mechanism of behavior or mechanism of failure that is referred to in engineering and scientific circles as a "boundary value problem" involving static, dynamic, steady state, thermal transient, stable, unstable, linear, nonlinear factors, and so on.

Closed-form Analysis

Closed-form analysis consists of solving the governing differential equations and boundary conditions in terms of failure-mode parameters listed in Table 5-1 (p. 134). Generally, closed-form solutions exist for problems involving objects with simple geometries (e.g., beams, bars, and cylinders), loadings, and boundary conditions; statically determined shell structures (no bending); axisymmetrically loaded and deformed shells of revolution with bending; and cylindrical shells with simple supports, or if the deformation is confined to a relatively narrow "boundary layer" in the neighborhood of the edge. Nonlinear problems are limited to relatively few and simple structures.

For this type of analysis a computer is usually used to perform parametric studies, and the core size and speed of the computer are usually not limiting values of problem tractability. A handbook such as the one by Roark and Young (1975) is often consulted for structural engineering formulas. Similar handbooks or textbooks are available for formulas regarding heat transfer, vibration, dynamics, fluid flow, and electrical circuits (e.g., Chu *et al.* 1983; *Handbook of Chemistry and Physics* 1989; Shaum's Outline Series of Handbooks).

Series-expansion Methods

Because most problems cannot be solved in closed form (particularly shell structures), the actual solution is assumed to take the form of an infinite series of power expansions or known functions with undetermined coefficients. The values of a finite number of undetermined coefficients are calculated from a set of simultaneous linear or nonlinear algebraic equations. The number of terms is dictated by the rate of convergence of the solution.

Series-expansion methods are generally used with either the minimum strain energy principle or the minimum complementary energy. When the minimum strain energy principle is used, displacements are the dependent variables. When the minimum complementary energy is used, internal stresses or forces are the dependent variables. Linear and nonlinear problems can be solved with these methods, and a high-speed computer is advantageous.

These methods are usually limited to simplified shell configurations; they cannot be readily applied to general shapes, material properties, and loads. (Examples of series-expansion methods can be found in the following sources: Amsbartsumyan 1964; Kraus 1967; Lukasiewicz 1979; Szilard 1974; Timoshenko and Woinowsky-Krieger 1959; Vlasov 1964).

Numerical Integration

The numerical integration method is a step-by-step forward-integration method in which the value of a function at a specific point is predicted on the basis of information obtained from neighboring points (Cohen 1968; Kalins 1964; Lestingi and Brown 1973; Mason *et al.* 1965). It has been used to solve both linear and nonlinear problems. The advantages of this method are that large core computers are not generally needed, the method is self-checking, it is relatively inexpensive. One of its disadvantages is that because it relies on series types of shells, intermediate supports are not allowed (this problem can be overcome by Gaussian elimination procedures). In addition, only shells of revolution have been solved successfully, and general shell configurations cannot be handled easily, if at all.

Finite-difference Method

In the finite-difference numerical method, the equations governing the solid or fluid problem to be solved are represented by forward, backward, central, and other difference simultaneous equations. The dependent variables are represented by piecewise continuous functions at mesh points in a spatial domain. This method has been applied to a large number of linear, nonlinear, two-dimensional, and three-dimensional solid and fluid problems. The advantages of the method are its versatility, its suitability for computer programming (larger high-speed mainframe computers may be required for large or complex nonlinear problems), and the simplicity of the difference equations involved. Most technical problems can be solved with an acceptable degree of accuracy by selecting the appropriate mesh or a suitable step-by-step solution. Among the disadvantages are (1) the difficulty of achieving compatibility between complex geometries and mesh requirements, (2) the deterioration of accuracy with higher order derivative equations, (3) the step-by-step size limitations, in transient problems, of the explicit or forward-difference method related to mesh size (numerical instability), and (4) the tendency of the implicit or backward-difference method to require carrying a large number of significant digits throughout the solution process. (Applications of the finite-difference method can be found in sources such as the following: Fenves *et al.* 1973; Kraus 1967; Szilard 1974; Wah and Calcote 1970).

Finite-element Method

The finite-element method is a numerical technique for solving problems involving solid and fluid (continuum) boundary values by assuming that (1) the solid or fluid can be partitioned into finite volumes or "finite elements," (2) the functions (usually polynomials) of unknown variables such as displacement, temperature, velocity, and pressure are assumed with undetermined coefficients within each domain or "element," (3) the undetermined coefficients are calculated by minimization of energy, (4) the element matrix that relates the variables are derived once the undetermined coefficients are solved, and (5) the element matrixes can be assembled into an overall structural matrix, such as the stiffness matrix in stress-analysis programs.

Unlike the finite-difference and numerical-integration methods, the finite-element method partitions the continuum rather than the governing equations. Two finite-element methods can be used in structural analyses: force method and displacement method. The displacement method is the most popular because it lends itself more readily to computer implementation.

In the finite-element method, extensive use is made of matrix algebra, and convergence of accuracy is generally achieved by increasing the number of elements until the approximation functions in each element or domain are sufficiently accurate. The advantages of this method are that (1) relatively simple equations are usually involved, (2) general structural shapes can be used to solve thermal, dynamic, inelastic, and static problems, (3) predictions of static or dynamic, steady-state or thermal, and elastic or inelastic responses are extremely accurate, (4) solid or fluid problems can be solved, (5) convergence is usually not a problem, and (6) various types of element formulations can be used in a solution.

The finite-element method has a number of disadvantages. First, the bigger and more complex the problem (geometrically and materially nonlinear), the greater the need for a large high-speed computer and interfacing pre- and postprocessing software. Second, a large number of significant digits is required for computational truncation purposes. Third, the solutions obtained with this method tend to be more expensive than those achieved with the finite-difference and numerical-integration methods. Fourth, automatic solution convergence schemes either are not generally known or are not available in current finite-element programs. Finally, the generality of the method can tempt one to obtain more information than is necessary.

Of all the methods previously discussed, the finite-element method offers the greatest ability to analyze structures that have arbitrary shapes, material properties, and boundary conditions and are subjected to arbitrary applied loads. This numerical technique is particularly well suited for high-speed digital computers, which permit a simulated response of solid and/or fluid bodies (such as valves and internal and external environment) that are subject to loading conditions such as forces, displacements, temperatures, and pressures. Conversely, the simpler the geometries, material properties, boundary conditions, and applied loads, the more likely that a less expensive, simpler method of analysis will be adequate (such as a closed-form solution).

Although the finite-element method can be organized into routine operations, the equations can be relatively simple, the method is relatively accurate, and it can be applied to many types of problems and disciplines, it can require the storage of a tremendous amount of data and involve a tremendous number of calculations. The digital computer made it possible to turn this disadvantage into an advantage by allowing information about (1) the geometry of a product by way of coordinates of the nodes of each element or zone, (2) the material properties in each element or zone, and (3) output data about the state of the material (stress, strain, temperature, velocity) to be represented visually on a monitor or videotape and in drawings or plots. Computer programs called preprocessors help engineers assemble the data required to describe a structure. These preprocessors include model builders, diagnostic programs, and on-screen model imaging and plots. Programs called postprocessors—such as data imaging (plots of displacements, stress, strain, temperature, stress functionals, and so forth), data interpretation (per performance criteria), and convergence and optimization algorithms—help them interpret and visualize the output. A third type of software includes solvers of basic scientific laws of mass, momentum, and energy; constitutive relationships (characteristics of materials); and boundary conditions. Since the finite-element method was first introduced in the aerospace industry in the 1960s, most structural design engineers have used it to assess products for which information is not readily available or is nonexistent.

Many good sources of information about the finite-element method are available (see, for example, Bathe and Wilson 1976; Belytschko and Greer 1977; S. J. Brown and Hsu 1977; Bushnell 1969; Connor and Brebbia 1976; Gallagher 1975; Gallagher *et al.* 1982; Lestingi and Brown 1973; Papadakas and Scarton 1978; Park and Gartling 1978; Strang and Fix 1973; Tong 1970; Vinson 1980; Wilson and Nickell 1966; Zienkiewicz 1977).

EXPERIMENTAL METHODS OF ANALYSIS

As discussed in Chapter 5, experimental tests are usually classified as:

- Parametric versus full simulations (some selected loads and conditions versus actual simulations)
- Scale model versus full-size simulations
- Prototype demonstrations
- In situ demonstrations with actual products or systems

An experimental simulation consists of two parts. First, the factors that define the mechanisms of failure (or failure-mode parameters) are recreated. In other words, some or all conditions to be considered in the test are simulated. Thus, this part involves the test specimen or product and the load fixture, apparatus, chamber, and control site. The second part involves the systems of gathering data: that is, the methods of measuring various conditions during the simulation.

Product simulations can include many of the loading and measuring techniques described in Chapter 8, particularly if the specimens are small. One must bear in mind, however, that those techniques evolved with standardized specimens and testing procedures to characterize the behavior of materials. As we have mentioned before, the standardized testing methods defined by the American Society for Testing and Materials provide engineers with a common reference for assessing the materials specified in a design or used in a product that failed.

No standardized tests per se exist for simulations, as there are available for material testing. Instead, the design, operational, and intended use, the emergency loads to which the component is to be subjected; and in the case of an accident investigation and failure analysis, the unexpected events are considered as the bases of performing an experimental simulation. Scale models of large products are often chosen because of the lower expense of fabricating them and the possibility of destroying them. In addition, no loading system, test chamber, or test facility may be large enough to accommodate the full-sized product unless an in situ or prototype test is possible. A variety of standardized test fixtures, apparatus, chambers, and the like have been developed over the years to evaluate the mechanisms of failure or "boundary value problems" involved in a failed product. A number of these are listed Box 10-1.

Data-gathering instrumentation is rapidly evolving and increasingly becoming more sophisticated, especially since the introduction of the microprocessor. The *Journal of Experimental Mechanics* is a good source of information about techniques used in experimental tests and in data acquisition (see, also, Kobayashi 1988). The techniques for measuring

BOX 10-1. Types of Test Fixtures and Facilities

Load frames or fixtures. Load frames used in the private sector and in universities and laboratories range in size from small ones (see Fig. 8-9) to those large enough to apply concentrated force or control displacements to large beam sections used in bridge and civil structures. Portable hydraulic jacks with small to large loading capacities are used in shop or laboratory environments as well as in the field and in situ tests of large civil structures.

Pressure vacuums. Pressure loads can be used for internal loading of objects such as pressure vessels, tires, and air compressors, or for external loading of items such as submersible vessels and submersible suits. Internal positive pressure above ambient or internal negative pressure below ambient tests are conducted with pumps capable of producing pressures on the

BOX 10-1. (*continued*)

order of 300,000 psi to vacuum conditions simulating space flight environments. External pressure or vacuum loads are accomplished by means of pressure chambers that are pressure vessels.

Force (or displacement) -time applied loads. Applied force as a periodic function of time is commonly categorized as vibratory and nonsinusoidal and the results are usually called time histograms. Shakers or electromagnetic coils are commonly used to apply vibratory force to both small and large structures. The frequency of vibration of the applied force can be varied. Programmable hydraulic units also are used to apply vibratory and time-histogram force or displacement as a function of time (Stroud 1987). See Figure 10-10(b).

Shaker tables or frames. Shaker tables are typically used to evaluate civil structures by means of testing of scale models of buildings, bridges, and so forth subjected to earthquakes and other ground motion. See Figure 10-10(f).

Wind tunnels. Wind tunnels permit the study of structural performance in a low- to high-velocity flow field of compressible gas (usually air), and come in a variety of sizes and capacities ranging from a localized part to full-size simulations of building components, automobiles, and aircraft. They are also used to study natural phenomena.

Thermal chambers. Thermal chambers are used to evaluate products exposed to thermal radiation, conduction, and convection, and they vary in size from small chambers to chambers large enough to test aerospace vehicles and power processing systems. Internally applied thermal conditions are typically associated with simulations of product processing. Internal types of tests are used in in situ or prototype simulations involving the actual source of thermal energy.

Environmental chambers. These chambers permit the study of controlled, simulated atmospheric effects on products, including corrosion, ultraviolet rays, moisture, wind, and chemical particles. One common type is a humidity chamber. In situ tests are relatively common.

Fluid-flow loops. These loops, which are similar to wind tunnels, are used to evaluate the flow of fluid hydrodynamically over components designed to operate in a fluid-flow field. Flow loops are often designed to test components through which fluid will flow such as heat exchanges, valves, and piping. These loops vary in size and capacity and can be used to test either small household-sized items or large industry-sized items (see Fig. 10-12(c)) (Chen 1985).

Hydraulic tanks. These tanks vary in size from the size of a small room to the size of a pond and are used to study surface and subsurface effects of water (or fluids). For example, they are often used to study the hydrodynamics of ship hulls, wave motions, the effects of waves on fixed structures, such as barriers and wave–structure interaction on flexible structures such as offshore structures, storage tanks, or process tanks (e.g., David Taylor Naval Research Facility, Shell Research & Development, West Hollow Facility, Houston, etc.)

BOX 10-1. *(continued)*

Ballistic impact facilities. These facilities vary greatly, depending on the phenomenon studied. In general, they are used to study one or all of the following: initial boundary conditions of the fragment or missile, trajectory, and impact-penetration mechanics. Tests vary from small changers to large remote sites. Missile formation also is studied in connection with the explosion process, and the impacts can range from grams to tons (S. J. Brown 1986a; Zukáš 1980).

Explosion test facilities. These facilities range in size from small chambers to remote site facilities used in large-scale high-energy explosions. Tests are generally evaluated with respect to the kinetic energy and degenerative effects of the contained media. Simulations are geared to the evaluation of explosives or of systems that present a risk of unintentional explosion. Product rupture or disintegration associated with rotating equipment also is studied using such facilities and fixtures (Brown and Baker 1992, Baker (1983), Chou (1977), Baker *et al.* (1983)).

Dynamometer test fixtures. Dynamometers are electric, mechanical, hydraulic, etc., test systems used in the laboratory to test the performance of power engines. Typically, they measure engine shaft torque at various rpms and other parameters (such as temperature, pressure, ignition characteristics) are usually measured in conjunction with these tests.

Rocket-thrust facilities. Rocket thrust, jetting, and blowdown facilities are facilities that test the solid or liquid "thrust" performance of actual rocket and jet engines and of chemical systems that, through failure, produce unintentional thrust. Test facilities vary in size from small low-thrust systems to large remote facilities used to investigate forces from large rocket motors, rupture disks, relief valves, and pipe breaching.

Fireball-testing facilities. These facilities tend to be large remote facilities that are used to investigate failures associated with vapor cloud explosions, breaching, and ignition of flammable materials (Burgoyne 1985, Hasegawa (1978)).

Particle dispersion. These tests are used to study the movement of solid, liquid, or gas particulates (such as deposition, migration, resuspension) via meteorological effects, migration, and resuspension. Most studies tend to be in situ and most are oriented toward predicting the dispersion of chemical, radioactive, and biological materials resulting from accidental release (Pruppecher *et al.* 1983, Green 1982, Bennett 1982). Refer to general source references from NOAA, EPA, and CCPS.

Chemical-reaction apparatus. This testing apparatus varies in size from small benchtop laboratory equipment to full-size in situ demonstration or remote testing facilities. The purpose is to determine the parameters that influence chemical reactions, especially runaway, hazardous reactions that preclude shutting down a process safely and instead require venting and containment. Refer to the DIERS Project Manual, AIChE (1989); Bartknecht (1981); survey articles (Burgoyne 1985; S. J. Brown 1985a); and text (Scott 1979); and Baker *et al.* (1983).

BOX 10-1. *(continued)*

Coefficient of friction test fixtures. These test fixtures are designed to determine the static and dynamic ratio of the initiating force (parallel to the surfaces) to the normal force. Most devices are benchtop or floor-unit size; some are portable. Because of injuries associated with falling on slippery surfaces, a number of devices have been developed to evaluate the coefficient of friction between pedestrian walkways and footwear. A number of studies have been reported by the ASTM and the National Bureau of Standards; Adler and Pierman (1975). Also refer to Bowden and Tabor (1964) for theoretical/experimental classic tests, and Rosen (1983) for an overview of the ergonomics aspect.

Foundation–soil tests. These tests are used to examine the mechanical properties of soil and its relationship to static and dynamic soil-structural responses. Static and quasi-static tests are used to assess soil-bearing capacities and subsidence, whereas dynamic tests are used to study the effects of vibration and seismic or ground shock. Foundation–soil tests vary from a size appropriate for a laboratory to a size used at remote sites. Ground motion can be induced hydraulically, electrically, or mechanically or by explosives (S. J. Brown and Chu, 1983, Richmond (1970)).

Anechoic chambers. These chambers are used to investigate sound levels or characteristics in products in the absence of reflective waves. They can range in size from a benchtop unit to a unit large enough to accommodate large aircraft (Ford 1970; *Journal of Sound and Vibration*).

Electric-charge apparatus. Electrical performance testing apparatus ranges in size from small benchtop units to large remote-site units capable of simulating the effects of lightening strikes. Standard testing procedures are usually promulgated by organizations such as the Institute of Electrical and Electronics Engineers, the National Bureau of Standards, the Underwriters Laboratories, and the NFPA (National Fire Protection Association).

Biomechanics and ergonomics tests. Biomechanics testing considers the human anatomy as an engineering or scientific system (electrical, chemical, structural, biological, and radiological), and ergonomics tests study the application of biological and engineering principles related to the humans interfacing with machines (or products). S. J. Brown (1986c) summarizes some of the studies performed in connection with the structural aspects of humans to sustain kinetic energy (blast and impact): see White, Bowen, and Richmond (1965) and Clare *et al.* (1976). Astrand and Rodahl (1986) and Carlson (1972); Fund and Anliker (1972); Eastman Kodak (1983) provide an overview of recent trends and methods in ergonomics. Examples of ergonomic tests are forklift studies (Waterman 1985; Melvin *et al.* 1982), endurance (Nadel and Bussolari 1988), perception on stairs (Carson *et al.* 1970), plant operation (Burgoyne 1985), mechanisms of head and spine trauma and tolerances (Sances *et al.* 1985). Gymnastic biomechanics is discussed by George (1980 and to be a published volume). Human Factors (see Box 4-2) analysis and testing considers optimal integration (or lack) of human, environment, and machine in the context of a particular machine.

structural displacement, stress, force, and so on, discussed in Chapter 8, are applicable to experimental simulations.

In the context of evaluating the global response of systems, equipment, and structures there are numerous experimental techniques that are used to measure boundary values (surface conditions), initial conditions (initial velocity, temperature, time, etc.), and body forces (internal conditions, such as (1) displacement (strain gauge, micrometers, eddy probe, fiber optics, acoustic, radiation, thermography, accelerometers, bifringic, Moiré coatings); (2) pressure transducers (gravity types, direct-acting elastic types, indirect-acting elastic types, piezoelectric [see Baker 1983]); (3) temperature measurements (thermocouples, pyrometry, thermoresistive elements, infrared [see Chapter 8], thermometry [pressure, thermoelastic expansion]); (4) flow measurement (obstruction, velocity, hot wire, magnetic, sonic, mass, bifringic, turbine types); (5) forces and torque (mechanical, hydraulic, pneumatic, magnetic, electrical) utilizes displacements and pressure-measuring techniques; (6) motion (still and motion imaging, photography, VCR, laser optics, stroboscope, tachometry, radar, radiography [tomography], acoustics); and (7) electrical, chemical, biological, and radiological measurements. A discussion of mechanical devices is provided in Kobayashi (1988) and Beckwith and Buck (1961), and two sources of equipment, facilities, and services suppliers are provided through periodicals such as *Experimental Techniques* (Society of Experimental Mechanics, Bethel, Connecticut) and *Test* (Mattingly Publishers, Oakland, California).

PREDICTING THE RESPONSE OF A PRODUCT

This section presents some examples of simulations that will acquaint readers with some of the experimental and theoretical concepts previously discussed for predicting the responses of products and the relationship between the modes and mechanisms of failure. Fifty percent of the accident investigations and failure analyses involved were associated with litigation. The other 50 percent were conducted to improve the reliability of predictions or the performance of products. In most examples, the finite-element method of theoretical simulation was used because of the attributes described earlier.

Static and Quasi-static Analysis

Figure 10-1(a) illustrates a finite-element representation of a bicycle frame. Note that the tubular frame is represented by approximately 200

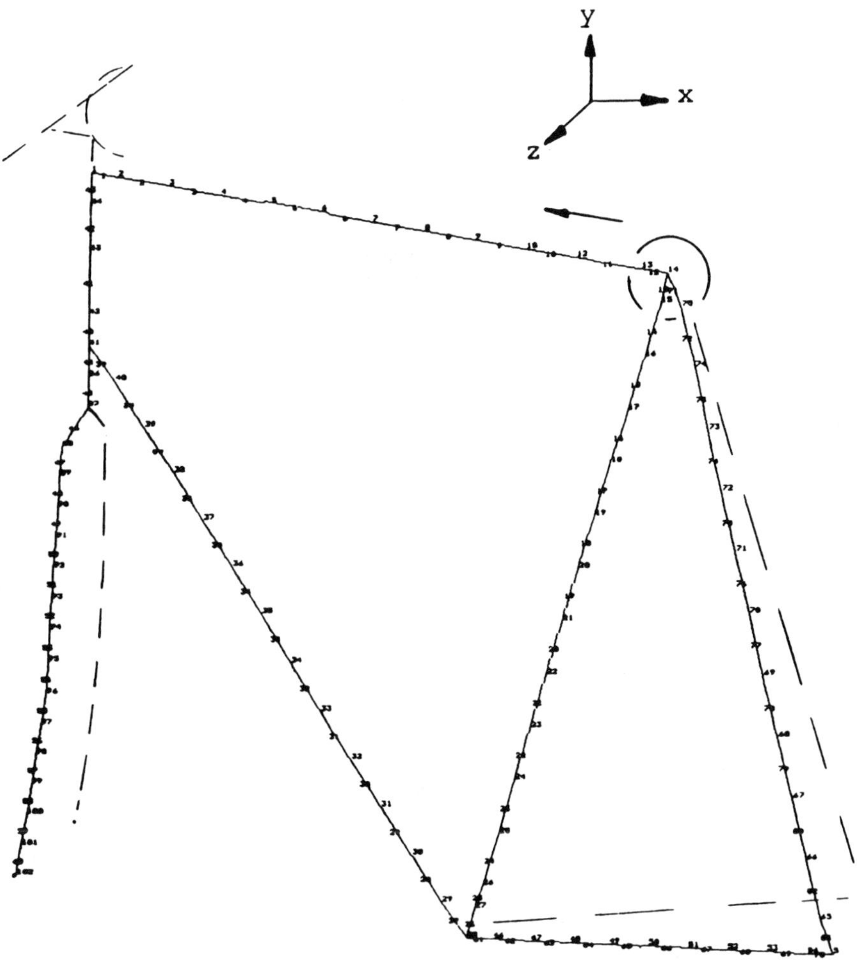

(a)

Figure 10-1. A dynamic analysis of an ultralight bicycle frame. (Courtesy of Quest Engineering & Development Corp.) (a) Computer finite-element simulation model. (b) An AUTOCAD drawing of the frame. (c) A comparison of force F_x versus displacement of the front steering fork determined by experimental and finite-element computer simulation. (d) Photograph of deformed frame after dynamic loading (front tire seizure against diagonal brake).

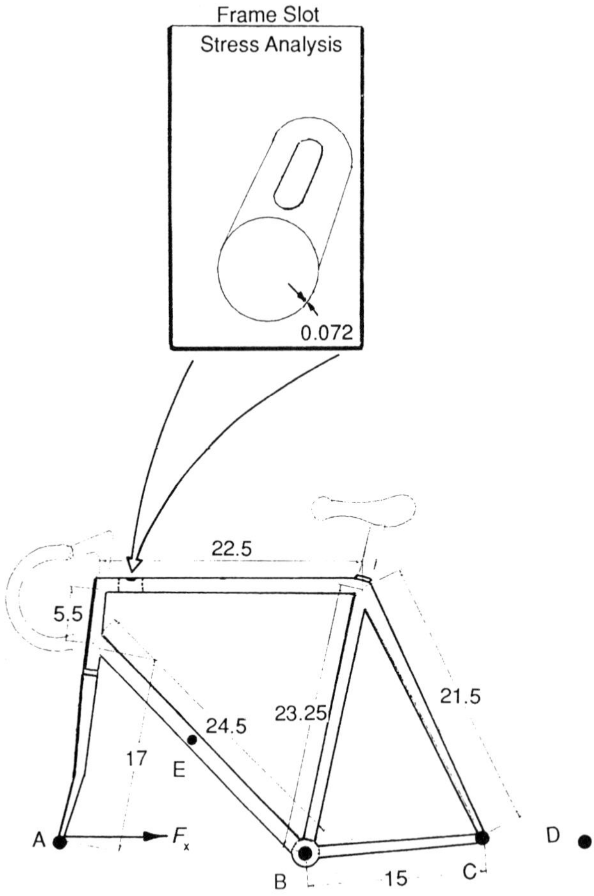

(b)

Figure 10-1. *(continued)*

beam-type elements, which appear in the figure as a straight line between each node point. However, an actual tube is simulated. The frame shown in Figure 10-1(b) was drawn by an interfacing drafting computer program. Experimental and theoretical simulations were performed to assess the static and dynamic deflection of the front fork (A) when subjected to a horizontal force (F_x). Figure 10-1(c) illustrates the measured relationship of force versus deflection. The experimental tests consisted of applying incremental loads and measuring the deflection (δ) with an extensometer (micrometer) between three pairs of points. The results of the finite-element computer simulation for the location A–C are plotted as a line in Figure 10-1(c); however, because displacements are com-

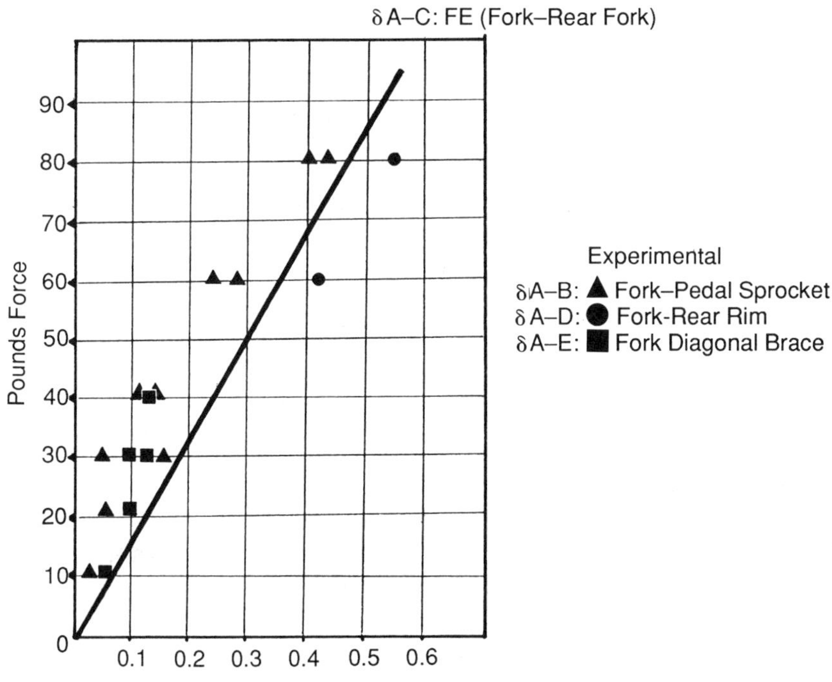

(c)

(d)

Figure 10-1. *(continued)*

puted at each node point, relative displacement throughout the entire system can be computed (see Fig. 10-1(b) for an illustration of the locations of the nodes along the frame). This simple model illustrates the ease with which the load versus displacement can be calculated by a computer simulation for a bicycle frame constructed of different alloys and having a tube of different thicknesses. In this case, the tests showed that an overly flexible front fork on the frame would cause the front tire to come into contact with the diagonal brace, which in turn would cause the front wheel to separate from the front fork.

Figure 10-2(a) illustrates a finite-element computer simulation of a coupler that is used to connect high-pressure flexible hoses (see Figure 10-2(d)). Unexpected uncoupling can result in a pneumatic or hydraulic thrust called a pipe whip that sends the hose flying at people and equipment. Figure 10-2(b) is a sequence of computer deformation plots, greatly magnified, illustrating that the rotation of the coupler sleeve is a function of where the roller bearing is located. Between frames 3 and 4, the locking surface of the sleeve on which the roller bearing applies load changes rotations from positive to negative. Figure 10-2(c) illustrates the plotted relationship between deflection of the coupler sleeve versus the position of the ball bearing that applies a force to the locking surface. In normal operation, the collar must be pulled back to release the male part of the connector. Figure 10-2(c) illustrates that the coupler is positively engaged between A and B, but tends to uncouple itself from C to D. The user cannot readily observe the small tolerance between the full-lock position and potential disengagement, and the spring that normally forces the sleeve to the engagement or locked position can be thwarted by twigs and other foreign material. Figure 10-2(e) illustrates a computer drawing of von Mises stresses as they occur in the coupler at t_3. The experimental test simulations consisted of hydraulically pressurizing a coupler and flexible hose system to specified pressure-gauge readings. A feeler gauge was used to calibrate the coupler position at different settings. The deflection and separation motion were videorecorded.

Figure 10-3(a) illustrates a computer simulation of a 300-psi-rated six-inch gate valve. Figure 10-3(b) is a computer drawing of the magnified deformation of the valve during a simulated pressure and flange-bolt load. Figure 10-3(d) is a color-coded plot of a von Mises stress distribution in the valve with internal pressure and bolt-torque loads. This simulation examines the variations of pressure on the gasket between the two flanges as a function of internal pressure, temperature, and various initial preloads (by applying a prescribed torque to the nuts of the bolt that hold the flanges together). Preventing leakage of chemicals past the gasket is the purpose of such a test. The manufacturer of the gasket recommended no less than 10,000 psi of gasket pressure to

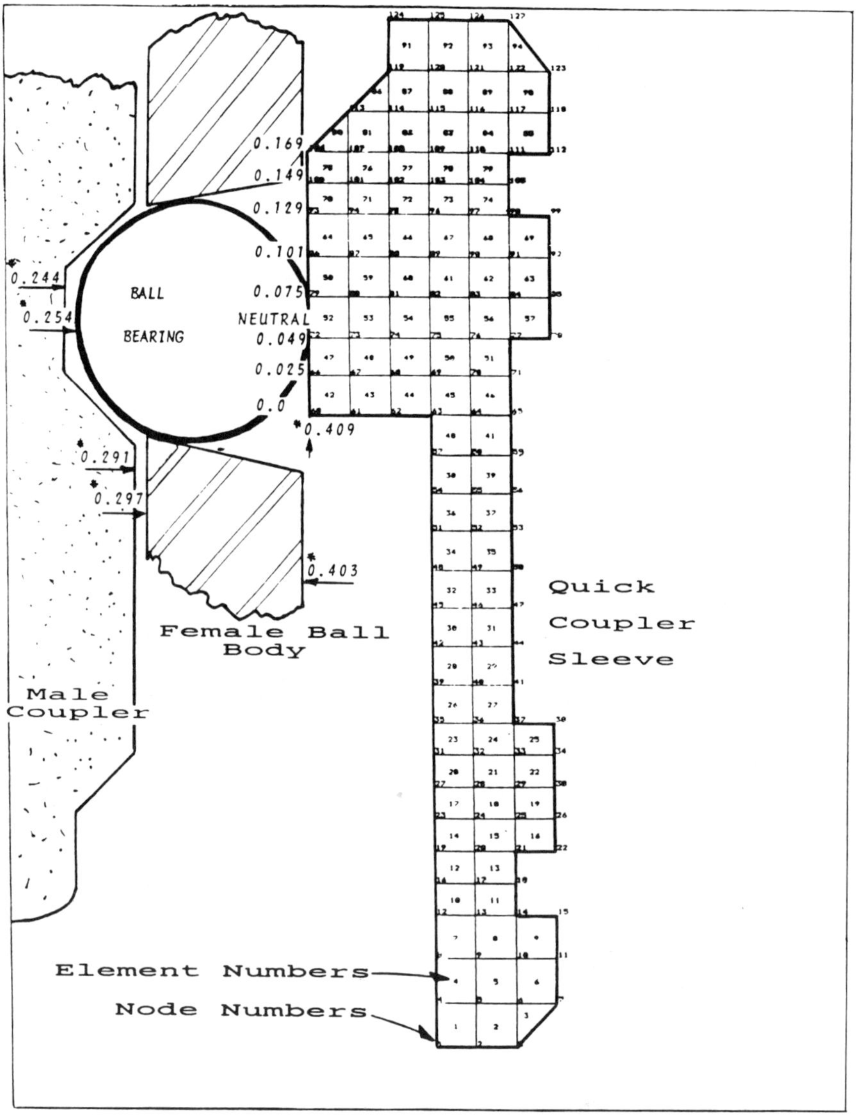

(a)

Figure 10-2. A dynamic analysis of a quick-release coupler. (Courtesy of Quest Engineering & Development Corp.) (a) Finite-element computer simulation of a quick-release coupler. (b) Time sequence of ball bearing displacement versus coupler sleeve deformation during disengagement of the quick-release hose coupler. (c) Location of the ball bearing pressure on quick-connect coupler sleeve versus deflection in the radial (Y) direction and versus node with greatest deflection in the radial (Y) direction. (d) Photograph of the quick-connect coupler assembly. (e) Computer plot of Von Mises stress (ksi) in the coupler sleeve at time (t) = t_2.

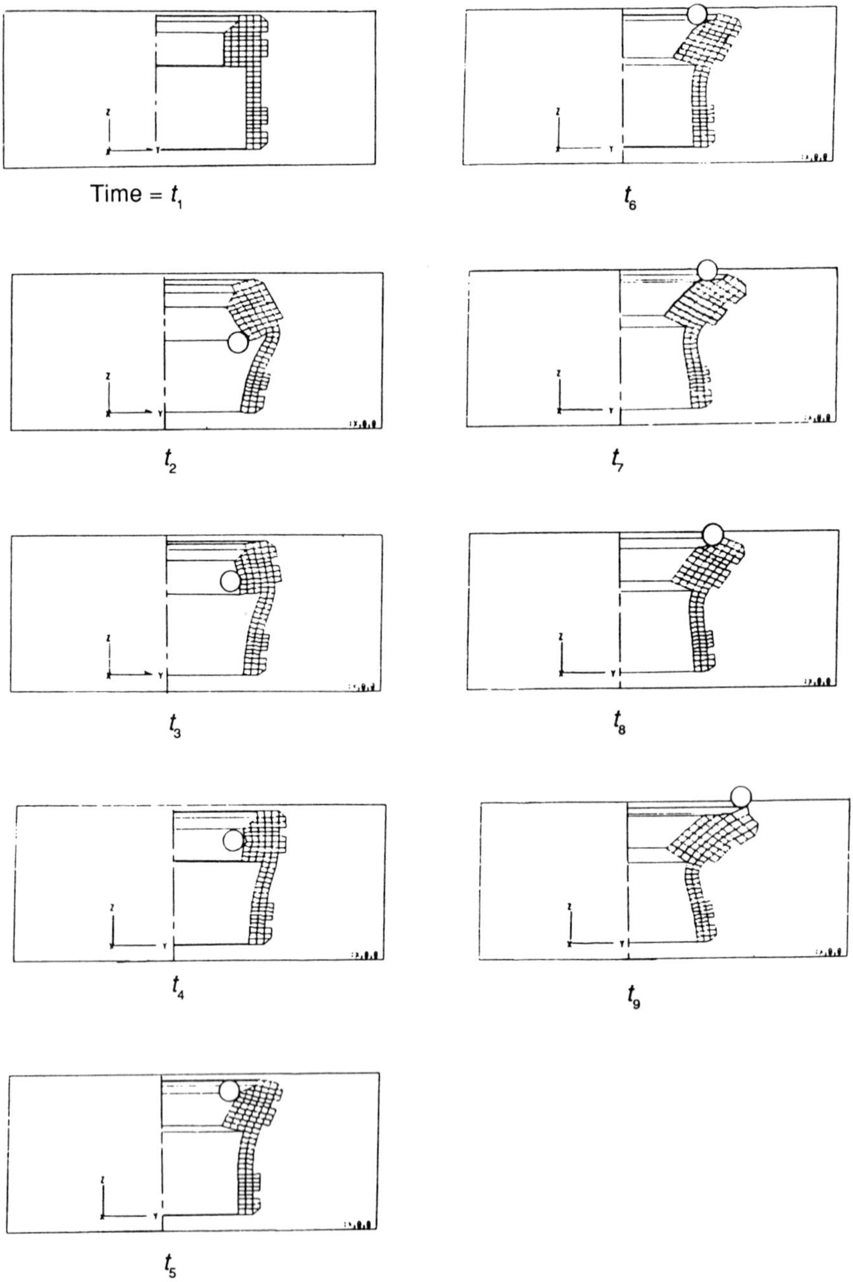

(b)

Figure 10-2. *(continued)*

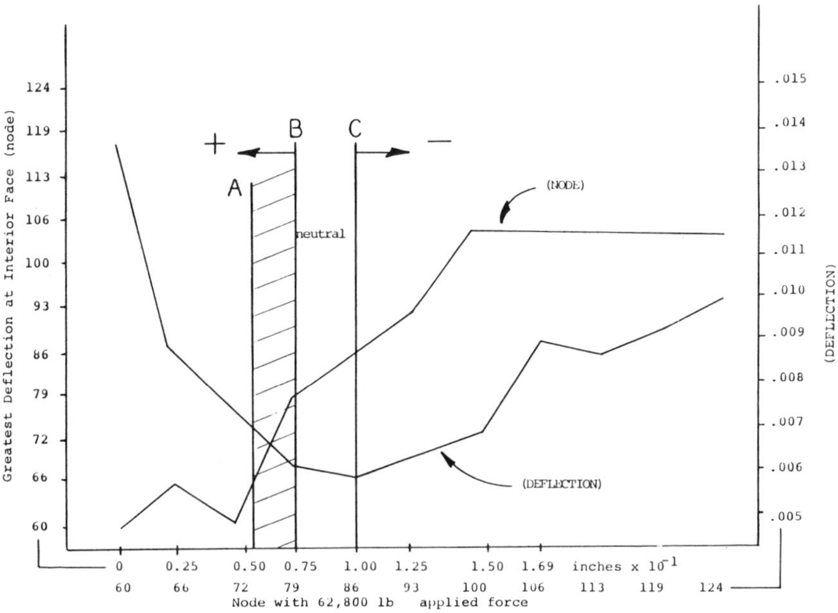

(c)

(d)

Figure 10-2. *(continued)*

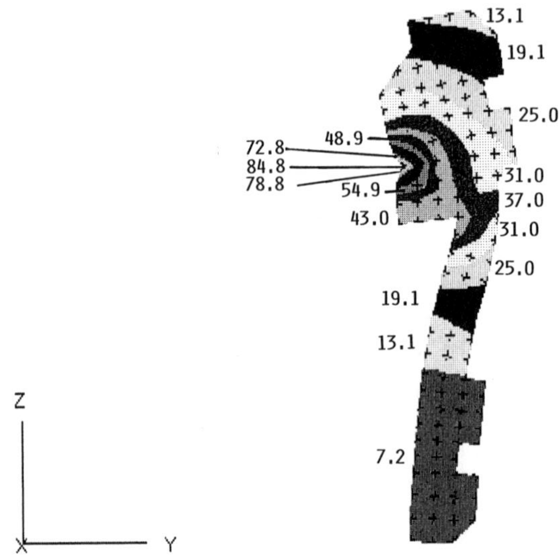

(e)

Figure 10-2. *(continued)*

ensure that leakage does not occur during operation. In Figure 10-3(c), curve I indicates that a bolt preload of approximately 60,000 pounds is the minimum load required to cause a pressure of 10,000 psi in the gasket; however, curve II indicates that a 70,000-pound preload is required to operate the valve at an internal pressure of 375 psi or less. Higher operating pressures require higher preloads. When the valve is heated to approximately 800°F (curve III), a preload of approximately 90,000 pounds is required to maintain a gasket pressure of 10,000 psi. The experimental tests consisted of preloading the bolts by torquing the nuts on the flange, then applying a hydraulic internal pressure load with saturated steam at a high temperature. Only qualitative data were gathered in this experimental test case; leak detection was performed. The thermal effects considered in the experimental and computer simulations of the valve are for steady-state conditions only (not time-dependent, temperature variations in the valve or "transient" thermal loading).

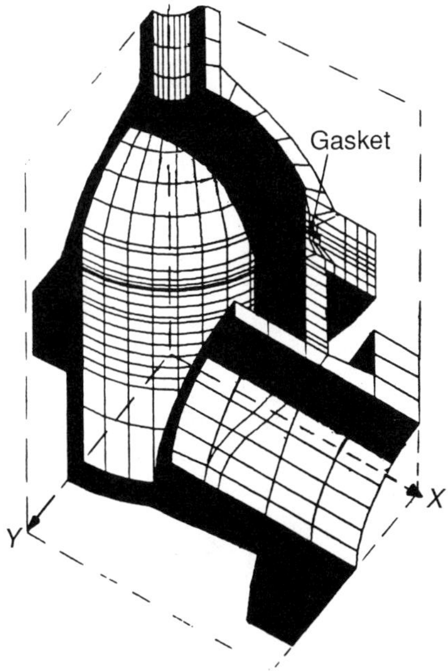

(a)

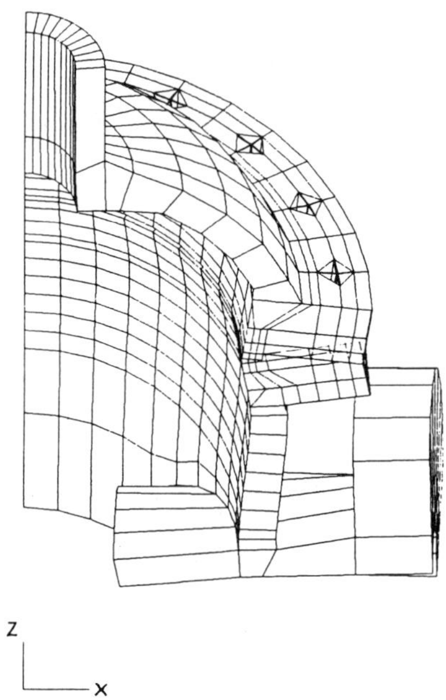

(b)

Figure 10-3. Thermal/mechanical analysis of a valve. (Refer to S. J. Brown and T. J. Brown 1990b) (a) Finite-element computer simulation of the valve body and bonnet, gasket, studs, nuts, and washers from the inside looking out. (b) Computer simulation of deformation caused by preloading of bolts and internal pressure. (c) Gasket pressure (ksi) versus valve preload (factor of 140,000#) (I), preload plus pressure (II), preload plus pressure and 800°F (III). (d) Valve Von Mises stress (ksi) computer plots for preload and pressure (case II). (Refer to Brown 1990a for example of Force and Moment Simulations.)

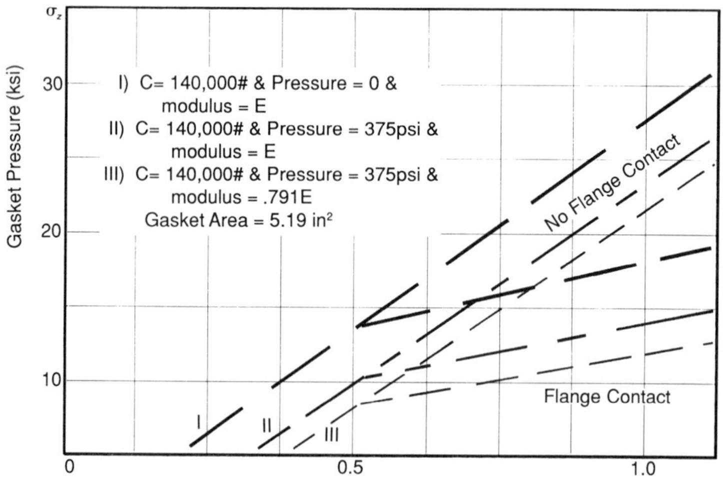

(c)

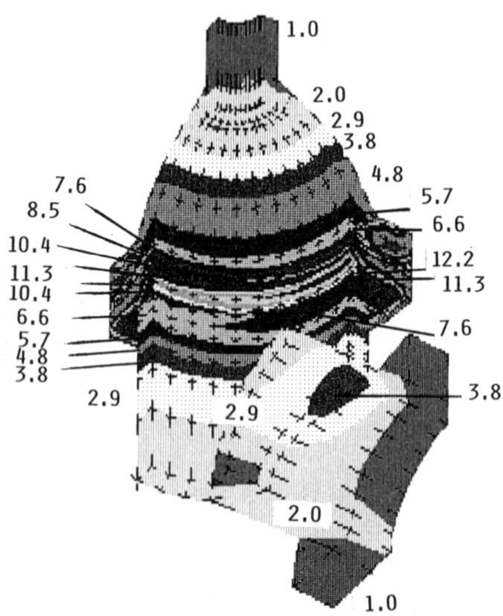

(d)

Figure 10-3. *(continued)*

Thermal–Mechanical Transient Analysis

Time-dependent pressure and temperature within a system are a common type of load found in power and processing equipment. Figure 10-4(a) illustrates an example of such simultaneous time-dependent loads. Figure 10-4(b) is a drawing of a photoelastic (bifringic) $\frac{1}{10}$-scale model of a large stainless-steel internally heated and pressurized component of a heat exchanger. Of particular interest in the experimental and theoretical simulations were the stresses in the transition region caused by tem-

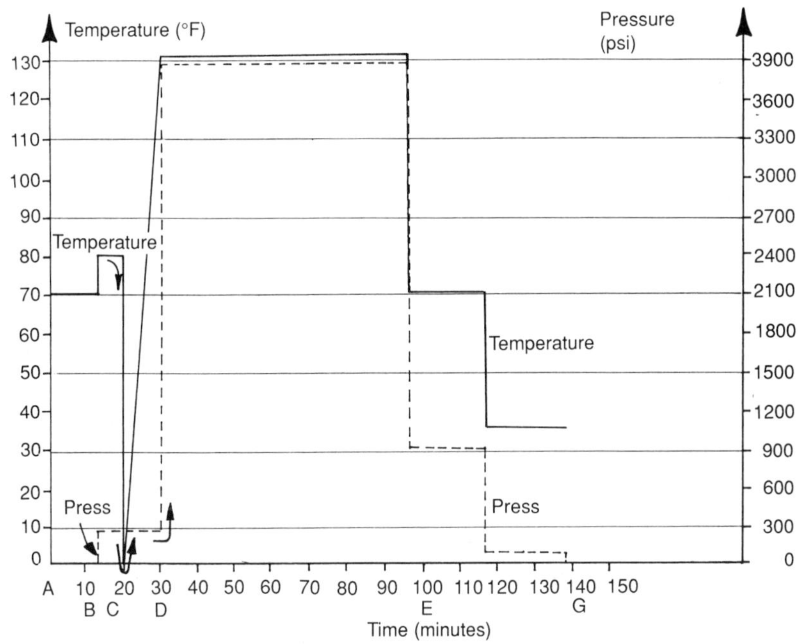

(a)

Figure 10-4. Thermal and pressure transient loads in a pressure-vessel "T". (a) Example of a temperature and pressure versus time (transient). (b) Illustration of complex "T" transition or cylinder-to-cylinder intersection (from S. J. Brown *et al.* 1977). (c) Surface stress distribution at "T" transition, inner and outer surfaces by finite-element (FE) computer simulation, experimental photoelasticity, and strain-gauge data (SG). ● = axial; · = hoop for FE. — = axial; –·– = hoop for photoelastic. ○ = axial; □ = hoop for SG data. (From S. J. Brown *et al.* 1977) (d) Three-dimensional finite-element mesh. (From S. J. Brown *et al.* 1977) (e) Computer drawing of temperature contours for transient analysis at time 0.11 (From S. J. Brown *et al.* 1977) (f) Contour map of effective (von Mises) stress for load case 1 of thermal transient. (From S. J. Brown *et al.* 1977)

DIMENSIONS	INCHES	mm
A	10	254
B	0.250	6.35
C	6.305	160.15
D	2.304	58.52
E	8.797	223.44
F	1.997	50.72
G	0.437	11.10
H	0.562	14.27

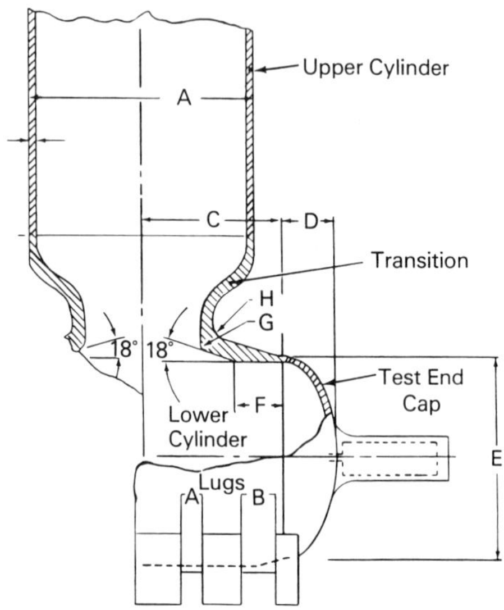

(b)

Figure 10-4. (*continued*)

perature and pressure. The scale model was constructed of bifringic material that can be pressurized, heated in an oven, then cooled—a process called stress freezing that "locks" in the stresses. By taking thin cross-sectional slices after the stress-freezing procedure and using photoelastic measuring techniques, stresses in the slice could be experimentally determined. Before the stress freezing began, however, strain gauges were mounted on the surfaces of the model in the transition region. The model was pressurized at room temperature, and surface stresses were computed from the strain-gauge data for later comparison to the photoelastic results and finite-element computer simulation results (see Fig. 10-4(c)). Figure 10-4(d) illustrates the three-dimensional finite-element representation of the model; only one-fourth of the model is shown because of the one-fourth symmetry in the transition region. (For a discussion of the results of the three-dimensional finite-element method and the photoelastic and strain-gauge pressure tests, see S. J. Brown *et al.* 1977; for a discussion about the thermal transient analysis, see S. J. Brown 1979.)

Because of the experimental difficulty of measuring the stress caused by thermal gradients (i.e., the change of temperature through the wall as the temperature of the interior fluid changes with time), this

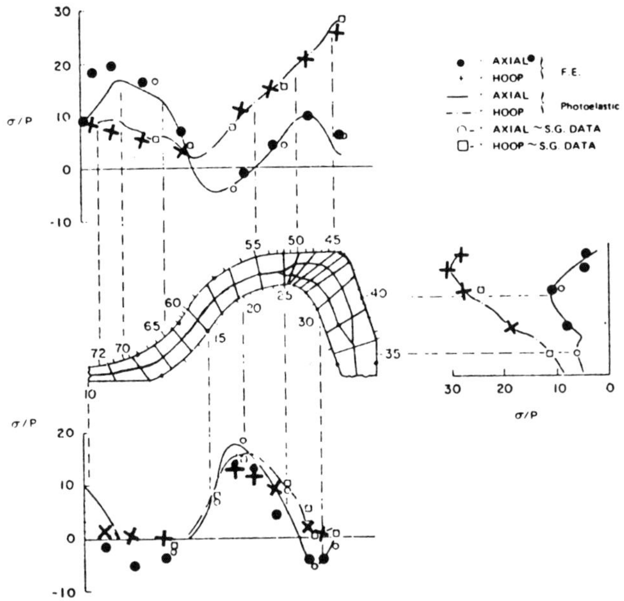

Normalized axial and hoop stresses (longitudinal section (0 deg))

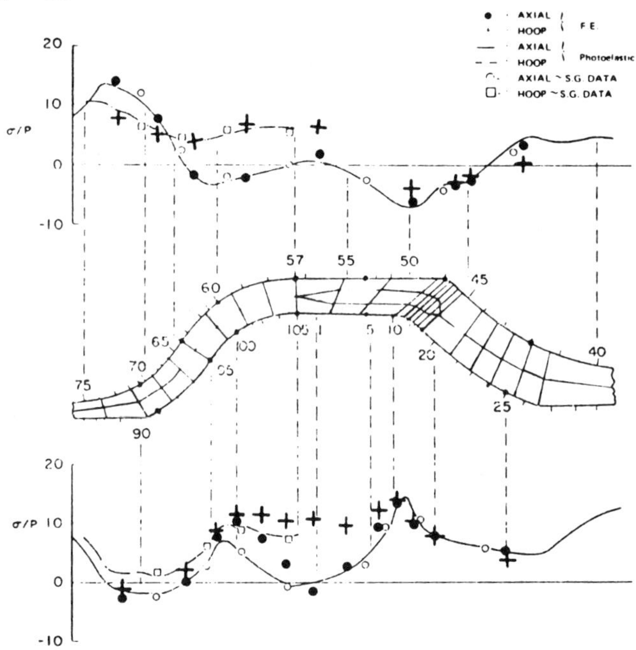

Normalized axial and hoop stresses (transverse section (90 deg))

(c)

Figure 10-4. *(continued)*

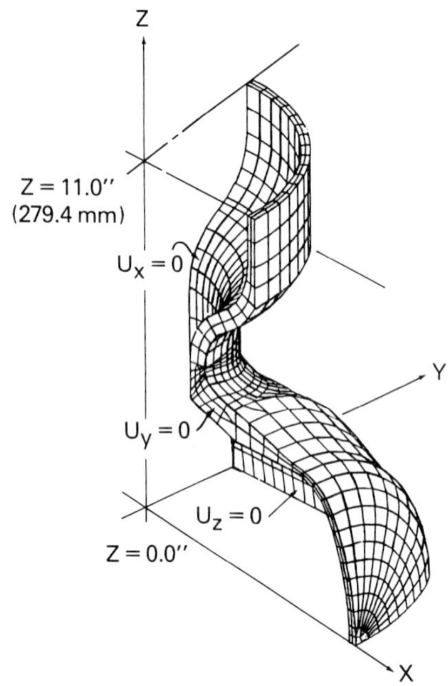

(d)

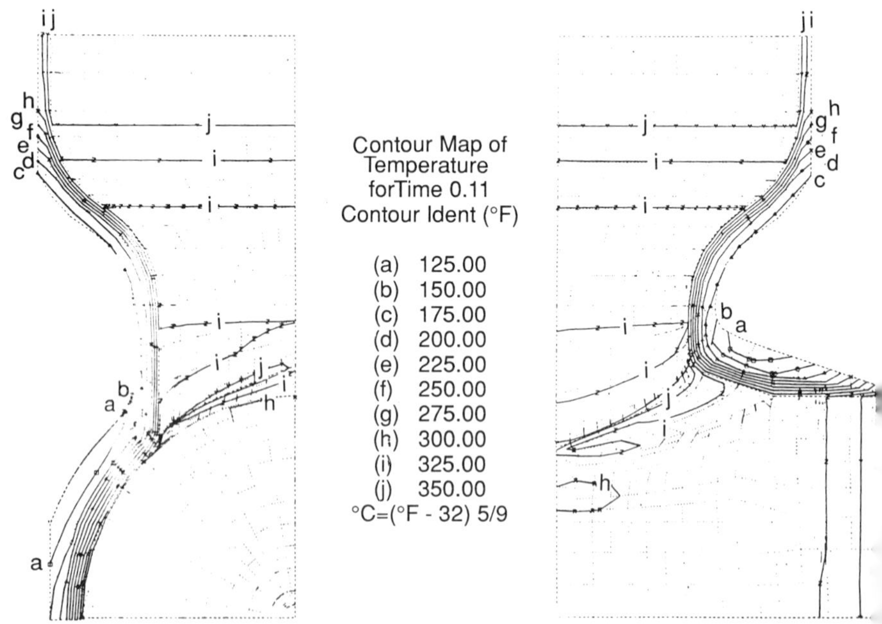

(e)

Figure 10-4. *(continued)*

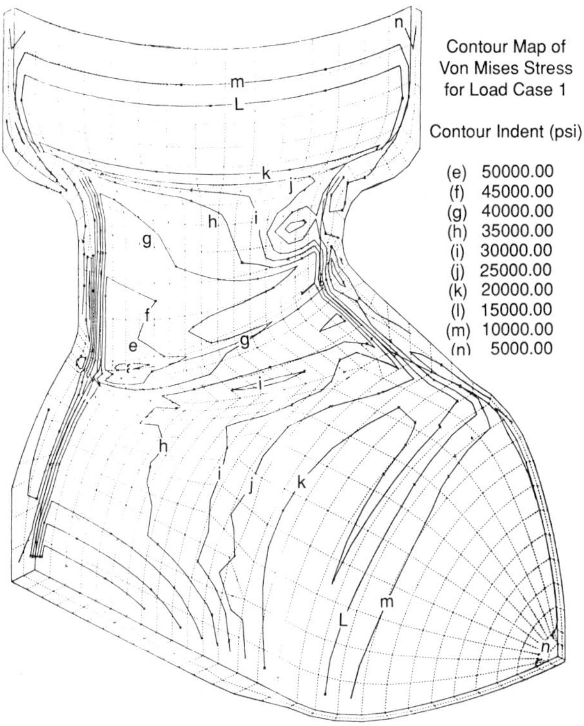

(f)

Figure 10-4. (*continued*)

part of the experiment was performed with a finite-element computer simulation. Figure 10-4(e) illustrates the thermal gradients with a computer drawing of constant temperature contours along the surface of the model. These gradients occurred at a specific point in the transient solution. Because the gradients varied in time, the computer solution was able to provide enough discrete drawings (or files) as a function of time to allow an animation of constant temperature contours moving spatially as a function of time. At any point, the equation-solving part of the computer program could calculate the stresses from the thermal, pressure, and any other applied loads. Figure 10-4(f) illustrates the computer-calculated constant stress contours on the surface and the interior of the component. Effective stress in this case refers to the von Mises stress state (see Appendix B for a brief discussion of von Mises and Tresca stress criteria). As shown in Figure 10-4(f), the highest stress occurred at point A on the structure; therefore, the evaluation of how the stress state complies with acceptable performance criteria (primary, secondary, shakedown [35 m], fatigue limits) focuses first on this location as being the likely location for breaching of the vessel wall.

Next, consider the case of a plate that is one-quarter inch thick and six inches wide, with a heat flux (Q_0) applied on the surface along its length (see Fig. 10-5(a)). The plate is composed of pyrolitic graphite;

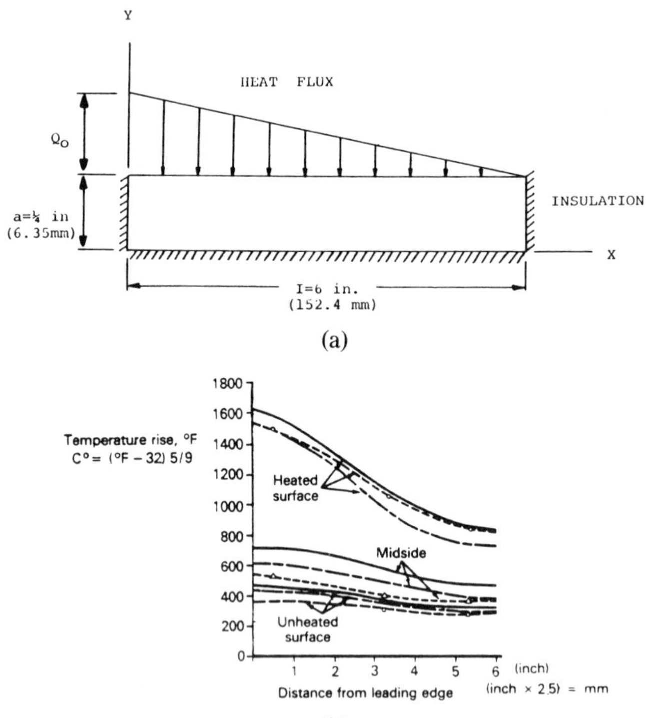

Figure 10-5. Theoretical and experimental thermal analysis of an orthotropic plate. (From S. J. Brown and Hsu 1977) (a) Two-dimensional orthotropic (pyrolytic graphite) plate subjected to linear heat flux over a single surface. (b) A comparison of the temperature distribution along a pyrolytic graphite plate (heated surface, midside, and unheated surface) by finite element, experimental, and series solutions. — = finite element; ---- = experimental; –·– = analytical.

thus, the properties are orthotropic (i.e., the thermal properties in the thickness direction are different from those in the width direction). The heat flux across the surface is applied by directing an oxyacetylene gas parallel to the surface to cause an uneven distribution of thermal stress. The temperature distributions along the heated surface, the unheated surface, and the midplane are determined by thermocouples. A comparison of the experimental results with the numerical finite-element method and a series solution is provided in Figure 10-5(b) (for the time of 20 seconds after the heating process began). (The theoretical and experimental results are discussed in S. J. Brown and Hsu 1977.)

Plasticity (Path-dependent Loads) Analysis

Figure 10-6(a) illustrates a high-heeled shoe (size 9) with the arch reinforced by a steel insert. The purpose of the experimental test was to determine the elastic limit of the shoe: in other words, the maximum allowable weight of the wearer before the heel buckles. The ergonomics of a person walking downstairs will be the maximum load expected

(a)

Figure 10-6. Plastic analysis of a shoe. (Courtesy of Quest Engineering & Development Corp.) (a) Dead weight load apparatus for measuring load versus deflection of a high-heeled shoe. (b) A slice view of the shoe with a metal-reinforced arch support. (c) Load versus deflection curve for the shoe.

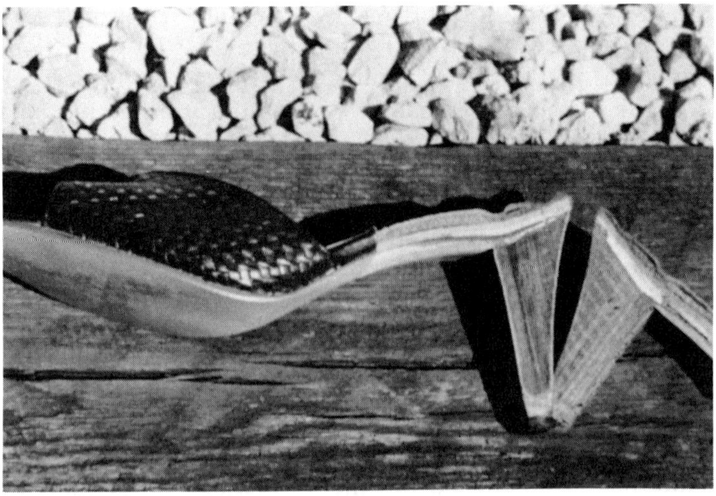

(b)

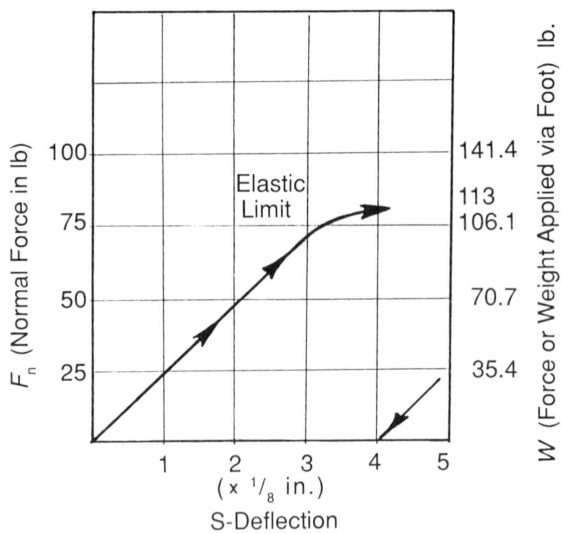

(c)

Figure 10-6. *(continued)*

during normal wear. As a person descends stairs, all the weight is put on the leading foot. The contact of the shoe on the stair is dynamic and will induce a greater force than does a static load. The function of the theoretical solution by the finite-element method is to correlate the theoretical and experimental results and then optimize the size of the metal insert to provide maximum performance and safety. Figure 10-6(b) shows a cross section of the shoe, and Figure 10-6(c) illustrates the experimental relationship of its load and deflection. The elastic limit is reached for a static load of approximately 106 pounds of force, which clearly is unacceptable unless the manufacturer is willing to place a warning on the shoe limiting the wearer's weight to 106 pounds. Obviously, the need for warnings could be eliminated by using a metal insert that can accommodate more weight; however, specifying a weight limit for a shoe with such an unstable configuration would not be farfetched.

Next, consider the computer simulation of a dome (Fig. 10-7(b)) that is clamped along its perimeter and loaded at the center by force *P*. Figure 10-7(a) illustrates the finite-element model. Figure 10-7(c) provides two comparisons: first, curve A, the actual elastic–plastic load versus deflection, is compared with curve B (a large deformation solution) and curve C (a linear elastic theoretical solution). Second, several numeri-

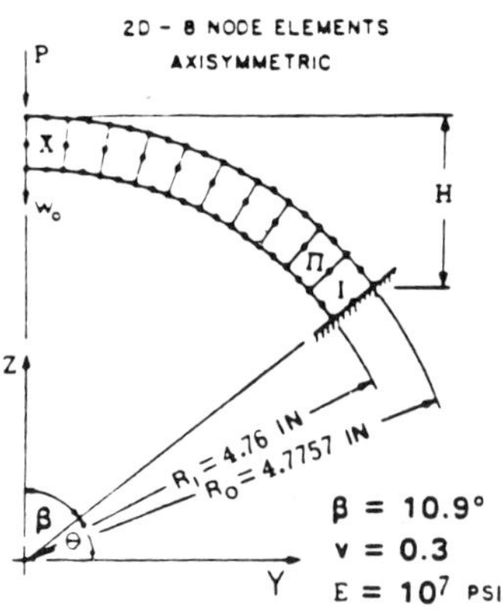

(a)

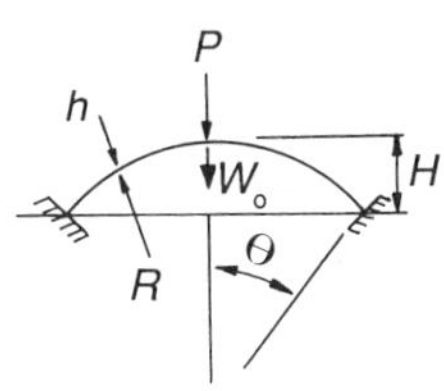

(b)

Figure 10-7. Comparison of linear versus nonlinear simulation of load versus deflection of a spherical cap. (a) Finite-element model of a cap clamped along its edge and loaded at the center (axis). (b) Geometric and material properties of the cap. Ten 8-node elements for axisymmetric analysis. $R = 4.76$ in.; $h = 0.01576$ in.; $H = 0.0859$ in.; $\theta = 10.9°$. (c) Load versus deflection curves for the spherical shell (A = elastic–plastic, B = large deflection, and C = linear solution). — = TL, UL solutions, no equilibrium iterations; -- = Stricklin (1970) (30 shell elements); ● = Mescall (1965) (finite-difference solution; -·- = elastic–plastic. (d) Elastic–plastic response of the spherical cap to central loading. (From Padovan and Tovichakchaikul 1982a, b)

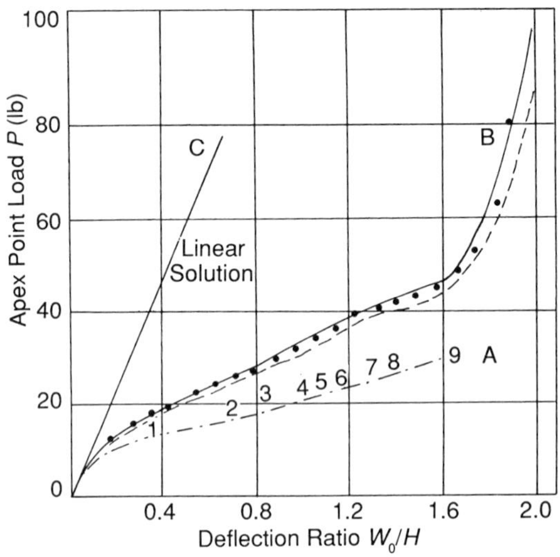

(c)

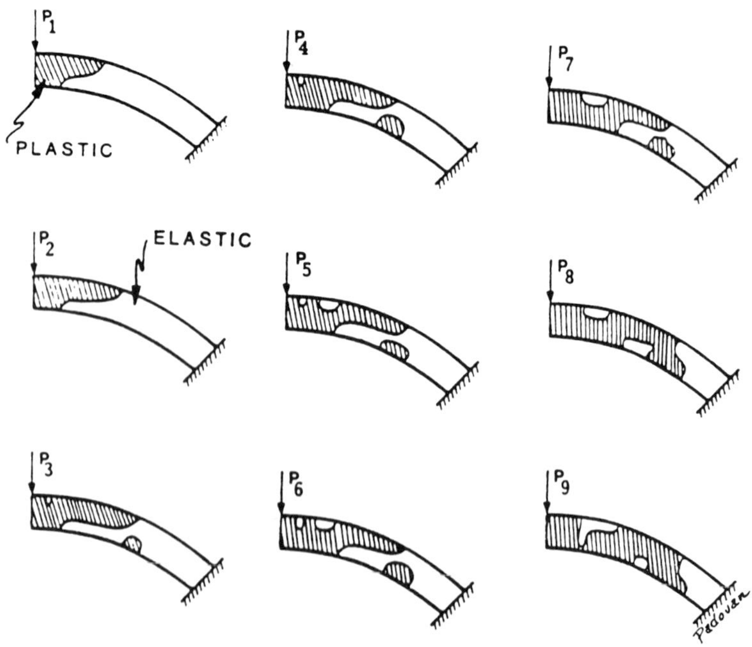

(d)

Figure 10-7. *(continued)*

cal solutions are compared for the nonlinear large-deflection solution illustrated in curve B. Figure 10-7(c) is a modified version of the original curve (Bathe, Ozdemir, Wilson 1974). The simulation shown in Figure 10-7(d) (performed by J. Padovan, a professor of mechanical engineering at the University of Akron) illustrates the regions of material in the dome that become plastic, that is, that exceed the initial yield surface. These theoretical computer simulations not only provide an extensive history of the strain in the structure but also demonstrate that the linear elastic simulation vastly underestimates the amount of deflection associated with a given load. Such structures are used in a number of applications, such as pressure-vessel access ports and as rupture disks (in which the dome is deflected against a knife edge to release the overpressure).

Vibration Analysis

Figure 10-8(a) is a schematic of a test layout for investigating the vibration of a thin dome. The effects of the boundary conditions at the support flange on the frequency of vibration has been discussed by S. J. Brown and Wu (1977). A loudspeaker causes the dome to vibrate, and the deflection (mode of deflection shape) is computed for each resonant

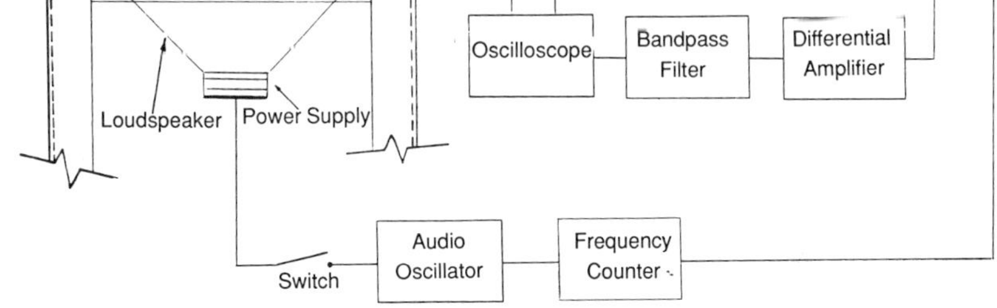

(a)

Figure 10-8. Experimental and theoretical vibration of a cap. (a) Schematic layout of test equipment for the free vibration of a spherical cap. (b) Finite element computer axisymmetric cap or dome. (From S. J. Brown 1977) (c) Comparison of experimental, finite-element, and series solution natural frequencies (Hz). (From S. J. Brown 1977) γ = rotational stiffness; K = tangent stiffness; G = normal stiffness; (M/M^*) = dimensionless mass; M^* = mass of shell above clamping ring.

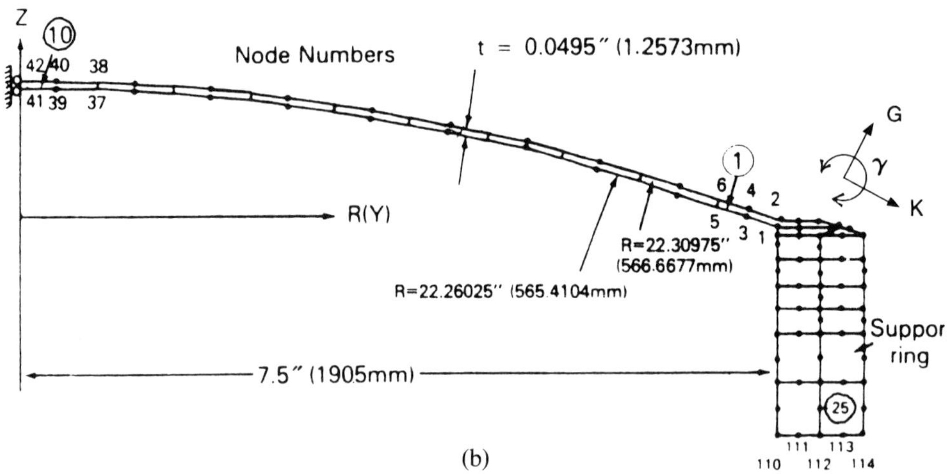

(b)

Natural Frequencies (Hz)

		$\gamma = \gamma_\infty = \infty$	$G = G_\infty = \infty$	
Frequency Mode Number	***Experi-mental***	***Series Solution about K = 1200 lb/in. M=0***	***Finite Element about K ≈ 1000 lb/in. M=0***	***Finite Element about K ≈ 3000 lb/in. M = (0.94)M****
ω_1	211	211	211	215
ω_2	446	389	392	393
ω_3	513	432	455	460
ω_4	649	530	630	605
ω_5	884	694	851	829
ω_6	1051	920	1145	1047

(c)

Figure 10-8. *(continued)*

frequency with a fotonic sensor (a fiber-optic means of measuring the deflection via light intensity). Figure 10-8(b) illustrates the finite-element representation of the dome and its support. In addition to the finite-element numerical solution, a series solution simulation of the frequencies of vibration and mode shapes also is calculated to provide an upper and lower bound solution. Figure 10-8(c) compares the first six experimentally determined frequencies by the two theoretical methods of analysis. The results (see Fig. 10-8(c)) provide estimates about (1) the accuracy of theoretical simulation in predicting the natural frequencies of vibration for the dome component, (2) the influence on the frequencies of the dome of the tightness of the flange bolts installed along the perimeter of the dome, and (3) the need to measure a series of experimentally

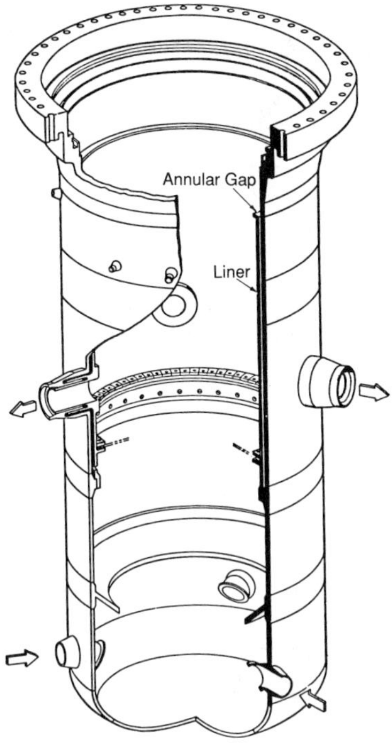

(a)

Figure 10-9. Vibrating structures in fluid. (a) A drawing of a reactor vessel and thermal liner, Clinch River Breeder reactor. (From Department of Energy) (b) A natural frequency test setup with acrylic coaxial cylindrical pairs Nos. 12 and 3. (From Chu and Brown 1985) (c) Finite-element three-dimensional mesh of Nos. 7, 8, and 9 cylinders. (Courtesy of Quest Engineering & Development Corp.) (d) Finite-element mesh of a cantilevered cylinder in a fluid-filled tank. (Courtesy of Quest Engineering & Development Corp.) (e) Three-dimensional finite-element cylinder, Mode 3, in a tank of water (water elements not shown). (Courtesy of Quest Engineering & Development Corp.) (f) Three-dimensional modal plot out-of-phase $n = 3$ of a coaxial pair of cylinders with H_2O filled annular gap (selected elements), thickness t_1 and $t_2 = 0.125$ inches, $\omega_3 = 8$ Hz. (Courtesy of Quest Engineering & Development Corp.) (g) Out-of-phase frequency versus annular gap (clearance between concentric cylinder) for four fluid viscosities. Outer cylinder thickness = 3.175 mm, inner cylinder thickness = 3.175 mm. (From Chu and Brown 1985) (h) Comparison of numerical solutions and experimental results for fluid wave profiles in a container. (From Su and Kang 1986)

(b)

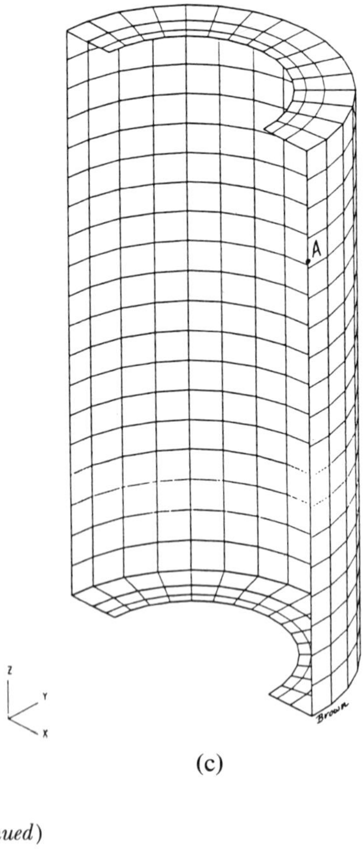

(c)

Figure 10-9. *(continued)*

derived frequencies to assess statistically the percentage variation of frequency in the actual dome component as produced.

An often-overlooked characteristic of the vibration or dynamics of structures is the effect of a gas, fluid, or two-phase medium. The frequency of vibration and response of internals in a process loop and of structures such as offshore platforms is influenced by the surrounding media. The data from scale-model testing of cylinders vibrating in air and emersed in a fluid can be found in Chu *et al.* 1979. The scale-model acrylic cylinders, which were approximately 26 inches long, 12 to 13.625 inches in diameter, and 0.125 to 0.25 inch thick were tested to assess the effects on frequency of cylinder thickness and gap thickness—the fluid-filled space between the inner cylinder and a concentric outer cylinder. Chu and his colleagues used these test data to verify or benchmark predictive theoretical methods. The theoretical methods in turn are used to evaluate the frequency and modal response of large expensive

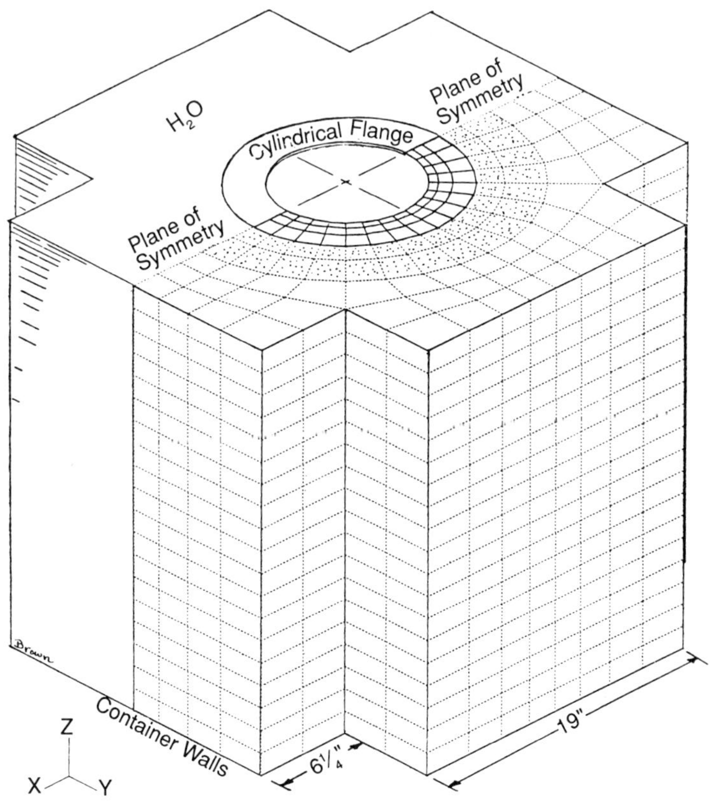

Figure 10-9. *(continued)* (d)

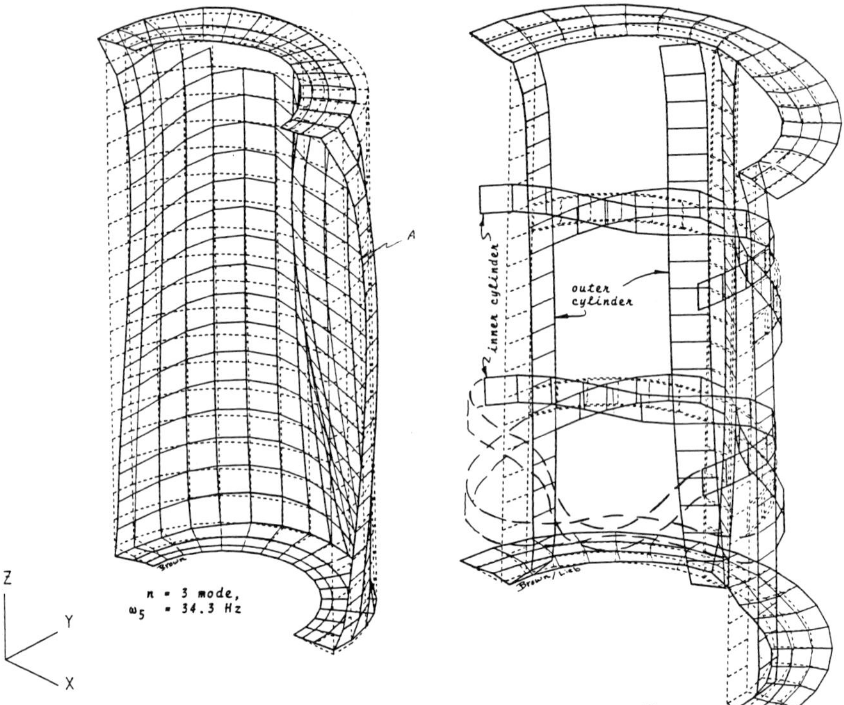

(e) (f)

Figure 10-9. *(continued)*

structures such as the vibration of the pipe liner in the simulation discussed in Chapter 5.

Another example is the thermal liners found in large chemical and nuclear reactors, as illustrated in Figure 10-9(a). Figure 10-9(b) shows the testing apparatus that is used to determine natural frequencies of the acrylic cylinders, individually and as concentric pairs with fluid in the annulus between the two. Chu and Brown (1981) summarized the experimental test procedure in which a shaker was used to excite the cylinders to a desired frequency, and an eddy-current technique was used to map the mode shapes (displacements). Figure 10-9(c) is a finite-element model simulation of one of the acrylic cylinders (only half is shown). The importance of the effect of the environment on the vibration of a structure can be illustrated best by considering the third natural frequencies of the cylinder shown in Figure 10-9(c): (1) ω_3 = 150 Hz in air, (2) ω_3 = 34.3 Hz in a large reservoir of water (see Fig. 10-9(d) and 9(e) for mode or deflected shape), and (3) ω_3 = 8 Hz for the acrylic cylinder placed within another flexible acrylic cylinder with a fluid-filled gap of 0.125 inch between the two (see Fig. 10-9(f)) (Chu *et al.* 1979, S. J.

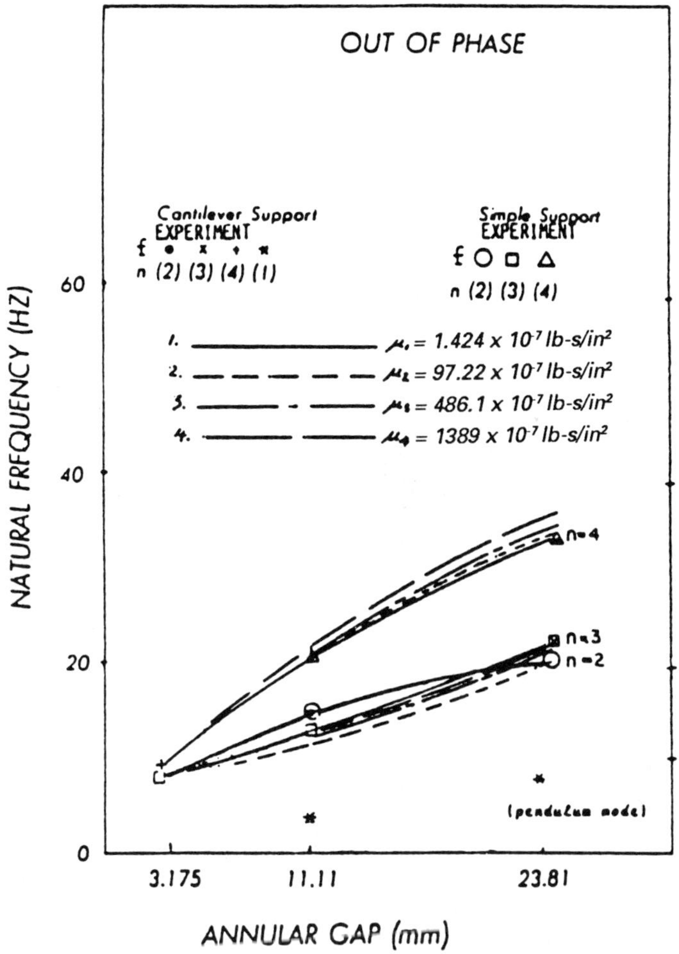

(g)

Figure 10-9. *(continued)*

Brown 1982). Chu and Brown (1985) reported the results of experimental simulations of coaxial cylinder vibration with various fluid viscosities coupling them (see Fig. 10-9(g)), and in Yan (1982) Brown discussed the effects of axial length and mode shape of coaxial cylinder vibration with a coupling fluid.

Figure 10-9(h) compares experimental and theoretical (finite-difference) simulations for fluid sloshing in a container that is sinusoidally vibrating (Liu *et al.* 1986). This type of test is of particular interest with respect to spills from large storage tanks that may be subjected to foundation motion.

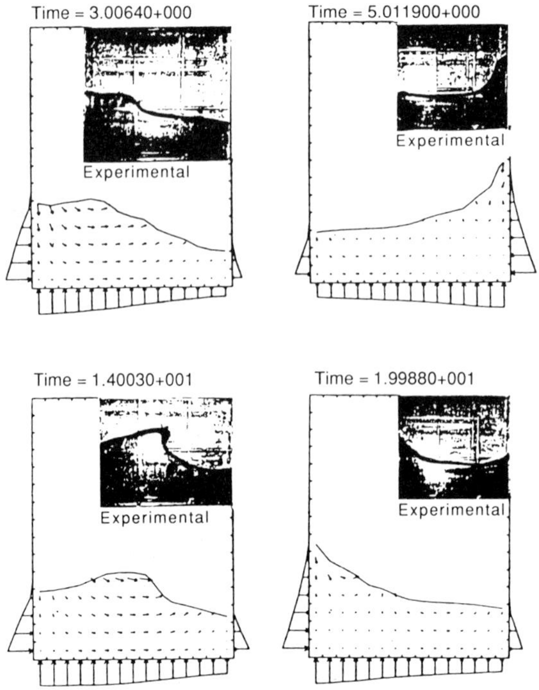

(h)

Figure 10-9. *(continued)*

Force-Displacement-Time Histograms

Figure 10-10(a) shows a test setup that permits the input of displacement to the test fixture as a function of time. The test fixture supports a pair of coaxial acrylic cylinders that are coupled by fluid in the annulus space between the cylinders. The purpose of the test is to correlate the results of the scale-model experiment with the results of theoretical predictive methods. Once the theoretical predictive methods are thus benchmarked for certain parametric ranges of seismic foundation displacements, they can be used to predict the responses of large components and the accompanying stress levels in the components. Figure 10-10(b) shows a schematic of the test setup. Figure 10-10(c) illustrates the input acceleration versus time histogram or curve to the frame to which the acrylic cylinders are attached (point C in Fig. 10-10(b)). Figure 10-10(d) also illustrates the output of relative displacement between points A and B shown in Figure 10-10(b). Global acceleration at points A, B, and C is measured with accelerometers; relative displacement between points A and B and points A and C are measured by eddy-current (small metallic

(a)

Figure 10-10. Fluid–structure dynamic coupling. (a) A seismic shaker test setup with cylinder pairs Nos. 12 and 3 (gap filled with fluid). (From Chu and Brown 1985) (b) Schematic of the seismic test setup. (c) Acceleration time–history curve sample (at point C) cylinder base (experimental). (d) Relative displacement of point A to B versus time (for three experimental tests). (From S. J. Brown and Chu 1983) (e) Three-dimensional finite-element (Nos. 12 outer with 3, 6, or 9 inner) cylinder pair idealization with fluid. (Courtesy of Quest Engineering & Development Corp.) (f) A seismic shaker table with a model of a building attached. (Courtesy of the University of California at Berkeley)

disks are glued to the acrylic surface). The table below compares the displacement at point A ($U_A \times 10^{-5}$ in.) determined experimentally (RMS) versus values computed by the theoretical finite-element time–history solution (deterministic method) and to the response-spectrum solution (quasi-statistical method) using the model illustrated in Figure 10-10(c).

$U_A \times 10^{-5}$ in. **Displacement at Point *A* (see Fig. 10-10(b), (e))**

	Finite Element Computer Simulation	
Experimental (RMS)	Time History (max)	P Spectrum
8.34 ($H_2 0$)	8.7 (20 percent damping)	9.1 (20 percent damping)
NA	21.0 (0 percent damping)	22.04 (0 percent damping)

When one assumes zero damping, the displacement at point A in the above table illustrates the advantages of the theoretical solution for ex-

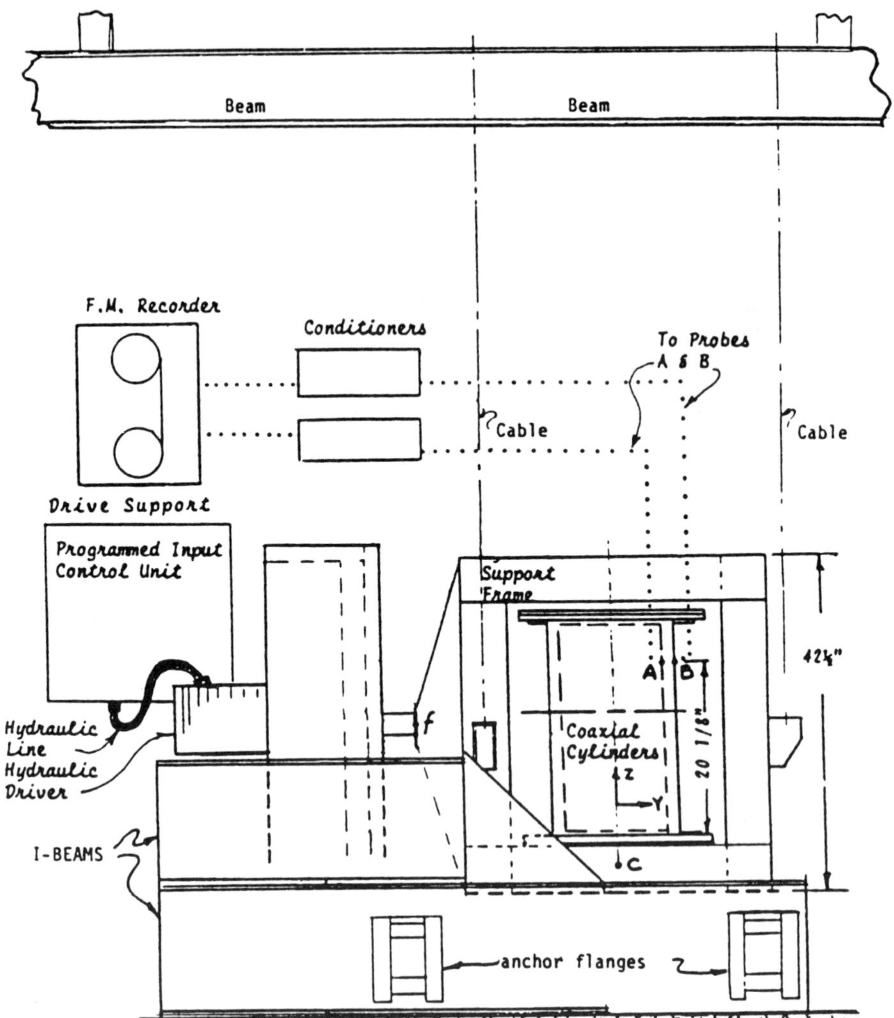

(b)

Figure 10-10. *(continued)*

amining the effects of phenomena that cannot be readily examined experimentally. The input displacement versus time histogram for this test was based on a measured seismic event at El Centro, California; however, the input could have simulated other dynamics to the components of the scale model, such as flow-induced vibration, foundation motion, and shock. Figure 10-10(f) illustrates the shaker table seismic simulation

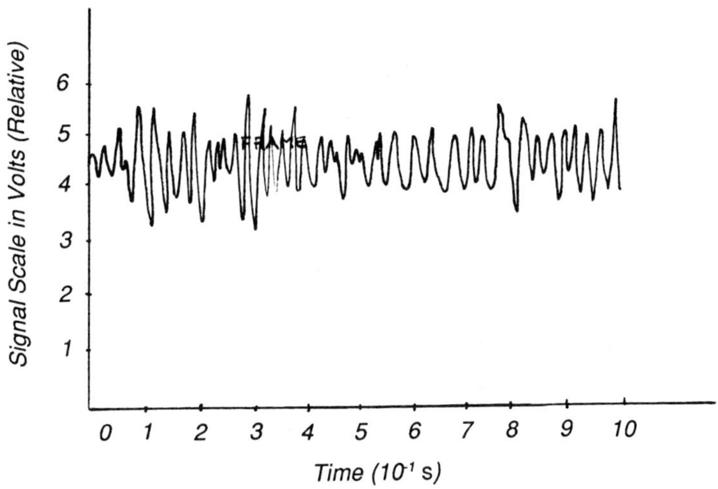

(c)

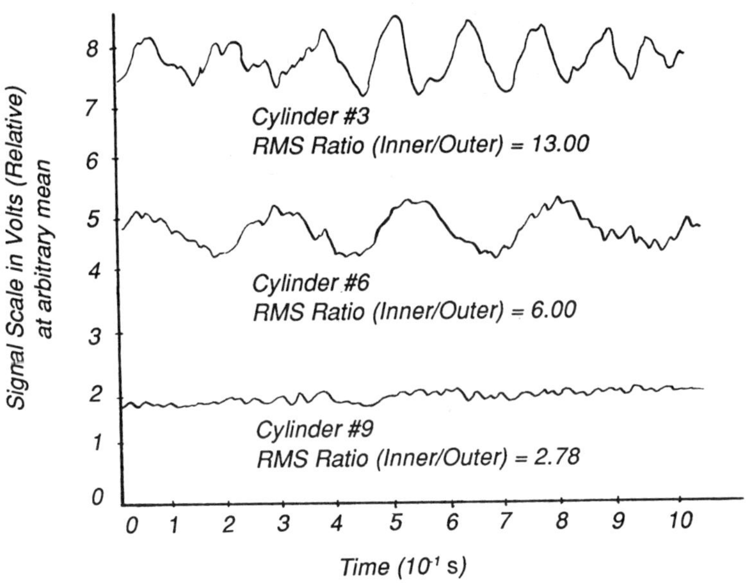

(d)

Figure 10-10. *(continued)*

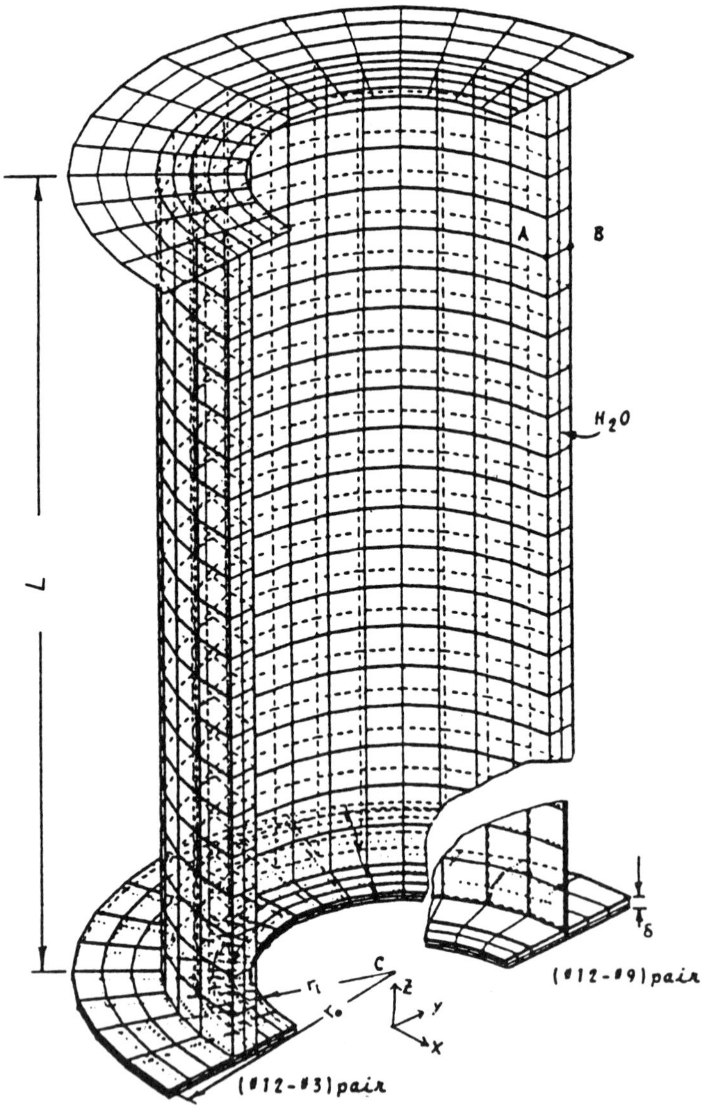

(e)

Figure 10-10. *(continued)*

(f)

Figure 10-10. *(continued)*

of a scaled building that was performed at the University of California at Berkeley. (For a more detailed description of the tests reviewed here, see S. J. Brown and Chu 1983.)

Flow-induced Vibration Analysis

Fluctuations in pressure associated with the flow of fluids and gases within or over the surfaces of a structure have been studied for decades as an excitation mechanism to induce vibration in the structure and thus cause metal fatigue. The recommended practice for designing structures that will operate in a flow field is to either change the resonant structural frequencies or the vortex shedding frequencies so that they do not coincide (Chen 1985). The effect of materials selected on the basis of fatigue properties (when considering a fixed design) means that the difference between two materials is not likely to be *if* a failure will occur, but *when*. For example, an impeller made of 304L stainless steel will fail much sooner than one made of a nickel–aluminum–bronze alloy (Fig. 10-11(e)). In this example, computer simulation (Fig. 10-11(a), (b), (c)) allows the engineer to economically examine changes in the impeller's configuration that will reduce or improve its performance before testing theories on an actual impeller. Figure 10-11(d) illustrates the fatigue mode of failure of the blades by the flow-induced vibration mechanisms of failure. In situ measurements of natural frequency are made experimentally using an accelerometer and a Fast Fourier Transform analyzer (S. J. Brown 1989; S. J. Brown and Moody in press). Figure 10-11(f) illustrates the effect of blade thinning by erosion.

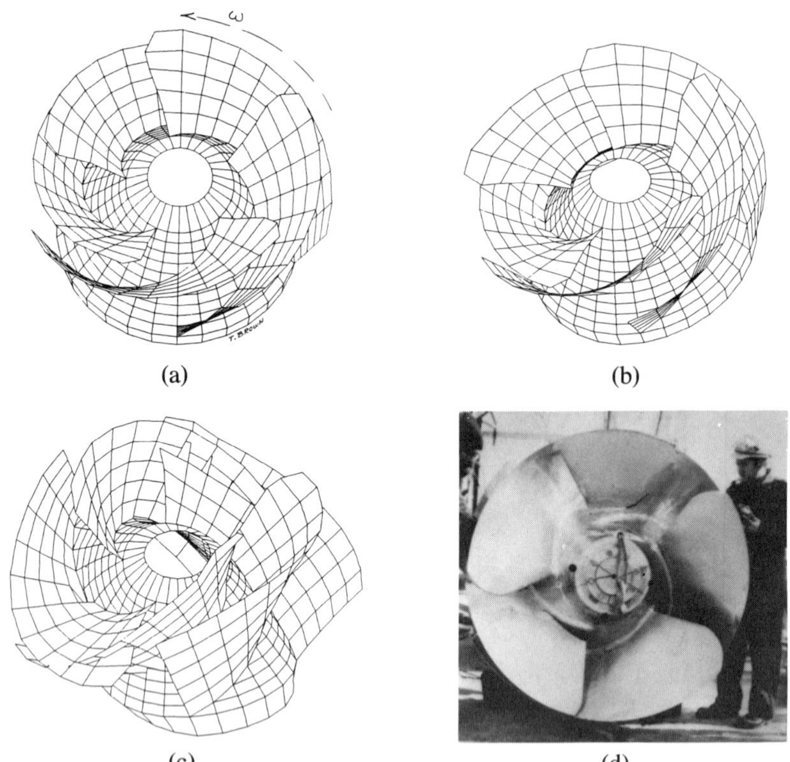

Figure 10-11. Flow-induced vibrational fatigue of an impeller. (Courtesy of Quest Engineering & Development Corp.) (a) Finite-element computer model of a pump impeller that is six feet in diameter. (b) Pendulum (first mode) of flow-induced vibration in water and air: ω pendulum = 173 Hz (in air); ω pendulum = 61.3 Hz (in water). (c) Flow-induced vibration of the impeller in water and air (n = 2, second mode). ω_2 = 161.5 Hz (in air); ω_2 = 73.5 Hz (in water). (d) Impeller blade failure at the hub. (e) Rotating beam fatigue curve for small impeller material. (f) Eroded impeller blade leading edge.

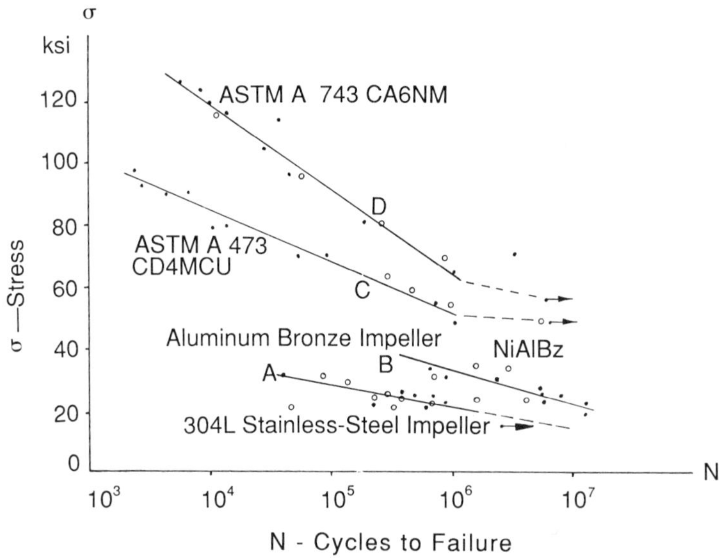

(e)

(f)

(Following page)

Figure 10-12. Parallel flow-induced vibration. (a) Schematic of flow path of an experimental test setup of fluid-coupled concentric cylinders. (From Au-Yang and Galford 1981) (b) Finite-element pendulum mode vibration prediction of the inner liner of the cylinders. Although included, the fluid and outer shell are not shown. (Frequency of vibration, 29.6 Hz.) (From S. J. Brown and Hsu 1978) (c) Flow induced vibration test facility (FIVTF). (Courtesy of Argonne National Laboratory)

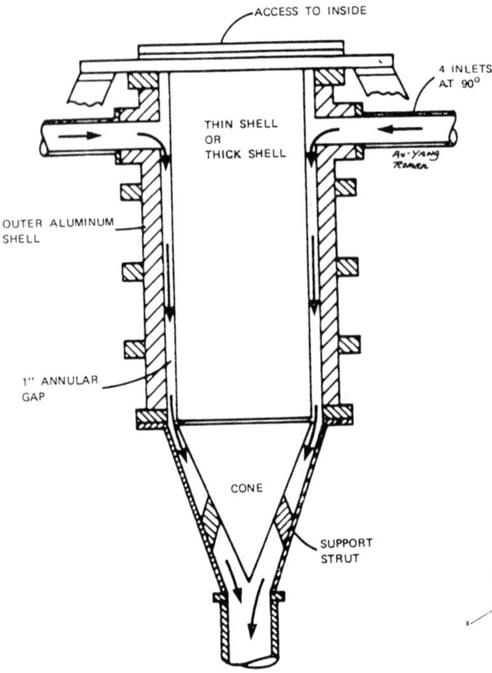
ACCESS TO INSIDE
4 INLETS AT 90°
THIN SHELL OR THICK SHELL
OUTER ALUMINUM SHELL
1" ANNULAR GAP
CONE
SUPPORT STRUT

(a)

Z
X
Y

(b)

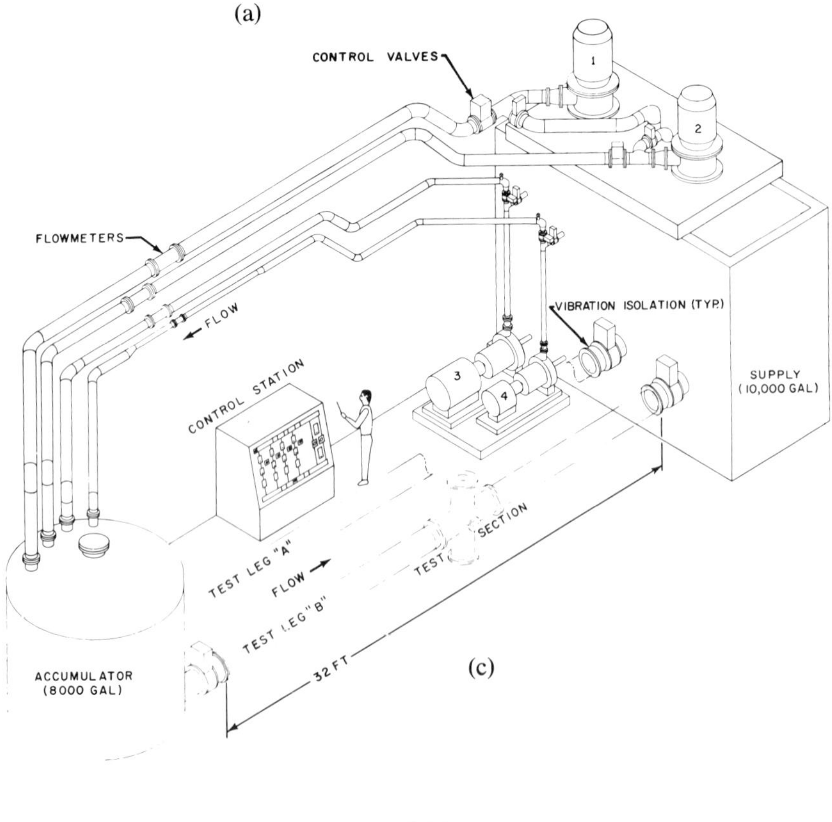
CONTROL VALVES
1
2
FLOWMETERS
FLOW
VIBRATION ISOLATION (TYP.)
CONTROL STATION
3
4
SUPPLY (10,000 GAL)
TEST LEG "A"
FLOW
TEST SECTION
TEST LEG "B"
32 FT
ACCUMULATOR (8000 GAL)

(c)

An example of a scaled cylinder model in a parallel flow-induced vibration apparatus is shown in Figure 10-12(a). Figure 10-12(b) shows a finite-element model used to correlate the experimental results (Papadakas and Scarton 1978). The natural frequencies of the liner are determined experimentally by means of shaker excitation and accelerometer measurements. Pressure fluctuations are measured by pressure transducers. (Au-Yang and Brown 1979 reported on an in situ study of dynamic pressure inside a similar but full-sized system.) Figure 10-12(c) illustrates a flow induced vibration test facility.

Figure 10-13(a) illustrates the flow-induced vibration mechanism of a vortex shedding off of a circular cylinder perpendicular to a flow field. The phenomenon is similar for the impeller, but a cylinder is used here

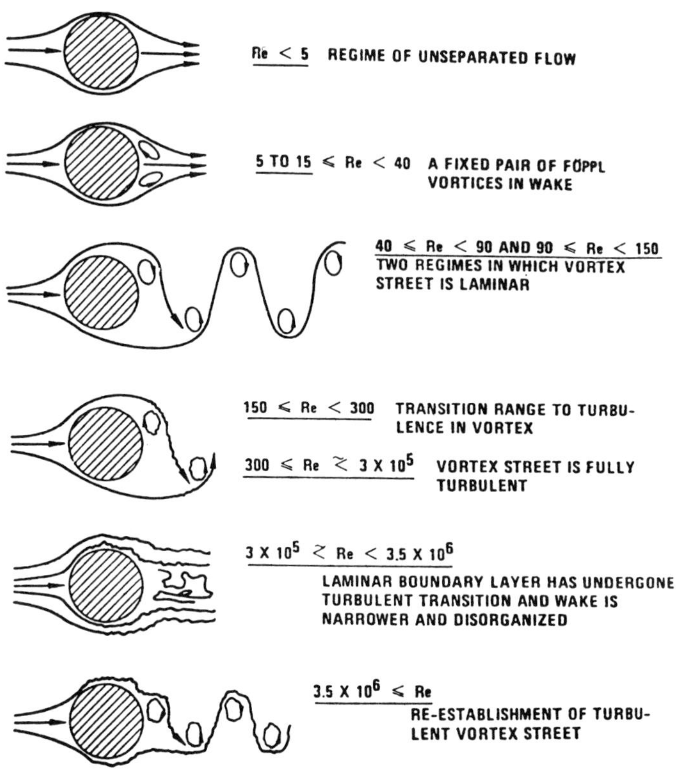

(a)

Figure 10-13. Perpendicular flow-induced vibration. (a) Regimes of flow across a circular cylinder. (From Lienhard 1966) (b) Envelope of Strouhal/Reynolds number relationship for circular cylinders. (From Jendrzejczyk and Chen 1985) (c) Actuator and valve assembly, flow-induced vibration. (Courtesy of Quest Engineering & Development Corp.)

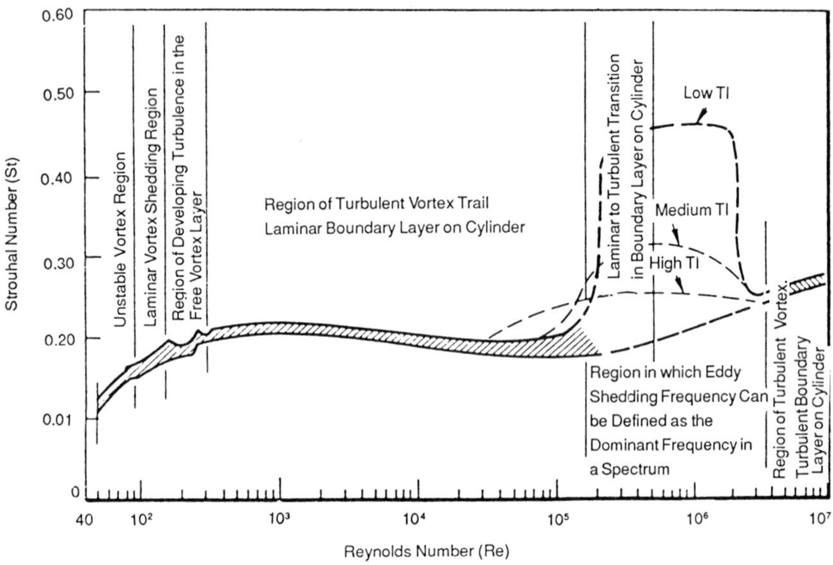

(b)

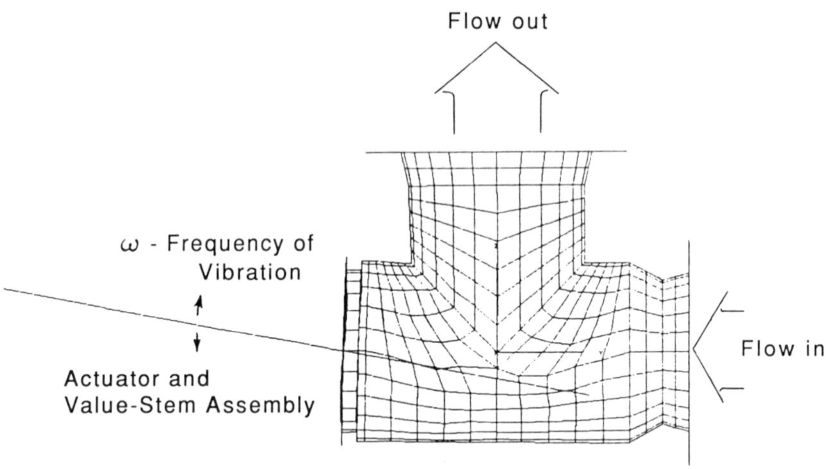

(c)

Figure 10-13. *(continued)*

because of its simplicity and because its geometry has been investigated so thoroughly. Figure 10-13(b) shows the relationship between the Reynold's number and a parameter called the Strouhal number, which is fairly constant over a wide range of Reynold's numbers, except in the region between 10^5 and 10^6. Figure 10-13(c) illustrates a finite-element simulation of an angle valve. The actuator guide tube and shaft were susceptible to fatigue-type failures because of flow-induced vibration.

Another application of fluid-structure dynamics is prediction of the effects of fluid motion on transport vehicles such as ships, aircraft, railroad cars, and trucks, which have large fluid reservoirs (see Fig. 10-9(h)).

Fatigue Analysis

In the previous section, we provided examples illustrating how structures undergo flow-induced vibration by vortex shedding and pressure fluctuations, only one of many mechanisms of induced vibration. The mode of failure that is of concern and is usually associated with induced vibration is material fatigue. Experimental cyclic loading of the actual

(a)

Figure 10-14. Fatigue failure of rotating composite disk. (a) A failed disk made of fiberglass fibers and epoxy resin. (b) Closeup of a crack in the disk. (c) Illustration of the composite test specimen and the loading fixture. Top: Tensile test specimen for Instron TTD test machine; middle: fatigue test specimen for Instron 1350 machine; bottom: jig used in fatigue testing. (d) Photographs of the specimen and posttest fragments. (e) S-N curve for the composite material. ▲ = experimental test of one-quarter-inch-thick disk; —·— = 4-ply specimen $\frac{3}{8}$-in. thick; △ = tensile test failed disk specimen; — = specimen data 3 ply—$\frac{1}{4}$ in. thick; ◇ = failed disk—specimen. (f) Finite-element simulation of point load on grinder disk (top and side view, respectively). (Courtesy of Quest Engineering & Development Corp. and Penn State University)

(b)

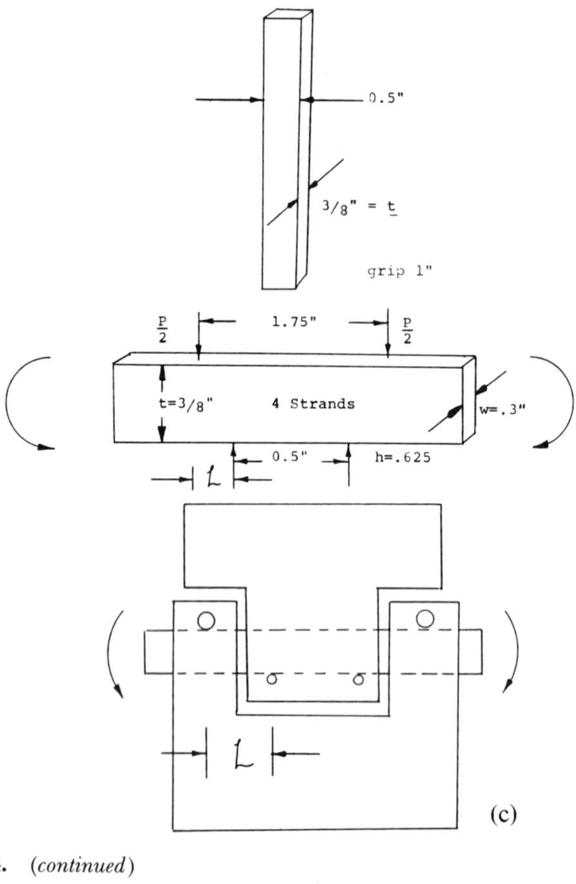

(c)

Figure 10-14. *(continued)*

product or system is usually an expensive method of assessing fatigue characteristics. For this reason, the fatigue characteristics of a material are determined by testing samples using methods such as those discussed in Chapter 8. Knowing the relationship between the stress level and the number of cycles to failure (having a fatigue curve for the material), fatigue life can be predicted by knowing the stress levels in the component. Figure 10-11e illustrates a series of fatigue curves that were developed in conjunction with the impeller simulations illustrated in Figure 10-11. Note that curve A (304L stainless steel) has poor fatigue properties compared with nickel–aluminum–bronze and duplex stainless-steel alloys. Thus, for the same geometry, the same loads, the same mass density, and the same elastic properties, greater fatigue life for the impeller can be achieved by using a duplex stainless steel (curve D), provided there are no problems with respect to maintaining material properties during fabrication and/or use. Figure 10-11(e) shows the results of a fatigue test on a rotating beam specimen.

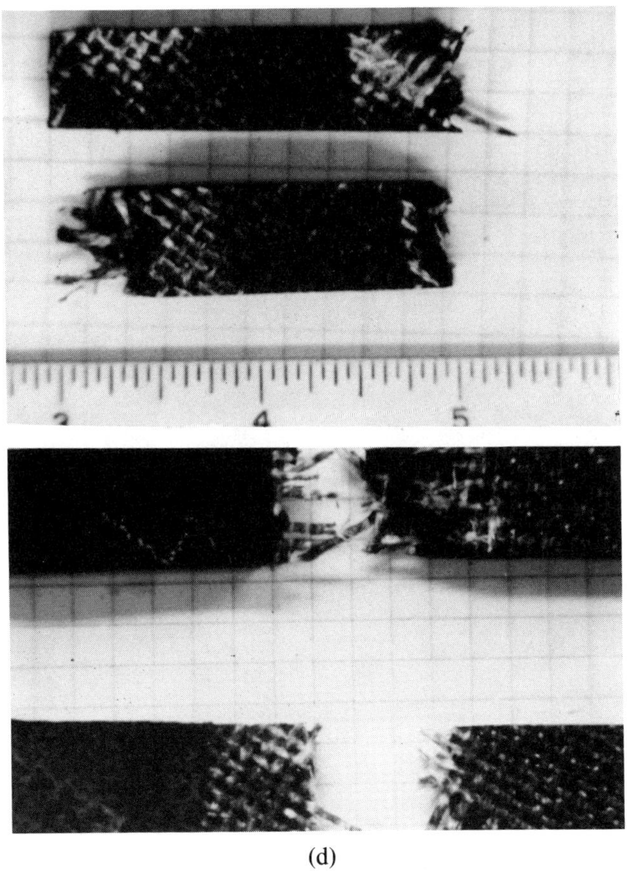

(d)

Figure 10-14.
(continued)

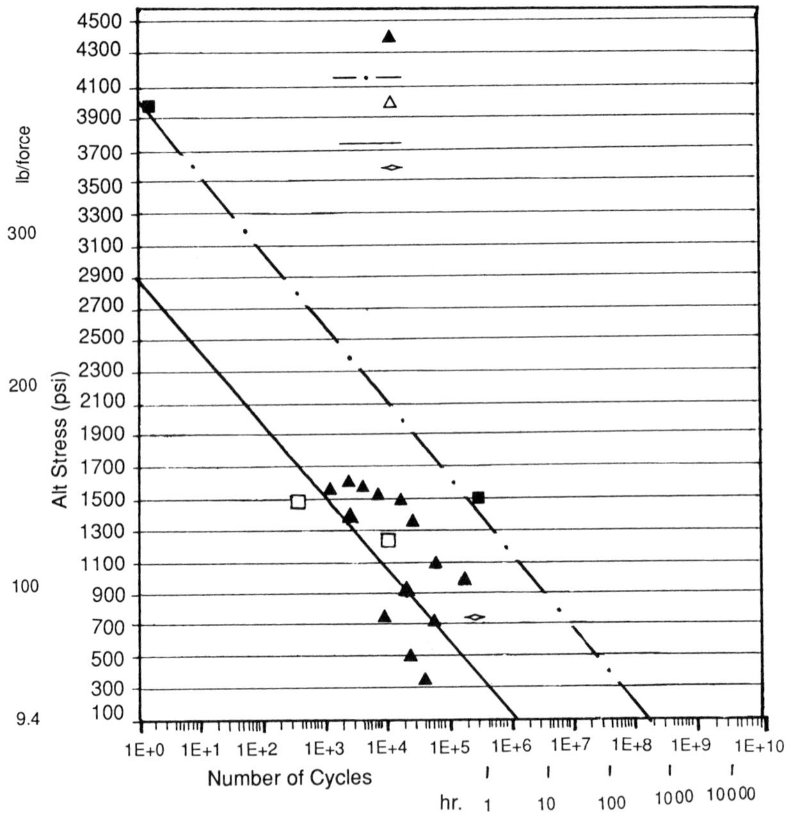

(e)

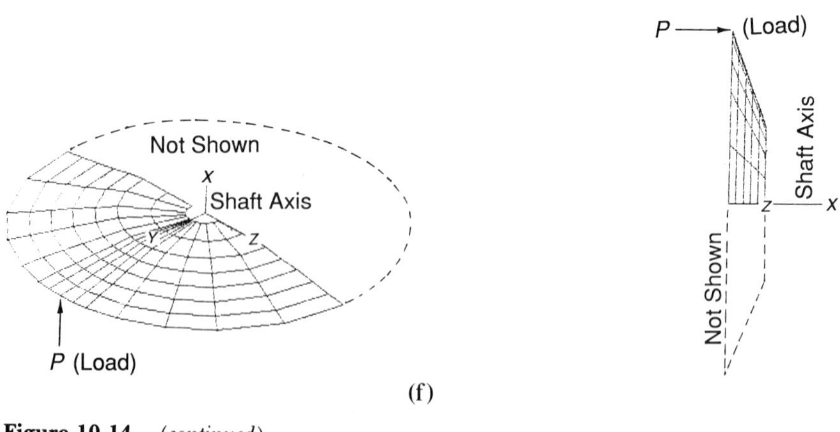

(f)

Figure 10-14. *(continued)*

Figures 10-14(a) and (b) show a global and macroscopic view of a failed composite disk. Figure 10-14(c) illustrates the fatigue testing fixture for a beam specimen from the disk, and Figure 10-14(d) provides a photograph of the specimen. Figure 10-14(e) illustrates fatigue data for several different tests. Data points ▲ are experimentally determined alternating stress versus the number of cycles to failure for a full-sized 3-ply disk. Curve A is a linear regression best-fit curve for these experimental data. Curves B and C are derived from sample specimens that are tested in the Instron-loaded beam-bending fixture. To correlate the stress versus the number of cycles from the specimen test data to applied loads on the disk, one must determine, either experimentally or theoretically, the load-versus-stress relationship. Figure 10-14(f) shows a finite-element model simulation that establishes the relationship between applied force on the disk and stress in the material. Note in Figure 10-14(e) that the relationship of applied force versus hours to failure can be computed based on the finite-element simulation.

Impact Analysis

Impact simulations are used to predict the response of a target (Fig. 10-15(a), (b)), a missile (Fig. 10-15(c), where the target is rigid), or the

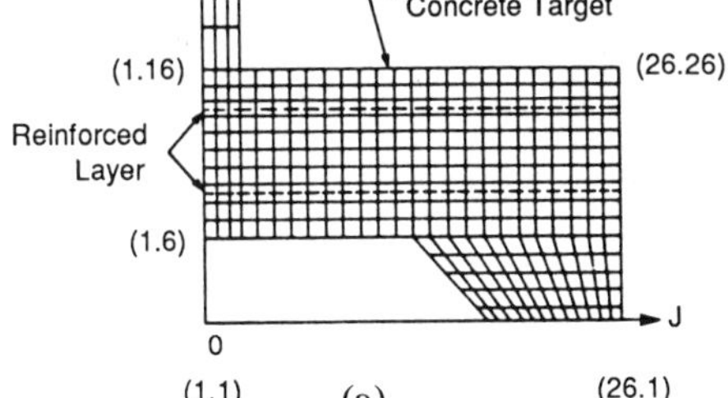

(a)

Figure 10-15. Illustration of experimental and theoretical impact analysis. (From S. J. Brown 1984) (a) Layout for a two-dimensional computer simulation of a steel rod forced into reinforced concrete. (b) Comparison of the computed and observed damage for a 22.3 m/sec penetration into reinforced concrete using the static set of concrete parameters. (c) The impact of a metal bar on a rigid wall. (1) Dyna3d bar impact on rigid wall; time word = 0. (2) Dyna3d bar impact on rigid wall; time word = 2.00097E + 01. (3) Dyna3d bar impact on rigid wall; time word = 4.00039E + 01. (4) Dyna3d bar impact on rigid wall; time word = 6.00023E + 01.

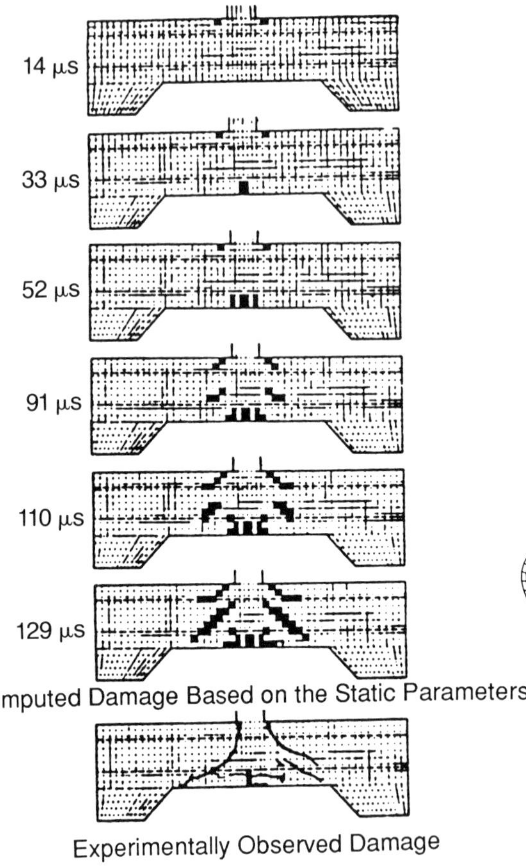

(b)

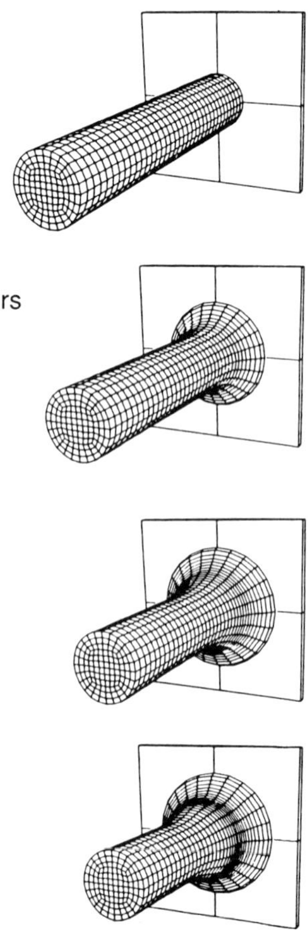

(c)

Figure 10-15. *(continued)*

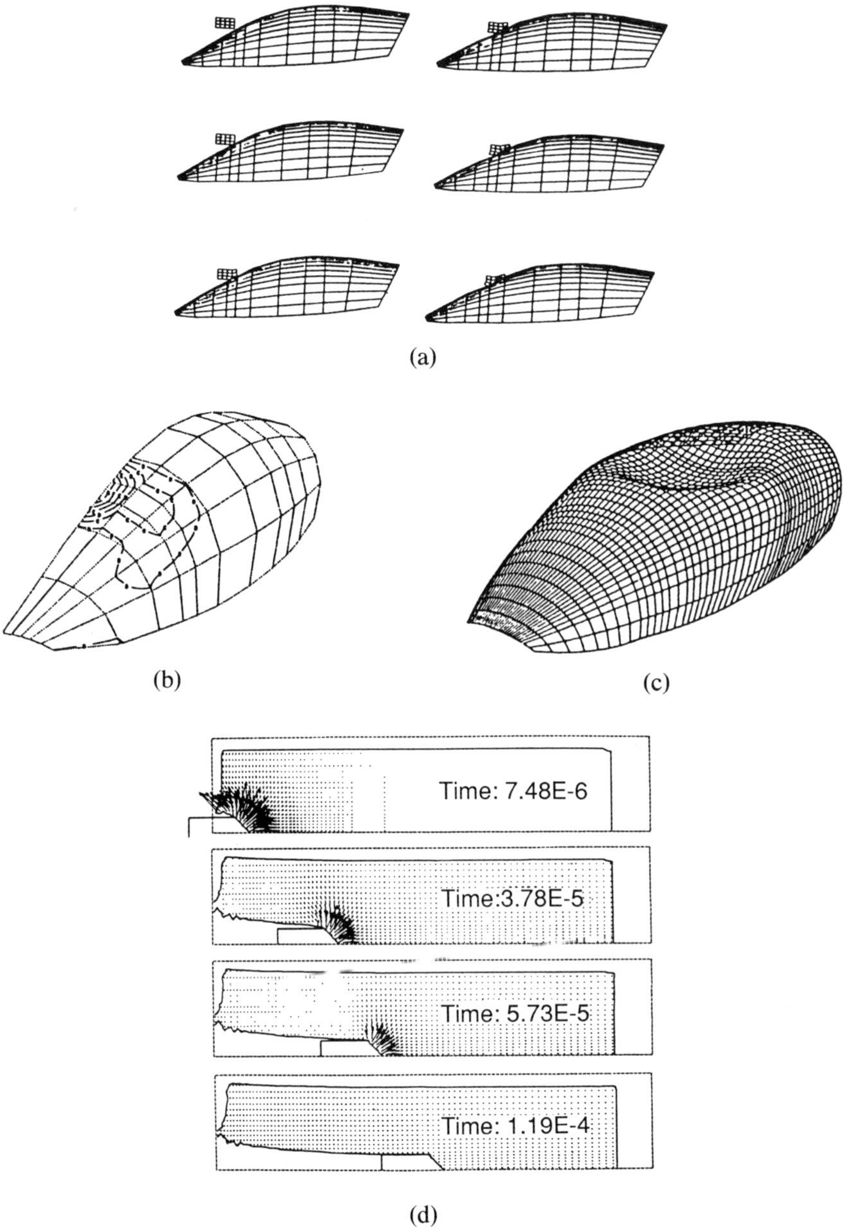

Figure 10-16. A computer-simulated collision of a bird into the windshield of a fighter plane. (a) The collision sequence in time. (b) Equivalent stress contours on the windshield. (c) The deformed windshield after the impact. (d) Rigid projectile impact and penetration into engineering material. (e) Test aircraft specimen in NASA Langley fixture. (f) Test aircraft specimen after impact.

(e)

(f)

Figure 10-16. *(continued)*

missile and the target, both of which are deformable (Fig 10-16). Impact and penetration simulations address both global and local material characteristics (Fig. 10-16(d)). The theoretical and experimental simulations shown in Figures 10-15 and 10-16 are discussed by Shaaban (1986). Figure 10-16(e), (f) illustrates an experimental full-size aircraft impact facility.

Figure 5-5 (p. 144) illustrated the influence of the deformation and dissipating effects that bauxite ore has on the response of the flooring in an ore crusher building. Figure 10-17 illustrates a computer simulation of a forklift truck performed by Clark Equipment Co. using the Automatic Dynamic Analysis of Mechanical Systems (ADAMS) computer program. The theoretical simulations were correlated with experimental tests summarized by McConville (1985).

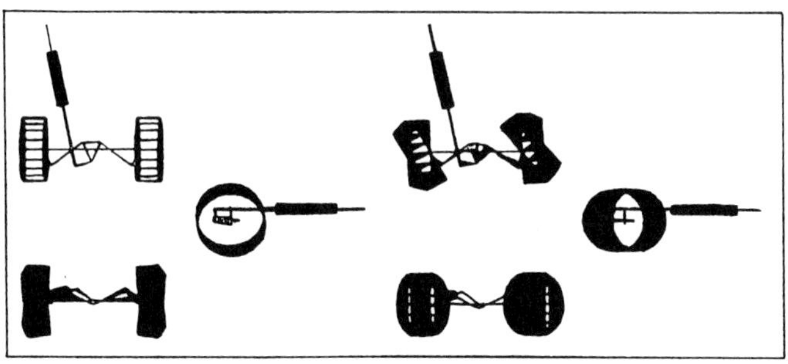

(a)

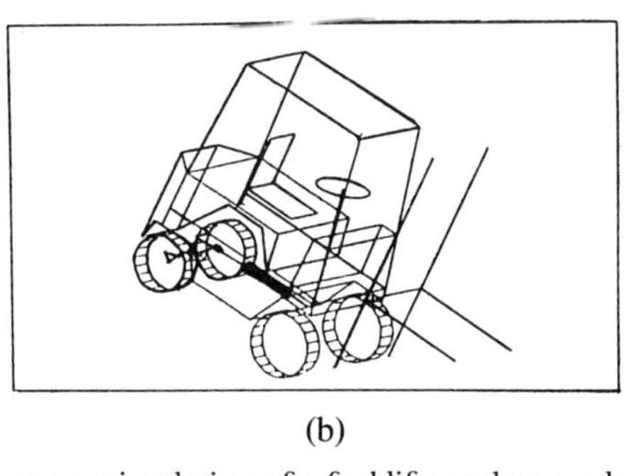

(b)

Figure 10-17. Computer simulation of a forklift truck to study its dynamics. (From McConville 1985) (a) Steering mechanism for the truck. (b) Computer-drawn model of the truck. (c) Elevation view of a simulation of the truck using the Automatic Dynamic Analysis of Mechanical Systems (ADAMS) program. (d) Plan view of computer simulation of forklift maneuver and its dynamics.

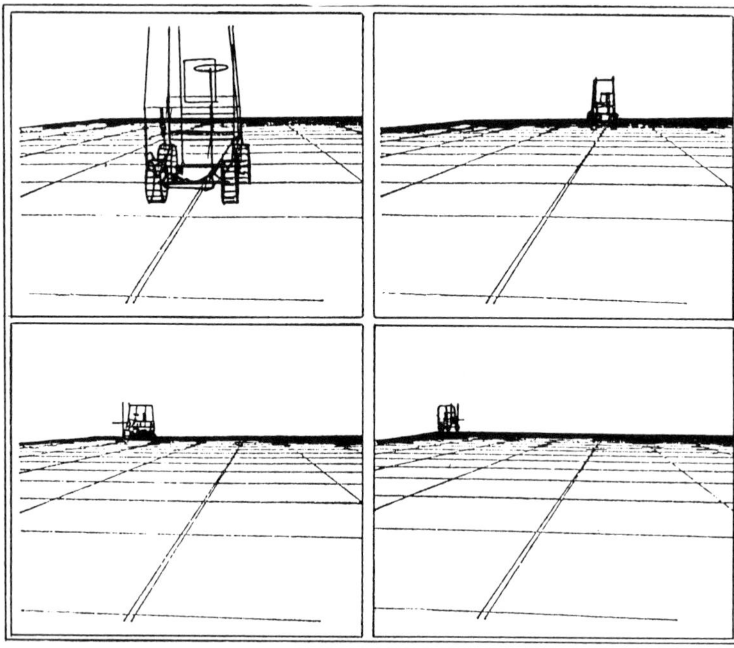

(c)

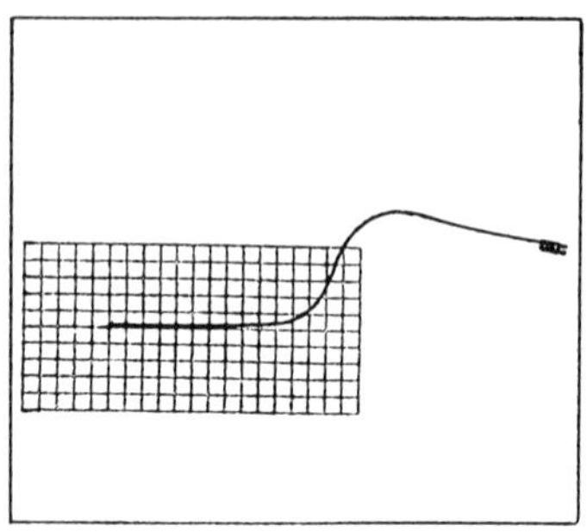

(d)

Figure 10-17. *(continued)*

Hazardous Release Analysis

Figure 10-18(a) illustrates a small-scale test chamber discussed by Fauske and Leung (1985), which can be used as a bench-scale apparatus to obtain data that can be extrapolated directly to full-sized process vessels for vent sizing: measurements of thermal data under nearly adiabatic conditions as well as a flow regime viscosity characterization under runaway conditions. Figure 10-18(b) and 18(c) illustrates tests versus model

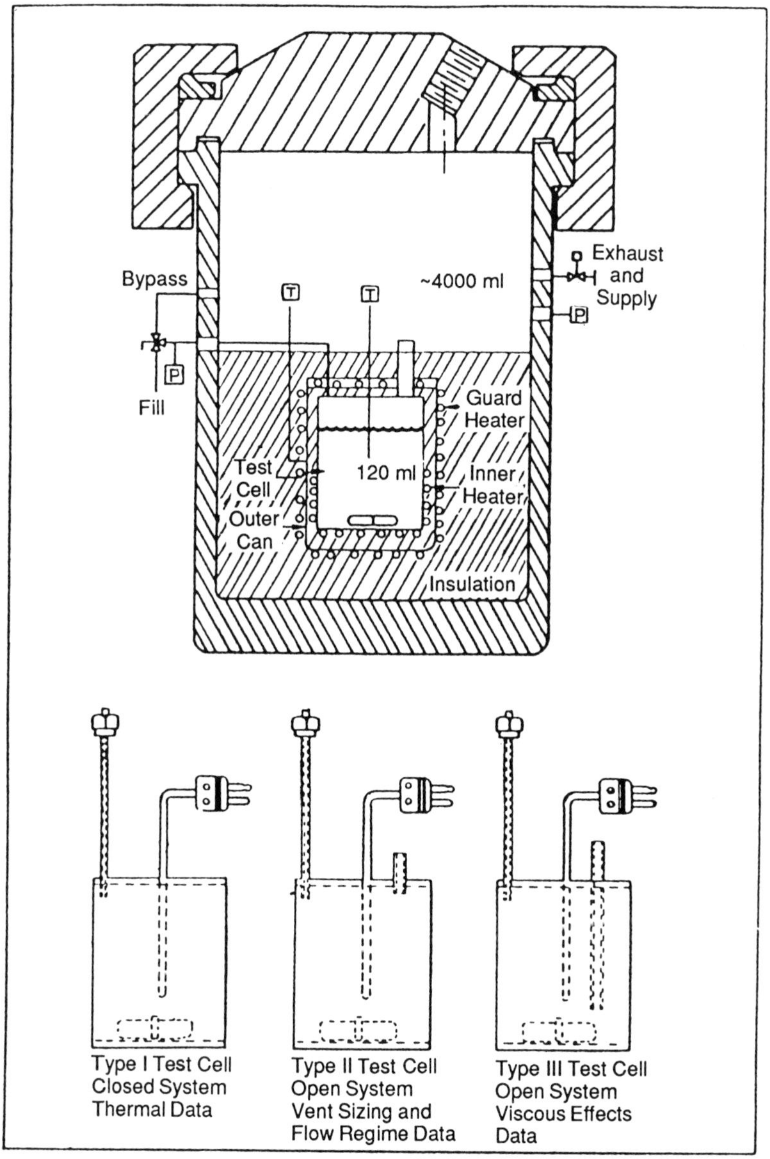

(a)

Figure 10-18. Small-scale test equipment to characterize runaway chemical reactions. (From Fauske and Leung 1985) (a) Small-scale testing equipment with closed and open test-cell designs. (b) A scale model predicting temperature reaction. (c) A scale model predicting pressure reaction.

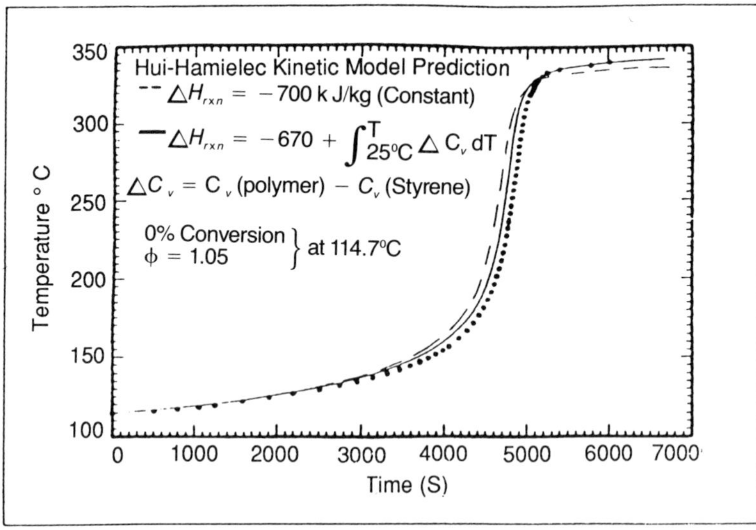

(b)

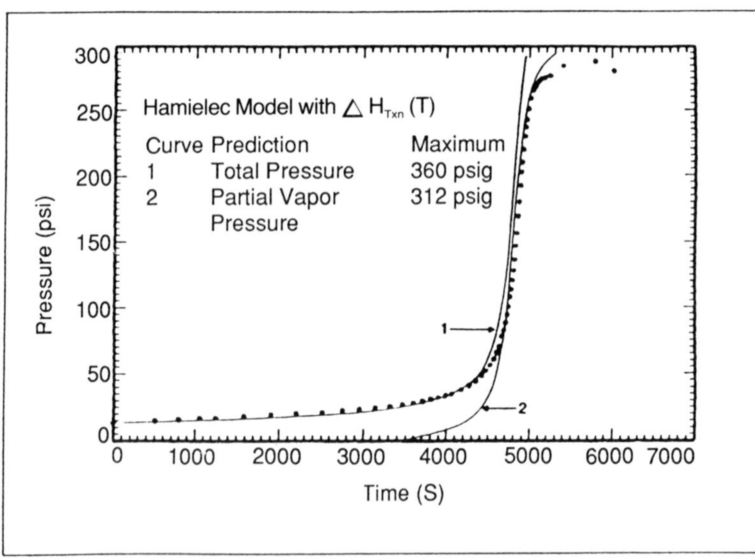

(c)

Figure 10-18. *(continued)*

predictions. The purpose of such apparatus is to provide data about the sizes of rupture disks or relief valves that will protect the pressure vessel or storage tank against runaway chemical reactions. These tests also can be used to predict ways of chemically or mechanically inhibiting the process in a controlled manner to prevent catastrophic failure and uncontrolled venting.

Figures 10-19 and 10-20 illustrate computer simulations of energy release in a reactor vessel. These simulations are conducted in conjunc-

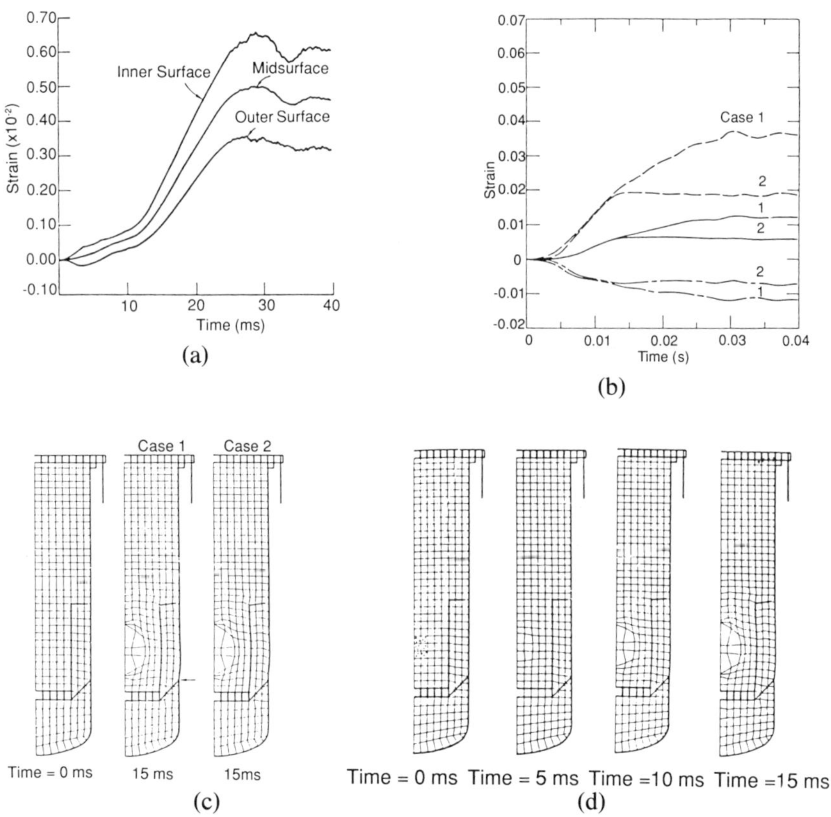

Figure 10-19. A computer simulation of a critical reaction event in a concrete vessel. (From Belytschko and Kennedy 1975) (a) Strain histories in the skirt for the complete model. (b) Comparison of strains for a reinforced (case 1) and a unreinforced (case 2) concrete vessel. --- = inner surface; —— = midsurface; -·- = outer surface. (c) Comparison of deformed shapes with and without reinforcement at skirt. (d) Complete model of primary containment and internals and deformed configurations.

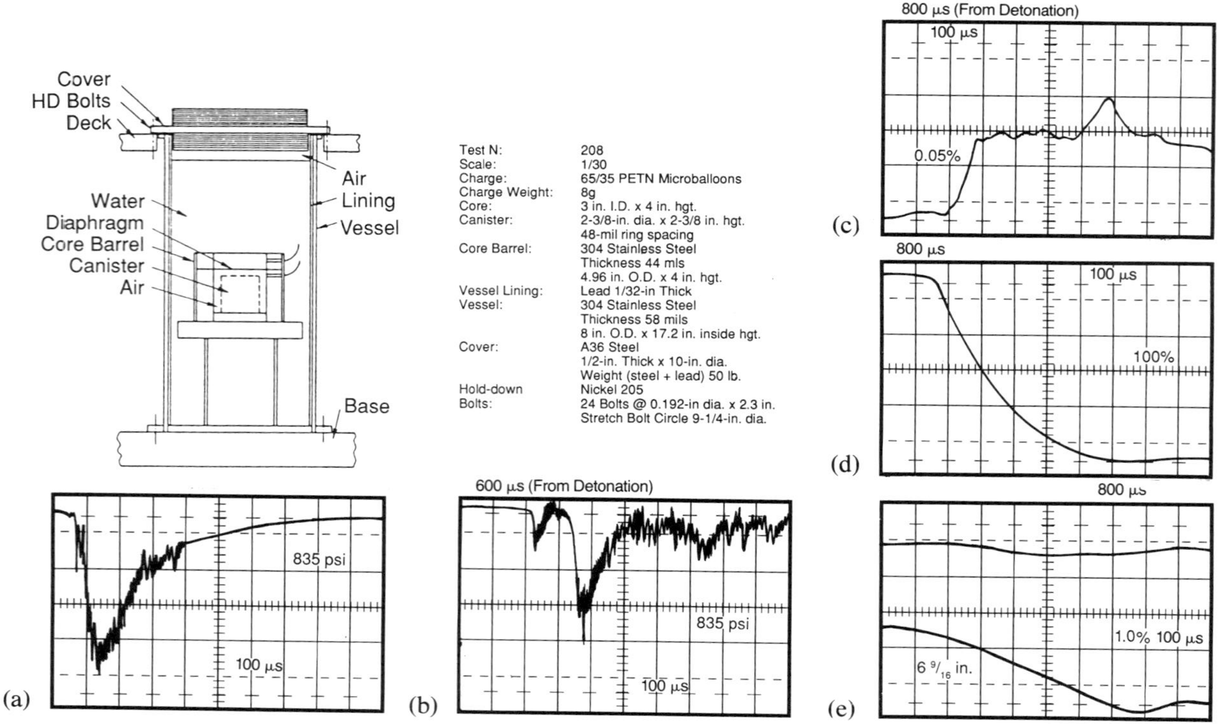

Figure 10-20. Illustrations of an experimental scale-model of a critical reaction in a vessel. (From Florence *et al.* 1975) (a) Core pressure. (b) Cover pressure. (c) Cover strain. (d) Vessel strain (circumferential strain $1\frac{1}{4}$ inches below the top of the vessel). (e) Vessel strain (circumferential strains 4 inches and $6\frac{9}{16}$ inches below the top of the vessel).

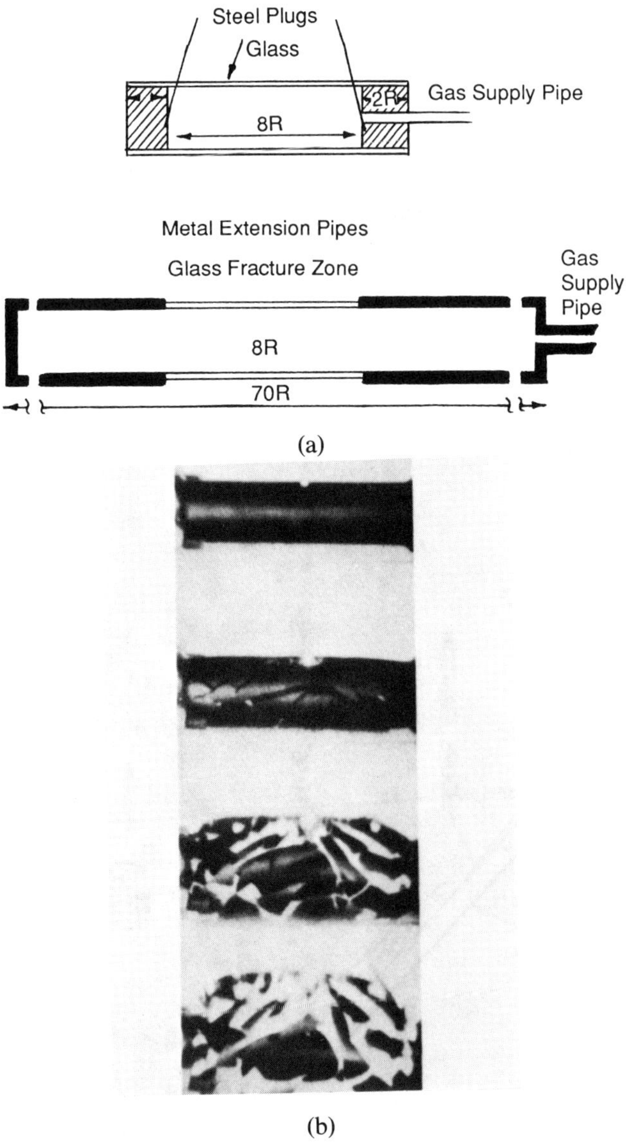

Figure 10-21. Fragments from bursting pressure systems. (From Baum 1984) (a) Top: Simulation of cylindrical vessel disintegration. Bottom: Simulation of the disintegration of a section of pipe line. (b) A side view of the section of pipe. (c) An experimental simulation of a pressure vessel rocketing caused by a circumferential failure. (d) Experimental simulation of a vessel closure becoming a missile. Release of an end-cap missile. (e) Drawing of how a ductile pipe-line missile is formed. (f) Film of the formation of a pipe-line missile. (g) A missile generated by the rupture of a pipe with a diameter of 0.6 meters. (h) The velocity of missiles generated by rupture of a pipe line. V = measured velocity; a_e = speed of sound in the atmosphere; V_p = velocity predicted assuming all the energy available in the rupture zone appears as fragment kinetic energy.

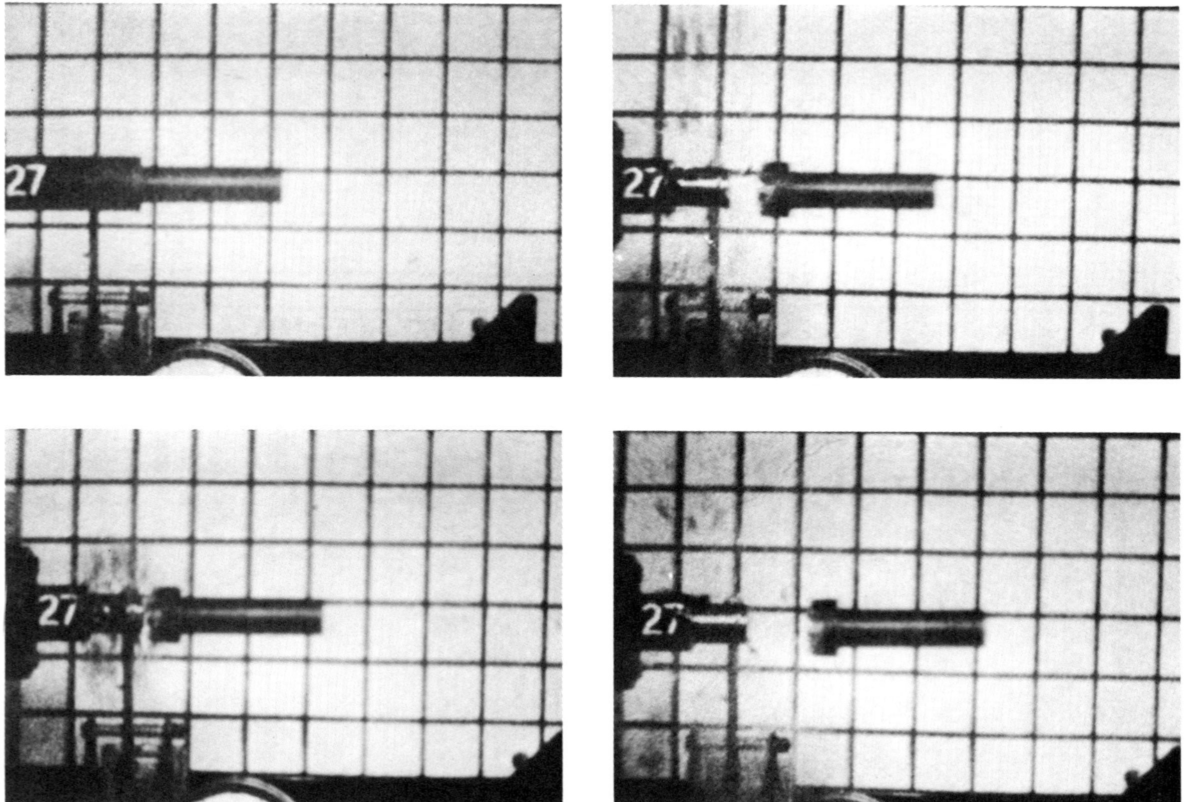

(c)

Figure 10-21. *(continued)*

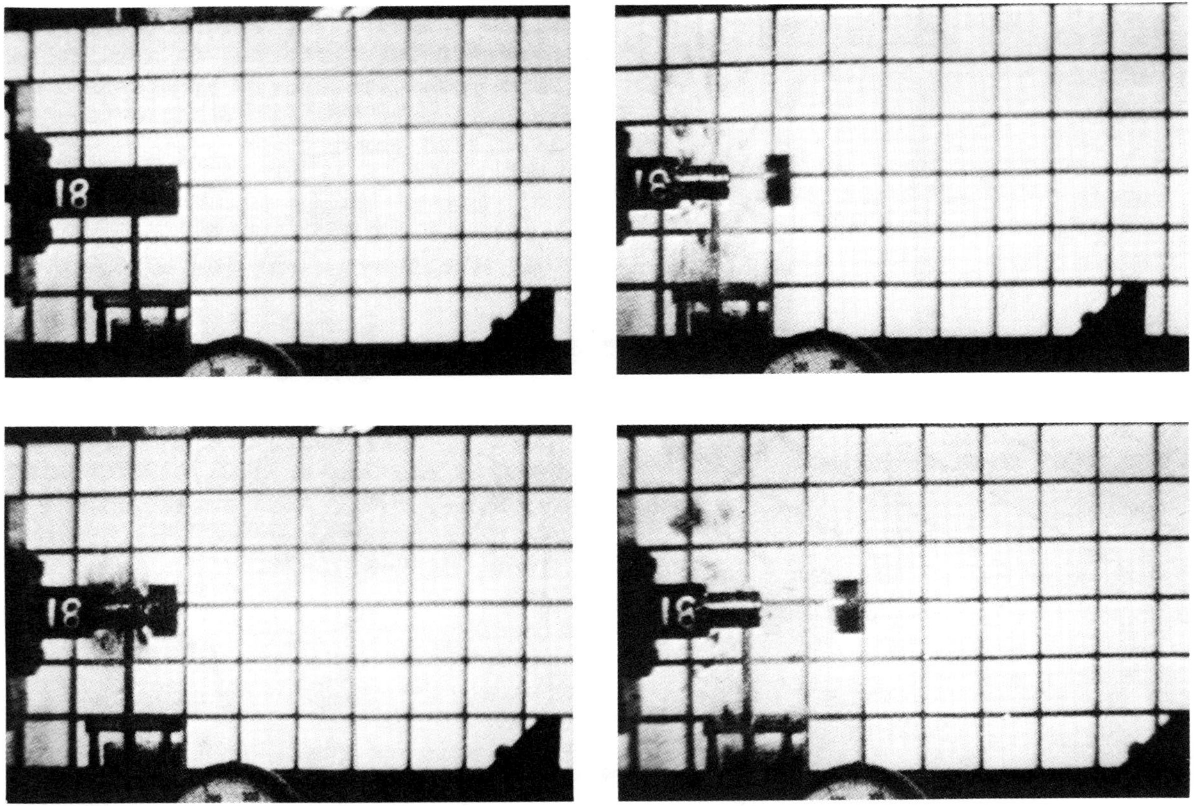

(d)

Figure 10-21. *(continued)*

tion with experimental simulations of a hypothetical core-disruption accident (HCA) test and are geared principally to assess the response of the vessel and its ability to maintain integrity during such an event (Florence *et al.* 1975). Figure 10-21 illustrates experimental burst and missile tests associated with theoretical and experimental studies conducted by Baum (1984). Baum (1989) also has provided data on experimental tests

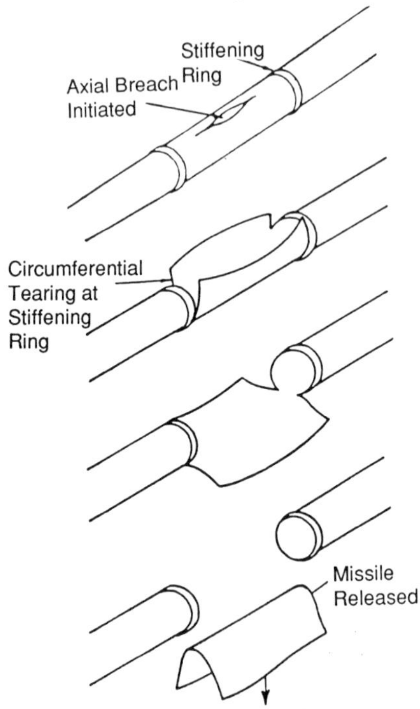

(e)

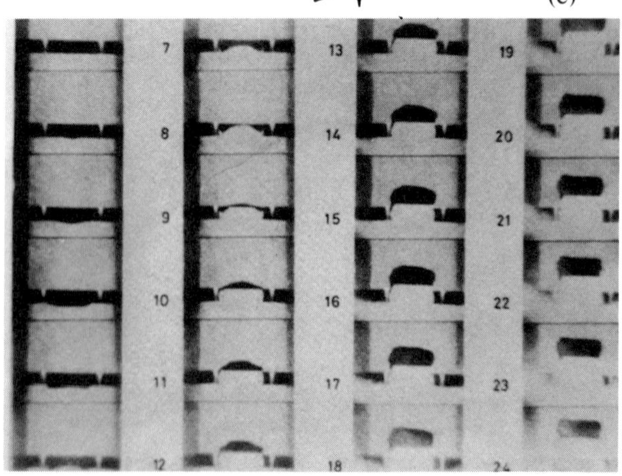

(f)

Figure 10-21. (*continued*)

(g)

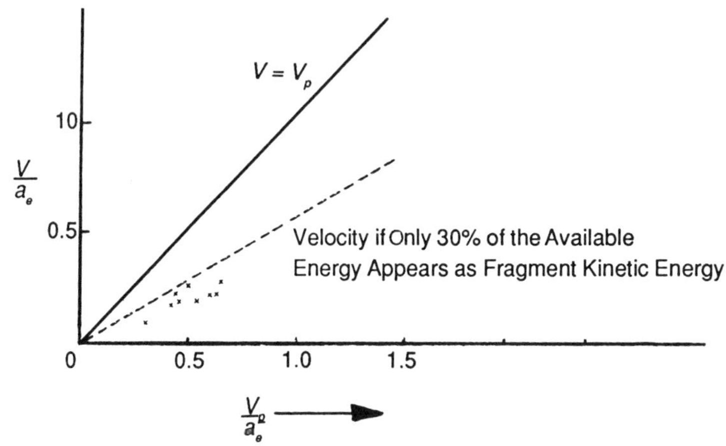

(h)

Figure 10-21. (*continued*)

on end-cap missiles ejected from vessels. Schroeder (1980) reported a burst test study for the Pressure Vessel Research Committee that is illustrated in Figure 10-22(a) and (b).

As was discussed earlier, not all hazards are associated with kinetic energy generated by breaching of a system. The contents pose a hazard regarding heat, chemical, biological, and radiological effects. Hooker (1990) reports the contamination of the drinking water aquifer by Chem-Dyne in Hamilton, Ohio, and the uses of computer models to analyze the problem and solution. In situ documentation of an event provides invaluable information (Bennett, Feates, and Wilder 1982; D.

(a)

(b)

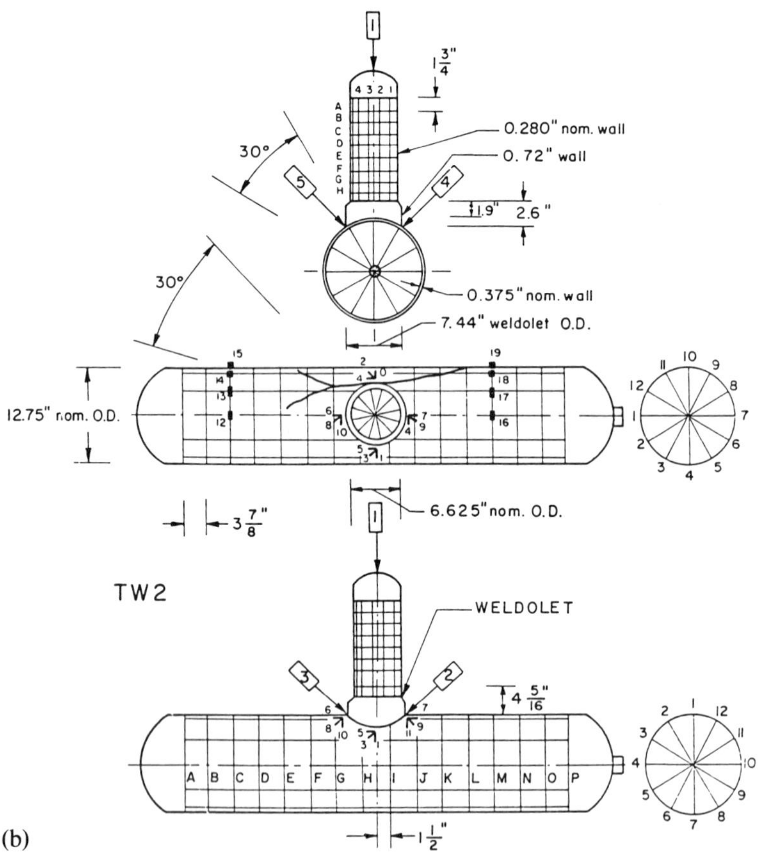

Figure 10-22. The rupture of a pressurized pipe branch. (From Schroeder 1980) (a) Experimental test of a pressure vessel bursting because of overpressure at the nozzle attachment. (b) Drawing of the T section of the vessel.

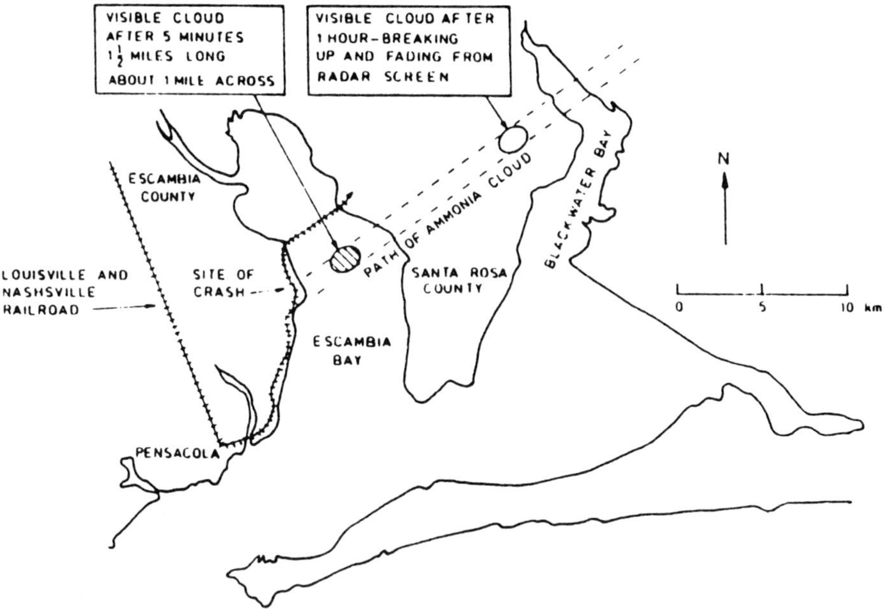

(a)

(b)

Figure 10-23. Vapor cloud release. (From Green 1982) (a) Path of an ammonia cloud released in Pensacola, Florida, in 1977. (From McQuaid (1985)) (b) View of the source tent being filled with refrigerant. (c) Collapse of the tent and release of a denser-than-air puff. (d) Collapse of the puff to a height of less than a meter when there is no wind.

(c)

(d)

Figure 10-23. *(continued)*

B. Brown 1976; Environmental Protection Agency 1972; Green 1982; Scott 1979). Figure 10-23(a) contains data collected by radar on an ammonia cloud released from a railroad train crash, and Figure 10-23(b)–(d) illustrates a vapor-cloud release test setup (McQuaid 1985). (See S. J. Brown 1984, 1985 for a summary of a number of testing programs on explosions and missile release developed in the United States, the United Kingdom, France, Germany, and Japan. See Green 1982 and Bennet, Feates, and Wilder 1982 for information about hazards with respect to high-risk technologies and spills.) The entire issue of *Chemical Engineering Progress, AIChE*, vol. 86/5, May 1990, provides an overview of regulatory and management issues regarding hazardous substances.

Creep Analysis

Figure 10-24(a) illustrates the load deflection of a dome. The effect of creep is illustrated by increasing the load at the center until the dome buckles in an elastic manner at room temperature versus the effect of heating the dome to some elevated temperature and observing the effect of time. Figure 10-24(b) illustrates that the buckling (instability) of the

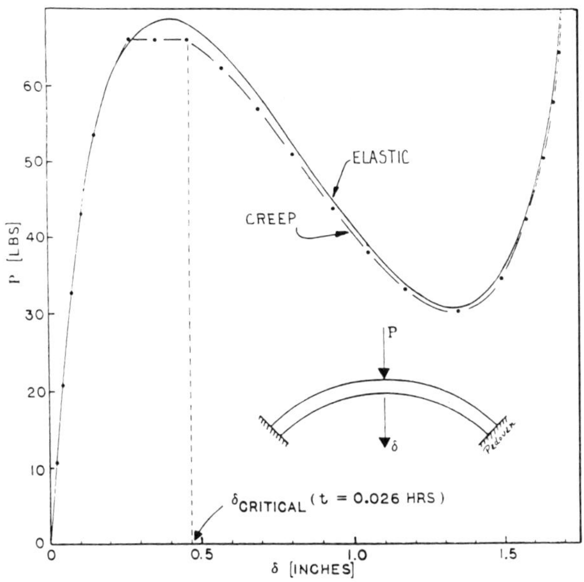

(a)

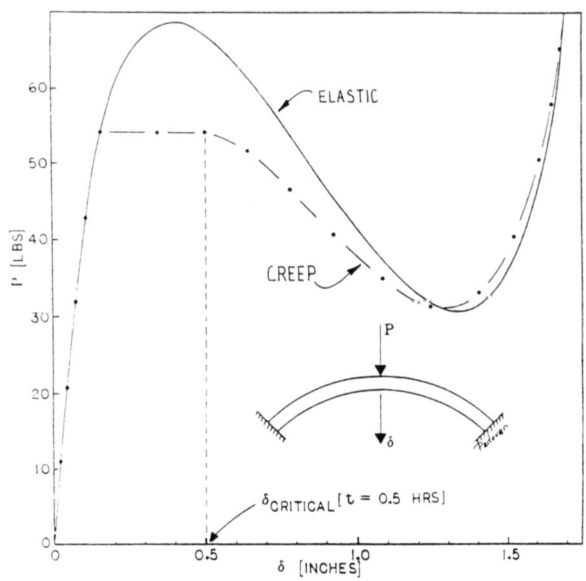

(b)

Figure 10-24. The effect of creep on buckling. (From Padovan and Tovichakchaikul 1983, 1984) (a) Comparison of buckling of the cap with creep at a fixed load versus perfectly elastic material. (b) Comparison of creep at a fixed load versus elastic buckling.

dome occurs at much lower loads with longer exposure to creep effects. The theoretical predictive solution was performed by Padovan at the University of Akron. The implications of such phenomena affect numerous applications: for example, rupture disks in processing equipment that are exposed to elevated temperatures can fail at lower loads.

Figure 10-25 illustrates bolt stress relaxation at elevated temperatures (*Criteria for Design* 1976). As we mentioned in Chapter 8, experimental testing to study the effects of creep on component design is difficult because of the long period required and because results from accelerated or scaled testing can be misleading. Thus, tests designed to correlate with theoretical predictive methods are especially important. Refer to Pugh (1982).

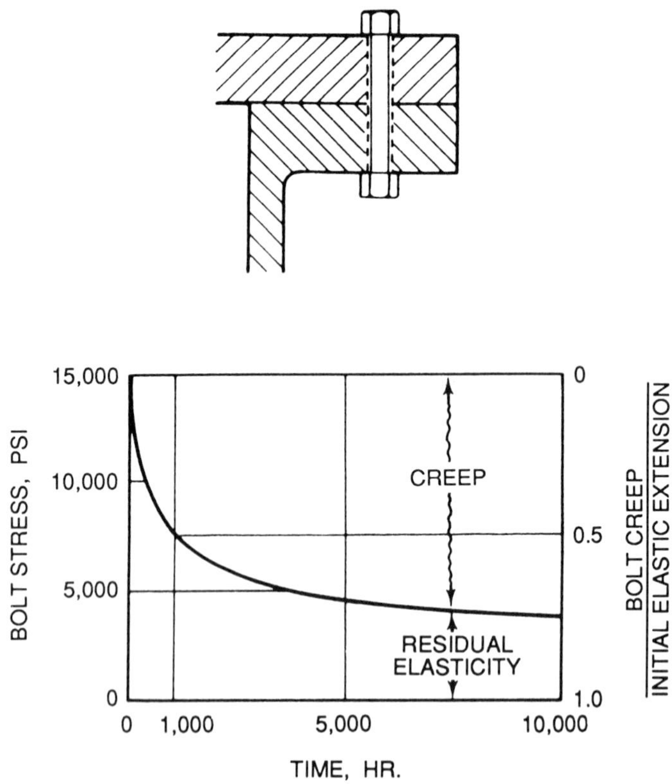

Figure 10-25. Creep relaxation behavior of a bolt in an unyielding flange. (ASME, N-47)

Fracture Analysis

Prediction of structural failures by fracture mechanics is relatively recent and therefore experimental, and theoretical simulation data are not as extensive as they are for other methods. Because the data are rather broad in scope, they are oriented to studying the fundamental factors that influence fracture: material defects, stress state (load rate, temperature, geometry), and material characteristics such as fracture toughness. In Chapters 8 and 9, we said that flaws, defects or cracks (caused by processing or crack growth) can take on many orientations and shapes that will influence the fracture process; thus, they are generally studied in small-fracture-specimen-type tests. In situ failures such as the experimentally ruptured vessel shown in Figure 10-22 are a source of information concerning how fractures develop in products.

Figure 10-26(a), (b) illustrates a ruptured pipe test reported by Kobayashi (1985). Figure 10-26(c) shows a finite-element model representation of Kobayashi's experimental test, and 26(d) compares theoretical

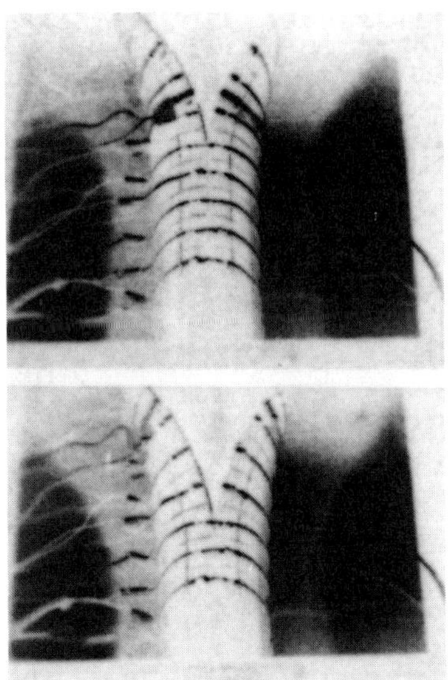

(a)

Figure 10-26. Experimental tests to investigate unstable crack propagation in a pipe. (From Kobayashi 1985) (a) The rupturing pipe. (b) The ruptured pipe after the crack is arrested. (c) Finite-element simulation of the crack propagation. (d) Pressure transient data for experimental and theoretical prediction.

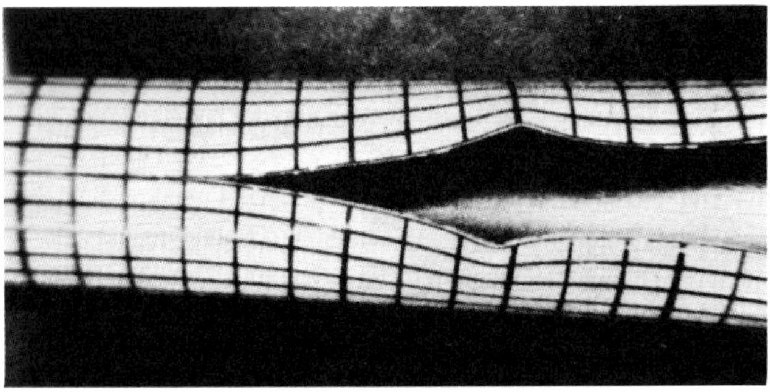

(b)

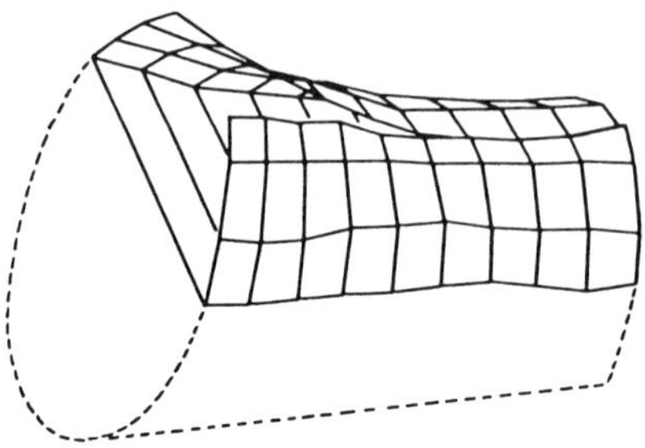

(c)

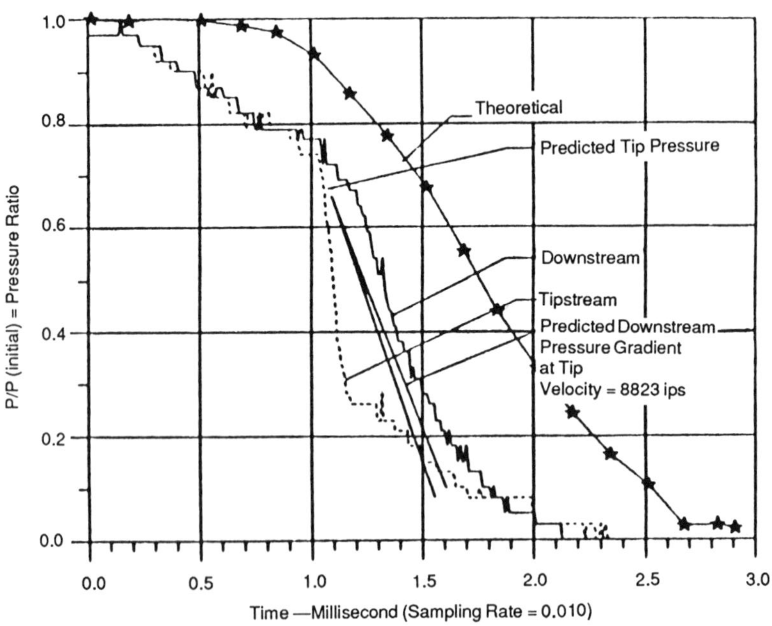

(d)

Figure 10-26. *(continued)*

data and experimental data with respect to pressure in the pipe system versus time.

Ando (1975) discussed an experimental and theoretical investigation of cracks in ship structures using the finite-element method and large structural testing (see Fig. 10-27). Linear fracture mechanics was used in

(a)

(b)

Figure 10-27. Test and analysis of ship structure by linear elastic fracture mechanics. (From Ando 1975) (a) Experimental K-value analysis for cracks in a local ship structure. (b) An illustration of the crash propogation in the one-half scale model of the local ship hull structure. (c) A comparison of Irwin's method for an infinite plate versus Isida's method of a finite plate for analyzing crack damage to ship structures.

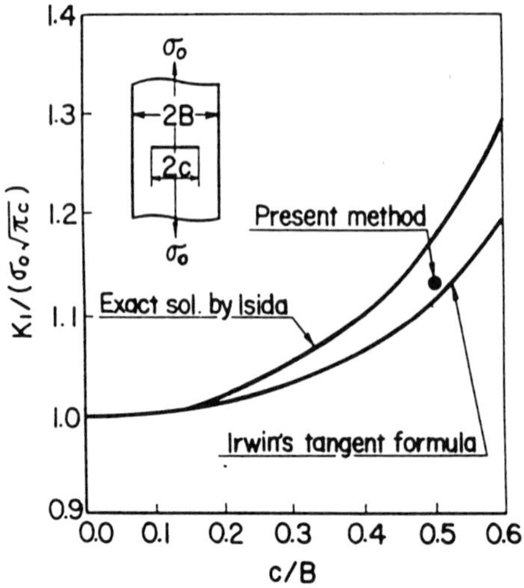

(c)

Figure 10-27. *(continued)*

this theoretical prediction. (See Garwood and Harrison 1985 for a discussion of a failure analysis of a ship that broke in half.)

Ergonomics Analysis

Figure 10-28(a) illustrates a lineman wearing a safety belt. As a result of numerous reports from safety-belt users who fell because they believed the safety strap attached to the belt was properly secured, studies were conducted into the ergonomics of safety-belt use and the causes of falls. Figure 10-28(b) illustrates one cause that was identified by fault-tree analysis: When left-handed users engaged the safety belt and safety strap on the left side, the belt tongue interfered with the strap. In situ tests demonstrated that the feel and sound of a properly engaged snap was present when the snap actually engaged the belt tongue rather than the D ring. This arrangement can support a lineman's weight for a considerable amount of time before disengagement takes place. Another hazard identified in the study was that the snap on the safety strap can open the catch of a large-diameter safety hook when the two are engaged.

Figure 10-29(a) illustrates a computer-simulated tip-over of a forklift truck, and Figure 10-29b represents the experimental simulation of the incident. Refer to Figure 10-17. Because of its short stance and high center of gravity, the forklift is susceptible to upsets, which subject the operator to side impact (Clark Equipment Co. 1983; McConville 1985). One purpose of these computer simulations and experimental tests was to assess the effectiveness of capsule- or wing-type seats (Severy, Brink,

(a)

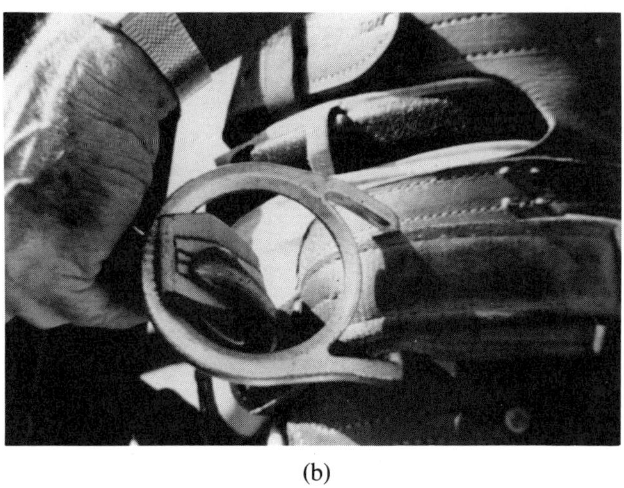

(b)

Figure 10-28. Example of lineman safety belt ergonomics. (Courtesy of Quest Engineering & Development Corp.) (a) A lineman wearing a safety belt. (b) A photograph showing what caused the belt to work improperly.

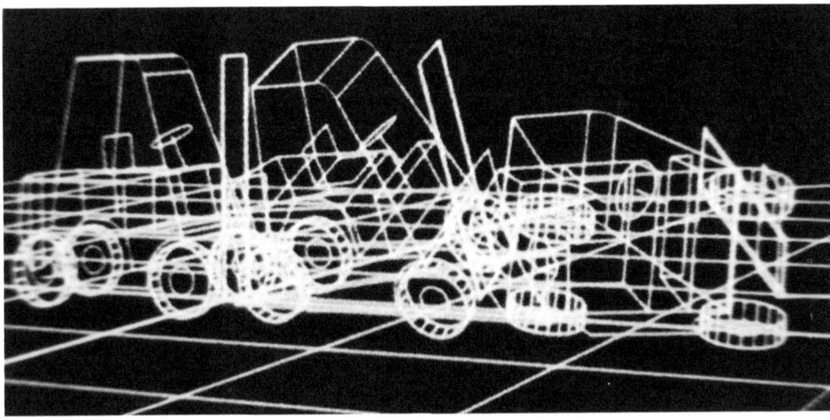

(a)

(b)

Figure 10-29. Driver safety during forklift tip-over. (a) Computer simulation of a forklift tip-over. (CASE) (b) An actual simulation of the tip-over. (c) The LM seat, which envelops the operator at the sides and rear, has an almost impenetrable structure.

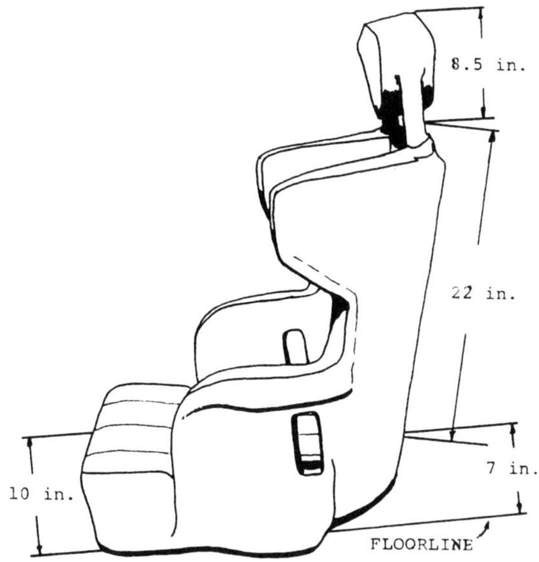

(c)

Figure 10-29. *(continued)*

and Baird 1968), which were shown to protect the operator of a vehicle subjected to side impacts. The effects of safety restraints and operator strength as means of preventing operator ejection were simulated analytically and experimentally.

SURVEYS WITH SIMULATIONS

As was discussed in Chapter 7, surveys, when combined with good statistical analysis procedures and methods, can be a valuable tool when gathering data on the incidence of events. They can also be useful for governments, industries, designers, manufacturers, and users in identifying needs for improvement. Figure 10-30 illustrates a laundry cart that was tested for potential hazards to personnel. Some of the hazards bear a striking similarity to the engineering principles involved in forklift overturns: that is, small-wheel stance, high center of gravity, and, in some instances, an overhead bar. The instability increased if heavy wet loads of laundry or infants were placed in the cart outside the support-leg axis. When the laundry cart overturned, nearby personnel were likely to be injured by the overhead bar. When the results of the laboratory simulations were compared with the results of questionnaires sent to distributors, the survey confirmed that the distributors were aware of the danger. Thus, the engineering firm that conducted the studies made recommendations for redesign and retrofit.

Figure 10-30. A commercial laundry cart.

CONCLUSION

This chapter has provided an overview of theoretical and experimental methods that are available to design engineers and forensic engineers. This overview has been cursory indeed. Simulation methods in a product liability and failure analysis context are extremely useful tools in establishing the mechanism of failure, but are only one step to be used in concert with the other steps in the investigation. This aspect of forensic engineering is important to keep in mind. Simulations demonstrate what is possible; however, investigation of the physical and documentary evidence, physical analysis and material testing, and identification of the mode and mechanism of failure all play a part in reconstruction of the incident, which provides a basis for determining causality.

Before moving on to the reconstruction of the accident and the determination of causality, we offer some final observations for the engineering investigator and attorney to consider. First, motion pictures, schematics, graphics, computer animations, and audiovisual media that are commonly used in the technical community to communicate concepts and observations to administrators, management, or other personnel should not be overlooked as tools to explain the results of experimental and theoretical simulations to the judge and jury. Thus, before performing potentially expensive experimental or theoretical tests, the investigative firm may find that it is advantageous to discuss with the client the possible uses of audiovisual documentation in preparing for trial.

Second, the engineer performing the simulations should discuss with the client the accuracy of the methods that will be used. This is important whether the simulations are performed at the design stage or during a forensic investigation. Economics frequently plays a significant role in the degree of accuracy of the predictive technique: that is, the greater the accuracy, the greater the cost is likely to be. And, typically, each successive increment of accuracy means an increased cost. Therefore, the client needs enough information to make a rational decision about the accuracy of the simulation versus its cost. In some product liability cases, an inordinately disproportionate amount of money is spent on depositions compared with the amount spent on developing scientific evidence. On the other hand, performing tests with little merit simply to dazzle the jury is another issue of concern.

Finally, having illustrated a number of experimental and theoretical methods of calculating product performance, we provide three example flow charts of performance criteria. Figure 10-31(a) and 31(b) are from the ASME Boiler and Pressure Vessel Code and are illustrative of design-by-analysis performance criteria. Figure 10-31(a) illustrates how structural stresses are to be classified, and the allowable stress values to which they are to be compared. Figure 10-31(b) is relevant when creep effects are to be considered. Both figures are criteria that are referred to as preventive. Figure 10-31(c) illustrates a flow chart to assess hazardous release protection and hence is referred to as a protective criteria. Protective systems have performance criteria that are a function of the hazard and the receptor that is to be protected (personnel, equipment, other hazardous material, property, or buildings, and protective systems). These illustrative flow charts are provided to give, particularly for the nontechnical reader, a feel for the relationship of performance criteria and the performance values that are calculated by the experimental and theoretical predictive methods, of which only a few have been reviewed in this chapter. Finally, the nontechnical reader should bear in mind that these flow diagrams provide only a simple schematic representative of the much more detailed and descriptive documents in which they are found.

Figure 10-31. Example of performance criteria to be used in the analysis of pressure systems. (ASME) (a) An ASME "Hopper" diagram for design by analysis of ASME Section VIII, Division 2, and Section III pressure vessels. (b) An ASME flow diagram for analysis of systems operated at elevated temperatures (Code case N-47). (c) A flow chart showing the proposed ASME SC6000 design-by-analysis criteria for hazardous-release protection. (S. J. Brown 1990)

Stress Category	*Primary Stress*			*Secondary Stress*		*Peak*
	General Membrane	*Local Membrane*	*Bending*	*Expansion*	*Membrane Plus Bending*	
Description (for examples see Table NB-3217-2)	Average primary stress across solid section. Excludes effects of discontinuities and concentrations. Produced by pressure and mechanical loads.	Average stress across any solid section. Considers effects of discontinuities, but not concentrations. Produced by pressure and mechanical loads, including inertia earthquake effects.	Component of primary stress proportional to distance from centroid of solid section. Excludes effects of discontinuities and concentrations. Produced by pressure and mechanical loads, including inertia earthquake effects. (Note 6)	Stresses which result from the constraint of "free end displacement" and the effect of anchor point motions resulting from earthquakes. Considers effects of discontinuities, but not local stress concentration. (Not applicable to vessels.)	Self-equilibrating stress necessary to satisfy continuity of structure. Occurs at structural discontinuities. Can be caused by pressure, mechanical loads, or by differential thermal expansion. Excludes local stress concentrations.	(1) Increment added to primary or secondary stress by a concentration (notch). (2) Certain thermal stresses which may cause fatigue but not distortion.

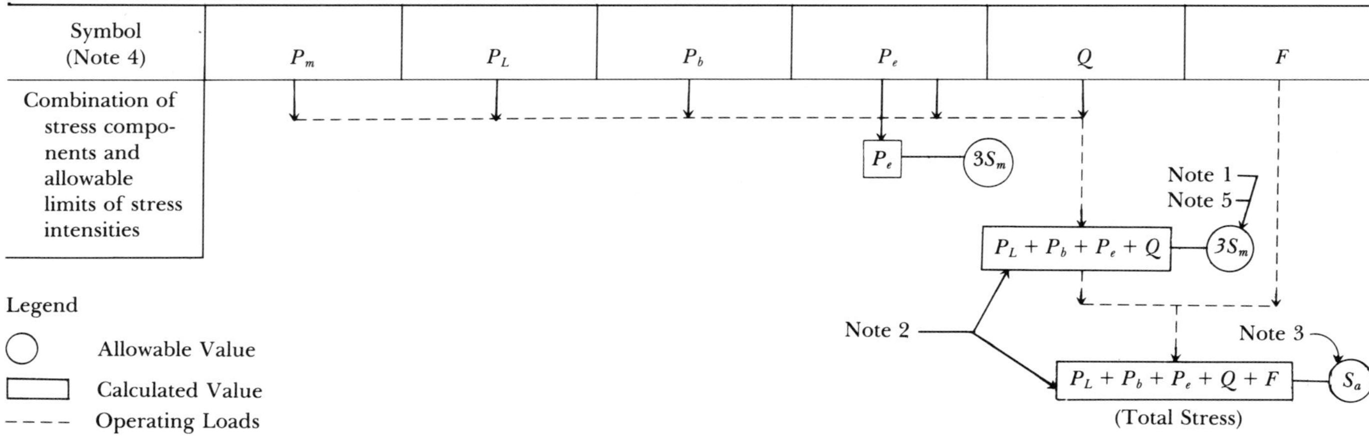

Note 1. When the secondary stress is due to a temperature transient at the point at which the stresses are being analyzed or to restraint of free end deflection, the value of S_m shall be taken as the average of the tabulated S_m values for the highest and the lowest temperatures of the metal during the transient. When part or all of the secondary stress is due to mechanical load, the value of S_m shall not exceed the value for the highest temperature of metal during the transient.

Note 2. The stresses in category Q are those parts of the total stress that are produced by thermal gradients, structural discontinuities, etc., and they do not include primary stresses that may also exist at the same point. It should be noted, however, that a detailed stress analysis frequently gives the combination of primary and secondary stresses directly and, when appropriate, the calculated value represents the total of $P_m + P_b + Q$ and not Q alone. Similarly, if the stress in category F is produced by a stress concentration, the quantity F is the additional stress produced by the notch over and above the nominal stress. For example, if a point has a nominal stress intensity, P_m, and has a notch with a stress concentration factor, K, then $P_m \leq S_m$, $P_b = Q$, $Q = 0$, $F = P_m(K-1)$, and the peak stress intensity equals $P_m + P_m(K-1) = KP_m$. However, P_L is the total membrane stress that results from mechanical loads, including discontinuity effects, rather than a stress increment. Therefore, the P_L value always includes the P_m contribution.

Note 3. S_a is obtained from the fatigue curves. The allowable stress intensity for the full range of fluctuations is $2S_a$.

Note 4. The symbols P_m, P_L, P_b, P_e, Q, and F do not represent single quantities, but sets of six quantities representing the six stress components σ_t, σ_l, σ_r, τ_{lt}, τ_l, τ_{lr}, and τ_{rt}.

Note 5. Special rules for exceeding $3S_m$ are provided in NB-3228.3.

Note 6. Bending component of primary stress for piping shall be the stress proportional to the distance from centroid of pipe cross section.

(a)

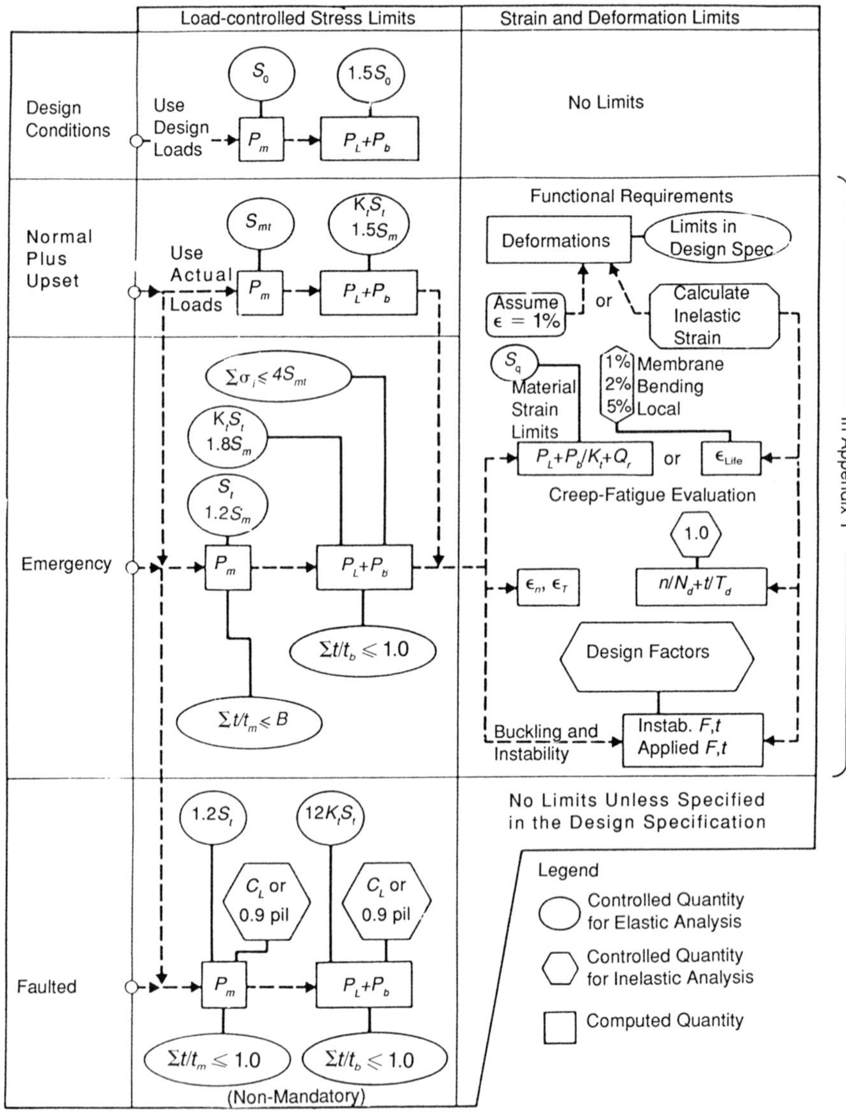

(b)

Figure 10-31. *(continued)*

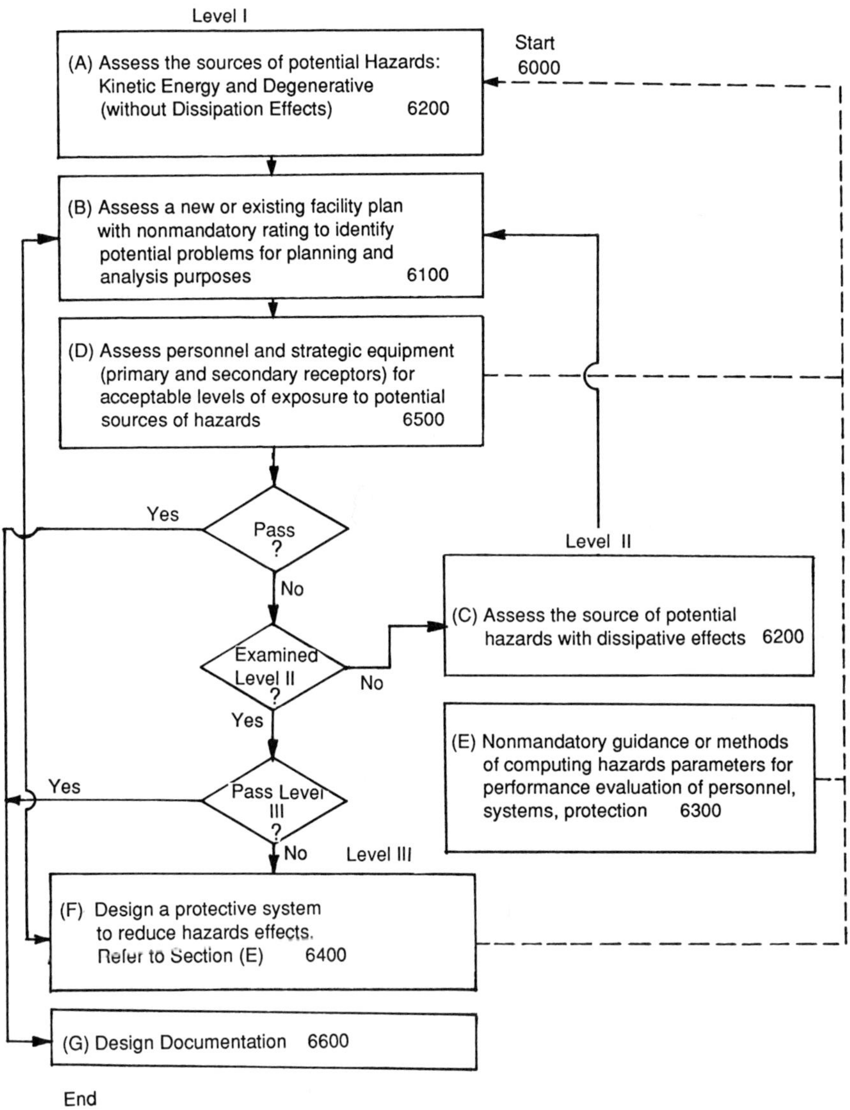

(c)

Figure 10-31. (*continued*)

11

Reconstruction, Causality, and Recommendations

Reconstructing an incident is one of the most challenging aspects of an accident investigation. It brings together the sequence of events leading up to and through the incident in which a product failed to serve its intended purpose and caused damage or injury. A reconstruction can be performed in a failure analysis while the investigator determines the mode and mechanism of failure. In most accident investigations, whether the outcome is a litigation or not, the reconstruction includes the determination not only of the modes and mechanisms of failure but of causality or responsibility as well. Each preceding stage of the investigation can play either a minor or major role in that determination. The motivation for an accident reconstruction is usually a desire to (1) determine how the incident occurred, (2) provide a basis for a legal opinion, (3) identify and assign responsibility, or (4) make recommendations for remedying the problem so that no similar incidents will occur (i.e., What lessons have been learned from the incident?).

As we have seen in previous chapters, some incidents involving personal injury or property damage are *not* the result of structural failure; in these cases, *no* failure-mode analysis is required, and the determination of the failure mechanism (how the failure occurred) focuses on the mechanics of the product, the personnel involved, and the relationship between the two. We also have seen that causality answers the question

"Who is responsible?" and that responsibility is usually divided into six general categories:

1. Design engineering—the responsibility of the design engineer
2. Selection of materials—the responsibility of the materials, metallurgical, or chemical engineer
3. Defective materials—the responsibility of the supplier of the materials
4. Improper methods of fabrication—the responsibility of those responsible for processes ranging from assembly to shipping
5. Improper service and operating procedures—ultimately the responsibility of the user-owner, but one that can involve subcontractors who perform these functions and the design engineer who supplied the service and operating specifications
6. Improper or inadequate maintenance and inspection—again, ultimately the user-owner's responsibility, but one that can also involve subcontractors who perform these functions and to the design engineer's specifications

Recognizing the scope, general nature, and overlap of these categories is extremely important. For example, fabrication, in addition to involving obvious processes such as machining and welding also includes assembly, quality control, inspection, and testing functions. In addition to loads and environment, the service or operating conditions may include storage, reshipment, and subcontracted maintenance and inspection services.

The six causes of failure can be the responsibility of a number of organizations or individuals or a single entity. Consider, for example, a cogeneration unit that is designed, manufactured, operated, maintained, and inspected by one organization and supplies heat, air-conditioning, and electricity and is leased to small businesses.

It bears repeating that, more often than not, the cause of a failure cannot be determined directly from the mode or mechanism nor failure but requires further investigation. For example, a component may fail because of fatigue at a defective weld caused by normal vibration during operation. One must investigate further, however, to establish causality—to determine whether the component was designed to sustain the amount of vibration it would be subjected to during its fatigue life and whether such a defect was anticipated in the design specifications. If the defect was not anticipated in the specifications, the fabrication defect may have been the site of the mode and mechanism of failure, but the underlying cause would be engineering design.

RECONSTRUCTION OF THE INCIDENT

The analogy used in Chapter 5 to describe the reconstruction of an incident is one of a person trying to correlate a series of picture puzzles of the scene, the pieces of which are not interchangeable. Rarely does one investigative method or technique allow one to reconstruct a complete picture of the incident—the sequence of events that led up to and through the incident. Frequently, the investigators must use two or more different experimental testing techniques on the same failed product to map defects or determine the properties of a material. Similarly, more than one simulation is often required to predict how a product will respond. In Chapter 10, we pointed out that both theoretical and experimental simulations are often used to correlate or benchmark the reconstruction predictive methods. This is particularly effective since the experimental simulation confirms the theoretical modeling. Let us consider several examples of reconstructions that involved product failures with or without material failure.

The Lineman's Safety Belt

In the previous chapter, we discussed the incident involving a lineman who fell from a utility pole because he believed the snap on the left side of his safety belt was engaged when it was not. He survived the fall but was seriously injured. During the deposition, he recounted that he had purchased the belt shortly before the incident. After wrapping the safety strap around the pole, he engaged the snap on the left side of the belt. He then began working, leaning away from the pole with his weight on the strap. Within a minute or two, he fell backward away from the pole. Photographs taken at the time of the incident showed that the snap on the left side of the belt had disengaged from the D ring. After an investigation of the accident site, eyewitness accounts, codes and standards, statistical information from databases, examination of the physical evidence, and experimental simulations (with videotape documentation) with the actual belt and an experienced line maintainer, the following information emerged.

- Only 10 percent of the population is left-handed
- Evidence at the scene of the incident showed that the snap was not engaged in the D ring
- The tongue of the lineman's belt protruded through the D ring and interfered with proper engagement of the snap
- Simulations demonstrated that the snap could engage both the tongue and the D ring while the tongue protruded through the D,

the arrangement could support the lineman's weight for long periods, and that the snap could be disengaged when the lineman moved (see Fig. 10-28)

- The snap could not be disengaged if it was correctly engaged about the D ring
- No warnings alerted users to this potential hazard, and the belt was not designed to prevent such incidents

A review of the data supported the following reconstruction. The experienced lineman was wearing a new safety belt as he ascended the utility pole. Once he had reached the desired height, he swung the safety strap around the pole from his right hand to his left hand. While holding onto the pole with his right hand, he engaged the snap on the D ring and the belt tongue, which interfered with the D ring. His old belt had been of a different design. Hearing the metallic click of the snap onto the D ring, he assumed that the snap had engaged the D ring when in fact the snap was not completely engaged. When he leaned into his belt, with his spikes set, and began working, the snap slipped off the ring after a minute or so of movement, and the lineman fell backward and down. This incident did not involve the failure of material.

A Faulty Ladder Foot Design

Some incidents are the result of a mechanism of failure involving a structural failure that occurs as a consequence of the event, but is not its cause. Consider the extension ladder with the faulty foot design discussed in earlier chapters (see Fig. 3-1) in which the user climbed the ladder to remove material from the roof of a single-story dwelling. Both ladder feet appeared to be resting on the ground, and the ladder felt steady as he ascended. After climbing six or seven rungs, the ladder suddenly shifted to the left. Impact marks caused by the ladder on the building confirmed the location of the ladder at the time of the accident. An examination of the ladder revealed the following.

- The ladder feet were constructed with a slot-and-pivot arrangement, which allowed the feet to remain horizontal with the support (see Fig. 3-1)
- The slot allowed the foot pad to travel one inch axially relative to the side rail
- The arrangement of the slot and sheet-metal foot permitted one ladder foot to rest firmly on the ground and carry all the weight of

the ladder while the other foot could also be resting on the ground but carrying no weight if the ground was uneven

- Experimental simulations with videotape documentation demonstrated this feature as well as the tendency of the foot assembly to bind and temporarily carry a force even when the foot was placed in a depression
- The simulations also demonstrated that one inch of unevenness between two ladder feet translated into almost one foot of horizontal motion at the top of an extension ladder when a user's weight shifted from one side to the other; thus, as the ladder shifted, the user was unable to stop the motion because of the uneven ground and the design of the ladder foot
- The dynamics of the fall were such that the ladder twisted on one leg as it shifted toward the unsupported side, and this twisting motion caused the deformation of the ladder foot.

In other words, the twisting of the ladder foot was *a consequence* of the incident, not the cause. The cause was a deficient design: flimsy light-gauge material used in the fabrication of the ladder combined with the slot-and-pivot arrangement of the ladder foot. (During the trial, the judge disallowed presentation of the videotaped simulations without explaining what parts were objectionable and could be edited out.)

Slippery Tile in an Entryway

Figure 11-1 illustrates an in situ coefficient-of-friction test using a Brungraber portable device. The investigating engineer was retained to reconstruct the sequence of events leading up to an accident in which a person was injured when she slipped on the tile entryway to a commercial building. Who was responsible? The designer of the tile, the architect who selected it, the factory that prepared it, the contractor who installed it, the person who walked on it, the shoe manufacturer, or the company who maintained and inspected the building? Field tests indicated that whether wet or dry, the tile had an unacceptably low coefficient of friction, and that this coefficient of friction could be influenced by the method of preparing the surface (waxing or sealing) that the maintenance crew used. The injury was the result of uneven installation of the tile, however, which allowed water to collect in pockets when the maintenance crew turned on the sprinkler system. Thus, the incident occurred as a result of hydroplaning rather than of a static or dynamic coefficient of friction between the injured party's relatively flat sandal-type sole and the walking surface.

Figure 11-1. A Brungraber test machine for determining in situ friction factors.

The Shattered Grinding Disk

Consider the failed composite grinding disk described in earlier chapters. An examination of the physical evidence showed that the mode of failure was a fatigue crack that was initiated at the outer edge of the disk and grew toward the center. When the crack reached the critical size, the disk shattered, severely injuring the 30-year-old operator. The reader is directed to Figure 10-14.

To reconstruct the incident and address the issue of causality, a number of questions about the mechanism of failure by the fatigue mode had to be answered: Was the fatigue design life of the disk significantly greater than the time it would take to consume the disk during normal usage? Was the optimum ply orientation used? Were there material defects, manufacturing defects such as inclusions, or assembly defects? Did the air-driven power unit on which the disk was mounted run faster than the rated rpms? Did the power unit have an adequate fragment guard? Could the guard contain all the fragments? Was it removable? If so, was an interlock system present that would prevent the power unit from operating? Was a maintenance and inspection procedure available for evaluating the presence of defects in the disk before operation? Did aging of the composite material influence the growth of cracks? Did the effects of aging change the material properties of the disk? Did the manufacturer provide instructions outlining the user's responsibilities? Were voluntary or statutory standards available concerning design, processing, performance protection, and warning criteria? Was the airline pressure to the power unit maintained at the proper level? Did the conditions at the work site contribute in any way to the failure? Did the user-employer or user-operator abuse the disk? Who had owned the disk between the time it was first used and the time it failed? Some or all of the organizations involved in the design, manufacture, and use of the disk could have been responsible for the failure; that is, the designers of the equipment, the supplier of the materials, the manufacturer of the composite grinding disk, the manufacturer of the air-driven grinder on which the disk was mounted, the purchaser of the equipment, the operator, the codes and standards, or the personnel who shipped and transported the equipment.

During the accident investigation and failure analysis, the investigator gathered the following information that enabled him to reconstruct the incident.

- The ply orientation of the fiberglass cloth could influence the strength and fatigue properties of the material, but the ply orientation was not considered in the design or specified in the manufacturer's assembly specifications
- The fatigue life of the disk in normal operation was relatively brief, as determined by the computer simulations and experimental fatigue tests illustrated in Figure 10-14(e) (p. 370)
- The strength and fatigue properties of the disk were influenced to a significant degree by the designer's choice of fiber and a resin base. The manufacturer, which was also the designer, used an inexpensive low-strength fiber. The strength and bonding capacity of the resin base was influenced by the abrasive used
- The disk did not contain any significant inclusions
- Some fiber-to-resin delamination was observed
- No manufacturing defects were evident in construction; however, the ply orientation was random (as determined with a sample of production disks)
- No overspeeds were found when the power unit was tested and disassembled
- No evidence of abuse or overload of the disk was found
- The owner of the power unit had removed the guard, which met the specifications of the American National Standards Institute (ANSI B1-7, 1978)

- The manufacturer of the power equipment did not inhibit removal of the guard
- The guard reduced but did not eliminate the operator's exposure to fragments
- The manufacturer did not specify to owners how to determine whether dangerous cracks were developing in the disk
- The manufacturer did not supply information about the relationship between the effect of the rate of consumption (grinding) on fatigue life and the safety factor
- The applicable codes and standards provided a considerable amount of safety information, but did not provide any performance criteria for designers of abrasive disks
- The manufacturer's batch records revealed that a number of years had passed between the time of fabrication and the time the disk was first used; thus, the history of the disk—from manufacturer, to distributor, to retailer, to purchaser—in the intervening years could not be traced
- A visit to the work site reveals whether the workstation in any way contributed to the incident.

Failed Retrofit Pump Impeller

A large regional utility (that will be called LRU) hired an architect/engineering firm to design an electrical generating facility (power plant) and assist the LRU in the selection of equipment and construction subcontractors from the contending bidders.

A pump manufacturer was chosen to furnish the four circulating water pumps (CWP) that would supply cooling water to the power plant from a nearby river. These CWPs perform a function similar to a water pump on a water-cooled automotive engine (pump water through a heat exchanger for cooling purposes). The pumps (costing approximately one-half million dollars each) were designed by the pump manufacturer's engineer, who also selected the materials. The pump manufacturer cast its own impellers, fabricated the pump system, prepared the operating procedures to be supplied to the utility company, and installed (tested and inspected during construction and at the commencement of the operation of the power plant) the pumps. The routine maintenance and inspection procedures that would be performed by the utility maintenance department were outlined by the manufacturer of the pump.

Just as an automobile's water pump is necessary for the functioning of the engine, the circulating water pumps (CWPs) are essential to the operation of the power plant, and their expense is relatively small compared to the overall cost of the facility.

During the first eight years of operation, cracks appeared in the CPW impellers, which were constructed of nickel aluminum bronze (NiAlBz). The cracks were weld repaired, but as time passed the impellers became thinner due to erosion (wearing away as a result of debris and sediments in the river water). Consequently, the frequency of blade cracking and failure increased. Figure 10-11(d) is a photograph of a failed impeller. When the frequency of failure became sufficiently high, the power plant's efficiency was threatened. The maintenance department of the (LRU) facility contacted the pump manufacturer for a cost estimate for replacement impellers and for estimates if a different alloy (304L stainless steel) was used in the fabrication of the impellers, in hopes of obtaining more than the eight years of service they had so far gotten from the impellers. Upon receiving the estimates from the pump manufacturer (approximately $60,000 per impeller), the utility decided that the prices were too high and that it was advantageous to manufacture the impellers themselves, since the LRU owned many power stations with similar pumps and impellers.

By examination, the LRU's maintenance department could see that erosion was thinning the impeller blades [for an illustration, see Fig. 10-11(f)]. Hence, they deduced that choosing a corrosion-resistant alloy with good weld repair properties would solve their problem. As for designing the impeller, the LRU's engineer decided that by simply copying the impeller and making the blades thicker (to account for the material that had been lost during the years of operation) they could circumvent their lack of experience in applying the fundamental design engineering principles of hydrodynamics, flow-induced vibration, structural integrity, and efficiency to the design of the impellers.

The LRU next approached a foundry to build a mold based on the eroded original impellers (however, the foundry was to add thickness to account for material lost in the blades). The LRU directed the foundry to cast four impellers. Since the LRU possessed neither the original impeller design drawings nor conceptual drawings, the foundry prepared "shop drawings" or "fabrication drawings" based upon their field measurements of the original, eroded impellers. The shop drawings were reviewed by the LRU, and thus became the basis for casting the impellers. After fabrication of the stainless-steel impeller by the foundry, the CWP systems were assembled by the LRU maintenance department, and operation commenced. Four 304L impellers were cast, shipped, and installed over several months. Within a year of operation, cracks began to appear adjacent to where the blades were attached to the hub of the new 304L stainless-steel impellers. It was at this time that the LRU and the LRU's insurer-inspector (with assistance from engineering consultants) began a critical look at the old NiAlBz and new 304L stainless steel impellers to determine why they were failing. The foundry that cast the impellers also retained an engineering design expert, since the LRU refused them their final payment for the castings. The foundry had cast the impellers in accordance with the drawings and ASTM procedures, and the chemical properties of the 304L alloy were typical (there were no anomalies that would contribute to the failure of the impellers). Why did the new impellers fail and who was responsible?

The following observations were outlined by the engineering firm retained by the foundry.

- The drawings that were the basis for manufacture and LRU's specification failed to provide tolerance and surface finish requirements that are particularly important to control hydraulic performance as well as strength.
- A computer analysis of the impeller with respect to natural frequency in water [for an illustration, see Fig. 10-11(a)–(c)] showed that the natural frequency of the impeller system was coincident with the vortex shedding frequency as originally designed by the pump manufacturer. The forward overhang of the leading edge promoted a torsional vibrational mode of the blade-to-hub attachment and concentrated stresses in this area. Typically, other designs either support this overhang with an integral outer ring or slope the leading edge backward instead of forward. The LRU was unaware of this fact. (Because it did not evaluate the adequacy of the original design, the LRU's engineers assumed the design was satisfactory.)
- The fatigue characteristics of the original NiAlBz impeller material were 10 to 10^2 on cycles to failure superior to 304L. This fact is readily available in a number of code documents (see ANSI standards and Fig. 10-11(e) for an illustration). A number of technical society publications (such as the American Welding Society, Klein, 1981) show that 304L is a poor choice for use as an impeller or propeller because of the difficulty in controlling the grain size in large monoblock impeller/propeller casting (in fact, the American Bureau of Shipping did not recognize 304L as an "official" material in propeller castings) and its susceptibility to stress corrosion cracking, particularly in the presence of chlorides. (The river water had a high chloride content.)

- 304L stainless steel is known to be susceptible to a more rapid failure of a component if it is welded upon in the presence of stress corrosion cracking. The LRU did weld repair the cracks as they appeared in the impeller blades. Many welds were performed in situ without benefit of controlled conditions. Lack of full-penetration welds was observed in the metallurgical examination of a sample specimen. The use of nonrecommended weld rod was also revealed in the metallurgical examination.
- The LRU performed its own fabrication and installation. This process contributed to improper alignment, and hence rubbing and impacting of the impeller blades against the CWP housing.
- Erosion was observed to intensify at the initiation site of the crack at the leading edge of the blade. The flushing of the fatigue crack with chloride-rich fluid also enhanced the rate of crack growth in the 304L impeller.
- It was discovered that the LRU operated the CWPs beyond the manufacturer's specifications. Velocities 120 percent above specification were observed.
- Weld buildup was added by LRU at the attachment of the blade to hub in order to thicken and build a root radius from 0.5 in. to 1.5 in. Crack initiation continued to occur 2 to 3 inches away from the hub, however, as occurred in the original NiAlBz impellers.

It was recommended by some of the LRU's consultants that monitoring devices be placed on the pumps in order to alert the maintenance department to improper operation or pump malfunction (so that prompt maintenance and inspection could be performed). One consultant suggested that the as-cast fillet radius of 0.5 in. to 0.625 in. at the blade-to-hub attachment along the circumference was the mechanism of accelerated failure versus a fillet radius of 1.25 in. to 1.50 in. in the eroded NiAlBz impellers. Failure did not occur along the circumference, however, but rather at the leading edge, even with a 1.5-in. fillet radius.

Who was responsible?

Clearly, neither the LRU nor the foundry was an expert in hydraulics, fatigue, material selection, or impeller design.

The LRU's engineer's and maintenance department made errors in judgment and hence, errors in the design and manufacture of the replacement impellers. Was it in the LRU's interest, in essence, to become a manufacturer of pump impellers? The answer, of course, is yes, if it was economically effective for the LRU to learn the technology of manufacturing pump impellers. Are the LRU's engineers, who were responsible for maintaining the CWP system, competent? The answer is, yes, in the context of their formal training and occupation; however, none of the maintenance staff had any formal training (and certainly no graduate training) in hydrodynamics, flow-induced vibration (FIV), or metallurgy sufficient to manufacture the CWP impellers without design and analysis calculations. Are there impeller scientists or impeller pump engineers? Not really! There are, however, engineers and scientists (by training and experience) who are qualified to assess the design adequacy of the impeller that is the subject of this case. Certainly the foundry did not have the expertise to design the impeller for FIV. Should the LRU have taken the long-term view that their experience was part of the learning curve for the engineering department and the development of a foundry to subcontract their casting work, rather than pushing along an expensive litigation?

All the information just listed was obtained from the technical literature, applicable codes and standards, the manufacturer's design and processing records and manuals, the user's shop records, interviews with the operator and eyewitnesses, the investigator's and manufacturer's test data, and computer simulations.

What picture emerges from this reconstruction? What was the causality? Who was responsible for the incident? Was more than one organization responsible? These questions will be discussed further later.

DETERMINATION OF CAUSALITY

In the engineering community it is the practice to assign the cause of failure or responsibility for failure to one or any of six categories. As we shall see in the following subsections, each causality category can be interrelated with the other five.

Improper Engineering Design

Design deficiencies are generally the result of the designer's failure to anticipate the mechanism by which a product will fail—the effect the failure-mode parameters will have on performance. Most important is the designer's failure to anticipate the failure mechanisms that pose the greatest risk and to consider whether the product can be manufactured or used safely. As we have stated throughout the book, one sometimes discovers during an accident investigation that the product was not evaluated by a qualified design engineer.

The lack of formal training in the principles of engineering often results in the inability to recognize that a design is likely to fail under certain conditions. Because even the best design engineer can err, however, evaluations by independent technical review committees of new and revised designs have become standard practice in many organizations.

Failures associated with design deficiencies are often related to the failure to appreciate the effects of elementary structural problems such as mechanical notches, improper radii, sharp corners, and so forth. For a large, complex fabricated structure or system, the design process is broad and comprehensive in scope. To summarize, the design phase includes (1) providing the required design analysis, (2) choosing relevant codes and standards, (3) incorporating passive protections, (4) specifying appropriate materials; developing relevant acceptable criteria regarding failure-mode and defect characteristics, (5) developing the proper fabrication and inspection specifications and procedures and preparing proper drawings for the fabricator, (6) determining the criteria for quality assurance procedures and documentation, (7) specifying the qualifications required to operate the equipment, and providing clear and complete procedures for operating the equipment, (8) devising relevant in-service inspection techniques, and (9) identifying uncertainties associated with equipment failure and human error via fault-tree, risk, and consequence analyses; providing protective devices when the risk of injury or damage is high; and determining whether a warning is a necessary feature.

If designers carry out these responsibilities carefully, accurately, and clearly, fabricators and users will be able to do their jobs properly. The failure to carry out any of these design responsibilities sets the stage for a failure as surely as any structural notch or concentration of stress. Because structural designers learn the cardinal rule of avoiding notches in engineering school, the problem usually does not occur in manufactured single-piece parts. Such notches do, however, occur at junctions between parts, either mechanical (bolted, etc.) or fabricated (welded), as the following examples illustrate.

Figure 11-2(a) shows a bicycle steering tube that suffered a failure at the threaded insert. The angle at the point of failure was too sharp, and the material in this region point suffered from notches caused by concentrated stress. The eventual result was fatigue failure (see Fig. 11-2(b)). In the dark, quarter-moon area, note the fatigue striations, which were present before the overall failure.

In the failure of a welded connection in an air compressor, the use of the slip-on flange illustrated in Figure 11-3(a) led to a concentration

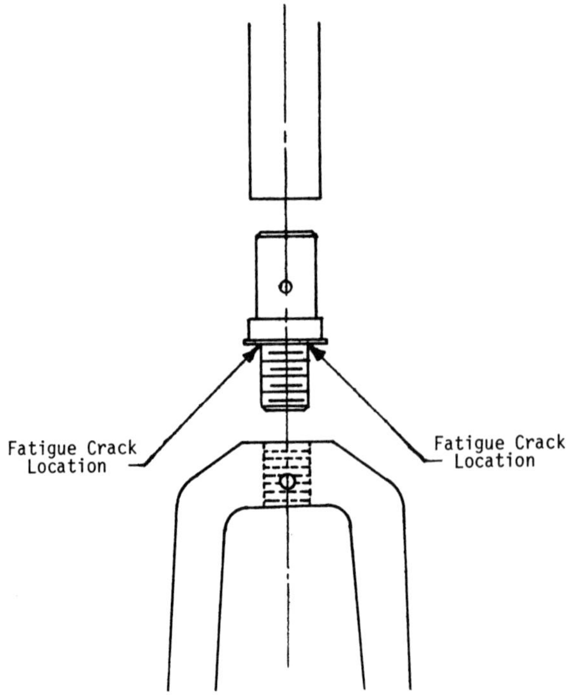

(a)

(b)

Figure 11-2. Fatigue failure of bicycle steering tube. (a) A bicycle frame steering tube. (b) The fracture surface of the tube.

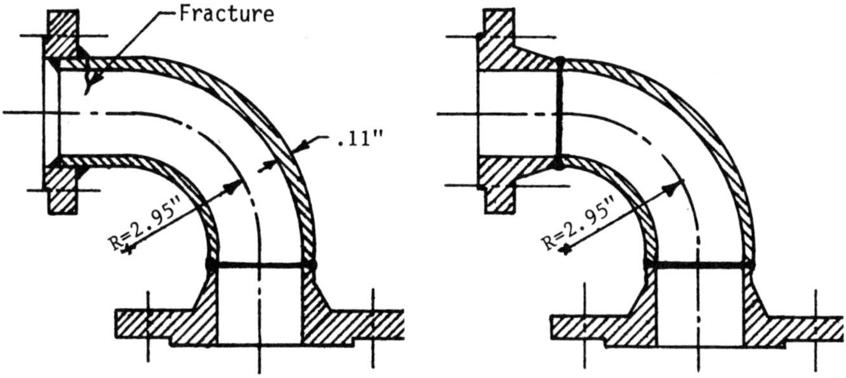

(a) (b)

Figure 11-3. Affecting fatigue failure by design. (From International Institute of Welding 1956) (a) A fracture in the flanged elbow of a compressor. (b) An improved design for the elbow.

of stress at the weld toe, which in turn caused the compressor to fail after 350 hours of service. Substituting a welded neck flange (Fig. 11-3(b)) proved to be a trouble-free alternative (International Institute of Welding 1956).

Notches that occur at welding junctions illustrate the need for a designer to be familiar with this fabrication technique in order to specify the proper weld design and inspection procedures in the fabrication specifications. The effect that weld "build ups" or notches can have on fatigue life is demonstrated in Figure 11-4. Differences in workmanship can result in differences in relevant design criteria (Newman 1960). Higher fatigue strength can be achieved with greater blending of the weld ($\theta = 180°$). Figure 11-5 is a schematic representation of this type of failure. The contour of a weld in a grinding mill (see the circular inset) caused the joint to fail. Crevices in the interbead surface of the weld resulted in a joint with a much greater notch stress-concentration factor than the designer had imagined. Because the designer had asked for a small weld fillet at that joint, which the shop had omitted, one could classify the failure as initiated during fabrication. But because the designer did not specify dimensions for the fillet or prepare any specifications for the welding or weld surface treatments such as grinding, the underlying cause of failure was associated with the design (*Metals Handbook* 1975).

A number of failures associated with design have been cited in previous chapters. For example, the failure of the flooring under an ore crusher described in Chapter 5 was the result of the "engineer's" failure

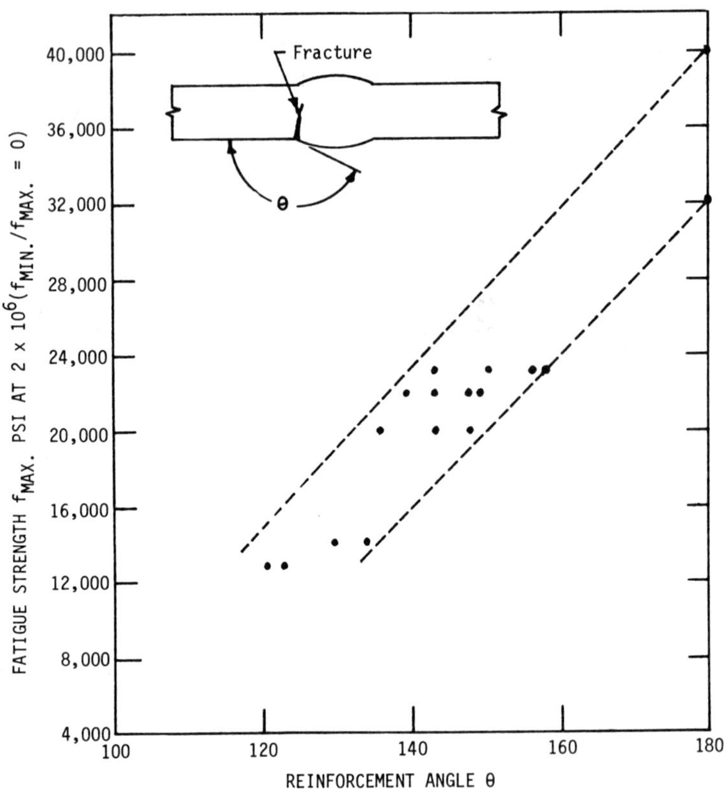

Figure 11-4. The effect of weld reinforcement on the fatigue life of steel butt welds. (From Newman 1960)

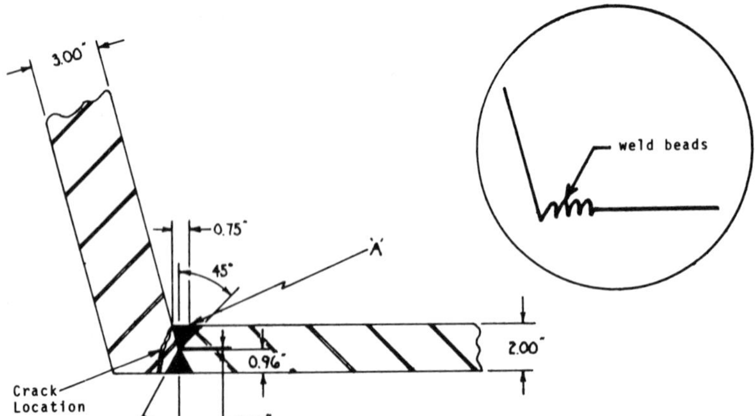

Figure 11-5. Cone-to-cylinder weld failures on a grinding mill. (From Svalbonas 1983)

to apply correct engineering and scientific principles of dynamic impact and his incorrect choice of structural material. And the failure of the pipeline liner was the result of inadequate specifications concerning testing procedures. In the case of the chemical processing vessel in Galveston, Texas, that exploded because of an overpressure, the owner had failed to specify to the designer the type of reaction that would take place in the vessel. Hence neither the vessel wall thickness was dimensioned to contain a reaction nor were the appropriate pressure-relief devices utilized.

Selection of Inappropriate Materials

Selecting materials is the design or metallurgical engineer's responsibility. To carry out this responsibility, the engineer must write proper material specifications and indicate any requirements of proper vendor-certified inspection and testing in preparation for the purchase of materials used in manufacturing or fabrication. Failures emanating from the selection of inappropriate materials, however, are more often the result of design unknowns rather than design deficiencies. Many such failures arise, not because the metallurgical engineer neglected to check the relevant criteria for a particular failure mechanism, but because that particular mechanism was not accounted for in the design engineer's or the user's specifications. These failures also can be linked to unknowns in the service environment.

One well-known series of incidents that occurred in the 1940s involved brittle fractures of welded ship plates, which caused these ships to break in half. These failures initiated research into brittle fractures and temperature-transition phenomena that eventually led to the studies of fracture mechanics discussed in earlier chapters. Today, failures associated with the selection of materials are often caused by overconfidence in the results of small-scale laboratory testing (see Chapter 8). The results of simulated service conditions in the laboratory (see Chapter 10) are often biased by the different specimen-to-prototype characteristics in machining, surface finishing, defect content, and possibly residual stresses. Many problems are associated with the effect of time in service (e.g., creep, fatigue, environmental factors) and the need to accelerate this variable in the laboratory or extrapolate test results. For example, when the effects of irradiation on materials were initially estimated for nuclear-reactor vessel designs, researchers sought to understand the dependence of the phenomenon on time. To accelerate the testing, the dosages were increased, that is, the researchers assumed that the effects of low doses of radiation over long periods could be simulated by exposing the metals to higher doses of radiation for shorter periods. Recent on-site investigations indicate that the early laboratory-based assumptions were not completely valid; therefore, predictions about the actual behavior of materials over time cannot be based entirely on the results of laboratory tests.

An example of a failure associated with an unfortunate choice of materials is illustrated in Figure 11-6. Because the hooks used with overhead cranes are subjected to high loads and long-term, repeated use, they must be fabricated of a steel alloy. The hook pictured in Figure 11-6(a) was made from a low-grade, nonalloy steel that lacked the properties specified in the engineering design. The fracture surface in Figure 11-6(b) shows the resultant propagation of fatigue cracks.

Figure 11-7 shows a failed linkage to a wine vat. The 304 steel used in the failed component became sensitized by welding: that is, the formation of chromium carbide depleted the chromium content of the steel along the grain boundaries, thus making the steel susceptible to intergranular corrosion. The result was stress corrosion cracking leading to a brittle failure. A satisfactory material in this case would have been a steel that is not susceptible to intergranular corrosion—perhaps a stabilized stainless steel such as type 321, and extra-low-carbon-grade steel such as type 316L, or a duplex stainless steel. The failure of the wine vat linkage

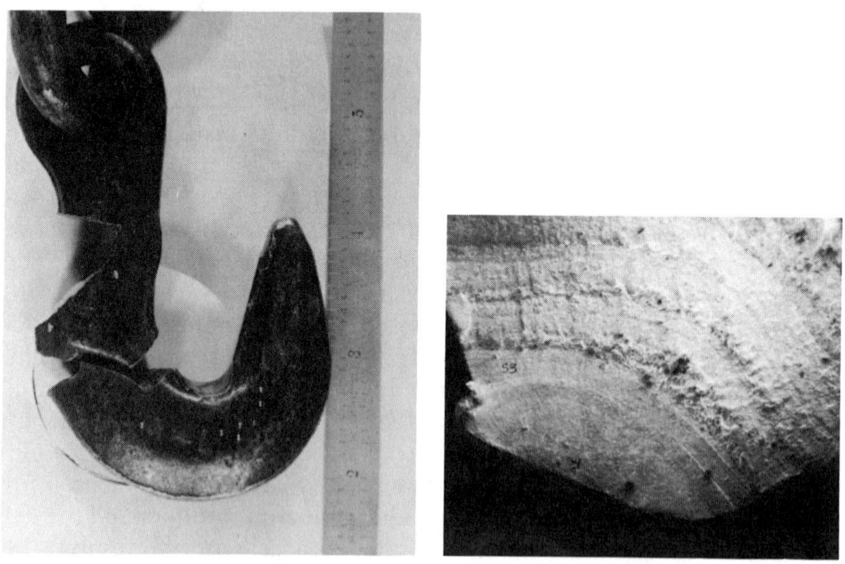

(a) (b)

Figure 11-6. Material selection affecting crane hook failure. (a) A failed hook on an overhead crane. (b) The fracture surface of the hook.

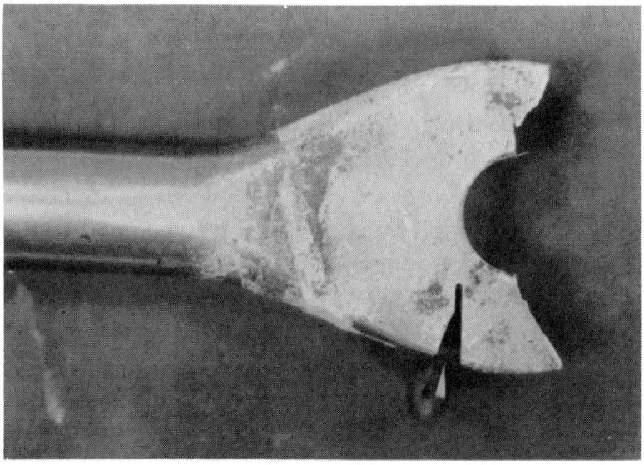

Figure 11-7. A failed linkage in a wine vat.

indicates the overlapping effects of material selection and fabrication (welding). The counterpart to stainless-steel sensitization by welding in ferritic steels is hot cracking, influenced mainly by carbon content and the manganese–sulfur ratios. Such cracking normally occurs during cooling after the deposition of weld metal at high temperatures. Thus, metallurgical engineers must be cognizant not only of the service environmental requirements for fabricated structures but also the proper sequence and type of fabrication.

Imperfections in Materials

Many failures originate from imperfections in materials that serve as a propagation path for failure. The imperfections may take the form of internal or external mechanical notches, sites of pitting or corrosion, or of metallurgical discontinuities that result in localized areas with high or low degrees of hardness. Forged products are often susceptible to defects such as laminations, laps, slivers, cracks, seams, or scabs. Castings are typically susceptible to gas holes, sand inclusions, shrinkage, chaplets, chills, hot tears, and other linear discontinuities. Rolled plates and extrusions are susceptible to defects. There are available steel mill certifications for ultrasonically cleared plates referenced to laminations or general freedom from inclusions.

One of the designer's responsibilities is to recognize the existence of such flaws, determine what levels are acceptable, and specify inspection procedures for detecting and evaluating them. Because a certain number of flaws are acceptable, inspection procedures should be in line with the "fitness for purpose" concept (Farrar, Charles, and Dolby 1971). The foundry or steel mill is then responsible for producing material that is free of the *detrimental flaws* specified in the designer's specifications. Ultrasonic and radiographic methods are the most popular methods of detecting internal flaws in forgings and castings, whereas magnetic-particle and liquid-penetrant examinations are most commonly used to detect imperfections on the surface (see Chapter 8).

Examples of imperfections found in materials are shown in Figures 11-8 and 11-9. Figure 11-8(a) shows a severe forging lap on the inside of a stainless-steel pipe section. Figure 11-8(b) shows unfused chaplets on the side of a cast-steel bearing wheel. Figure 11-9(a)–(c) illustrates shrinkages of varying severity found in gray-iron castings, while Figure 11-9(d) is a positive of a radiograph of severe shrinkage.

An example of a failure associated with material defects is illustrated in the photograph of a grinding mill in Figure 11-10(a). The mill experi-

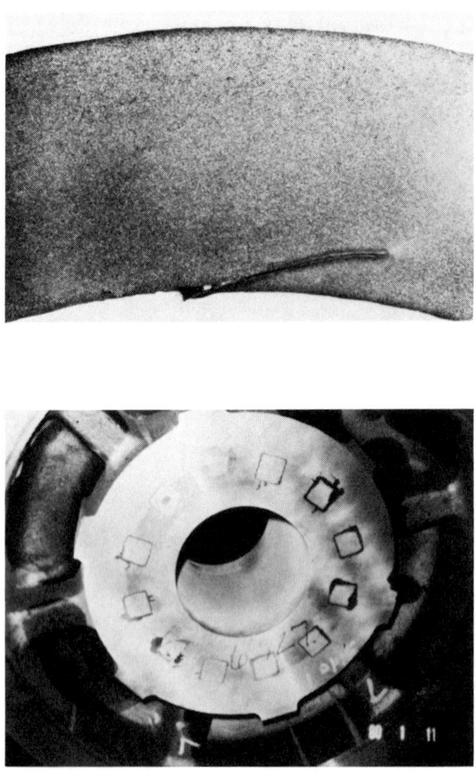

(b)

Figure 11-8. Flaws in a cast-steel bearing wheel. (a) A forging lap. (b) Unfused chaplets.

(a)

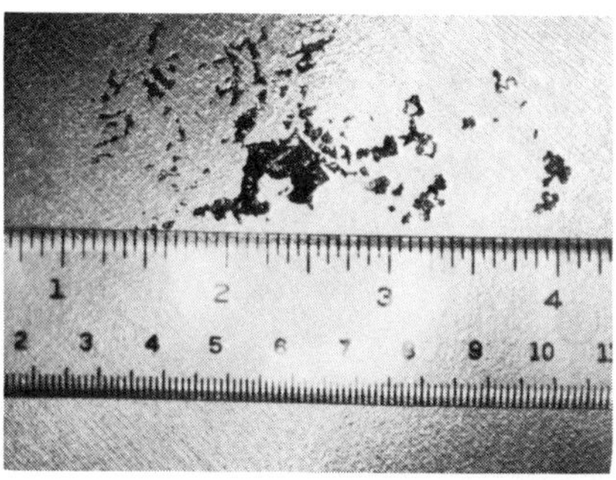

(b)

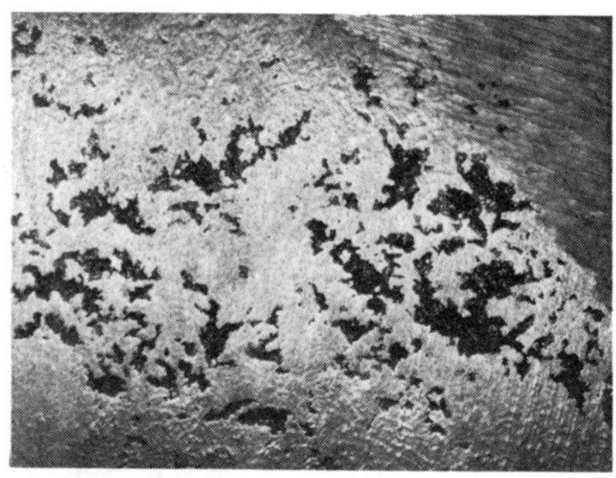

(c)

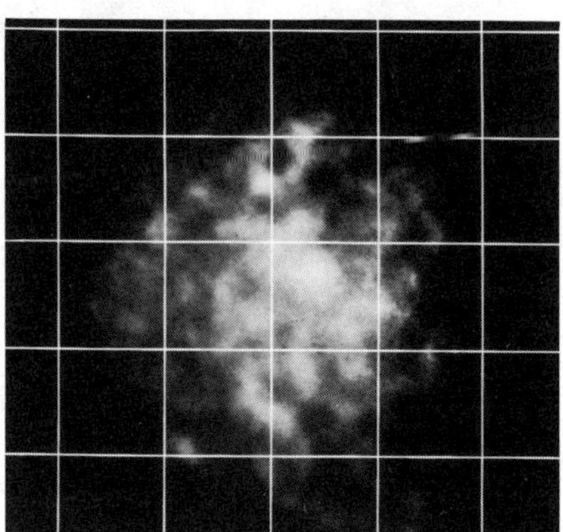

(d)

Figure 11-9. Shrinkage in gray-iron castings. (a) Macroview of surface effects. (b) Closeup view of surface. (c) Microscopic (low power) view of surface. (d) A positive of the radiographed interior.

enced severe lamellar tearing upon welding of certain joints (see Fig. 11-10(b)). Although the steel plating was thick and the joint configurations were restrained, the tearing was unexpected because a special low-sulfur rare-earth composition had been used to make the steel plates. A microscopic examination revealed that the lamellar tears had formed in dense inclusion planes that existed in some areas of the plate perpendicular to its thickness. When the investigator opened the cracks (Figure 11-10(c)) to reveal the fracture surface (Figure 11-10(d) and (e)), he noted that the density of the inclusions was extremely high in these planes. X-ray analysis showed that the inclusions were cerium and lanthanum, rare-earth elements used to control the sulfur content, and that they appeared to be concentrated in certain regions of the plate, essentially giving those re-

(a)

Figure 11-10. Failure of a mill for grinding autogenous iron ore. (a) Photograph of the mill, which is 36 feet in diameter by 15 feet long (10.97 × 4.57 meters). (b) Optical micrograph of cracking oriented normal to the edge of the plate (6×). (c) Scanning electron micrograph of cracks and inclusions on the as-polished surface (2300×). (d) and (e) Scanning electron micrographs of the fracture surface.

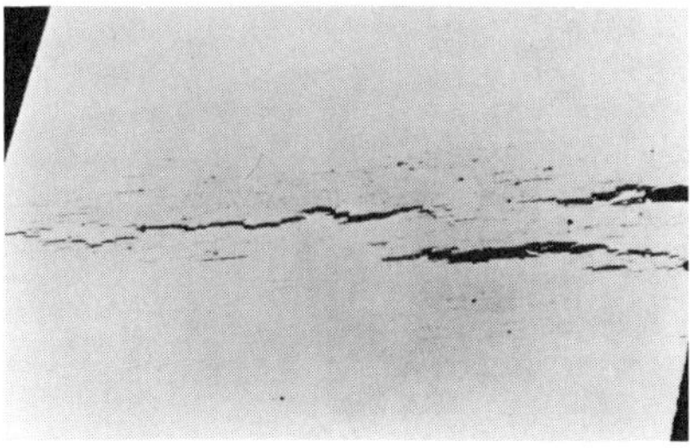

(b)

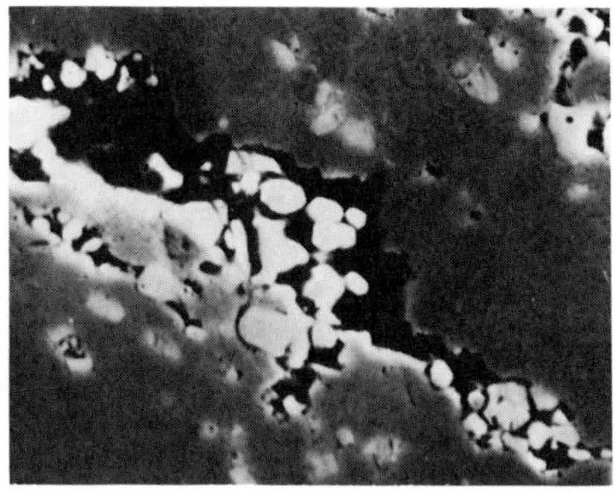

(c)

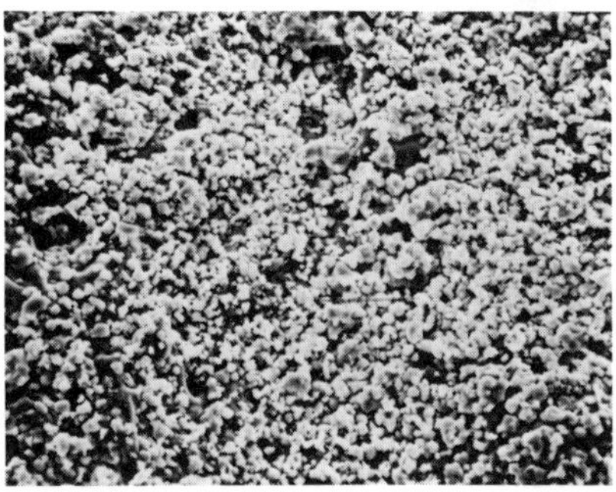

(d)

(e)

Figure 11-10. *(continued)*

gions zero through-thickness ductility and causing the severe welding failures (Farrar, Charles, and Dolby 1971; Glodowski 1983; Svalbonas 1983; Wilson 1981). Thus, the cause of the failure was not the fabrication process but the milling process.

The metallurgical properties of a material may constitute an imperfection if an error was made during chemical or thermal treatment at the foundry or steel mill. These are usually discovered during certification tests of chemical and mechanical properties. However, errors associated with spot areas or individual heat may still occur. An example of such a failure is shown in Figure 11-11. Some A193 Grade B-7 bolts were put in service as erection bolts of a grinding mill (as specified by the American Society for Testing and Materials, ASTM). When installed, the bolts were prestressed (preloaded) to 70 percent of yield. After some time in service, the mill was shut down for several months, and two bolts fractured in one area of the mill flange during the shutdown. Note in Figure 11-11(b) that the fracture appears to indicate a minor amount of fatigue

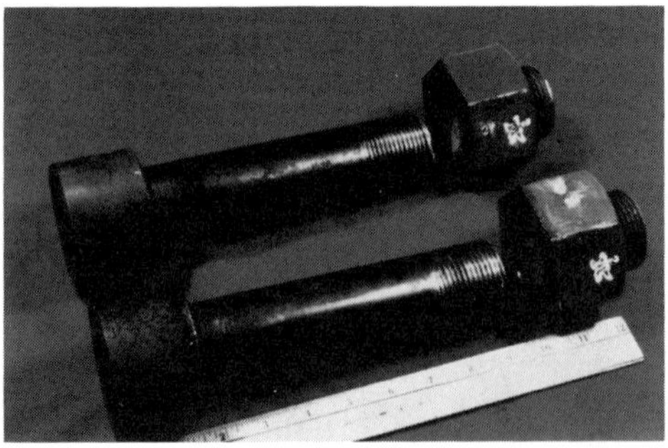

(a)

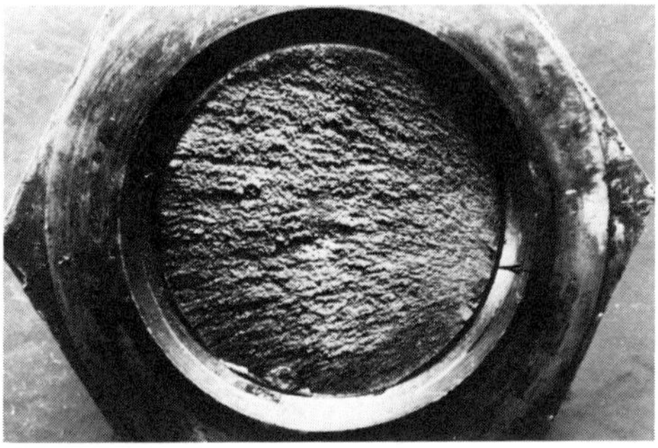

(b)

Figure 11-11. Failed improperly heat-treated grinding mill bolts. (a) Photograph of the fractured bolt. (b) The fracture surface of the bolt.

growth, followed essentially by a brittle failure. Samples taken from the bolt revealed an average of 161,000-psi yield and 190,000-psi ultimate strength, compared with the ASTM expected values of 105,000 psi and 125,000 psi, respectively. Brinell hardness readings of 375 were consistent with the greater strength, whereas the bolt chemistry was consistent with the ASTM's requirements for A193 B-7 bolts. Thus, it appears that incorrect heat treatments provided unusually high strength and corresponding brittleness. In addition, the nominal preload of 70 percent of yield was actually only 45 percent when compared with the actual yield values, which further reduced the expected fatigue life.

Fabrication Deficiencies

The fabrication process can be divided into the following subprocesses: handling of materials; storage of materials and consumables; manufacturing or construction processes such as cold forming, machining, grinding, heat treatment, acid pickling, electroplating, and welding; and inspection, reworking, assembly, storage, and preparation for shipping of the finished product. Any handling of the product can cause gouges, notches, or impact marks. Poor methods of storing weld wire can cause subsequent weld problems. Cold forming or overgrinding can produce high residual stresses. Improper heat treatment can cause decarburization or undesired hardening or softening. Acid pickling can cause hydrogen damage. Incorrect or inadequate quality control procedures at each step of the fabrication process can allow defective parts to pass. Monitoring the fabrication process is the responsibility of the manufacturer's quality assurance department, and the responsibility of setting the criteria for this process rests with the design engineer.

Theoretically, if the design engineer has chosen the appropriate design codes and criteria and has prepared appropriate fabrication documents, the only way a fabrication deficiency can slip through as if quality assurance personnel violate a procedure or miss an inspection. In practice, however, design engineers simply state inspection personnel should "follow good shop practice," and unless such practice is actually good, much can be missed. In this regard, understanding the limits of accuracy achievable with any inspection technique and operator qualification is important. Attempts to define these limits have been underway in several countries for more than a decade (Forli 1979).

The methods of nondestructive examination used in fabrication and maintenance fall into two categories: (1) surface investigation and (2) volumetric investigation. The tests most commonly used to investigate surfaces are the visual, liquid-penetrant, magnetic-particle, and eddy-current methods, and the most commonly used volumetric tests are the ultrasonic and radiographic methods (see Chapter 8).

An example of the mix of nondestructive methods of testing used to find fabrication flaws in power plant components is shown in Table 11-1 (Reinhard 1984). Such flaws can take the form of mechanical notches, as in weld defects or handling gouges, changes in metallurgical properties that occur during heat treating or cleaning operations, and unanticipated residual stresses. Figure 11-12 illustrates the types of notches that may be introduced by welding. Figure 11-12(a) shows a rather large crater pit, and Figure 11-12(b) shows a weld radius that has been ground smooth to expose areas of undercut (the dark areas). Figure 11-12(c) shows an excavated area in a cracked weld that was being repaired; note the large void that has been uncovered. Figure 11-12(d)–(f) demonstrates the accuracy of ultrasonic testing of welds. Figure 11-12(d) shows a drilled-out weld seam, exposing some slag inclusions at the weld root found by ultrasonics. Figure 11-12(e), (f) shows how ultrasonic testing

TABLE 11-1. Applications of Nondestructive Testing Methods to Power Plant Components

	Testing Method					
Component	*Visual*	*Liquid Penetrant*	*Particle*	*Ultrasonic*	*Radiographic*	*Eddy Current*
Boiler heater I.D.	Fiber optics and closed-circuit TV through access windows	Not used	Not used	Used for steam and mud-drum penetrations for measuring wall thickness	Not used	Not used
Transfer tubes O.D.	Eye examination for sagging or distortion of tube wall	Used to detect cracks	Use limited to detecting cracks in welds	Used extensively to measure wall thickness and detect cracks	Use limited to detecting cracks and inspect repair welds	Not used
Condenser tubes I.D.	Usually not used	Not used	Not used	Usually not used, but can supplement particle testing in or near tube support areas	Normally not used	Used extensively to detect cracks and thinning walls

Feedwater heater tubes I.D.	Usually not used	Not used	Not used	Use limited to carbon-steel tubing and as a supplement to particle testing for flaw analysis. Slow	Normally not used	Used extensively to detect cracks and thinning walls in noncarbon-steel tubing
Turbine and generator rotors (outside)	Eye examination for cracks	Use limited to supplement to particle testing	Major surface examination; draped and bucking field method	Peripheral examination as supplement to bore or when bore is inaccessible	Normally not used	Not used
Turbine and generator rotors (boreside)	Rigid borescope	Use limited to remote systems	Major examination; central conductor and yoke methods	Used extensively to detect service-induced subsurface cracks	Normally not used	Use limited to detection of surface defects
Turbine blades	Use limited to mirror-aided examination for defects or fiber optics through access ports	Used for nonferromagnetic areas of blades and as supplement to particle testing	Major method of examining surfaces; draped coil, bucking field, and some yoke	Use limited to detection of cracks in shroud area	Limited use	Limited use, but highly sensitive method of detecting small surface cracks

Table 11-2. Application of Nondestructive Testing Methods to Processing and Power Plant Components

	Testing Method				
	Hardness	***Acoustic Emission***	***Tomography***	***Infrared or Thermography***	***Interferometry and Surface Instrumentation***
Information	Used to determine tensile strength. Inexpensive method. Surfaces only	Used to locate crack growth and cracks for ultrasonic and radiographic testing	Used to define crack configuration by two- and three-dimensional CT	Used to locate high surface stresses by metal change in temperature	Used to determine strains on a surface by strain gauges and bi-fringic coating
Locations	Most surfaces that are accessible by hand	Most surfaces in operation or shut-down	Limited to 2 × 2 × 2-foot volumes, but portable units are possible	Limited to components at ambient (possibly steady-state) temperature	Most surfaces that are accessible to apply the material. Temperature can be a problem
Use or Application	Used to check properties of most components: e.g., studs, nuts, welds, vessels, pipes	Usually used in vessels, cylinders, and pressurized components	Usually used to inspect small components, fittings, instrumentations that are removed from the system	Usually used to examine structural elements such as supports or moving elements at steady-state temperatures	Usually used to examine structural elements for steady-state or transient mechanical or thermal loads

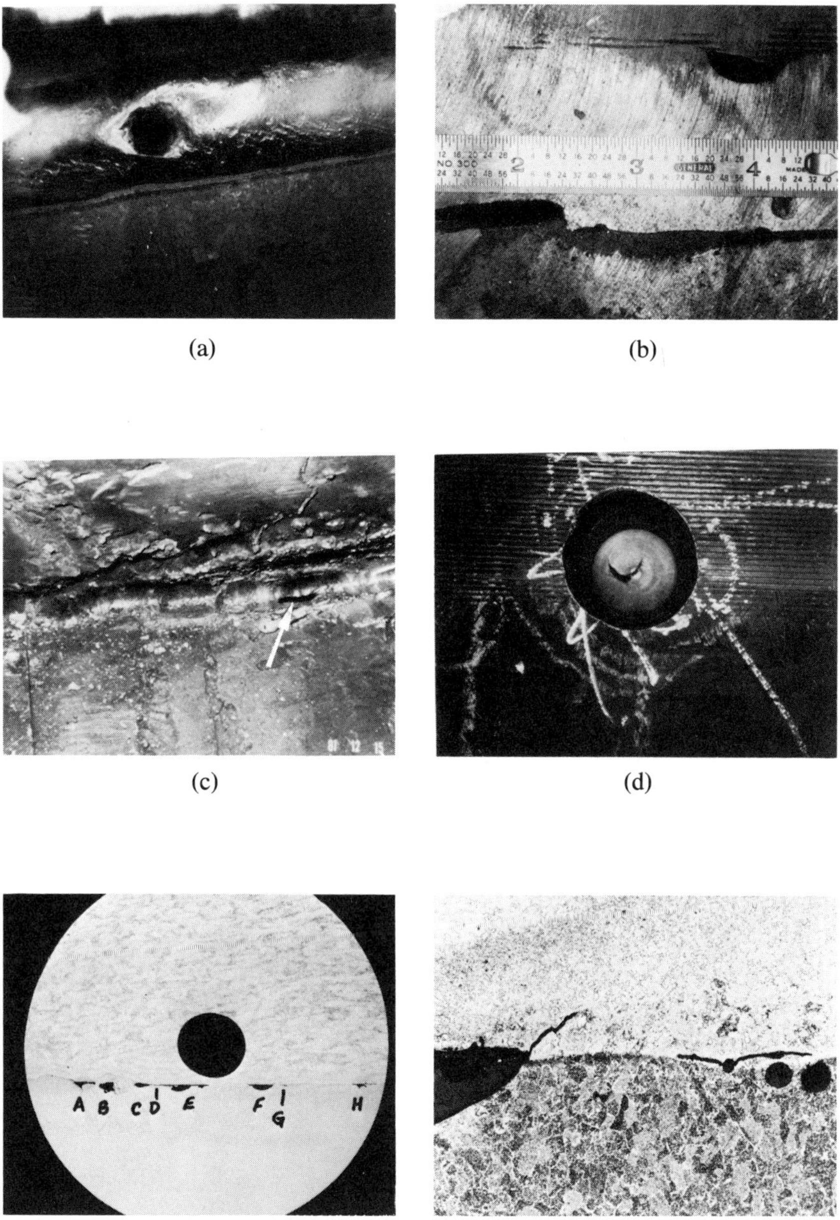

Figure 11-12. Void exposed in drilled-out weld seam. (a) A weld crater pit. (b) Weld undercut areas. (c) A weld excavation showing a possible source of cracking. (d) Subsurface weld slab at a weld root. (e) Slag inclusions at the fusion line. (f) Cracks growing from the inclusions.

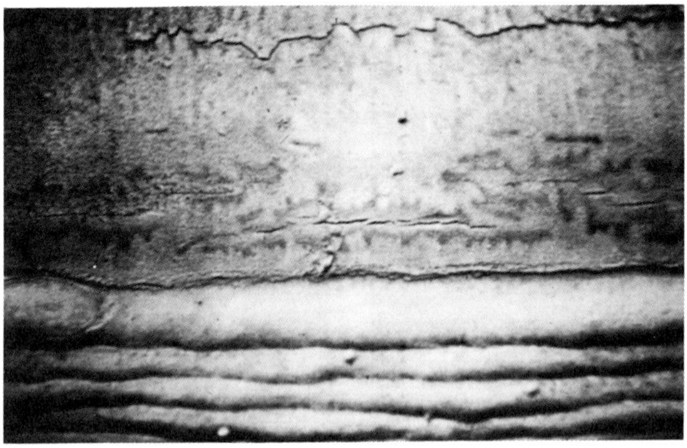

Figure 11-13. Lamellar tears after welding of a large plate.

immediately detected small cracks growing from individual weld slag inclusions shown in Figure 11-12(e) over a period of nine months of operation. Figure 11-13 shows lamellar tearing found in the plate adjacent to a large weld. This was probably caused by poor material or by poor heat control during the welding process.

Figure 11-14 illustrates a metallurgical notch introduced in a truck steering linkage by the fabricator. The results of microinspection showed that incorrectly applied chromium plating caused cracks that subsequently propagated into the main shaft and culminated in failure (Figure 11-14(b)).

(a) (b)

Figure 11-14. Crack initiation in improper plating of truck steering linkage. (a) Photograph of the steering linkage. (b) Microsection of the linkage.

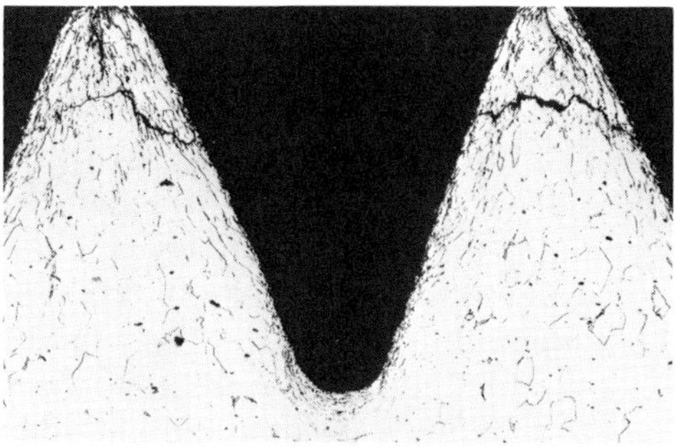

(a)

(b)

Figure 11-15. Defects introduced into threads during machinery operation. (a) A bolt crack in a cold-rolled thread. (b) Heat checks on top of a gear tooth.

Figure 11-15 shows defects that are introduced during machining operations. Figure 11-15(a) shows cracks that occurred during rolling of threads on a bolt, and Figure 11-15(b) reveals the characteristic network pattern of grinding checks, which were highlighted in the figure by using dry magnetic powder on a developer background. In some instances, these defects may not be detrimental (as is true in this application because of tooth-tip relief); however, the heat-checked area eventually may accelerate a failure associated with misalignment or wear.

Unanticipated Service and Operating Conditions

Normal service or operating conditions include forces, temperatures, time, and environment. The forces and temperatures can be steady or time dependent, and the environment can involve nuclear, biological, electrical, or chemical factors. If all these conditions are properly defined in the owner-user's performance specifications and the manufacturer's operation and maintenance specifications, overloads would occur only during transient fluctuations caused by the failure of all responsible parties to appreciate the consequences of equipment failure, by intentionally operating the product beyond its specified or intended use, or by human error during operation.

When time is a factor, one often finds that the design engineer, although aware of environmental factors, did not adequately model or understand them, especially when time, varying loads, and residual stresses are part of the equation.

The failure to perform adequate preoperational tests to identify the effects of complex loadings leads to equipment failure, for example, by the following.

1. *Vibration* [such as flow-induced vibration; sloshing (gravity waves)]
2. *Transient phenomena* (such as chugging, water hammer, and blowdown)
3. *Coupling between parts and/or the environment,* or a combination of the two.

Because the environment can affect all loading mechanisms, data obtained from *in vacuo* tests should be adjusted to account for environmental influences. Some relatively common environmental problems, such as hydrogen embrittlement, pitting, and stress corrosion cracking, were discussed in Chapter 9.

Although the damaging effects of these phenomena can be, and often are, considered in design, one usually finds that the rates of deterioration have not been fully anticipated, especially when influenced by changing operation and maintenance and the effects of aging. This is especially true in cases involving corrosion caused by various interrelated mechanisms (LaQue and Copson 1963). The lead-processing kettle picture in Figure 11-16(a) is an example of such a failure. Weakened by progressive liquid-metal embrittlement, the bottom of the kettle eventually collapsed. The findings of a through-thickness metallurgical study of the failed area are illustrated in Figure 11-16(b). Another example is shown in Figure 11-17, which documents a failure of steel piping associated with stress corrosion cracking caused by exposure to soil and some degree of cathodic protection. Figure 11-17(a) shows the secondary stress corrosion cracks revealed on the outside surface of the pipe by a

(a)

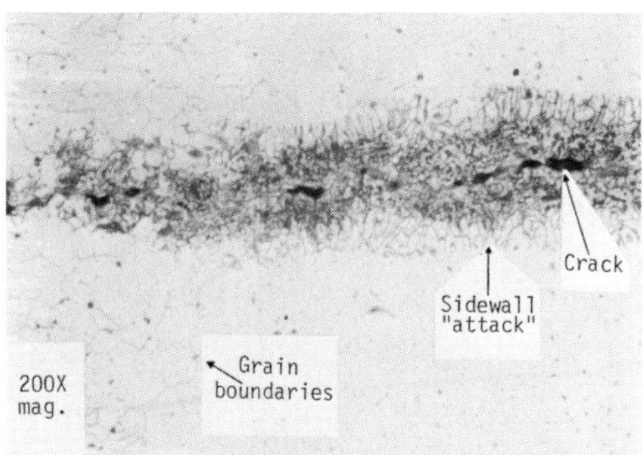

(b)

Figure 11-16. Metal imbrittlement failure of a processing kettle. (a) Photograph of the kettle. Note the missing bottom. (b) Through-thickness closeup of the embrittled kettle (200×).

dye-penetrant test. Figure 11-17(b) is a cross section through the cracks in an as-polished specimen, whereas Figure 11-17(c) shows the area with a picryl etch. Note the intergranular nature of the cracks.

The examples previously cited involved actual service and operational conditions. Other service overloads can be attributed to misuse or to unauthorized modifications such as those made in the hot tapping equipment and the buildings damaged by unexpected ground shock discussed in Chapter 5. Another example of such problems was the spectacular collapse at the Hyatt Regency Hotel in Kansas City (Petroski 1982).

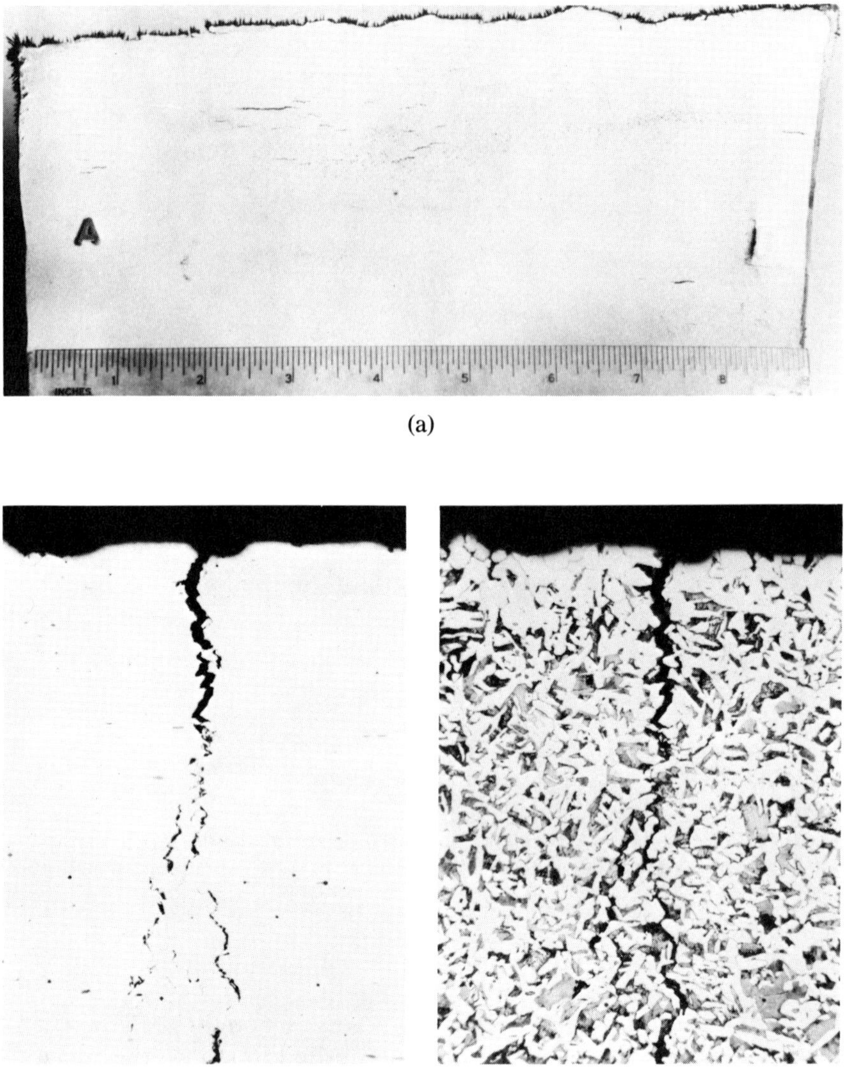

(a)

(b) (c)

Figure 11-17. Stress corrosion cracking failure of a steel pipe. (a) Secondary stress corrosion cracks on pipe O.D. (b) Cross section of one of the cracks, as polished (250×). (c) Cross section of the picryl etch (250×).

Improper Maintenance

Because of their importance in ensuring the safe operation of products and systems, maintenance and inspection are major factors in affixing responsibility for failure. If all potential causes of failure involving design, processing, and the service or operational environment have been ruled out, inadequate maintenance or inspection should be investigated. Once owners accept a fabricated product, they are usually responsible for maintenance and periodic inspection unless they have subcontracted the job to a separate service entity or the manufacturer, fabricator, processor, assembler, or retailer maintains and inspects the product as part of its sales package, as was the case of the co-generation equipment mentioned earlier.

Figure 11-18 shows a pinion that ran for approximately one week without lubrication. Note that the tooth profile has been completely destroyed—it no longer has an involuted form; instead, the teeth are worn to a point. The left edge of the teeth shows the original profile. Because the pinion and gear were axially mismatched, severe wear caused by scoring or scuffing appears on each tooth, and the tip contains jagged pieces of metal. Further investigation revealed that the alarm which should have warned the user about the lubrication failure had never been connected; however, daily visual inspections should have revealed the problem.

Another gear problem related to improper maintenance is depicted in Figure 11-19. Here, the problem is a combination of gear misalignment and poor initial cutting of the teeth. The dark areas on the teeth (Fig. 11-19(a)) show the typical one-sided, triangular pattern of contact

Figure 11-18. A damaged pinion.

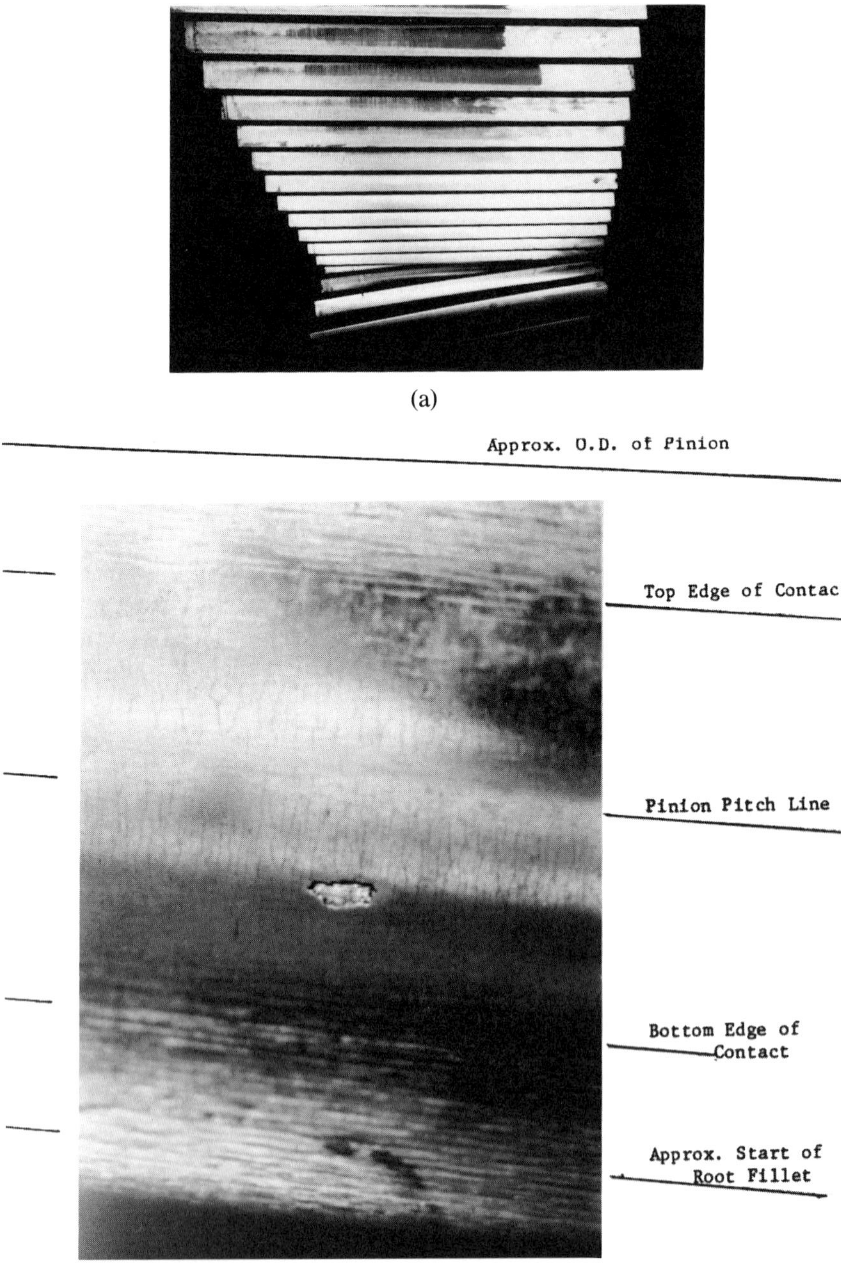

Figure 11-19. Gear teeth failure due to improper fabrication and maintenance. (a) Gear teeth showing the contact area. (b) Gear teeth showing heat checks.

associated with misalignment. The light vertical "brush marks" are cutting-tool feed marks indicating teeth-cutting problems. The overloaded areas shown in Figure 11-19(b) already show signs of surface damage by heat checking and the cross-linking of cracks that initiate surface failures. In this case, the cause of failure was linked to both fabrication (inaccurate tooth profiles and spacing) and maintenance (improper alignment of gear and pinion). Whether one or the other was the sole cause would be difficult to determine on the basis of technical reconstruction.

Figure 11-20(a) shows a corrosion fatigue crack that resulted from poor maintenance procedures. The inner surface of a cylindrical steel trunnion (pivot) on a grinding mill is usually protected by wear plates. When these wear liners are not properly maintained, ore slurry gets behind them and contacts the surface of the trunnion, causing erosion. These erosion grooves in an otherwise smooth machined surface then become sites of corrosion as well as notches, which amplify the alternating stresses on the trunnion and lead to the growing corrosion fatigue cracks shown in the figure. Figure 11-20(b) demonstrates another maintenance problem associated with grinding mills that are erected using prestressed erection bolts. Occasionally, these bolts need to be checked for the proper prestress or the need for replacement. The bolts in the photograph were overtorqued during replacement. Improper calibra-

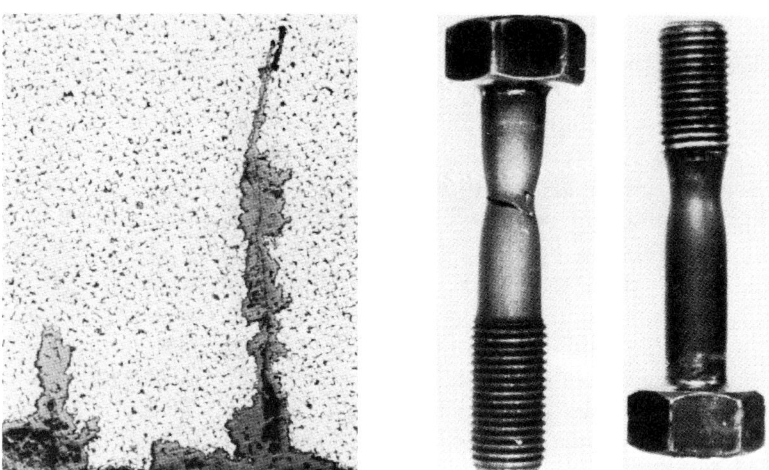

(a) (b)

Figure 11-20. Grinding mill failures associated with poor maintenance. (a) Corrosion fatigue crack in a steel trunnion on a grinding mill. (b) Overtorqued bolts on a grinding mill.

tion of a hydraulic torque wrench or the use of a hand wrench with a pipe extender handle (a "cheater bar") are possible reasons for the problem.

In Chapter 5, we discussed the separation of the flanged sections of a valve in the control room of a power station. The failure occurred because the owner's maintenance and inspection personnel did not notice the significant creeping of the flange bolt nut that had occurred during 27 years of operation. As a result of this oversight, some personnel were seriously injured, one employee died, and the utility sustained substantial economic losses. Figure 5-8(d) (p. 151) illustrates that because of the significant creep deformation (readily measured with a micrometer) of the nut threads, the washer end of the nut was *not* carrying any of the stud load. It is doubtful that the deformation could have been detected by visual inspection alone unless the deformation had reached the point of incipient failure. In such facilities, critical parameters should be monitored continuously at strategic locations and rigorous inspections should be conducted during scheduled shutdowns.

THE TECHNICAL REPORT

As we said in Chapter 5, the stages of an accident investigation and failure analysis are the same whether the motivation for the analysis is redesign or litigation; however, more formal recordkeeping is required for a litigation. Having discussed the first six stages of the accident investigation, we can now talk about the formal report that is usually prepared at this juncture. In this report, the investigating engineer outlines the findings obtained during the first six stages of the investigation and, in addition, offers *opinions* about who is responsible for the incident and makes *recommendations* for remedying the problem. The scope and amount of detail included will be influenced by the client and the purpose of the report (see Fig. 11-21).

Although the style of a report written for engineers and scientists usually differs significantly from one written for attorneys, clients, and the courts, the conclusion does not. Reports written in anticipation of upcoming litigation should teach and inform in uncomplicated language and contain helpful illustrations. The underlying scientific and engineering theories should be available in appendixes, references, or both, but the thrust of the report should be to make clear to the expert's client and attorney the expert's opinions about causality and the basis for those opinions.

If the purpose of the investigation was the redesign of a product (see Fig. 11-21, III), the report provides input for the redesign cycle of the

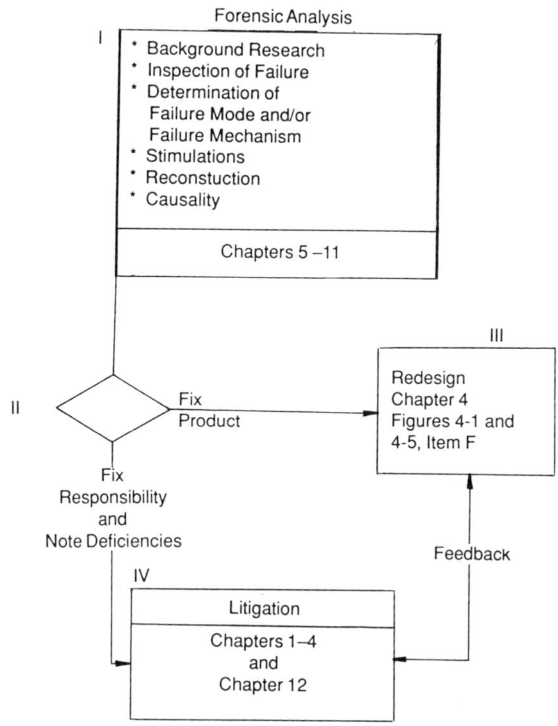

Figure 11-21. Flow diagram of a forensic analysis leading either to a redesign or to litigation.

product, as was discussed in Chapter 4. If the expert is preparing opinions and recommendations for use in litigation (see Fig. 11-21, IV), then the type of report is dictated by the client's legal strategy, a topic that will be discussed in the next chapter. Whether redesign or litigation is the purpose of the investigation, the technical report contains sections and subsections that correspond to the eight stages of the investigation. It should contain an introduction that summarizes the events and principal players involved in the incident. Usually a preliminary report is prepared based upon the preliminary findings of the first three stages, especially before proceeding to perform material tests, simulations, and extensive research. At this point, we want to reemphasize that an analysis of the mode of failure alone cannot be the basis for determining causality. To affix causality, the report also must address the mechanism of failure. This is why the investigator must have a comprehensive understanding of engineering functions in addition to paper credentials (Vance and Gould 1985; Weinstein *et al.* 1978).

In many incidences of product failure and litigation, the client's economic situation dictates whether the investigator will provide a preliminary or formal report. In either case—and whether the purpose of the report is redesign of the product or use in litigation to assess economic responsibility—a draft of the report should be reviewed with the client. If the report will be used in litigation, the draft should also be reviewed critically by the legal team that will handle the trial before the investigator completes the final report.

The final version of the report should contain the following:

1. A table of contents (for a sample table of contents, see Appendix D, p. 582–583)
2. An executive summary, especially if the investigation involved a considerable number of reconstructive stages
3. An analysis of the information on which the expert's opinion is based
4. A review of the methodology or rationale underlying the expert's analysis (Ladino 1986)

The report can be a particularly effective mechanism for encouraging an out-of-court settlement by the contending parties.

As was discussed in Chapter 5, a final technical report is frequently not prepared if the legal team anticipates that the case will go to trial ("the contending parties have dug in their heels") because they do not want to help the opposition (if the report must be produced to them) by (1) giving them time to prepare to attack the report, (2) giving them expensive data for which one's client paid, and/or (3) outlining a legal or technical position that may change. This is especially true if the opposition's resources are limited (i.e., to have their expert perform the same or similar tests) in their preparation. As we mentioned in Chapter 5, "losing" data that are useful to the opposition has a similar effect.

Just as it is important to conduct independent reviews during the design phase of a product, it is equally important to obtain independent reviews of the report of an accident investigation and failure analysis. Recall that two types of independent reviews are typically used in an accident investigation and failure analysis: a peer review (one by a disinterested party) and a blind review of the first six stages of the investigation (one in which the expert's conclusions, opinions, and recommendations have been left out).

To illustrate the usefulness of a technical report, consider the four accident reconstructions discussed in this chapter: (1) the lineman's unsafe safety belt, (2) the poorly designed ladder foot, (3) the slippery tile, and (4) the shattered composite disk. A technical report was prepared in all four cases, a settlement was extended in all cases, and the offer of a settlement was accepted in three of the four cases (Cases 1, 3, and 4). And in all cases, the opposition's legal team cited the technical reports as the reason for the offer. In Case 2, the plaintiff's attorney refused the opposition's offer, failed to get the videotaped simulation data into evidence, and lost the case, and the plaintiff received no compensation. In all four cases, the plaintiff's attorney commissioned the significant investigative tests and simulations just before the end of the discovery period.

In each instance, the plaintiff's legal team worked on a contingency basis, whereas the defendant's legal team worked on an hourly basis. In the four instances, the plaintiff's law firm consisted of one or two attorneys who had little or no product liability experience and limited financial resources, whereas the defendant's firm was a large established one whose attorneys were experienced in product liability. A word of caution, however: large, established law firms should never underestimate the technical or financial resources available to the small firm that handles personal injury cases. Some of these small firms have established a successful practice in product liability law.

CONCLUSION

Although the causes of product failure have been grouped by category in this chapter, responsibility is usually associated with a complex array of interacting, causative actions. The science involved in a failure analysis and the art of conducting an accident investigation require the investigator to unravel the individual contributions of the design engineer, the manufacturer, and the user-consumer, then reconstruct the overall root cause of the problem. In this way, improvement of products can be achieved in the future on the basis of the lessons learned from the product failures of today. Equally important, the courts will be better able to assess responsibility for repairs and compensation for injury and damages and to determine what punitive measures are appropriate.

12

Product Liability Litigation

The intent of the preceding chapters was to (1) illustrate the continuity of the processes leading from an incident involving economic or personal loss through an investigation to determine causality, (2) examine how the primary influencers of product safety (see Chapter 2) are interwoven in litigation, and (3) enhance the reader's appreciation of each aspect that contributes to the overall complex pattern of the product liability and forensic engineering framework and the production of high-quality, safe products. The litigation process is not the preferred method of controlling product quality, but by the same token, it is the centerpiece of a free-enterprise system and a democratic society, where disputes over causality or responsibility are resolved in the context of restitution, compensation, and punitive damages.

This chapter explores the legal framework in which the attorney and the engineering or scientific expert work with one another and their client to assess the merits of their own position and that of their opponents through technical reports and depositions in preparation for a trial. Financial arrangements for conducting an accident investigation and failure analysis are discussed, and guidelines for technical experts and attorneys when preparing for a deposition and trial are provided.

THE EXPERT–ATTORNEY–CLIENT RELATIONSHIP

Who are the principals in a product liability suit? The principals are the law firm, the experts, the client, the opposition, and the court, which may or may not include a jury. The attorneys convey information presented by their experts and clients to the court, which weighs the merits of the case on the basis of the information presented by each side. The combatants may be one or all of the primary influencers of product quality or any of the six instrumentalities discussed in Chapter 2, including an organization that writes voluntary codes (see, for example, the case of *Hydrolevel v. American Society of Mechanical Engineers* 1982). In many instances, insurance carriers for both sides are involved in the litigation. Both defendant and plaintiff can be a large organization or an individual. When compensation is the primary thrust of a suit, the best defendant—from the plaintiff's viewpoint—is a large organization, an insurance carrier, or both.

The Technical Expert

As we have said many times, the role of the technical expert is not only to interpret data such as the mode and mechanism of failure but to affix responsibility for the failure as well. Because a product failure can be related to improper engineering design, selection of inappropriate materials, material processing (to eliminate the effects of defects), manufacturing and construction, operation and use, and maintenance and inspection, the engineering expert must be knowledgeable in one or all of these six areas "beyond the ordinary experience of the layman." Medical and actuarial experts also may be involved in a case, particularly when an injury or fatality has resulted in economic loss to the victim's family or employer. The duty of an attorney is to demonstrate to the court that experts are qualified to testify because of their (1) formal education and occupational and research experience, (2) self-education effects that represent the legitimate application of their basic skills, and (3) qualifications that are consistent with the technical issues they have addressed as a result of the thoroughness of their self-education efforts (Weinstein *et al.* 1978).

Under common law and Rule 702 of the Federal Rules of Evidence, an expert who is qualified by training, skill, education, or experience may testify in the form of an opinion on the particular matter at issue (Ladino 1986). The rule states that if "scientific, technical, or other specialized knowledge will assist the trier of fact to understand the evidence or to determine fact in issue, a witness qualified as an expert by knowledge, skill, experience, training or education, may testify thereto in the form of an opinion or otherwise."

Some states hold that the expert's testimony must be *necessary*. For example, in *Keefer v. C. R. Bard, Inc.* (1981), a Michigan court set the following guidelines for the admissibility of expert opinion: The expert must be qualified in his field, facts that require interpretation or analysis must be involved, and the witness's knowledge must be peculiar to experts rather than to laypersons. Federal Rule 703, "Basis of Opinion Testimony by Experts," states that

> the facts or data in the peculiar case upon which an expert bases an opinion or inference may be those perceived by or made known to him at or before the hearing. If of a type reasonably relied upon by experts in the peculiar field in forming opinions or inferences upon the subject, the facts or data need not be admissible in evidence.

Federal Rule 704, Opinion on Ultimate Issues, states that "testimony in the form of an opinion or inference otherwise admissible is not objectionable because it enhances an ultimate issue to be decided by the trier of fact."

The limits of an expert's opinion need to be clearly defined. A judge or jury must know the underlying facts that form the basis of the expert's opinion and what facts have been excluded. An expert's conclusions are often based on hearsay, but courts do not limit the expert's testimony because of that fact. As we shall see, any weaknesses in the expert's reasoning can be explored on cross-examination or challenged by the opposition's expert.

The expert's expertise generally falls into one or all of the following categories:

- Knowledge of and training in the basic scientific principles (underlying material science, dynamics, fatigue, computer science, or chemical analysis) that are independent of the product or equipment such as found in the six causes of failure.
- Product experience, such as applied engineering, manufacturing, marketing, and other occupational skills involving products such as valves, impellers, forklifts, and grinders.
- Experience in an industry infrastructure, practices, codes and standards, and the like that influence the quality or safety of the product (e.g., pressure vessels in the process industry versus pressure vessels in the nuclear or transportation industries).

The Manufacturer

The manufacturer's primary motivating force is to make a profit by producing a product that is in demand, not to advance the state of the art. As we discussed in earlier chapters, the manufacturer's role in producing a finished product can vary considerably—from affixing its name to a product produced by another manufacturer to assembling subsystems produced by other manufacturers to manufacturing all parts and assembling the parts in-house. The manufacturer can be viewed as an organizer and customer of the many professions, skilled trades, and technologies that must come together to produce a product. The people who provide professional, skilled trade, and technical services can be either employees or contractors. In many instances, a manufacturer needs only one or a few skills to produce a product. High productivity and cost effectiveness are its guiding principles or goals. Manufacturers assign occupational titles according to the employee's function, and a title may or may not correspond with the employee's formal training and experience, or lack thereof.

Many products are "engineered" by copying a competitor's design. It is not unusual for the president or owner of a manufacturing company to acknowledge during a deposition that no engineering evaluation was ever performed on the product, that the product has been used for X years without any reports of serious injury, and that this is the basis for estimating its margin of safety. Some courts and members of the public fail to recognize that engineering a product and designing a product are entirely different activities. This lack of understanding is prevalent, not only in many small manufacturing companies, but in Fortune 500 corporations whose technical research and development activities consist primarily of reviewing a competitor's new products and patent applications and assessing the desirability of acquiring a new business. On the other hand, a number of manufacturing companies have set the standards for high-quality research and development, design analysis, manufacturing, and inspection because it makes good economic sense. One common misconception found among attorneys and juries is that the manufacturer or employee "expert" who has many years of experience in a particular area is automatically knowledgeable about and has expertise in all applications of basic scientific and engineering principles. The bottom line, however, is this: Although courts may have preconceived or ill-conceived notions about what qualifies an expert, it is the attorney's duty to (1) understand the implications of the underlying engineering principles, product characteristics, and industry infrastructure associated with the product, (2) determine which expert will help develop the client's case, and (3) prove to the court that the expert is qualified. In addition, the attorney should be prepared to prove the expert's qualifications by reviewing the expert's general credentials and those that are relevant to the case. As Vance (1988) suggested, qualified experts can be found among university professors who have worked in or with industry, among consultants who have both academic and industry credentials, or among in-house experts who have the requisite technical and occupational skills. However, by using an in-house expert, one risks perceptions of bias or intimidation by the employer, which the opposition may explore or infer, fairly or unfairly. As we have pointed out before, the courts' ignorance concerning the relationship between actual expertise and actual experience with the specific product at issue is often used effectively as a perverse weapon during trials. In the Highlift forklift case, for example, the court did not allow the plaintiff's eminently qualified expert to testify because the defense attorney convinced the judge that the expert was not a "multifunction control pedal" engineer. Thus, the attorney must sort through the occupational experience and professional training of both the plaintiff's and defendant's experts that are relevant to the case.

Judges and juries are impressed by paper credentials such as degrees and memberships in professional, product, and trade associations (*The Expert and the Law* 1988), although the only requirement for membership in many cases is the ability to pay dues. This is also true of many organizations that write codes and standards. Even degrees cannot be taken at face value because baccalaureate, master's, and Ph.D. degrees in management and engineering can be obtained by mail order from non-accredited colleges and universities.

The Attorney

The law firm in a product liability case can vary in size from one attorney to a large law firm that resembles any other large professional organization. Large law firms are organized into groups that specialize in various aspects of the law such as product liability, personal injury, and mineral rights. A small firm is likely to work on a contingency basis when representing personal injury plaintiffs (typically 30 percent of the amount of the settlement). A large firm, which is likely to handle large clients such as corporations, usually charges an hourly rate. Clients who retain a law firm must fend for themselves when assessing the training and experience of an attorney or law firm in the legal and technical aspects of product liability. No legal screening process analogous to that of the technical expert's deposition exists.

Some law firms and attorneys attempt to litigate a personal injury suit despite their lack of experience with product liability cases; the incentive is to win a large percentage of a particularly large settlement. However, there are monetary risks. Attorneys who work on a contingency basis must bear the cost of the expert's fees, laboratory tests, and technical reports during the development of the case, and they may lose the case. Some attorneys try to reduce this risk by trying to settle the case before obtaining an expert's evaluation of the incident; by waiting to retain an expert, or by having the expert delay the necessary tests and the technical report until the end of discovery is in sight or until the judge or the opposition prods them to obtain this information; or by relying on the opinions of established experts rather than on scientific proof. Some law firms only handle personal injury product liability cases, whereas others handle such cases sparingly. Some scholars attribute the increase in the number of product liability suits, particularly personal injury, to the desire to achieve the "get-rich-quick" or "big-bonanza" settlements that are possible for the law firm. Some argue that these attorneys are the ones who minimize their investigative expenses, request outlandishly high settlements, and refuse to pay their experts' fees, especially if they lose the case. The fee problem will be discussed in the next section. The counterargument is that personal injury attorneys are on the increase because of the inadequacies of workers' compensation and the failure of employers to provide safe equipment and work places, to evaluate their products adequately, and to train their employees to use and maintain equipment properly.

A large law firm represents a microcosm of the legal profession, just as an engineering consulting firm or engineering research institute represents a microcosm of the engineering profession. Some members of the firm are legal and legislative scholars who do research, publish articles, and participate in seminars that define and interpret the laws and technical principles that form the basis of the law (similar to the function of engineering scientists). Others specialize in a specific industry (similar to professional engineers who work in a specific industry). Still others are legal clerks who have a baccalaureate degree (similar to the engineering technologist or baccalaureate engineer who lacks advanced training and professional registration). One finds that more and more attorneys who practice product liability law have a baccalaureate degree in engineering, physics, chemistry, or the biological sciences; some of these attorneys have a doctorate in engineering. This background can be disadvantageous as well as advantageous. Obviously, attorneys who understand engineering and manufacturing principles are more likely to understand what is involved in an accident investigation and failure analysis. On the other hand, they may be tempted to become their own expert and try to impose their interpretation of the technical facts on their technical experts. This is a frequently cited pitfall confronting technical experts when they interview a technical client and the client's technical employees.

An attorney who understands technical language and the product and the industry involved in the case is advantageous from a technical expert's viewpoint. Some examples of technical material geared to give the attorney this "big picture" or overview are Blundell (1985), Frumer and Friedman (1988), Frumer, Benoit, and Friedman (1988), Hall (1985), R. L. Miller (1986), Peters and Peters (1984), Rosen (1984), and Sances *et al.* (1986). In addition, there are general overview texts that are prepared for industry personnel, which can be recommended by the expert, and are equally informative to the attorney (i.e., Bennet, Feates, and Wilder 1982; Boley 1977; *Dow's Fire & Explosion Index* 1981; Green 1982; Hammer 1972; Meidl 1970, 1978; and Sax and Lewis 1986).

Should a case require several technical experts, the attorney handling the case should discuss this with the "lead" or "coordinating" expert who has a broad knowledge of the type of incident involved. No purpose is served by hiring experts in different areas only to discover that they have been on opposite sides in other legal disputes, have little professional regard for one another, and regard their own area of expertise as the most significant. For example, a metallurgical expert may cite defects in the materials as the proximate cause of an incident, whereas the mechanical engineering expert may insist that the proximate cause was the mechanism of loading, when in fact the actual proximate cause was, and usually is, a combination of the two. As pointed out in previous chapters, if more than one technical expert must be retained, one should serve as coordinator unless the expertise of each expert is unrelated to the expertise of the others. Similarly, if a team of attorneys is handling the case, one attorney should be designated as coordinator. If more than one engineering firm has been retained, no work should be performed without a clear agreement as to who must authorize work done by the experts—the client or the law firm. Furthermore, each phase of the work should be covered by a separate authorization.

FINANCIAL ARRANGEMENTS

As we have pointed out previously, a financial agreement between the expert and the attorney is as important as the financial agreement between the attorney and the client. In both instances, the agreement should specify the services to be provided, the tasks to be performed, the scope of the work, and the provisions for checks and balances—that is, the agreed-on milestones for reviewing the work and paying for work in progress. All agreements should be in writing, particularly if tests and analyses will be performed. Detailed invoices that describe the work being done help to minimize misunderstandings. The client and attorney, as well as the expert, are entitled to have the expert's costs and activities spelled out in advance, when possible.

Problems Experts Encounter Regarding Their Fees

Because the tests and analyses described in earlier chapters can be expensive, a number of legal strategies have evolved concerning the acquisition of technical data. Some attorneys believe that data derived from expensive tests and analyses should not be provided to the opposition, whose own technical expert could use the information as a basis for his or her opinions at the expense of the client who paid for the work. As we stated in Chapter 5, the legal tactics for keeping data out of the opposition's hands include (1) delaying the performance of such tests until just before the period of discovery ends, to (2) the attorney not "officially" requesting or the expert not finalizing or transmitting the data (by giving the expert some latitude to use his or her own discretion), and (3) to "losing" the data temporarily—all of which are within the limits of the law. These tactics are usually directed by the defense against the personal-injury law firm that works on a contingency basis. Such tactics also are used occasionally by the defense and plaintiff attorneys in cases that may or may not involve personal injury.

When a law firm works on a contingency basis, the risk of nonpayment for technical services increases significantly. The most common reasons for a law firm's failure to pay experts are as follows: (1) the case involves a personal injury suit, (2) the case is lost without compensation, often because of the law firm's or client's go-for-broke strategy, (3) the negotiated settlement is insufficient to cover all expenses, or (4) the settlement is significant but the law firm tries to force the expert to reduce the fees on unpaid services. Law firms that try the fourth strategy use the revolving-door approach to experts: they hire different experts for each case. Evidence suggests that the incidence of these tactics in the guise of a business disagreement are fairly widespread. As the National Forensic Center (an organization founded to bring attorney–expert issues to a forum of discussion) points out, the expert is on the attorney's home turf when a real or imagined disagreement occurs. Insisting on work authorizations is a relatively effective method of screening for or reducing such unsavory practices. Not allowing the unpaid balance to go too high is another. The safest method, however—suggested by Judge Lucius Bunton of the Federal District Court in Midland, Texas—is to require payment in advance. This practice puts the attorney and client on notice about the monetary risk of going to trial versus accepting a settlement.

Another complaint one hears from experts is that they appear at a deposition at the client's request, only to discover afterward that the client and the opposition disagree over which side should pay for the expert's travel, time, and services. In some instances, an expert finds that the opposition is ultimately responsible for the payment, and, of course, they are in no hurry to pay. The following is another common occurrence: after months of no contact with the law firm that retained them, experts suddenly receive a panickly request from the firm for the tests that were discussed long ago. "We need to have them performed in a week or two because the judge wants a report." The experts oblige, spending many hours in overtime to perform the tests and prepare a report, and ship the report to the law firm with a feeling of satisfaction. This feeling can turn to dismay when an expert finds that the client has settled the case and refuses to pay the bill.

Consider the following example reported during a survey of consulting engineers and scientists from major universities and consulting firms across the country. A law firm asked a metallurgist at the University of Florida to perform physical tests on some evidence and determine whether a suit was warranted. The professor completed his analysis and issued a report stating that, in his opinion, the client did not have a case. When he wrote a letter to the law firm asking why he had not been paid, the firm responded that because he said a suit was unwarranted, he should not be paid.

A forensic engineering firm reported that a law firm representing a plaintiff retained them to perform an accident reconstruction. When the technical report and depositions were completed, the expert and the attorney decided that the most effective way to demonstrate the results in court was to videotape a simulation. The expert verbally quoted a price for the simulation, videotaping, editing, and narration. The day he was to deliver the videotape to the law firm, one of the firm's junior attorneys told him that the case had been settled and that the report and the videotape had been instrumental in this outcome. The senior attorney then stalled payment for a year on the pretext that the case had been settled before the videotape was delivered. The consulting firm eventually turned the unpaid balance over to a collection agency, which ultimately collected by "hounding the hell out of them."

Even work authorizations, written agreements, and letters of responsibility do not guarantee the collection of fees. For example, a Connecticut coordinating technical firm refused to pay a forensic engineering firm because the client had not paid them. "It is only business." Collection proceedings went on for years.

Consider the case involving the shattered grinding disk that seriously injured a 30-year-old employee of a steel fabrication shop in Midland, Texas, in April 1985. On October 24, 1985, a suit was filed on the plaintiff's behalf in the U.S. District Court for the Western District of Texas. The presiding judge was Lucius D. Bunton, and the plaintiff's attorney, whom we will call Mr. White, was a member of a local three-man firm. The defendant was a Delaware corporation whose subsidiary manufacturer was located in Massachusetts. Mr. Green, the defense attorney, also was a member of a local law firm. The same day, White contacted Dr. Smith, a forensic engineering consultant in Houston, and described the incident. Smith outlined the procedures for performing a forensic analysis and said that to reconstruct the incident and determine its cause, a number of questions needed to be answered concerning the mode and mechanism of failure. On December 3, 1985, White sent a retainer check to Smith to act on the plaintiff's behalf after receiving a description of the engineering firm's capabilities, a fee schedule, and Smith's curriculum vitae.

On December 17, White forwarded the following items to Smith accompanied by a letter of transmittal: the fragments of the failed grinder disk, the air-driven grinder, a new grinder wheel, the work piece the plaintiff was grinding at the time of the accident, a series of photographs (some of which were of good quality, most of which were not), a copy of the plaintiff's deposition, the date of the trial (March 28, 1986), a retainer fee, and a request to begin testing the materials as quickly as possible. The chronology and cost of the forensic investigation that ensued appear in Appendix E, which contains 13 invoices totaling $14,600.

In early January 1986, Smith did a preliminary inspection of the failed grinder disk, performed some preliminary calculations, interviewed the plaintiff's plant supervisor, and submitted a strategy for the forensic investigation to White (see Item 1 in Appendix D). White asked Smith not to begin testing the materials at that time.

Between mid-January and late March, Smith interviewed the principals in the case, reviewed the technical literature on composite materials, and performed other preliminary examinations on the failed disk and new disks (see Item 2 in Appendix D).

On April 5, Smith flew to Midland to give his deposition in the presence of White and Green, the defense attorney. While speaking to White on April 3 about the deposition, Smith had suggested that the visit would be a good opportunity to visit the accident site. However, when Smith arrived in Midland, White decided to cancel the site visit (see Item 5, Appendix D). The following is an excerpt from Smith's deposition.

> **Smith.** We have sought to obtain information from the defendant with respect to this specific product—the specific composition of the materials in the grinding wheel. Having not done so, we are using some literature that gives some physical data.

Green. What's that literature?

Smith. We have been using a number of sources. I believe it's primarily from the American Society of Testing and Materials. . . . [Green asks, What types of testing on the wheel in question have you done?] We have done basically a microscopic examination of the wheel—the grinding surfaces and the failure surfaces.

Green. You have looked at it through a microscope?

Smith. Right. Also, I believe we have done a pull test on the fiber.

Green. Have you made any changes in that piece of evidence since you received it?

Smith. We have cut a sample out of the wheel . . . .

White. Both sides talked about that at our meeting in Houston—keeping a sample.

Smith. Yeah. At least that was my understanding of our discussion with [the defendant's in-house expert] . . . for destructive testing we would take and cut a piece off . . . .

Green. How are you going to measure the strength of either wheel: the new one or the old one?

Smith. This isn't totally resolved. One way is to make some coupons or a test specimen to run in a tensile (compression) testing machine.

Green. What other kinds of tests are you going to do?

Smith. Another one that we are considering is a possible fatigue test . . . . What I stated here was that these are some possible tests that we are contemplating, but we have not been given the go-ahead to do them.

Green. Who are you seeking this go-ahead from?

White. From me.

Green. You intend to do the fatigue tests in your lab. Is that correct?

Smith. We don't have any fatigue testing equipment at our lab. It would either be done at a lab in Houston, or we might send it up to Penn State University.

Green. How does that work?

Smith. [He described the procedure.] It can be a tedious operation since it requires a lot of data points. Load points versus number of cycles define the curve . . . .

Green. Do you know if White has made a request? (for test data performed by the defense's expert).

Smith. For a fatigue data?

Green. Right.

White. Are you talking about a formal request? I want the record to reflect that we have asked for it, perhaps not on a formal basis.

Green. I think we can take that up later.
[At this point, White stated that he was currently authorizing the expert to perform the necessary tensile and fatigue tests on the composite material to complete the calculations.]

Green. Did you do any testing of the grinder itself?

Smith. Yes, I did.
[A discussion followed about the expert's examination of the grinder, pressure versus speed tests, and use of the manufacturer's technical manuals. Green then asked whether any other consultants are involved.]

Smith. Dr. X of Intertech is also working on this case.

Green. What type of computer analysis are you doing?

Smith. We are using the finite-element method, which allows us to simulate the composite material—its anisotropic properties.

Green. How do you do that stress analysis?

Smith. This goes back to the computer analysis . . . .
[Smith notes that he has cut some wheels apart to examine the fiber ply orientation.]

Green. What calculations have you done with regard to the guard?
[Green then asked questions about the analysis of codes and standards, labels and warnings, equipment guarding, and the like. He also examined and discussed equipment the expert had brought with him for the planned visit to the accident site.]

White. Referring back to your test of the grinder, you mentioned that you put a tachometer on the grinder, and you also used a pressure gauge on the compressor you used to pump the air into the grinder. Correct?

Smith. There are actually two gauges. One is on the compressor reservoir. The other is a line pressure gauge.

White. How did you know that your pressure gauges and your tachometer . . . gave you accurate readings on the day you tested this material? . . .

Smith. We used several calibration pressure gauges to verify the system gauges. . . . we checked it (tachometer) against an electrical motor (with a known calibration speed). This is a recommended procedure in the Biddle tachometer manual.

White. As far as what your records of data indicate, does anything that you have so far developed, either by computer analysis, observation, any facts, lead you to believe that the plaintiff improperly used this equipment?

Smith. There is no evidence that I have seen that would lead me to believe that is the case.

During the depostion, White authorized Smith to have the material testing done at Penn State and to have a computer analysis performed by Intertech Service, Inc. (see Items 7 and 8 in Appendix D). On May 7, 1986, White advised Smith that Judge Bunton had scheduled the case for May 27. "Your report was fine. I forwarded it to the attorney for the defendant. Enclosed please find a copy of your deposition taken May 1, 1986." On May 27, Judge Bunton dismissed the case without prejudice. The settlement was reported to be $400,000.

On June 3, White sent a letter to Smith stating that "the case has been settled and we will be receiving the money in the next few days. At that time, I will be dispensing checks for expenses. I cannot get our settlement check unless I exchange the check for [the manufacturer's specifications], so please send them to me immediately."

On July 17, 1986, White informed Smith that he would refuse to pay $5805 of the total bill because

> we feel that certain charges should not have been made. In particular, I refer you to the March and April bills (Items 2, 3, and 4) [see Appendix D] for evaluation hours totaling 31 hours and then 14 hours (Item 5) for a deposition in which the total bill is in excess of $3100. First of all, the deposition did not last more than two hours and you admitted at the deposition that you had not done any testing and could not give any real opinions because you said I had not authorized you to do the testing. Then in reviewing your bill, we find the $4000 (Item 7) for tension tests, but then find another $2000 for "grinder speed tests and

grinding wheel samples" and for "computer analysis" (see Item 8 and Item 3).

Your technical report was very good and really redeemed the deposition you gave.

Note the inconsistencies in White's statements, which are highlighted in the above chronology and by the items in Appendix D. The technically astute person will recognize that the properties of the material were determined by comparing specimens taken from the failed disk and the new disk and that this information was used in the computer simulation to predict the responses of both disks. The computer model permitted data to be gathered on various thicknesses and orientation of fibers. Performing experimental simulations with the failed disk was impossible because the disk had shattered.

Why did White delay requesting the material tests until the deposition? Why did he fail to formally request tensile and fatigue data on the materials from the manufacturer? (In fact, he did not know very much about legal procedure for a product liability case and was afraid to take it to trial because of his inexperience.) Did he actually believe that Smith should charge for the two-hour deposition, but not for the day and half he spent in travel and at White's office? For whom did he believe he was writing his letter? Should forensic experts require payment in advance from attorneys who are working on a contingency basis?

Smith filed a law suit in Houston against White's Midland, Texas, law firm, and the judge ruled that the case would be heard in Midland, where the plaintiff's suit was heard. After a considerable amount of legal maneuvering and legal harassment by White's law firm, he paid the engineering firm shortly before the trial date in Midland, Texas.

Situations such as this indicate that experts should require payment in advance when they agree to take such a case (plaintiff personal injury contingency funded). This is unfortunate because most attorneys are ethical and serve their clients well. Some attorneys seem to believe, however, that they can get away with such practices—"that it's only business."

During research for this book, it became clear from interviews with professional and academic experts that this problem needs to be resolved because clients and the judicial system as well as technical experts tend to be ill-served by attorneys who engage in such practices. Interestingly, attorneys and legal scholars recommend that one should insist on a signed agreement before work is performed.

Another interesting facet of attempting to collect fees from a law firm is the common experience of having considerable difficulty finding another law firm that is willing to go against fellow attorneys. Clients who want to file a suit against their attorney are likely to have the same experience. In small towns, we have found that one must go to a nearby town to find another attorney. Even in large cities, plaintiffs may have difficulty finding an attorney who feels comfortable about filing a suit against a fellow attorney, a situation that is characteristic of other professions as well.

The following section reviews recommendations regarding fees for technical services. It also offers guidelines concerning who pays for written reports and depositions and who determines whether fees are reasonable.

The Expert Witness's Fees and Expenses

First, the relationship between experts and those who hire them is a matter of agreement. Fees should be discussed when the expert is first retained and a personal services contract or work authorization should be drawn up (Ladino 1986; Branton and Lovett 1987; Poynter 1988). (A sample fee schedule and work authorization appears in Box 12-1.) If the consulting firm does not have a standard fee schedule, the attorney may want to check with a source such as the National Forensic Center or an appropriate professional society.

The attorney should ask the expert questions such as the following: Do you charge by the hour, by the day, or some other basis? Do you provide fixed-fee estimates for projects? Do you have different charges for different services? Do you require payments in advance for the investigation and for travel expenses? If so, when? Will you need an initial retainer to participate in a deposition or trial? If so, will the retainer be held until the end of the project? What will the billing schedule be? Will there be late charges, administrative fees, fees for rental of equipment, and laboratory fees? Do you have a sample work authorization?

Other matters that should be agreed on at the outset are whether a written report will be required and how detailed the report should be, whether the expert will testify in person at a trial or whether his or her deposition will be used at the trial, and where the deposition will be taken. Finally, difficulties can be avoided if both sides clearly understand and agree on what the expert's specific assignment is (and is not).

Who Pays When the Opposition Requests the Expert's Report? Federal Rule 26 (b)(4)(A)(ii) provides for "discovery by other means" than through interrogatories upon a motion to the court. This "means" could include an expert's written report. The discovery of the expert's opinions, grounds for opinions, and so forth through means other than

BOX 12-1. Standard Proposed Engineering Services Form

RE: ______________________________ FOR: ______________________________

ITEM #

1) Perform engineering service specified in attachment(s) ____________ at $ ____________ (fixed fee, not to exceed cost, or hourly rate)

2) Review technical material, depositions, specifications, etc. at $ ______ (fixed fee, not to exceed cost, or hourly rate—principal engineer).

3) Office visit, plant visit or visit to failure/accident site as requested by client—hourly rate—including travel time (principal engineer).

4) Travel—air and ground transportation, parking, meals, lodging, (round trip from the office to site).

5) Technical preliminary report preparation at $ ____________ (fixed fee, not to exceed cost, or hourly rate),—the minimum approximate time required is 10 to 16 hours (principal engineer).

6) Audio-Video Documentation at hourly rate plus equipment use which is charged on a lease basis. Photographic, VCR, audio dubbing, etc. at $ ____________ (fixed fee, not to exceed cost, or hourly rate).

7) Background data and research at $ ____________ (on a fixed fee, not to exceed cost, or hourly rate). Technician to consult as required.

8) Retain the Company as Expert Consultant. It is understood that a retainer of $1000 is due prior to the initiation of work; the company agrees to refund any unused portion of the retainer in the event of early termination of the project.

FEE SCHEDULE

Service	**Project Basis***	**Hour-by-Hour Rate**
Principal Engineer	3.20 x/hr	9.5 x/hr
Consultant	2.50 x/hr	7.5 x/hr
Senior Engineer	2.00 x/hr	6 x/hr
Engineer	1.65 x/hr	5 x/hr
Senior Technician	1.25 x/hr	4 x/hr
Technician	x/hr	3 x/hr
Testimony at Trial	12.5 x/hr	12.5 x/hr
Testimony at Deposition	12.5 x/hr	12.5 x/hr

x = base technician dollars/hour rate

***Fixed fee costs (excluding overhead) will be provided on a project basis with a lower estimated hourly rate dependent upon project length and scope. For example—on a large fixed fee project costing $50,000, the fees estimated above can be used for comparison.**

BOX 12-1. *(continued)*

WORK AUTHORIZATION

The above prices, specifications, and conditions are satisfactory and are hereby accepted. The client, ______________________ has authorized the company to perform the above following items: ______________ ______________________ and omit items: ______________ ______________________. It is understood and agreed that payment will be made at street, city, state upon receipt of invoice and that there is a 1.5%/mo. charge on accounts over 30 days. The client agrees that any outstanding balance will be paid upon the settlement of this case including any costs incurred to collect unpaid balance.

This work authorization expires on ______________________
(indicate date and/or "completion of requested items")

______________________ ______________________

Client Date

interrogatories is specifically made subject to Rule 26 (b)(4)(c), which requires that the party seeking discovery must pay a reasonable fee.

The rules of the specific state also should be checked. For example, Texas Rule 166b.2.e(4) gives the trial judge discretion to order one's expert to reduce his or her findings and opinions to "tangible form," but does not address the issue of payment. If the person who retained the expert has not requested a report, one could argue that the person moving for a report under the Texas rule should pay the expert a reasonable fee for preparing the report.

Who Pays When the Opposition Requests the Expert's Deposition? The Federal Rules of Civil Procedure do not specifically grant permission to the opposition to take the deposition of the other side's expert witness. Indeed, Rule 26 (b)(4)(A)(i) specifically provides that such expert testimony is to be obtained through interrogatories. However, Part (ii) of the rule provides that, upon a motion, the court can order "further discovery." Consequently, experts should anticipate having to produce all documents and provide an extensive description of everything they have done up to the time of the deposition.

Because state rules vary, they should be checked. For example, Texas Rule 200 provides for the taking of a person's oral deposition but does not address the subject of who pays.

Who Pays When the Expert Witness Gives a Deposition? Federal Rule 26 (b)(4)(c) specifically provides that the party seeking discovery shall pay a reasonable fee for expenses the expert incurs during the discovery process. This rule stems from the philosophy that one side should not obtain free what the other side has paid for.

State rules vary. For example, Texas has no rule that addresses the issue of who must pay the expert witness's deposition fee. Under the basic rules of contract law, the party who wants to take the deposition should be responsible for paying the fee. The rules have essentially left the opposing parties to work out the agreement. In the event that no agreement is reached and the opposition refuses to compensate the other side's expert for a deposition, the attorney can move for a protective order (before the deposition) under an amendment to the Texas Rules of Civil Procedure dated April 1, 1984.

Under this amendment, the trial court had the authority, by way of a protective order, to require a person seeking an expert witness's deposition to deposit in the court's registry a reasonable fee before the deposition is taken. In *Mahan v. Stover* (1984), a case of first impression, the defendant attempted to depose the plaintiff's two handwriting experts by tendering a $1.00 witness fee and $0.06 per mile travel fee under Texas Revised Civil Statute (TEX.REV. CIV.STAT.) Article 3708 (Vernon 1926). Holding that the trial judge did not abuse his discretion in issuing a protective order refusing the deposition until the parties could agree on the amount to be paid and later clarifying the order to require that the defendant deposit $600 in the court registry for each day of the expert's deposition, the court ruled that Texas Rules of Civil Procedure (TEX.R.CIV.P.) Rule 166b governed rather than Article 3708. The court pointed out that although the Texas rules did not have a specific provision that permitted the trial court to assess the party seeking discovery for the expert witness's fees, it could be argued that the Texas rules, through silence, gave the Texas courts even more power under Rule 166b(4) by providing that "the court may make any order . . . to protect the movant from . . . unnecessary expense" and further "ordering that the discovery be undertaken . . . upon such terms and conditions . . . [as] directed by the court."

Who Determines Whether the Fees are Reasonable? Under Federal Rule 26 (b)(4)(c), which permits the court to assess the expert witness's fees against the party seeking discovery, the fact that the court can determine what is a reasonable fee is well established. In *Anthony v. Abbott Laboratories* (1985), a U.S. district court in Rhode Island held that an hourly rate of $420, which a plaintiff's medical expert attempted to charge the defendant, who wanted to take his deposition, was unreasonable. This decision was based on a relatively in-depth analysis of fees charged by expert witnesses and what constituted a reasonable charge. The court considered the following factors in assessing the reasonableness of the physician's fee: (1) the location of the deposition, which was selected to suit the witness's convenience, (2) the objection to the amount of his fee, which was made in a timely manner, and (3) the witness's credentials.

State rules vary, but they may implicitly endorse the federal rules. In *Mahan v. Stover,* the Texas trial court's assessment of the reasonableness of the fee was not challenged. However, the trial judge's authority to assess any amount was challenged in the appellate court, which sustained the trial court's authority to assess the fee of $600 per day, which happened to be the amount the expert was asking. It seems reasonable that the authority to assess a fee implies the authority to assess a specific amount and to determine what a reasonable fee is under the prevailing circumstances. The Mahan–Stover case leads to the following interpretation: it sustains the trial court's authority to assess the fees of expert witnesses, it assumes that the trial court has the authority to set the amount, and it implicitly endorses the philosophy espoused in the federal rules that the party seeking discovery from an expert witness must pay a reasonable fee.

PREPARATION FOR TRIAL

Preparation for trial should begin after the expert has correlated the gathered data and has formulated an opinion about the mode and mechanism of failure and causality that is sufficient to prepare a preliminary technical report. As the end of discovery approaches, the legal and technical positions of both plaintiff and defense ideally have been established and defined. Reports and depositions are read and scrutinized, and available documentary audio-video data are reviewed for possible use at trial. Weaknesses in one's legal or technical position will be strengthened, and the opponent's weaknesses will be assessed (refer also to the section on advice for expert witnesses at the trial).

Before we discuss depositions, consider the recommendations that technical experts offer to attorneys involved in product liability litigation. These recommendations, which are listed in Box 12-2, represent a distillation of information gathered during a survey of expert witnesses for this book (also, Finn 1987) and data obtained from published articles (Briem 1987; Burgher 1988; Jur *et al* 1981; *Kellogg Corporation News* 1986; Peters *et al.* 1977; Poynter 1988; and Talbot 1975).

Preparation of the Expert's Deposition

Obviously, a technical expert who is well prepared knows far more than an attorney about the technical complexities of a product liability case.

BOX 12-2. Advice that Experts Offer to Attorneys

- Don't wait until the last minute to retain an expert.
- Ask the expert to help you formulate questions for depositions, obtain information during discovery, and search for relevant documents.
- Don't wait until the last minute to have an expert perform basic procedures such as examining the product and visiting the scene of the incident.
- Don't attempt to do any engineering investigative work without direction from an expert.
- Don't be influenced unduly by previous cases.
- Don't use an in-house jack-of-all-trades to coordinate a technical product liability suit.
- Don't withhold data from the technical expert.
- Have written agreements with the client and the expert.
- Look for an expert who has certified training and experience, including experience with depositions and trials.
- Request a preliminary report on the expert's findings and review it carefully to ensure that you understand the technical issues.
- Solicit the expert's input about technical material to ask the opposition versus build an offense.
- Obtain the expert's estimate of the technical costs as soon as possible. If you can't afford the necessary expertise, turn the case over to another law firm or collaborate with the firm.
- Critique the content and style of the expert's testimony (in the deposition and at the trial). Experts appreciate this, but attorneys rarely provide this information (see Finn 1987).
- Tell the expert early about the need to control expenses; don't let the expert spend hours on an investigation if you don't have the resources to pay for these activities if the case is lost.
- Tell the expert early in the planning stages how much money the client is willing to settle for (an estimate of the operating budget).
- Discuss with the expert the strategies you and the opposing attorney may use; for example, the "smoke-screen" tactic—diverting attention to technical issues that are not the real technical issues but are within the letter of the law; the "empty-seat" tactic—heaping blame on a party that was not named in the suit; and the "discredit-the-expert" tactic, usually directed toward someone who can damage the opposition's case.
- Ask the expert about the cost of an accident investigation and failure analysis, what it entails, and what information it will provide for the case.
- Look for an expert who is *interested* in product liability and trial law—that is, one who is interested in the strategy, planning, and mechanics of the case itself.

The strategies an attorney uses during a deposition will depend on the purpose of the deposition, the strategies the opposition uses, the client, and the characteristics of the case. Some attorneys advise experts to relax and tell the opposing counsel everything he or she wants to know. Others may admonish the expert to offer as little information as possible. In some cases, the opposing counsel asks the expert only a few questions to avoid giving the other side the opportunity to prepare questions that will be asked at the trial.

When preparing for the deposition, the expert and the attorney should review the checklist shown in Box 12-3 together. Doing so will

BOX 12-3. Checklist for Joint Expert Witness and Attorney Review Prior to Deposition

The Expert should review the following with the Attorney:

- The expert's background, qualifications, experience, achievements, and publications
- Unfavorable facts about the expert's background that might tend to discredit his or her opinions
- Publications that might appear to conflict with the expert's opinions in this case
- Information on which the expert has based his or her opinions
- Assumptions the expert has made and why
- A defense of the expert's assumptions and opinions
- Why the expert was hired to testify in the case: that is, what specific issues the expert was asked to address or what he or she was asked to do
- The expert's fees and expense arrangements

Attorney review with the Expert Witness:

- The expert should remember that his or her opinions must be based on reasonable probability
- The expert should avoid contact with the opposition's attorneys; they are not his or her friends
- The expert should answer the attorney's questions as positively as possible and correct any mistakes as they occur. Generally, an attorney cannot ask leading questions; therefore, the expert should be as responsive and helpful as possible in helping the attorney convey information to the jury
- The expert should not answer yes or no when cross-examined by the opposition's attorney, unless that clearly is his or her answer.
- The expert should explain his or her answers if necessary, but avoid seeming argumentative or appearing to be a smart aleck. Avoid exaggerations

BOX 12-3. *(continued)*

- The expert should avoid getting involved in arguments between the attorneys over objections. The expert should remain silent until instructed on what to do
- The expert should answer difficult questions if they conform to the rules, and rely on his or her client's attorney to judge the difference between abuse and difficult questions and make the necessary objections
- The expert answers all questions he or she is able to answer; if the expert doesn't know the answer or can't remember it, he or she should say so
- The expert should review the checklist of deposition questions with the client's attorney
- The expert should be prepared to explain how much or what percentage of his or her consultation work has been for the plaintiff and how much has been for the defendant
- The expert should be willing to admit that he or she is unfamiliar with obscure technical terms the interrogator may use to catch him or her off balance. The expert should ask the interrogator to define what the term means
- The expert should review his or her relevant credentials with the attorney
- The expert should remember that the judge and jury may be unable to formulate an opinion about matters such as the mode or mechanism of failure and are likely to be impressed by paper credentials. The expert should let the client's attorney demonstrate the expert's expertise
- The expert should beware if he or she is asked whether he or she would have done what the defendant did. The interrogator is asking the expert to ignore all the information he or she has gathered during the investigation; the expert is being asked to speculate
- The expert should be aware that being asked to assess or criticize the plaintiff's or defendant's conduct is unfair because the expert has the benefit of hindsight
- The expert should be prepared to perform calculations or derivations during the trial, and take the time to be as accurate as possible. This is an opportunity for the expert to inform the court
- The expert should discuss all pertinent technical points with the client's attorney, who will then be able to develop or avoid developing points for the record, depending on what information the opposition wants to develop or suppress
- The expert should be fresh and alert: that is, he or she shouldn't take the "red-eye" special the night before the deposition or stay up late working on the case unless necessary

help both to avoid any nasty surprises during the deposition. Appendix E provides a checklist of questions the attorney may use when deposing opposing counsel's expert. This checklist may be a useful supplement to Box 12-3.

Attorneys should always remember that they are dealing with people: the opposition's clients, attorneys, and experts and the judge and jury as well as their own colleagues, clients, and experts. How they deal with all these people can play an important role in bringing the case to an effective conclusion. As we said earlier, establishing trusting and professional relationships with clients and experts is important. The following paragraphs offer guidelines for attorneys to consider when preparing for a deposition (based on Johnson 1984).

1. *Do not go to the deposition armed with an exhaustive list of predetermined questions.* Some attorneys go to a deposition (or trial) with a laundry list of questions and charge ahead, regardless of what a witness may say that makes the list useless. Because they do not listen, they are oblivious to a witness's lies or evasions; their main goal is to get through the list. They may learn something new, but if they have not done their homework and do not listen, they will not learn much. The purpose of deposition is to help one's case.

Rule 1 of the Federal Rules of Civil Procedure states that the purpose of depositions, along with other discovery devices, is to promote "the just, speedy, and inexpensive determination" of lawsuits. Despite the innocent hope expressed in this ruling, the words do encourage attorneys to make every witness for the opposition look like a fool or a liar. When attorneys are on the offensive, they must expose an opposing witness's biases. When they are on the defensive, they must protect their own witnesses against the opponent's maneuvers and manipulations. If they expect the real contest to take place during the trial, they can lose the case during the deposition. Thus, they should prepare their expert witnesses as carefully as possible before the deposition and listen carefully to what they have to say during it.

2. *Being rude with advocacy.* Attorneys can be congenial until they adopt the role of "advocate," at which time they can become incredibly rude and vicious. Nowhere is this so apparent as at a deposition, because no judge and jury are present to ensure that the attorneys' behavior remain within the bounds of civility. An attorney can turn a good case into a bad one by being rude. Too many clients have paid dearly for disputes that could have been resolved before a trial were it not for the hostility between opposing attorneys. Videotaping the deposition can be quite effective in keeping the proceeding civil.

3. *Do not fail to establish rapport with neutral or opposing witnesses before the deposition.* Most witnesses are anxious about being deposed, and they

resent the fear generated by the exercise. If the attorney is the one who insisted on the deposition, he or she should contact neutral or opposing witnesses, explain why their depositions are needed, and try to accommodate their schedules rather than select an arbitrary date for a subpoena. Witnesses from out of town should always be subpoenaed, however, even if they agree to appear voluntarily, to avoid having to pay the opposition's travel costs [Rule 10(g)(2)] and attorney's fees should a witness fail to appear.

Assuming that the witness is not a party to the law suit or a party's expert, the attorney should arrange to have his or her own expert interview the witness, if possible, if the witness's information is relevant to the expert's opinion and arrange for the attorney to meet with the witness to discuss the areas of testimony the attorney expects to cover. Attorneys should not go through a list of questions that will be repeated at the deposition because the opposing attorney can make an issue of this.

Attorneys should identify the party they represent and the claims of their case. If the witness wants to know what will happen at the deposition, the attorneys should make no promises regarding other or opposing attorneys, and they should refer the witnesses to their own attorneys.

Finally, witnesses tend to give more complete, less guarded answers to questions during depositions if good rapport has been established. By meeting a witness ahead of time, the attorney will have some idea of how the person is likely to behave under oath. For example, will the witness tell the same story without hesitation in the deposition room? Is the witness afraid that his or her employer, professional colleagues, or future clients will retaliate for harmful testimony? If a witness refuses to meet with the attorney before the deposition, it may indicate bias against the side the attorney represents.

4. *Do not let a witness go into the deposition unprepared.* Because attorneys have been trained to handle verbal confrontations and probably have never been deposed themselves, they may forget that no matter how sophisticated witnesses are, they will be anxious about the interrogation that is about to ensue. Thus, attorneys should not only prepare their witnesses for the kinds of questions they are likely to be asked (note the word "prepare," not "rehearse") but also share some of their perceptions about the personalities of the other attorneys. For example, if Ms. X is likely to ask verbose, convoluted questions, the attorney should explain that the witness does not need to answer questions he or she does not understand. If Mr. Y is likely to ask personally embarrassing questions, the witness should be reassured that he can consult with his attorney before answering. If the witness is the attorney's client, the attorney should tell the witness during preparation for the deposition that he or she will be instructed not to answer irrelevant questions and is free to

interrupt the deposition at any time to confer with the attorney. In other words, the attorney should try to discover what concerns the witness, especially if the witness is a client, about the forthcoming ordeal. Provide as complete a picture of the deposition process as possible, show the witness the room where the deposition will take place, and where he or she will sit. Explain that the attorneys will make objections from time to time "for the record" and that the witnesses should ignore those objections as being "business" between attorneys. The witness should be advised, however, that an attorney may object to a question that seems misleading or incomplete.

Finally, the attorney should explain Federal Rule 30(e), which gives witnesses the right to change their testimony after a deposition is completed. Although witnesses may have to explain during a trial why they gave an incorrect answer at the deposition, they will understand that their answers are not immutable. Prospective witnesses visibly relax when they hear that good news.

5. *Do not allow the opposition to run roughshod over the documents of your own witness.* Many witnesses will calmly answer outrageous or embarrassing questions, only to fall apart when an attorney begins to place exhibit stickers on their original business records, scientific calculations, and so on. Attorneys tend to forget that what is merely documentary evidence to them may be the lifeblood of a witness's business—for example, the minutes of a board of directors meeting ("Our only copy!") or a secret chemical formula embedded in a laboratory test report ("If our competitors get their hands on that, we'll lose our market!"). Thus, if the witness is a client, the attorney should obtain protective orders for sensitive documents before the deposition begins. If the attorney wants nonparty witnesses to be favorably deposed, the attorney should explain the right of the witness to petition for a protective order for documents they do not want to produce and suggest that they seek the advice of their own attorney.

Attorneys seldom need to turn over a client's original documents. Copies are sufficient as long as the originals are available for comparison [Rule 30(f)]. Finally, attorneys should examine their client's records before giving them to the opposition. Some clients have strange filing practices; for example, an attorney's letters may be mixed in with invoices and sales brochures.

6. *Do not be derailed by a hostile expert witness.* Expert witnesses, especially technical experts, resent the passive role of being deposed, particularly if they believe that an attorney does not understand the complex issues they have been called on to explain. Do not allow yourself to be baited or interrogated by a frustrated witness who feels he or she has the right to take a few shots at you.

Do not allow a witness's editorializing to stand on the record if it is irrelevant to your question. Simply say, "Move to strike the last remark of the witness as nonresponsive," and move on. Some expert witnesses believe they can gain the advantage by making you define your words. On the other hand, your choice or usage of a word may be unclear to the witness. For example, attorneys often use the word design in a manner that is unfamiliar to engineers: "Have you ever *designed* a nuclear power plant, a telephone pole, a ladder, a road?" To an engineer, the term means to create a concept. Rarely do they actually create a product, whereas an attorney may use the term to mean calculate, to prepare a drawing, or to apply a scientific law—meanings that are redundant to an engineer. Use the word "engineer" or "analyze" in place of design.

If you use a clearly understood English word such as rapid, and the witness says, "That all depends on what you mean by the word," ask the witness to define the word as he or she would use it. Then ask whether that definition is sufficient to answer the question fairly. If not, ask the witness to explain why the question is unfair. If the witness prefers to use a scientifically more precise substitute, ask for one. Keep the discussion on the witness's turf. If the witness becomes tedious or silly about definitions, however, ask a question such as the following: Have you had a difficult time deciding, on any other occasion, what the word rapid means? The witness will get the message. If the opposing witness's opinions are highly technical, consult your own expert witness before or during the deposition about what terms to use.

7. *Do not load the deposition with garbage to fill the time allotted.* Johnson's Law—"Depositions expand to meet the time allotted for them"—should stand next to Murphy's Law and the Peter Principle (Johnson 1984). Our corollary to Johnson's law is: "The indicator to the method of payment (contingency or hourly) to the law firm that requested the deposition is a function of the duration of the deposition and the number of attorneys involved." If the issues covered in the deposition are simple, the attorneys are well organized, and they get the testimony they want in two hours during a deposition scheduled to last two days, why waste time? Dazzle the opposition and your client with your confidence. Go home!

8. *Do not neglect the court reporter.* Attorneys tend to forget that the court reporter must record all testimony and objections completely and correctly. Because the final transcribed record is only as good as the reporter's opportunity to hear everyone clearly, let nothing get in the way of that opportunity.

Find a good court reporter and stay with that person. If the case is complex and involves a large amount of money, have the reporter travel with you. And maintain a clean record of the proceedings by avoiding the following pitfalls.

- Do not talk while someone else is talking. The reporter can take down the statements of only one person at a time
- Do not mumble, chew gum, cover your face with your hand, use foreign words you cannot pronounce, or talk too fast
- Do not engage in marathon interrogations. The reporter needs breaks to be effective
- Do not let witnesses answer questions by nodding or shaking their head
- Do not allow the witness to point to documents such as maps or photographs or refer to places as here, there, this, or that. Ask the witness to make clear references to the exhibit by drawing on it with different colored pens and labeling significant features with letters, words, and arrows

Ask the court reporter to read back an answer or a question occasionally to ensure the accuracy of the record. Some attorneys ask a reporter to read back what they regard as a wonderful answer. That is a judgment call. An attorney may want to know whether the answer was recorded correctly; on the other hand, drawing attention to it may cause the witness to recant or backtrack at the next opportunity.

9. *Do not risk becoming burned out.* You cannot do a good job for your client if you come to the deposition straight from an all-night flight. This is also true of your expert witness. In a long, drawn-out case, let other attorneys shoulder part of the burden. The jousting that takes place during depositions takes an emotional and physical toll.

10. *Save your "prize cow" for the jury.* For strategic reasons, it may be wise to avoid taking an expert witness's deposition altogether or taking the deposition only until you get a sense of the impression the witness will make at trial. In other words, do not give the opposition an opportunity to prepare its own witnesses, especially if they lack the depth of knowledge to determine causality.

Advice for Expert Witnesses for Trial

The purpose of the expert's testimony at a trial is to communicate conclusions about the causality or responsibility for personal injury or property damage caused by an incident and the basis for arriving at these conclusions. Thus, the attorney's job is to elicit information from the expert on the results of the accident investigation and failure analysis and the reconstruction of the incident. As we have mentioned many times before, audiovisual aids can be extremely effective in conveying this information to the jury.

Chapter 6 provides a discussion of the uses of videotape to illustrate documentary data, and the nine suggestions by Brown should be reviewed for guidance in preparation for trial. When considering the use of videotape to inform the court, the user must understand the difference between material and equipment scientific "simulations" (Chapters 8 and 10, respectively) and "reenactments." Edited sequences have been favorably received by the courts, as long as the raw footage is available. Audio dubbing should be limited to the actual audio track and the narration of the appropriate expert witness; however, most experts suggest that no audio dubbing be used and that narration be provided by the expert witness at trial.

The section on visual documentation in Chapter 6 alerts the forensic engineer to the value of photographic and videotape documentation of the incident site for possible use in an audio-video presentation at trial that will make the accident site real to the jury. Audiovisual documentation of material testing (as discussed in Chapter 8) and simulations (as discussed in Chapter 10) may also be included in a presentation by the forensic expert. These can help the expert explain relevant technical issues to the judge and jury (as well as the attorneys and clients before trial). Video, photographic, commercial art/graphics (Fig. 12-1), and models (Fig. 12-2) can supplement or be added to the edited sequences

Figure 12-1. Use of posters and videotape aids for expert use in instructing the court in technical issues of the forensic analysis.

Figure 12-2. Use of Styrofoam model representation of heavy steel vessel for use in courtroom demonstration.

of the technical audiovisual documentation to produce a story for the judge and jury.

The videotape format has the appeal of (1) familiarity to the jury, and (2) being a means of ensuring that the viewer/listener follows a planned and controlled narration of the reasoning and visual effects. (Refer to the section on visual documentation in Chapter 6 for a discussion of the uses of videotape and other visual documentary methods.) Without question, independent of the issues of legal strategy, a technical report, particularly in a complex case involving complex technical issues, provides the technical team an opportunity to explain the implications of their analysis and the basis for their conclusions to the attorneys, the clients, the jury, and the judge. The technical report provides the opposite appeal of videotape. With a report, the reader can not only follow a logical flow presented by a writer but can concentrate on any areas of the report at his or her leisure. A number of aspects regarding the preparation of the technical report are discussed in Chapter 11.

Like the report, the deposition will be scrutinized by the attorneys, experts, and courts because it can play an important role in the case. Hence, the importance of preparation for a deposition by the expert and attorney.

Most experts prefer to use the presentation format whenever possible because it allows them to develop the technical basis for their conclusions in a logical and sequential manner, progressing from relatively simple information to more complex issues. This technique is particularly effective when the expert narrates a videotaped presentation or a series of photographs and illustrations. As Briem (1987) notes, "the question and answer format is confusing and awkward. In no instance has the judge objected to this presentation format." Engineers constantly use narrations in lectures, presentations, and symposia.

In addition to the basic qualifications of knowledge and experience in his or her field, an expert witness must have the ability to communicate clearly and objectively. Cantor (1985) offered suggestions on topics ranging from manners to mannerisms that will help experts communicate in this manner and therefore help the judge and jury reach fair and proper conclusions based on their testimony. Cantor's suggestions, to which a few have been added, are listed in Box 12-4.

A Checklist for Cross Examining the Opposition's Expert Witness

The function of the cross-examination under the adversary system is to permit the opposing counsel to inquire into the source of information, the basis for opinion, the specific material on which the opinion is based,

BOX 12-4. Tips for Expert Witnesses Giving Testimony at Trial

- Walk to the witness stand with even steps
- Hold your right arm high with fingers straight when taking the oath, and look the officer administering the oath in the eye
- Say "I do" in a loud voice when the officer finishes the oath, "So help you God," so that everyone in the courtroom can hear you. Do not act timid
- Think before you speak
- Stop speaking immediately if an attorney says "Objection" or if the court interrupts. Wait until the court gives its ruling
- Be fair and frank; don't be too anxious to please or too eager to fight
- Admit it if you make a mistake or contradict yourself and correct it
- Keep your temper, and don't let an attorney trap you into arguing over trivial or even important points. Be firm but flexible
- Say so if a question cannot be answered with a yes or no; modify your reply by saying "under the circumstances . . . ."
- If an attorney asks you an illegitimate question, say you do not know or cannot remember. Such answers are appropriate in this situation
- Avoid speech mannerisms: for example, prefacing all your replies with "I can truthfully say that . . ." may cast unwarranted doubt on your

BOX 12-4. *(continued)*

entire testimony. If you were not telling the truth, you certainly wouldn't say so

- Do not get trapped by questions such as "Did you ever discuss this with anyone?" Of course you did, so name people with whom you talked
- Stop and ask yourself whether you are positive about a statement if an attorney asks: "Are you as positive about this as you are about the rest of your testimony?"
- Listen closely to questions such as "Do you want this jury to understand that . . .?" If you do not, clarify what you do want them to understand
- Never act like a smart aleck
- Indicate this to the judge if the opposition's attorney interrupts you before you have completed your answer
- Do not volunteer any information
- Express yourself well: Use simple technical language that the judge, jury, and attorneys can understand
- Be brief; answer the question and stop
- Do not converse with other witnesses or parties to the controversy during recesses. Speak only with the attorney who retained you
- Do not chew gum
- Do not memorize your testimony
- Never nod your head to indicate yes or no
- Speak up so that the judge, jury, counsel, stenographer, and others can hear you
- Wait until the attorney has asked the entire question before you answer
- Do not look at your attorney when the opposition is cross-examining you
- Keep your hands from your mouth and face
- Address the judge as "Your honor" or "Judge Jones," and address the attorneys by their names
- Ask, after concluding your testimony, if you may be excused, then leave the courtroom immediately
- Remember that the witness stand is not a comfortable place for a person who is not telling the truth, the whole truth, and nothing but the truth
- Be prepared to use audiovisual aids when necessary to bring the court to the scene, demonstrate scientific principles, and document simulations. But be brief, organized, and direct
- Acknowledge the judge's and jury's presence by occasionally speaking to them directly when answering an attorney's question (*The Expert and the Law* 1988)

the amount of compensation received, testimony given at other trials, pretrial depositions, and other matters that would tend to rebut inferences drawn during direct examination (Colangelo and Thornton 1981). The opposing attorney attempts to neutralize the impact of the opposing expert's direct testimony by cross-examination. The opposing counsel concentrates on the negative aspects of the expert's qualifications and other weak points. As Ladino (1986) indicates, however, if the expert is neutral and objective, an antagonistic attorney can actually lose points with the jury.

Practically speaking, many attorneys choose to do little or no cross-examination, and instead use the testimony of their own expert to neutralize the testimony of the opposition's expert. Those who do choose to cross-examine the opposing witness should find the checklist of questions in Box 12-5 helpful. Appendix E provides a checklist that may be used as a supplement to Box 12-5.

BOX 12-5. Checklist of Questions Attorneys Should Ask when Cross Examining the Opposition's Expert Witness

- What is your education, employment history, and professional certification?
- What is your experience as an expert in litigation and related matters and in consultation?
- What percentage of your experience as an expert has been for the plaintiff or defense in litigation and consultations?
- When was your first contact with opposing counsel, and what were you asked to do?
- How much time have you spent on the investigation and what have you done?
- What information, documents, depositions, exhibits, and so on have you been furnished?
- What assistance have you received from others?
- What are your tentative opinions: Please list each one without any explanations (i.e., just give your opinion)
- State each of your opinions and give the basis for it, including any support in the literature
- What data are inconsistent with each opinion?
- What work, if any, have you performed since your deposition? Has this changed your opinion in any way? And (for a trial) if so, how?
- What work, if any, do you contemplate doing before you are ready to testify (for deposition)?

THE TRIAL

The deposition is viewed as an extension of a trial. As Ladino (1986) points out, however, during a trial the opposition's attorneys tend to use their own experts to rebut an expert's conclusions and opinions rather than rely on cross-examination. On the other hand, as Irving (1980) indicates, opposing attorneys who know their trade will detect, during cross-examination, an expert witness's failure to perform a sound forensic investigation. In the case of the exploding gas range described below (Irving 1980), the testimony of an expert witness revealed that the witness had not personally examined the evidence, had not visited the scene of the incident, and was not prepared to provide the basis for his calculations or to perform those calculations at the trial. (All names and dates have been changed.)

The injured parties in the case were Mr. Swartz and his 9-year-old daughter Heidi. Swartz and his family moved to a large American city from North Dakota in 1969 with their seven children. In 1971 they purchased a gas range from the Smith Range Co., which installed it in the Swartz's kitchen. Two years later, the kitchen caught fire, and Swartz and Heidi suffered severe burns. It was generally believed by the parties interested in this incident that the fire was caused by a natural gas explosion. Just before the explosion, Swartz had placed a pot of hot dogs on the left-front burner. Heidi was seated about three feet away with her back to the stove. Swartz turned the burner on, but nothing happened. After about 30 seconds, he turned it off, then turned it on again. A ball of fire immediately engulfed him and his daughter.

Four years later, the case went to trial. The gas company and the range company were codefendants. (The jury was not aware that the range company had filed a cross-claim against the gas company.) The participants were Mr. Jones, the range company's attorney; Mr. Willingham, the gas company's expert witness, a mechanical engineer in private consulting practice with experience in investigating fires; Mr. Heintz, the Swartz's attorney; the judge; Mr. Cunningham, the gas company's attorney; and Dr. Smith, the range company's expert witness, who had a doctorate in mechanical engineering and specialized in product evaluation and testing. The following are excerpts from the actual trial.

Jones. When this suit was stated against the gas company in July or August of 1975, the gas company did not have the benefit of your opinion. Is that right?

Willingham. That's correct.

Jones. When this trial began and Mr. Cunningham stood before this jury [on January third] and represented that he had a defense and was going to prove it in this case, he did that without the benefit of your opinion. Did he not?

Willingham. Yes.

Jones. And when they called you . . . on Wednesday of this week and asked you to become involved on behalf of the gas company, they didn't know that you would be able to become involved favorably to them as a result of the phone calls. Is that correct?

Willingham. Yes, that's correct.

Jones. And yesterday, when you sat in the office and waited for Mr. Cunningham while he was finishing this session in court, they still didn't know whether your participation in this case would be favorable to them. Isn't that right?

Willingham. As far as I know.

. . . .

Heintz. Have you ever testified in court before, sir?

Willingham. Yes.

Heintz. Have you ever been offered as an expert witness in any case and been rejected . . . by the court?

Willingham. Only in one case involving a—a case in federal court that I can recall.

Heintz. Was that before Judge Herman, sir?

Willingham. Yes.

Heintz. Were you offered as an expert witness and rejected?

Willingham. On the basis of how I was offered, yes.

Heintz. Was that because you were offered as a witness to give an opinion . . . in a matter that it was determined you were not qualified to testify as an expert?

Willingham. The matter involved a steel mill. Somehow the offering as far as what I was to do was, in my opinion, not presented properly.

Heintz. Oh, you mean the lawyer made a mistake?

Willingham. I'm not saying who made the mistake. For what I was asked to do, I felt—qualified. Safety engineering was what I was supposed to be doing. How it was presented, I have no knowledge of that.

Heintz. But the judge ruled that you were not qualified to testify.

Willingham. Along with Dr. Tomes of the university, who's head of their mechanical engineering department.

Heintz. You were supposed to testify as a *safety engineer* in that regard, sir?

Willingham. I don't know. All I know is what I was asked to do and what [information] I had developed on my own. Since I never got on the stand—

Heintz. May I ask you this, sir? What was the nature of the expertise required in that particular matter?

Willingham. It involved the movement of steel—hot steel bars from the mill to a tabling device next to a loading area.

Heintz. Did it have anything to do with fires?

Willingham. Not a thing.

Heintz. Have you attempted to, or have you, testified as an expert witness relative to things other than fires in the last years?

Willingham. Yes.

Heintz. How many different things have you testified to relative to these aspects?

Willingham. Gas-gases, explosions, fires, and mechanical matters.

Heintz. What mechancial matters, sir?

Willingham. Machinery of different types.

Heintz. All different types of machinery?

Willingham. Yes. Printing—printing machinery, steel-making machinery, any number of machines. I majored in *machine design.*

Heintz. Have you ever testified relative to component parts of automobiles?

Willingham. Yes.

Heintz. How about mobile homes?

Willingham. Mobile home fires, yes.

Heintz. Have you ever testified relative to any other items that you haven't told us about?

Willingham. I—I don't know. I didn't make a study of that. We open files on anywhere from 80 to 100 cases a year, which involves, over the years since 1955 to the present, literally hundreds and hundreds of different events. Relatively few get into final litigation. But they cover the marine field—

Heintz. Marine?

Willingham. Yes, I have the—

Heintz. Ships?

Willingham. The ship that blew up at the dock. I have several oil refinery fires.

Heintz. Other than fires, what other things have you testified about, sir?

Willingham. Machinery failures of one kind or another.

Heintz. That's the point I'm trying to get from you, sir, if you can tell us. What machinery *have you* testified about?

Willingham. (No response.)

Heintz. You've told us you've testified about printing machinery, steel-making machinery, automobiles, and the components—

Willingham. Punch presses—

Heintz. Of automobiles?

Willingham. Punch presses, *die presses.*

Heintz. Die presses.

Willingham. Various automotive parts. Sometimes for and sometimes against.

Heintz. Various automotive parts. Like the brakes?

Willingham. Brake systems.

Heintz. Yes? Brake systems. *Transmissions?*

Willingham. Yes, I think I had one or two transmission cases. And—

Heintz. Steering mechanisms?

Willingham. Steering mechanisms. Well, that's generally anything that's within the—

Heintz. Anything to do with an automobile?

Willingham. No. Well, not everything, but— I mean, I haven't testified in every phase of the automobile, but positioning of gasoline tanks on occasion.

Heintz. Have you ever testified in any airplane cases?

Willingham. No, I haven't. I've done some work—

Heintz. I'm sorry. Have you investigated and advised counsel in any airplane cases?

Willingham. I have one that's—made it into litigation, so I won't be priviledged to tell you anything about that at this time.

Heintz. I wasn't going to ask you. I just want to explore—

Willingham. There's just one.

Heintz. —your qualifications and let the jury know *how many different things* you're qualified in. Are there any other things you're qualified in?

Willingham. Well, let me—I think I could answer your question by saying—

Heintz. Let me ask . . . [the question]. I haven't asked it.

Willingham. I know. But you are asking me the questions, and I do want to give you an answer.

Heintz. I would like that, sir.

Willingham. But if we could lump it up together and say that mechanical engineering is the wheel and axle in the incline plane, and how it's put together doesn't matter. It's just like you understand the law and all its concepts and I understand mechanical engineering.

. . . .

Jones. Isn't it a fact, Mr. Willingham, that you make yourself available to lawyers and others involved in litigious situations . . . to advise them from an engineering standpoint about products—

whether they're safe or not safe, whether they're defective or not defective?

Willingham. Absolutely.

Jones. And in situations where a company which specializes in a particular product—whether it be an automobile, a gas range or whatever—is in a litigious situation, you come in as an engineer, whatever . . . the product, to comment either favorably or against the product.

Willingham. Well, I comment on how I see it. That may be favorable or it may not be.

Jones. All right. And in that capacity, would you admit at least that you're somewhat of a generalist?

Willingham. No. I wouldn't say I'm a generalist.

Jones. All right. Have you *ever* worked for a company that manufactures gas ranges?

Willingham. No.

Jones. Have you *ever* designed and manufactured a gas range?

Willingham. No. But I think we're getting into the same category that I explained: the wheel and the axle in the incline plane. It doesn't mean that I don't understand them or I don't understand the standards under which they are manufactured.

Jones. I see. But in other words, you hold yourself . . . [up] as a person who is qualified to comment upon any product that involves any concept of mechanical engineering.

Willingham. Well, that's the reason I'm a registered professional engineer in this state.

Jones. All right.

Willingham. They have agreed that I am competent to do that.

Jones. Well, . . . could you answer this: If a cubic foot of gas was present at a pressure of 300 pounds per square inch in a gas pipeline at a temperature of 60 degrees Fahrenheit, how many cubic feet of gas would this produce if it was released into a person's kitchen at 75 degrees Fahrenheit and atmospheric pressure?

Willingham. No, I certainly wouldn't answer that from here.

Jones. That's beyond your competence?

Willingham. You're talking about— No, it isn't. There's two gases: one's Boyle's—Charles's—Law, and the other is Boyle's Law. One deals with temperature; the other deals with pressure. And the volumes change inversely with the change of those particular constants.

Jones. Then what's the answer?

Willingham. The answer is that if you release a—a cubic foot of gas under 300 pounds' pressure to a lower pressure, it will expand until it reaches the low pressure.

Jones. No. No. I mean the numerical—

Willingham. I don't understand what you're—

Jones. Is it beyond your competence to answer the question that I've put to you in specific numbers?

Willingham. I don't think you've answered my question.

Jones. That answers my question, sir. What was your grade in math at college: your first course in mathematics in 1929 or 1930, applied math?

Willingham. My first— When I went to college the first time?

Jones. Yes.

Willingham. I failed my first course in mathematics, and I would end up getting an—an "A" for the whole year in calculus and higher mathematics.

Jones. How about your course in applied physics?

Willingham. I don't recall that—what the grades were. I don't have an idea, and I don't know where you would get this information.

Jones. I have your transcript in front of me, sir. You *flunked physics*, didn't you, Mr. Willingham?

Willingham. That's high school physics, yeah. I can't—I graduated from a lousy high school. That's why I flunked.

Jones. I'm looking at your record here, and I don't see the "A" in calculus—not that year, anyway. Might it have been some later year?

Willingham. If it wasn't that year, you wouldn't have the calculus the same year you flunked in—

Jones. I thought you said it was later that year.

Willingham. No, I didn't.

Jones. (Jones asked Mr. Willingham if he had testified in a series of trials in 1970, including cases involving an electrical appliance, a lathe, a high lift loader, an automobile, and a conveyor belt.) Would you agree, sir, that you testified 60 times in the five-year period from 1970 to 1975 in the various courts in this area expressing opinions on *almost as many products* as there were times that you testified?

Willingham. Yes. And I think my track record is pretty good, too, honestly.

Jones. Can you write the chemical reaction equation for the combustion of methane and air: balance the equation and compute the ratio of fuel to air on a volume basis for a chemically correct mixture?

Willingham. I could work it out, but I'm certainly not going to do it from here.

Jones. Well, would you give me the formula?

Willingham. I don't recall the formula offhand. They're all in the gas handbooks. I don't keep all the formulas in my head.

Jones. Do you have a gas handbook with you?

Willingham. No.

Jones. Now, Mr. Willingham. What would you say is the ratio of cases in which you're employed to the cases in which you actually come to court and testify?

Willingham. The ratio is— I don't know the percentage, but it certainly is a very low percentage of the number of cases that we do.

Jones. In other words, a very low percentage actually goes to court?

Willingham. That's true.

Jones. So, if you testified 60 times in five years, could you give this jury some idea of how many times your services were retained during that period?

Willingham. In five years? Maybe 250 times, roughly.

Jones. Do you consider yourself a scientist?

Willingham. A scientist?

Jones. Yes.

Willingham. I don't know exactly what the definition of a scientist is. I think I understand scientific terms, and I think I have a—a knowledge of what science is supposed to be all about.

Jones. Do you approach your opinion making and opinion forming in a scientific manner?

Willingham. I think I do.

Jones. Have you done that in this case?

Willingham. Yes, I think I have.

. . . .

Judge. Mr. Willingham, you mentioned the David Engineering Company.

Willingham. Yes.

Judge. Is that the company that you head or are associated with?

Willingham. Yes, it is.

Judge. What's your position with that company?

Willingham. I'm president of the company. It's a corporation now; I'm president of it.

Judge. What does David Engineering Company do?

Willingham. We are consulting engineers, and we're also manufacturers' representatives.

Judge. How many employees are there of David Engineering?

Willingham. At present, it's just myself. I have a secretary, and I have two men under project contract. I hire people when I need them.

Judge. So, do I understand—and correct me if I'm wrong—that David Engineering Company is you?

Willingham. At present, it's—I'm the—and my secretary are the full-time employees. Yes.

Judge. That's what I asked. Can you give me . . . an estimate of the percentage of your work that entails consulting work where there is no litigation involved at all—just consulting to a contractor or to an architect, or whatever—as compared to the work that you're engaged in lawsuits, whether they come—

Willingham. I see.

Judge. —to trial or not?

Willingham. I had a contract for three years with a company in the south and their groups just to aid them in their process—times of process—what to do in the event of fires and safety. They deal in toxic materials. That was strictly a consulting job.

Judge. All right. I didn't ask you about anything specific. I just wondered if you could give us an approximation of— What is the main thrust of your business? Is it consulting engineer that has nothing to do with . . . litigation, or is it mostly that has to do . . . with litigation?

Willingham. I would say it's about 50-50.

. . . .

Cunningham. What is your opinion as to what did cause gas to be released?

Willingham. Something failed at the range—within the range. I didn't see it. I didn't make any examination. It's been removed from the premises, so I have no idea. But something failed that would release a sufficient amount of gas in that time frame to cause exactly what happened.

Cunningham. Now, Mr. Willingham, do you have an opinion as to the amount of time between the—the amount of time the gas would be in the room to cause this type of ignition?

Willingham. No more than a half minute to a minute. That would be sufficient to cause this amount of—this little puff that caused the burns and set fire to the room.

Cunningham. Could it have been more time than that in your opinion?

Willingham. No.

. . . .

Heintz. Now, sir, you have now admitted that you can make an estimate as to the size of the leak; you have admitted that the leak was between a quarter of an inch and a half inch; and you know the size of the room; and you know the gas pressure was six-inch water column pressure. Can you tell us, sir, how much time it would take, [considering] the leak that you think was there . . . [and] the gas pressure that you know was there, . . . [to fill] the size of that room?

Willingham. It can be calculated, certainly.

Heintz. Would you *calculate it* for us?

Willingham. I can't do it from the stand. That takes—

Heintz. Would you go to the board?

Willingham. No, I will not do it that way.

Heintz. Why won't you?

Willingham. Because you have to refer to the gas formulas and the pressures, and it's not anything you can do in a few minutes.

Heintz. I told you the pressure.

Willingham. I know what the pressure is; it's roughly three inches of—six inches of water column.

Heintz. You told us it was about a pound per square inch.

Willingham. I told you, if it's three ounces per square inch and the number of square inches—32 inches—I mean 32 ounces to a pound, . . . it's roughly—

Heintz. I thought there's 16 ounces to a pound, not 32.

Willingham. Sixteen ounces to a pound. You're correct. So, it's roughly only a few it doesn't take much to get a gas flow of sufficient quantity, because you're not just dealing—

Heintz. Sir, would you please answer the question?

Willingham. I said it can be calculated, certainly.

Heintz. Would you calculate it for us?

Willingham. No. I will not do it from the stand.

Heintz. Aren't you able to?

Willingham. Not from the stand. I don't—

Heintz. Where would you be able to calculate it from?

Willingham. I would have to go to my office, refer to my textbooks, and make the calculation.

Heintz. Why is that, sir?

Willingham. Because that's the way it would be done, as far as I'm concerned, to be completely accurate.

Heintz. Well, why don't you just do the *best you can right here*, sir?

Willingham. I—I wouldn't guess at it.

Heintz. I don't want you to guess. I just want—

Willingham. I can't—

Heintz. Unless you will admit you're not able to.

Willingham. I will not. I am able to.

Heintz. Do you know that they hired another expert besides you?

Willingham. No.

Heintz. They didn't tell you that?

Willingham. I know there is another man as an expert, a Mr. Clayton. Is that—is that who you're referring to?

Heintz. Did you speak to him?

Willingham. Yes.

Heintz. Did they hire him to be an expert, too?

Willingham. I don't know for what reason they hired him. I didn't discuss that with them. That's up to Mr. Cunningham, what he chooses to do with Mr. Clayton.

Heintz. Did you find out Mr. Clayton was from Ohio and was going to be an expert?

Willingham. I knew he was from Ohio.

Heintz. But you didn't know that he was an expert witness or a proposed expert witness or might be an expert witness.

Willingham. Yes, I knew that he could be used as an expert witness.

Heintz. Did you *ask* Mr. Cunningham if he could arrange for you to go and visit 1819 North Sherman Street?

Willingham. *No.*

Heintz. Did you do any type of mathematical calculations whatsoever in order to come here and give an opinion to this jury that the fire happened that fast?

Willingham. No. I didn't think that it was necessary on the basis of the facts as I see them.

Heintz. Are you always able to reach your conclusions and opinions this quickly, or—

Willingham. No.

Heintz. —does it depend on the fee?

Willingham. No. Sometimes it takes weeks and sometimes longer than that. And it doesn't depend on a fee and never has.

Heintz. You charge whether you testify or not, don't you, sir?

Willingham. Absolutely.

Heintz. As soon as they talk to you, it costs them money.

Willingham. I imagine it's the same as [for] you.

Heintz. As soon as they talk to you, they have to pay you. Is that right?

Willingham. That's exactly right.

Heintz. And, in fact, if they settle a case after you get involved in it, they have to pay you a fee—

Willingham. No.

Heintz. —whether you testify or not, don't they?

Willingham. Absolutely not.

Heintz. Oh. Have you changed that from the way you used to do it?

Willingham. Yes, I did.

Heintz. You used to charge a fee—

Willingham. There was a reason for that, if you'd like to know it. It was because—

Heintz. I know the reason, sir.

Willingham. Do you?

Heintz. Yes, thank you.

. . . .

Jones. Sir, would you please tell his honor, the ladies and gentlemen of the jury, what the significance of these photographs was?

Smith. These photographs, as I illustrated, depict the condition of the range at the time I saw it. They show me conclusively that the cause of the fire in that range was the connection with the flexible connector attaching to the manifold and that this was brought about by the lack of an adapter between the flexible connector and the manifold.

This was supported by the fact that there is carbon . . . soot on the underside of the range top at exactly that location, and it [the soot] extends back from that location toward the rear of the range toward the left rear burner, which is exactly where you would expect to find soot from gas burning in a very rich mixture from a gas leak at that connection to the manifold.

And evidence that I pointed out about flash tubes—the ones in the back being in good condition, the ones in front being melted—is consistent with having gas near the rear of the range too rich to burn.

I talked about the cup for the pilot light being burned on the side opposite where the manifold connection is, and this is also consistent with flooding the area with gas having an extremely rich mixture, so that when it burns, it produces a lot of carbon but doesn't produce a very high temperature. The flame is cooler than [it] would normally be . . . if the mixture were correct.

In other parts of the range, where there's more oxygen available and you get gas consumption, you'll get a hotter flame, and you can expect more damage but no soot. We find that in other areas of the range.

Jones. Dr. Smith, when manufacturers of gas ranges such as Smith . . . sell their ranges, do they sell them with or without the flexible connector?

Smith. They're sold without a flexible connector. This is done for a very good reason. The reason is that installation of ranges varies from kitchen to kitchen. In some kitchens, the gas riser comes out of the floor, in which case you need a long flexible connector to come up to the top of the range. In other installations, the gas riser comes right out of the wall behind the range, which . . . would require . . . a very short flexible connector to connect it up to the range.

Jones. Are these gas ranges that are sold by Smith and . . . various other manufacturers sold with or without this adapter bushing that you have discussed?

Smith. The adapter bushing is part of the flexible connector. It's not part of the range. It's sold with the flexible connector.

Jones. Do the diameters of the flexible connectors vary?

Smith. Yes. There are half-inch flexible connectors and three-quarter-inch flexible connectors.

Jones. The flexible connector then and the adapter bushing are purchased separately from the range. Is that correct?

Smith. They're purchased from a hardware store or Sears, Roebuck or anybody that sells plumbing . . . hardware.

Jones. Now, when you examined the range parts that were laid out for you, did you find any evidence of any gas leak in any of the parts of the Smith range itself?

Smith. I went through that range and looked at every gas connection and component in the range, and I found no evidence whatsoever of a gas leak anywhere else in that range except where the flexible connector connects to the manifold.

Jones. Will you tell the . . . jury how many hours you spent there poring through those parts?

Smith. I spent a whole day, from early in the morning till late in the evening.

Jones. The gas that was in the Swartz home at the second that it exploded, as Mr. Swartz described. Should the odor of that gas have been detectable by Mr. Swartz and by his daughter?

Smith. Yes.

Jones. Would a person . . . [with an] ordinary, reasonable sense of smell detect gas in that room at that time under those conditions?

Smith. Yes.

Jones. You were in court on Friday when Mr. Willingham testified, weren't you?

Smith. Yes, I was.

Jones. Do you recall Mr. Willingham testifying that, in his opinion, this leak existed for 30 to 60 seconds?

Smith. Yes, I heard that.

Jones. Based upon your evaluation of this range and your having listened to the description of this accident as described by Mr. Swartz, in your calling upon reasonable engineering and physics principles, is it possible for the explosion described by Mr. Swartz to have occurred with a gas leak of 30 to 60 seconds?

Smith. It is physically impossible for that to have occurred.

Jones. Will you tell . . . the jury why a gas leak that only lasted 30 to 60 seconds would not have caused the damage, the injuries, that we've seen in this courtroom?

Smith. The description of the accident provided by Mr. Swartz indicated that the flame had extended out approximately three feet at least around the range and up and that when he turned around, he saw the entire kitchen was engulfed in flame. This indicates that the gas had been distributed . . . through the room at the time that it was ignited and not just concentrated right in the range.

Jones. Dr. Smith, Mr. Willinham also testified that, in his opinion, this explosion, or whatever he described it as, occurred in the burner box itself.

Smith. Yes.

Jones. If Mr. Willingham said there was a half a cubic foot of gas in that box and the rest of the box was filled with the range components and air, do you have an opinion as to whether or not this explosion, as described by Mr. Willingham, was (a) possible or (b) was the cause of the damage sustained by the Swartz family?

Smith. It is physically impossible—I've said that before—because the amount of gas required in there to reach the explosion limit would require that there be at least another $3\frac{1}{2}$ cubic feet or so of air, minimum, and there just isn't enough volume in that box.

Jones. All right. You heard them describe [that] they spray a liquid odorant into the natural gas that's passing by at high pressure in big pipes.

Smith. Yes, I heard that.

Jones. Are you familiar . . . with the general type of equipment that they use to do this type of work?

Smith. Yes. They use positive-displacement metering pumps to pump the odorant into the spray assembly in the pipeline.

Jones. These are machines, are they not?

Smith. Yes, they are.

Jones. Now, are you familiar with the use of machinery such as this type . . .?

Smith. I'm very familiar with positive-displacement pumps of that type.

Jones. Does this machine ever fail?

Smith. It does quite frequently.

Before the closing arguments began, the gas company offered to settle for $250,000. The plaintiff's attorney rejected the offer, insisting that anything less than $1 million was unacceptable. The jury awarded the Swartzes $835,000: $10,000 to Mrs. Swartz, $200,000 to Mr. Swartz, and $625,000 to their daughter.

CONCLUSION

The journey from the beginning of an accident investigation that concludes with a trial is often a long one. Between 90 percent and 95 percent of suits are settled before a trial begins. The American legal process is an expensive one that encourages early out of court settlement. Of the small percentage of product liability cases that do go to trial, a modest percentage are settled during the course of the trial. Thus, by the time the case has proceeded to trial, the contesting parties have staked out their economic, philosophical, and legal claims.

Some experts argue that the judicial procedures for presenting complex technical issues and the jury system for evaluating those issues are cumbersome and inefficient. They point to the fact that in a product liability suit, a trial by one's peers should be interpreted as a panel composed of qualified experts, particularly when the issues are extremely technical in nature.

The counterargument is that such a system would be prohibitively expensive, either to taxpayers or to the contesting parties. Furthermore, if the issues are so complex that they cannot be explained to a jury composed of ordinary citizens rather than experts, how can the nontechnical decision makers, such as manufacturers, users, and the instrumentalities that influence the quality of products, be expected to make decisions about product quality and safety?

We end this book with the observation that the quality and safety of products is not only about the law and science but about the people who are capable of the best and worst that we value. The universal desire of humankind to hold in high regard the Golden Rule—Do unto others as you would have them do unto you—is manifested in all cultures and all religions. However, the reality that in some cases this ideal principle must be enforced with the authority of law has been recognized since ancient times. The courts can uphold the best and the worst that we value because they are a human creation and humans, after all, are imperfect. Imperfect as our legal system may be, we cherish it because it provides a forum for both honorable and dishonorable people to be heard and to seek protection. We cherish it because it is a living and growing instrumentality, because it simultaneously reflects national, special-interest, and individual values.

Because engineers, manufacturers, users, and the other instrumentalities that affect product quality and safety also are not perfect, we must conclude that, in a democratic free-enterprise system, maintaining a balance between these entities is the best way to ensure safe and high-quality products.

Appendix A

Requirements for Registration as a Professional Engineer

STATE REQUIREMENTS FOR REGISTRATION AS A PROFESSIONAL ENGINEER

	1 Minimum Age, E.I.T.	2 Minimum Age, PE	3 Degree Accreditation ABET School or Board approved (BA)	4 Experience Credit for EIT No Degree, Years	5 Experience Credit for PE with Degree, Years	6 Credit Teaching Experience	7 Credit Military Experience	8 Credit Contracting Experience	9 Board Provides Licensure by Endorsement	10 Month Written Exam Held: Apr = 4, Oct = 10	11 Written Examination: E.I.T. or Fundamentals, Hours	12 Written Examination: PE, Branch, or Specialty, Hours	13 Reciprocity or Comity Practices for PE	14 Certificate of Verification Recognized	15 Temporary Practice Permit Granted	16 Time Limit on Permit, Days	17 Register in Branches or PE	18 Seal Required on Drawings, Plans, Specs	19 Time Limit on E.I.T. Certificates, Years	20 Reinstatement Policy
Alabama	—	—	ABET	8	4	Asst prof	Eng	Design	No	4, 10	8	8	Yes	Yes	Yes	Bd rev	PE	No	10	—
Alaska	—	—	ABET	8	4	Eng	Eng	Eng	No	4, 10	8	8	Yes	Yes	No	—	Branch	Yes	5	—
Arizona	—	25	ABET	8	4	5 yr max	Yes	Eng	Mt req	4, 10	8	8	Yes	Yes	No	—	Branch	Yes	8	Yes
Arkansas	—	—	ABET	8	4	Yes	Eng	Ind bas	Yes	4, 10	8	8	Yes	Yes	Yes	No limit	PE	Yes	None	—
California	—	25	BA	6	2	—	Eng	Eng	Yes	4, 10	8	8	Yes	No	No	—	Branch	No	11	Yes
Canal Zone	—	21	Deg	12	4	Yes	Eng	No	No	4, 10	8	8	Yes	Yes	Yes	1 yr	PE	No	None	—
Colorado	—	25	Deg	8	4	Yes	Eng	Eng	No	4, 10	8	8	Yes	Yes	No	—	PE	Yes	None	Yes
Connecticut	—	—	ABET	10	4	Yes	Eng	Eng	Yes	4, 10	8	8	Yes	Yes	Yes	30	PE	Yes	10	—
Delaware	—	—	ABET	8	4	Eng	None	None	Yes	4, 10	8	8	Yes	Yes	Yes	30	PE	Yes	None	None
District of Columbia	21	25	BA	8	4	Yes	No	Yes	Yes	4, 10	8	8	Yes	Yes	Yes	1 proj	PE	Yes	None	Yes
Florida	—	24	ABET	10	4	Yes	None	None	Yes	4, 10	8	8	Yes	Yes	Yes	1 yr	PE	Yes	12	Yes
Georgia	—	25	ABET	12	4	—	—	—	No	4, 10	8	8	Yes	Yes	Yes	90	PE	Yes	10	None
Guam	—	21	BA	8	4	B.A.	B.A.	Eng	Yes	4, 10	8	8	Yes	Yes	No	—	PE	Yes	None	Yes
Hawaii	—	—	Deg	12	3	1 yr	Yes	Yes	No	4, 10	8	8	Yes	Yes	Yes	1 yr	Branch	Yes	12	—
Idaho	—	—	Deg	8	4	Eng	Yes	No	No	4, 10	8	8	Yes	Yes	Yes	30	Branch	Yes	None	Yes
Illinois	—	—	BA	8	4	1 yr	—	—	Yes	4, 10	8	8	Yes	Yes	Yes	30	PE	Yes	12	Yes
Indiana	—	—	ABET	5-12	4	Yes	Eng	None	No	4, 10	8	8	Yes	Yes	Yes	30	PE	Yes	10	Yes
Iowa	—	—	ABET	12	4	Eng	Eng	No	Yes	4, 10	8	8	Yes	Yes	Yes	Bd rev	Branch	No	None	Yes
Kansas	—	—	BA	8	4	Eng	Eng	—	No	4, 10	8	8	Yes	Yes	No	—	PE	Yes	10	Yes
Kentucky	—	—	ABET	8	4	Yes	Yes	No	Yes	4, 10	8	8	Yes	Yes	No	—	Branch	Yes	None	Yes
Louisiana	—	—	ABET	8	4	Yes	Maybe	Yes	Yes	4, 10	8	8	Yes	Yes	Yes	30	Branch	Yes	10	Yes
Maine	—	25	ABET	12	4	B.A.	B.A.	B.A.	Yes	4, 10	8	8	Yes	Yes	Yes	30	PE	Yes	12	Yes
Maryland	—	—	ABET	12	4	Yes	Eng	No	Yes	4, 10	8	8	Yes	Yes	Yes	1 yr	PE	Yes	None	Yes
Massachusetts	—	—	ABET	8	4	Asst prof	Eng	Eng	Yes	4, 10	8	8	Yes	Yes	Yes	30	PE	Yes	None	Yes
Michigan	—	21	ABET	8	4	Eng	Eng	No	No	4, 10	8	8	Yes	Yes	Yes	60	PE	Yes	None	Yes

Minnesota	—	25	ABET	13	4	1 yr Eng	Eng	1 yr Eng	Yes	4, 10	8	8	Yes	Yes	Yes	1 yr	Branch	No	10	Yes
Mississippi	—	—	BA	8	4	Yes	Eng	Eng	Yes	4, 10	8	8	Recog	Yes	Yes	30	PE	Yes	10	Yes
Missouri	—	21	BA	8	4	Yes	No	Eval	Yes	4, 10	8	8	Yes	Yes	No	—	PE	Yes	8	Yes
Montana	—	—	ABET	8	4	Yes	Yes	Yes	Yes	4, 10	8	8	Yes	Yes	No	—	PE	Yes	10	Yes
Nebraska	—	25	ABET	8	4	Yes	Eng	Eng	Yes	4, 10	8	8	Yes	Yes	No	—	PE	Yes	None	Yes
Nevada	—	—	BA	8	4	Yes	Lim	Yes	Yes	4, 10	8	8	Yes	Yes	Yes	Bd rev	PE	Yes	8	Yes
New Hampshire	—	—	BA	8	4	Eng	No	No	Yes	4, 10	8	8	Yes	Yes	No	—	PE	Yes	10	Yes
New Jersey	21	25	ABET	8	4	Yes	Yes	Eng	Yes	4, 10	8	8	Yes	Yes	Yes	30	PE	Yes	10	Yes
New Mexico	—	—	BA	8	4	Yes	No	Policy	Yes	4, 10	8	8	Yes	Yes	No	—	PE	Yes	None	Yes
New York	19	19	ABET	12	4	Yes	Eng	Policy	Yes	4, 10	8	8	Yes	Yes	Yes	30	PE	Yes	10	Yes
North Carolina	—	21	BA	10	4	4 yr max	Policy	Policy	Yes	4, 10	8	8	Yes	Yes	Yes	30	PE	Yes	None	Yes
North Dakota	—	21	ABET	—	4	4 yr min	Eng	Eng	No	4, 10	8	8	Yes	Yes	Yes	1 yr	PE	Yes	12	Yes
Ohio	—	—	ABET	8	4	Eng	Eng	Eng	Yes	4, 10	8	8	Yes	Yes	Yes	60	PE	Yes	10	Yes
Oklahoma	—	—	ABET	8	4	Yes	Yes	Eng	Yes	4, 10	8	8	Yes	Yes	Yes	1 yr	PE	Yes	7	Yes
Oregon	—	—	BA	8	4	Eng	Eng	Eng	Yes	4, 10	8	8	Yes	Yes	Yes	30	PE	Yes	10	Yes
Pennsylvania	—	25	ABET	12	4	Yes	Yes	Some	Yes	4, 10	8	8	Yes	Yes	Yes	30	PE	Yes	None	Yes
Puerto Rico	—	—	ABET	—	4	B.A.	B.A.	B.A.	Yes	4, 10	8	8	Yes	Yes	Yes	Bd rev	PE	Yes	Indef	Yes
Rhode Island	—	—	ABET	12	4	1 yr	None	None	No	4, 10	8	8	Yes	Yes	Yes	30	PE	No	None	Yes
South Carolina	—	—	BA	8	4	Yes	Eng	None	Yes	4, 10	8	8	Yes	Yes	Yes	30	PE	Yes	10	Yes
South Dakota	—	21	ABET	8	4	Yes	Eng	Eng	No	4, 10	8	8	Yes	Yes	Yes	1 yr	PE	Yes	None	Yes
Tennessee	—	—	BA	8	4	Yes	2 yr	B.A.	Yes	4, 10	8	8	Yes	Yes	No	—	Eng	Yes	None	Yes
Texas	—	—	Deg	8	4	Asst prof	Eng	No	No	4, 10	8	8	Yes	Yes	No	Bd rev	PE	Yes	12	Yes
Utah	—	—	ABET	8	4	2 yr	No	No	Yes	4, 10	8	8	Yes	Yes	Yes	30	PE	Yes	10	Yes
Vermont	—	—	BA	12	4	Yes	Eng	Eng	Yes	4, 10	8	8	Yes	Yes	Yes	30	PE	Yes	12	Yes
Virginia	—	21	ABET	10	4	Yes	Eng	Lim	Yes	4, 10	8	8	Yes	Yes	No	No	PE	No	10	Yes
Washington	—	21	ABET	8	4	Yes	Eng	Eng	No	4, 10	8	8	No	No	No	—	PE	Yes	None	Yes
West Virginia	—	—	BA	8	8	Yes	Eng	Yes	No	4, 10	8	8	Yes	Yes	Yes	60	Branch	No	12	Yes
Wisconsin	—	—	ABET	12	4	Eng	Eng	Eng	Yes	4, 10	8	8	Yes	Yes	Yes	60	PE	Yes	10	Yes
Wyoming	—	—	ABET	8	4	Yes	Yes	Yes	Yes	4, 10	8	8	Yes	Yes	No	—	PE	No	None	Yes

*The exhibit was developed from data furnished by boards of registration of the states and represents the latest information available. Because boards retain the right to change conditions, and the exhibit contains space-saving simplifications, engineers are cautioned to verify answers to questions with their boards of record. Addresses of state boards may be obtained from the National Council of Engineering Examiners (NCEE), Box 1686, Clemson, South Carolina 29633-1686.

Most states have no minimum age requirement for Engineers in Training. Upon graduation from an accredited program in engineering or science or upon completion of the board-approved equivalent in engineering experience, the graduate is permitted to sit for the first of two parts (Fundamentals of Engineering) of the written examination. After serving his or her internship (4 years or more) in engineering practice, he or she may qualify to sit for the second part of the examination (Principles and Practice). In all states, each of the two parts of the written examination runs 8 hours long.

†Some states require an engineering degree from a school accredited by ABET, and some are satisfied with a degree from a board-approved school.

(Courtesy of Hicks and Mueller *Standard Handbook of Consulting Engineering Practice,* 1986, New York: McGraw Hill.)

SCHEDULE OF EXAMINATIONS ON THE PRINCIPLES AND PRACTICES OF ENGINEERING

The specifications listed below became effective in the spring of 1983. They reflect the findings of the Task Analysis Survey conducted by the National Council of Engineering Examiners to determine the specific tasks that licensed engineers perform on the job.

Group I Examinations Administered in Spring and Fall

Discipline	Category	Subject	Approximate no. problems/ category
Chemical	1	Thermodynamics	3
	2	Process Design	4
	3	Mass Transfer	4
	4	Heat Transfer	2
	5	Chemical Kinetics	4
	6	Fluids	2
	7	Engineering Economics	1
		Total	20
Civil/Sanitary/ Structural	1	Transportation	5
	2	Structural	7
	3	Sanitary	5
	4	Hydraulics	4
	5	Soils	2
	6	Engineering Economics	1
		Total	24
Electrical	1	Power	8
	2	Electronics	6
	3	Control Systems	5
	4	Computers	4
	5	Engineering Economics	1
		Total	24
Mechanical	1	Mechanical Design	8
	2	Management	1
	3	Energy Systems	6
	4	Control Systems	1
	5	Thermal and Fluid Processes	3
	6	Engineering Economics	1
		Total	20

(Continued)

Group II Examinations Administered in Spring and Fall			
Discipline	Category	Subject	Approximate no. problems category
Group II Examinations Administered in Fall Only			
Aeronautical/ Aerospace	1	Structures	7
	2	Aerodynamics	4
	3	Flight Mechanics	4
	4	Propulsion	4
	5	Engineering Economics	1
		Total	20
Agricultural	1	Irrigation & Drainage	3
	2	Soil & Water Conservation	3
	3	Machine Design	2
	4	Power, Control & Systems	2
	5	Structures	2
	6	Environmental	2
	7	Crop Handling & Processing	3
	8	Food Engineering	2
	9	Engineering Economics	1
		Total	20
Ceramic	1	Phase, Microstructure, Mechanical & Electrical Properties	4
	2	Processing	4
	3	Glass Technology	2
	4	Refractories, Corrosion, Heat Transfer	3
	5	White Ware, Glazes, Enamels, Coatings	2
	6	Structural Clay, Plant Operation, Plant Design, Mineral	4
	7	Engineering Economics	1
		Total	20
Fire Protection	1	Hydraulics	4
	2	Suppression systems	2
	3	Fire Behavior	2
	4	Fire Communications	2
	5	Hazards	4
	6	Structures	4
	7	Fire Protection Management	1
	8	Engineering Economics	1
		Total	20

(Continued)

Group II Examinations Administered in Spring and Fall			
Discipline	Category	Subject	Approximate no. problems/ category
Industrial	1	Facilities Planning	3
	2	Management	2
	3	Manufacturing	3
	4	Operations Research/Systems	3
	5	Production & Inventory Control	3
	6	Quality Assurance & Applied Statistics	2
	7	Work Measure & Ergonomics	3
	8	Engineering Economics	1
		Total	20
Manufacturing	1	Processes	4
	2	Operations	3
	3	Production Facilities	6
	4	Manufacturing Management	4
	5	Systems	2
	6	Engineering Economics	1
		Total	20
Metallurgical	1	Fabrication & Processing	2
	2	Metallurgical Thermodynamics and Corrosion	3
	3	Selection of Material	2
	4	Thermal Treatment & Solid State Processing	3
	5	Structure/Property Relationships	2
	6	Failure Analysis	2
	7	Mineral Processing	2
	8	Extractive Metallurgy	3
	9	Engineering Economics	1
		Total	20
Mining/Mineral	1	Exploration & Geology	3
	2	Mine Planning	6
	3	Mine Operations	3
	4	Ground Control	3
	5	Mineral Processing	3
	6	Environment & Governmental Regulations	1
	7	Engineering Economics	1
		Total	20
Nuclear	1	Thermal Power Plant Operation	4
	2	Radiation Protection & Instrumentation	3
	3	Neutronics & Core Physics	2
	4	Reactor Engineering	2

(Continued)

Group II Examinations Administered in Spring and Fall			
Discipline	Category	Subject	Approximate no. problems/ category
	5	Nuclear Fuel Cycle Analysis	2
	6	Safety and Licensing	2
	7	Shielding	2
	8	Waste Management	2
	9	Engineering Economics	1
		Total	20
Petroleum	1	Drilling and Completion	6
	2	Production	4
	3	Reservoir & Improved Recovery	6
	4	Formation Evaluation	3
	5	Engineering Economics	1
		Total	20

NOTE: Above listing was obtained from the *Registration Bulletin*, December 1982, courtesy the National Council of Engineering Examiners. Table reflects the findings of NCEE in their Task Analysis Survey to determine specific task areas that licensed professional engineers perform on the job.

Source: Reprinted through the courtesy of the *Registration Bulletin*, December 1982. National Council of Engineering Examiners.

Appendix B

Elastic and Inelastic Behavior of Materials and Fundamental Physical Laws

STATIC (QUASI-STATIC) MECHANICAL PROPERTIES

Stress–Strain Diagrams of Typical Engineering Metals

In all cases, the nominal tensile stress

$$\sigma = \frac{P}{A_0} \qquad \text{(force/area)} \tag{1}$$

where

A_0 = original cross-sectional area
P = force applied to the specimen

and the nominal strain

$$\varepsilon = \frac{L - L_0}{L_0} \qquad \text{(change in length/length)} \tag{2}$$

where

L_0 = original unload gauge length
L = the gauge length at some load (P)

are plotted (see illustration in Fig. B-1) based on a standard tensile test (2-in. specimen loaded at 10^{-5} to $10^{-3}\ s^{-1}$) at room temperature. Reversing the direction of the load ($-P$) produces a compression test that is used particularly for masonry material, such as concrete, that usually has a higher compression stress range than a tensile stress range.

Ductility

The property of a material that represents its ability to deform plastically is called *ductility*. The three measures of ductility are (1) total elongation, (2) uniform elongation, and (3) reduction in area. Usually, these values are given as the percentage of geometry and size of the specimen that will affect these measures of ductility if all other factors are similar.

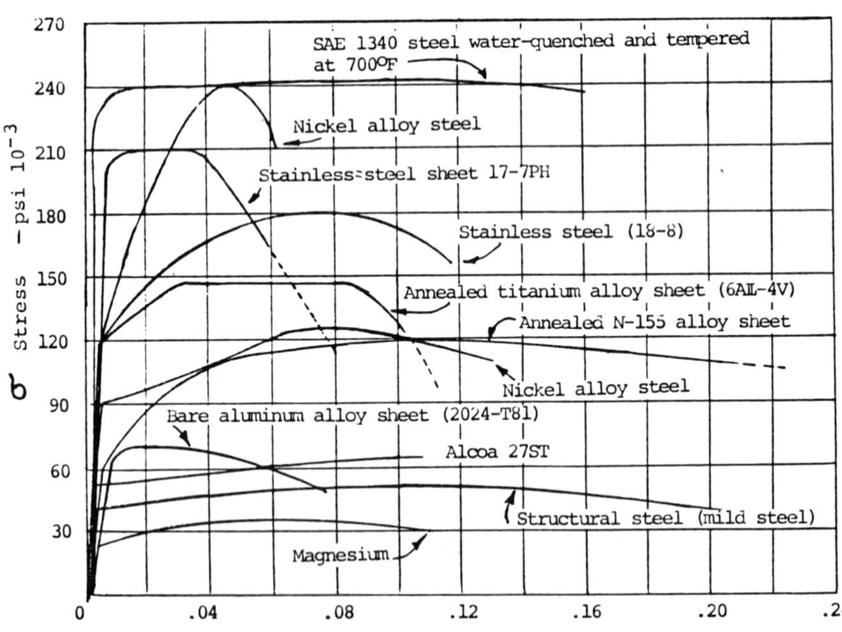

Figure B-1. Nominal tensile stress (σ)–strain (ε) diagrams of selected metals. (From J. Marin, *Mechanical Behavior of Engineering Materials*, Englewood Cliffs, NJ: Prentice-Hall, 1962)

Large specimens tend to be less ductile than small specimens, and round specimens are more ductile than rectangular ones. Thus, to compare the ductility of two materials, their geometry and size must be similar.

The percentage of total elongation is defined as

$$D_T = \left[\frac{(L_f - L_0)}{L_0}\right] \times 100\% \qquad (3)$$

where L_0 = original gauge length, and L_f = final gauge length at fracture.

Uniform elongation (D_u) is defined as

$$D_u = \left[\frac{(L_u - L_0)}{L_0}\right] \times 100 \qquad (4)$$

where L_u = gauge length at σ_{ult} (see Fig. B-2), and where $L_u = (1 + \varepsilon_u)L_0$ or exact

$$L_u = \left(1 + \varepsilon_u - \frac{\sigma_{\text{ult}}}{E}\right)$$

and σ_{ult} = ultimate stress (maximum stress in the specimen before or at failure).

The reduction in area is defined as

$$R_A = \left[\frac{(A_0 - A_f)}{A_0}\right] \times 100 \qquad (5)$$

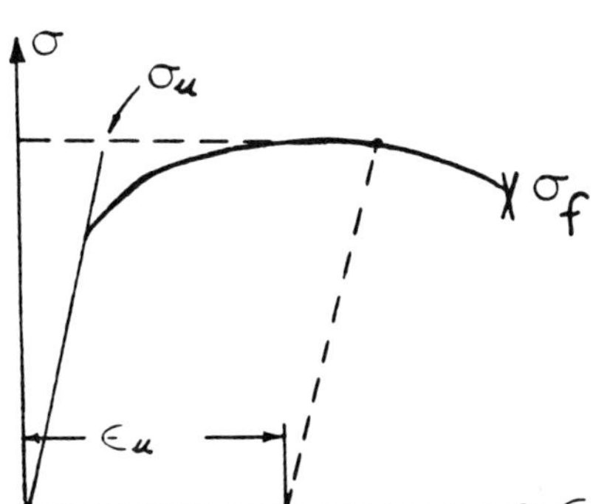

Figure B-2. Ductility curve.

where A_0 = cross section of the original specimen, and A_f = cross section of the fractured area.

The total elongation (D_T) depends on the original gauge length (L_0) because of the local deformation during "necking" of the specimen. Thus, as L_0 decreases, D_T increases in magnitude.

Uniform elongation (D_u) and reduction in area are not affected by L_0. The gauge length (L_0) should be reported with D_T. The next section, "Properties Determined from Tensile Tests," will show that the reduction in area can be related to be *true* strain at fracture.

In addition to geometry, ductility is a function of the state of stress. Under conditions of high hydrostatic compressive stress, the ductility of all materials increases significantly (Bridgeman and Von Karman tests). In fact, materials such as marble, which are usually regarded as brittle, may become ductile (Fig. B-3). On the other hand, materials that are usually regarded as ductile will act in a brittle manner when subjected to triaxial tension that may be caused by notches.

Because of the many factors that affect ductility, classifying material as ductile or brittle can be extremely difficult. At times, metals with

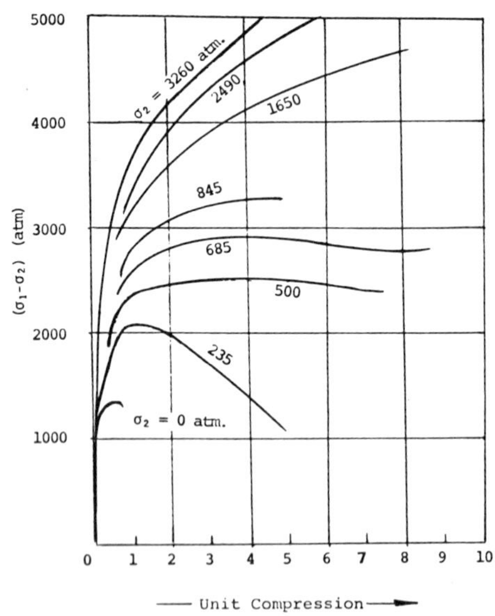

(a)

Figure B-3. Von Karman test illustrating that a material that appears brittle in a uniaxial test is ductile in a triaxial test. (a) Results of Karman's tests on marble cylinders under combined axial (σ_1) and lateral (σ_2) hydraulic compression. The lateral pressure σ_2 was kept constant in each test. (b) Schematic showing Karman's test of marble cylinders.

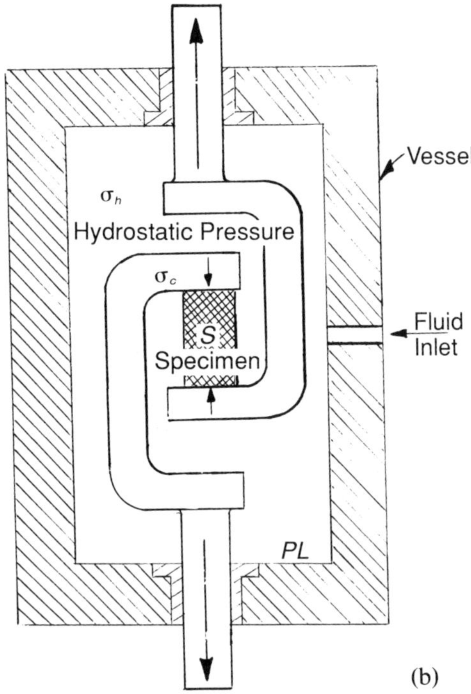

Figure B-3. *(continued)*

nominal tensile strains in excess of 0.05 in./in. at rupture are considered ductile and those with less strain are considered brittle (Marin 1968).

Properties Determined from Tensile Tests

Many mechanical properties of materials are determined from the standard tensile test. A number of these properties are defined as follows.

- *Proportional limit:* The stress at which stress is no longer proportional to strain (Fig. B-4(a), point *a*).
- *Elastic limit:* The maximum stress that can be applied to a material in a standard tensile test without causing permanent deformation. The elastic limit usually exceeds the proportional limit (Fig. B-4(a), point *b*).
- *Upper and lower yield points:* The higher and lower yield points. For example, mild steel is one of the few materials that has a higher yield point and a lower yield point. These points are local maxi-

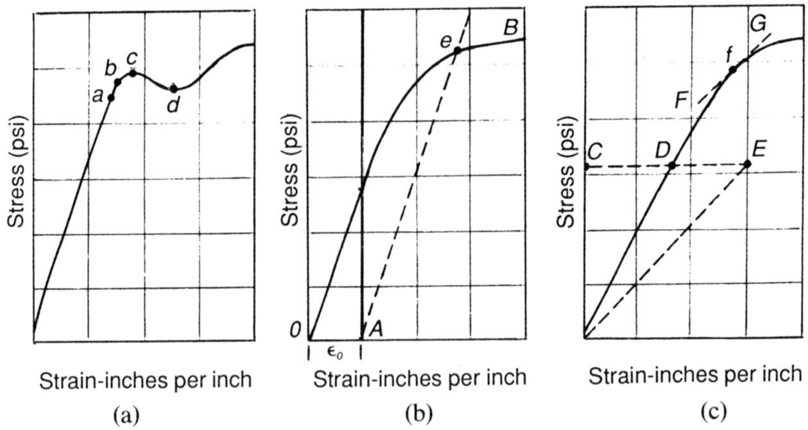

Figure B-4. Material characteristics and properties determined by tensile tests.

mum and minimum (Fig. B-4(a), points *c* and *d*, respectively). If this same steel is heat-treated to increase hardness or is work-hardened, these natural yield points are not observed in the σ–ε diagram, which is similar to the diagram in Figure B-4(b). Also, if the tensile test machine is controlled for load rather than strain, these points are not observed.

- *Offset yield strength:* For materials that do not have a well-defined yield point (see Fig. B-4(b) and (c)), the offset method is usually used to define a yield strength. An offset strain (ε_0) on Figure B-4(b), which is usually 0.2 percent, is measured from the origin of the σ–ε diagram. The point at which the σ–ε curve is intersected by a line through ε_0 and parallel to the elastic portion of the curve is the 0.2 percent offset yield strength (point *e*).
- *Tangent modulus* (E_t): The slope of the σ–ε curve at any point (Fig. B-5, point *a*).
- *Secant modulus at a point:* The ratio σ/ε at a point E_s is in Figure B-5. In the plastic portion of the curve, the secant modulus is the slope of a line connecting point *a* with the origin; hence at point *a*

$$E_s = \frac{\sigma_a}{\varepsilon_a}.$$

- *Young's modulus:* The slope of the σ–ε curve in the elastic (or linear) portion of the curve in Figure B-5. Young's modulus (E) is

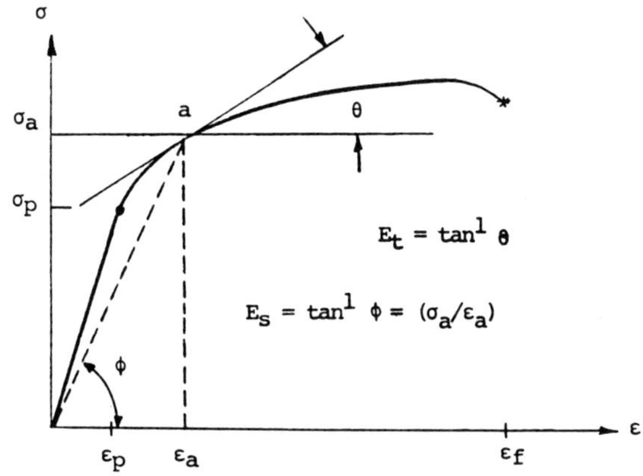

Figure B-5. Curve showing tangent modulus, point *a*.

also called the elastic modulus. In the elastic region of the σ–ε curve, the three moduli are equal:

$$E_s = E_t = E \qquad \left(\text{if } \sigma \text{ is less than } \sigma_p\right). \tag{6}$$

- The elastic modulus also is used as a measure of the stiffness of a material. Examples of some materials are listed in Table B-1.
- *Resilience:* The measure of the ability of a material to absorb energy elastically (Fig. B-5):

$$U_R = \frac{1}{2}\frac{\sigma_p^2}{E} \tag{7}$$

TABLE B-1. Young's (or Elastic) Modulus (*E*) for Typical or Common Materials

Material	*E (psi)*
Steels	$28–30 \times 10^6$
Aluminum alloys	$10–10.5 \times 10^6$
Magnesium alloys	6.5×10^6
Cast iron	$14–16 \times 10^6$
Wood	$1–2 \times 10^3$
Concrete, rock	$5–10 \times 10^6$
Titanium alloys	16.5×10^6

- *Toughness:* The measure of the ability of a material to absorb energy plastically. Thus, the total area under a σ–ε curve minus the resilience is toughness:

$$U_T = \int_0^{\varepsilon_f} \sigma \, d\varepsilon - U_R \tag{8}$$

- *Poisson's ratio:* The ratio of the strains that occur in the axial and transverse directions when a tensile specimen is loaded in the axial direction. Note in Figure B-6(a) that the specimen elongates axi-

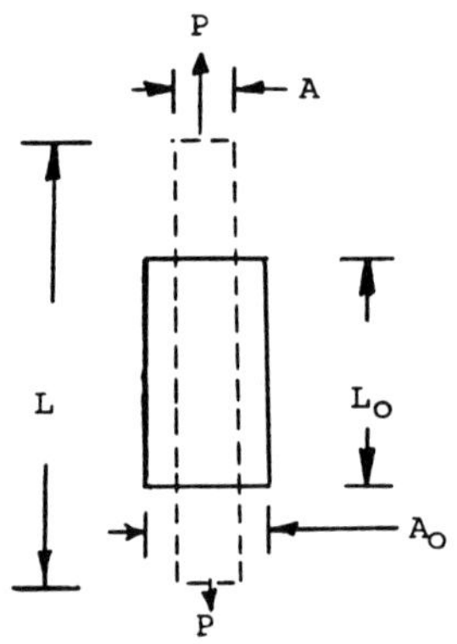

(a)

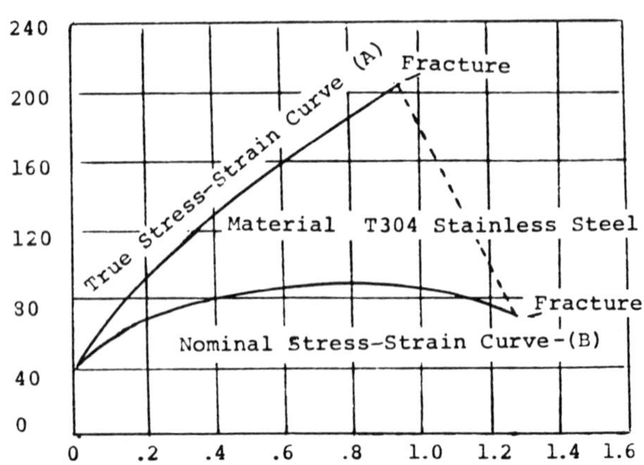

(b)

Figure B-6. (a) Tensile specimen of true stress–strain measurements. (b) Comparison of nominal versus true strain (inches per inch).

ally and contracts in the transverse direction. The Poisson's ratio is

$$\nu = \frac{\varepsilon_t}{\varepsilon_a} \tag{9}$$

where ε_t = transverse strain, and ε_a = axial strain. In the elastic region, $0.25 < \nu < 0.35$ for most metals; in the plastic region, $\nu = 0.5$. For this value, the volume of metal, and therefore the density, is unaffected by stress. In an isotropic material, only two elastic constants (e.g., E and ν) are needed to define elastic behavior. More constants are needed for nonisotropic and inelastic material.

True Stress–True Strain. True stress in a tensile specimen is defined as

$$\bar{\sigma} = \frac{P}{A} \tag{10}$$

where $\bar{\sigma}$ = true stress, P = applied load, and A = cross section area for load P. As a basis of comparison, the nominal stress (σ) is defined as

$$\sigma = \frac{P}{A_0} \tag{11}$$

where A_0 = original stress area.

True strain (ε) in a tensile specimen is defined as

$$\bar{\varepsilon} = \int_{L_0}^{L} \frac{dL}{L} = ln\left(\frac{L}{L_0}\right) \tag{12}$$

where ε = true strain, L_0 = original gauge length, and L = final gauge length.

As a basis of comparison, the nominal or engineering strain is defined as

$$\varepsilon = \left(\frac{L - L_0}{L_0}\right) \tag{13}$$

A true-stress–true-strain curve is compared to a nominal σ–ε curve for material T304 in Figure B-6(b).

True Fracture Strain. True strain at fracture is related to the reduction in area (Dieter 1961). It is an experimental fact that the density of metal remains almost constant during plastic flow. Thus, in the necked area of a tensile specimen

$$LA = L_0 A_0 \tag{14}$$

where L_0 = original length, A_0 = original cross section, L = length at load, and A = cross section at load. Thus, $L_f A_f = L_0 A_0$ at fracture and reduced area (R_A) is written as

$$R_A = \frac{A_0 - A_f}{A_0} = \frac{L_f - L_0}{L_f} \tag{15}$$

However,

$$\bar{\varepsilon}_f = ln\left(\frac{L_f}{L_0}\right) = ln\left(\frac{A_0}{A_f}\right) = ln\left[\frac{1}{(1 - R_A)}\right] \tag{16}$$

For this reason, the reduction in area is an important measure of ductility.

True Stress–Strain Curves. For many metals, true stress–strain curves are approximately linear on log-log paper. As a result, these curves can be approximated by the equation $\bar{\sigma} = \bar{\varepsilon}^n K$, where K = strength coefficient, and n = strain-hardening coefficient. Constants for common metals are listed in Table B-2.

TABLE B-2. True Stress–Strain Material Constants *n* and *K* for Different Sheet Materials[a]

Material	*Treatment*	*n*	*K* (*psi*)	*Thickness* (*in.*)
Steel				
0.05 percent carbon rimmed	Annealed	0.261	77,000	0.037
0.05 percent carbon killed	Annealed and temper rolled	0.234	73,100	0.037
0.05 percent carbon killed, completely decarbonized	Annealed in wet hydrogen	0.284	75,500	0.037
0.05/07 percent phosphorus low-carbon steel	Annealed	0.156	93,330	0.037
SAE 4130	Annealed	0.118	169,400	0.037
SAE 4130	Normalized and temper rolled	0.156	154,500	0.037
Type 430 stainless (17% Cr)	Annealed	0.229	143,000	0.050
Aluminum				
Alcoa 24-S	Annealed	0.211	55,900	0.040
Reynolds R-301	Annealed	0.211	48,450	0.040

[a]Tests were conducted by J. R. Low and F. Garafalo.

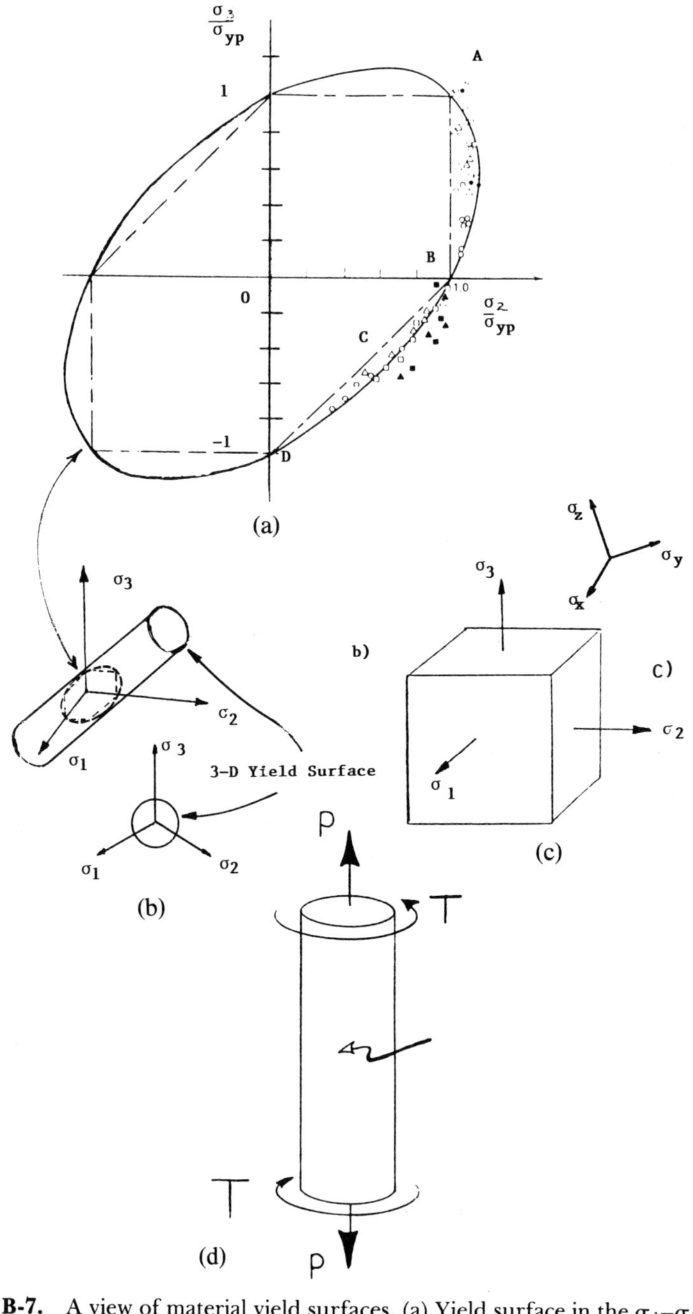

Figure B-7. A view of material yield surfaces. (a) Yield surface in the σ_1–σ_2 plane. Comparison of biaxial yield strength data with theories of failure for a variety of ductile materials. *Solid line:* Von Mises distortion energy theory; *dotted line:* Tresca maximum shear theory. (□ aluminum, △ copper, ● nickel, ○ steel, ▲ mild steel, ■ carburized steel). (b) Yield surface in σ_1–σ_2–σ_3 space. (c) Principal stresses. (d) Biaxial load specimen (tension (P), torque (T), and internal pressure (p)).

Yield Surface (Multidirectional Loading). Figure B-4 illustrates the yield point of a material when irreversible strains, permanent deformation, or permanent set occur in a *uniaxial tensile test.* Figure B-7(a)–(c) illustrates a yield surface defined by *multidirectional* stress states that can be determined in a tensile testing machine with a thin cylindrical specimen that also can be internally pressurized and axially torqued (see Fig. B-7(d)). In other words, yielding of a material does not depend on a single value of stress, but rather on the stress state. Note in Figure B-7(a) and (b) that point A corresponds to equal stress in the principal directions (axial and circumferential), point B corresponds to uniaxial load only (P), point C corresponds to torque (T) or shear stress only, and point D corresponds to compressive stress and load ($-P$).

Two theories are commonly used to describe the multistates of stress and material yielding in mathematical terms: the von Mises theory (or distortion energy theory) and the Tresca theory (the maximum shear theory). The von Mises theory is usually preferred in theoretical and computational mechanics with the Prandl–Reuss equations because of its ability to describe the phenomena of material hardening and loading–unloading.

Baushinger's Effect

For an isotropic homogeneous material, the tensile yield (σ_{yt}) and the compressive yield (σ_{yc}) are equal. If the material is strained plastically to point A on Figure B-8 and unloaded, the material will unload elastically to point B. On reloading in tension, the yield occurs at A. Strain hardening increases the yield in tension. If, however, the specimen is strained to point C and then loaded in compression, yielding occurs at D. The compressive yield D is less than the original compressive yield. This reduction is called the *Baushinger effect* and is described in the context of kinematic hardening. (See Lam *et al.* 1983 for a more detailed discussion of this phenomenon.)

DYNAMIC PROPERTIES FRACTURE AND CREEP

Effect of Rate of Strain

As the rate of loading increases, the mechanical properties of a test specimen are affected, and both the yield strength and the ultimate strength increase. Results of tests conducted by Manjoine are shown in Figure B-9.

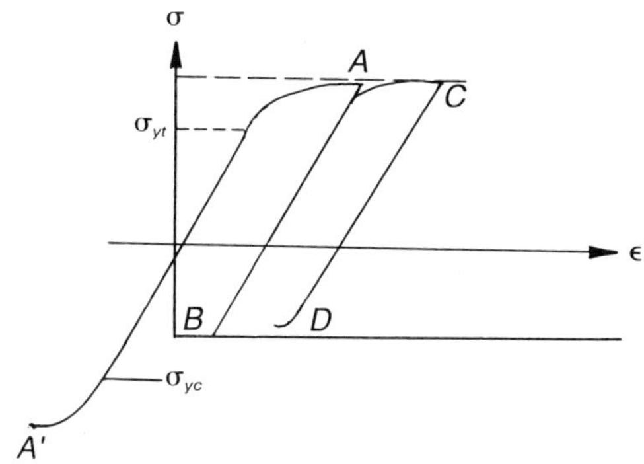

Figure B-8. Curve showing Baushinger's effect.

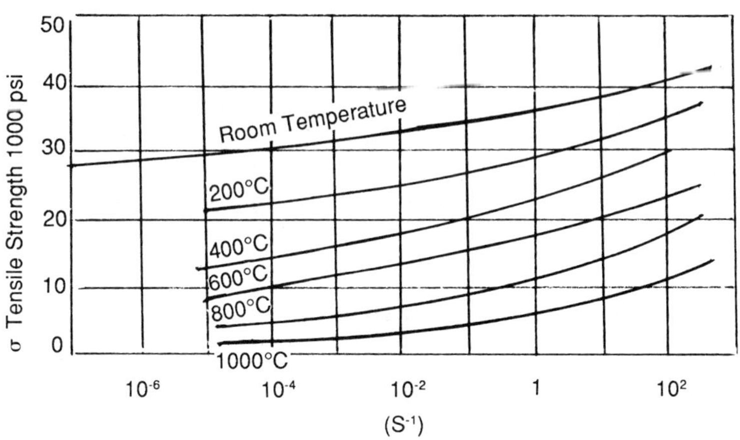

(a)

Figure B-9. The effect of strain rate (the rate of elongation) on materials. (a) Effect of strain rate on the tensile strength of copper for various specimen temperatures. (b) Influence of average strain rate on mechanical properties.

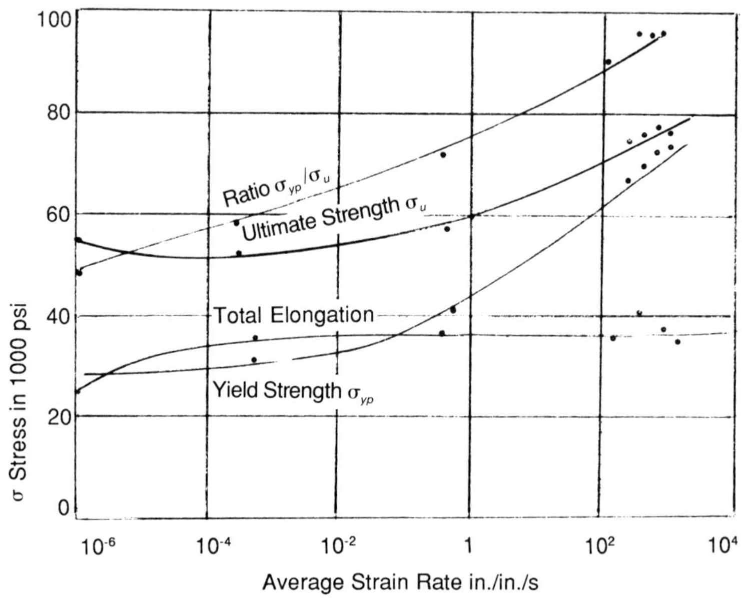

(b)

Figure B-9. *(continued)*

Impact Testing

Many tests have been developed to investigate brittle fracture of "ductile" materials caused by impact loading. The most common tests are the Charpy and Izod tests. The test machine, Charpy specimens, and Izod specimens are shown in Figure B-10. Charpy results are shown in Figure B-11, which also plots the energy required to fracture the specimen.

Critical Crack Size and Crack Growth

Fracture is predicted by the maximum normal stress theory if any one expression of Figure B-12 is satisfied:

$$\begin{array}{ll} \sigma_1 \geq \sigma_t & \sigma_1 \leq \sigma_c \\ \sigma_2 \geq \sigma_t & \sigma_2 \leq \sigma_c \\ \sigma_3 \geq \sigma_t & \sigma_3 \leq \sigma_c \end{array}$$

(a)

(b)

(c)

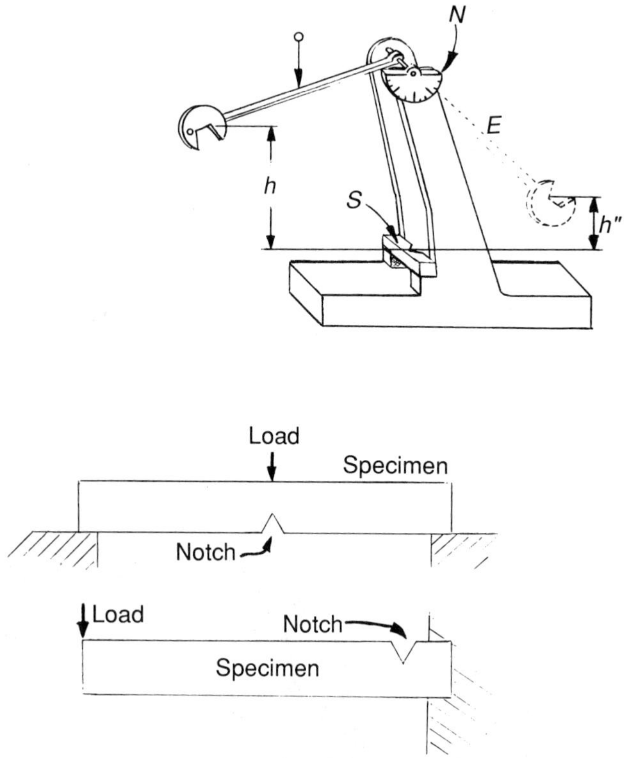

Figure B-10. "Notched bar" impact test. (a) Impact testing machine. (b) Charpy specimen loading. (c) Izod specimen loading.

where σ_1, σ_2, and σ_3 are the principal normal stresses; σ_t is the uniaxial failure strength in tension; and σ_c is the uniaxial failure strength in compression. Comparison of this failure theory with experimental data on brittle fracture for different biaxial states of stress is shown in Figure B-12.

Experience shows that ductile materials also fail by a brittle-fracture response in the presence of cracks or flaws if the combination of crack size, geometry of the part, temperature, loading rates, or all of these lie within certain critical regions. The combined use of higher strength structural alloys, welding, and thicker sections has reduced the capacity

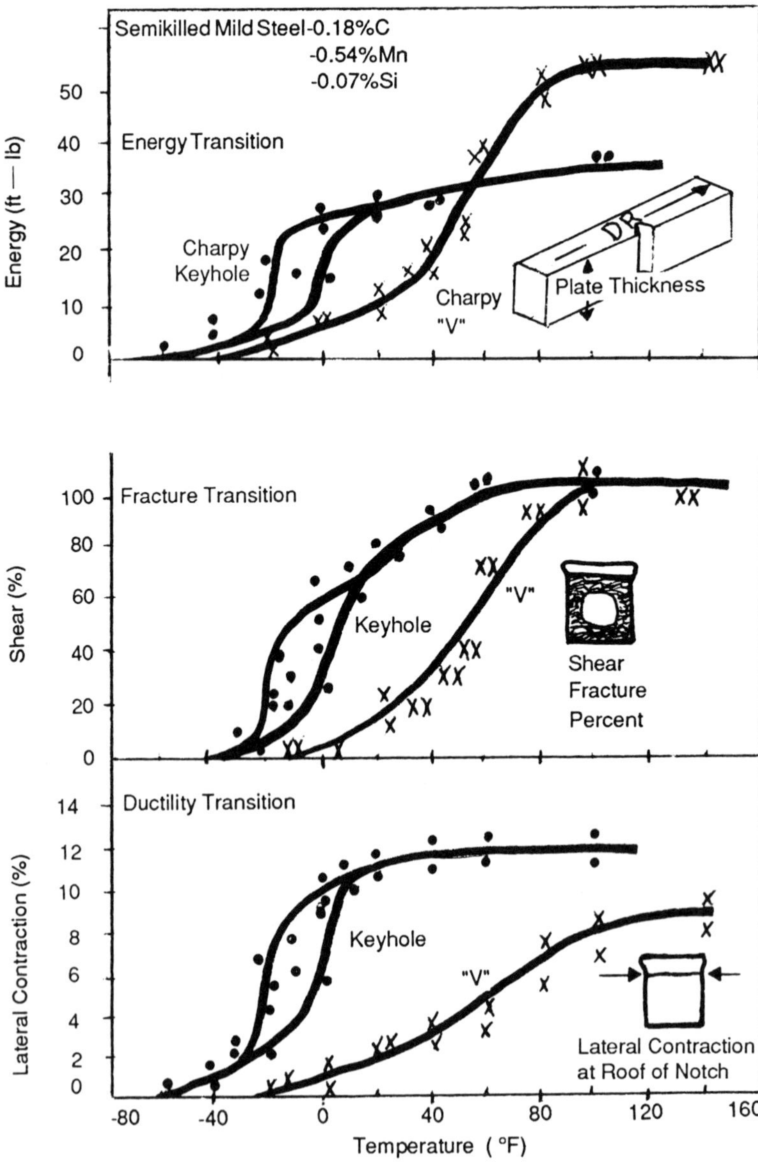

Figure B-11. Transition-temperature curves versus on energy absorbed, percent shear fracture, and percent lateral contraction (for Charpy "V" and "Keyhole" specimen).

(a)

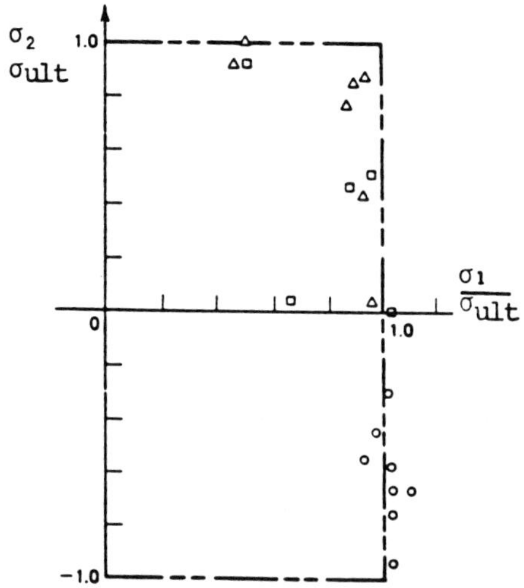

(b)

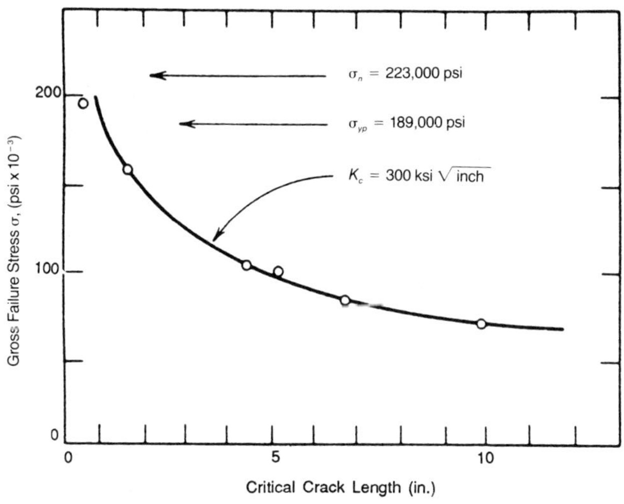

Figure B-12. The effect of cracks and stress on brittle fracture. (a) Comparison of data on biaxial brittle-fracture strength with normal stress theory for several brittle materials (□ = brass, △ = cast iron, ○ = cast iron). The broken line = maximum normal stress theory. (b) Influence of crack length on gross failure stress for center-cracked 4330 M steel plate 36 in. wide and 0.14 in. thick at room temperature. (Copyright: American Society of Testing and Materials. Adapted with permission.) (c) Influence of crack length on gross failure stress for center-cracked aluminum plate 24 in. wide and 0.1 in. thick (2219-T87 aluminum alloy) (longitudinal direction). (After Dieter (1961). Copyright: American Society of Testing and Materials.) (d) Coordinates measured from the leading edge of a crack.

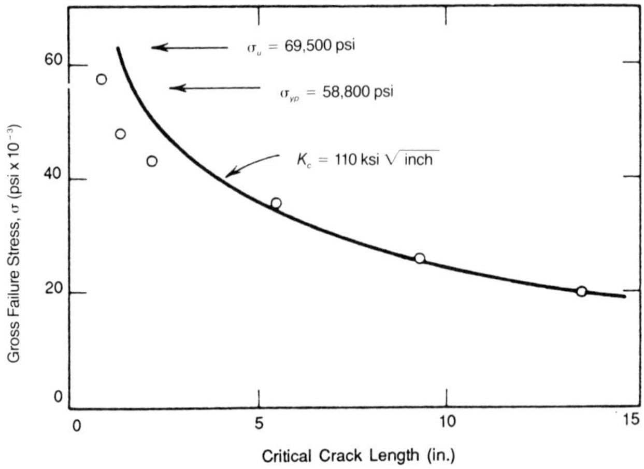

(c)

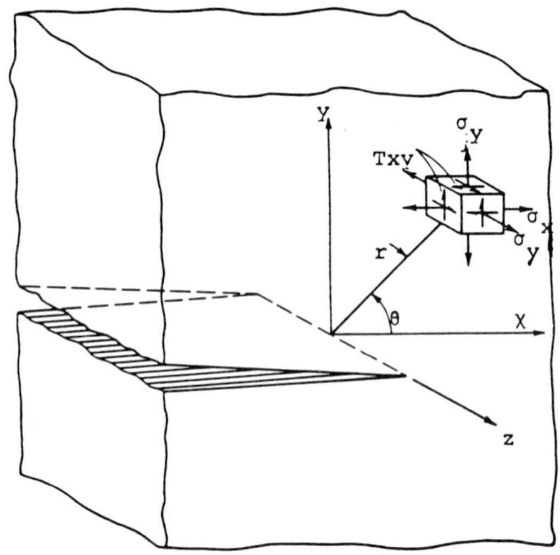

(d)

of some structural members to accommodate local plastic strain without fracturing. In addition, residual stress (due to fabrication by welding, machining, and assembly mismatching in production), fluctuating service loads, and aggressive environments also contribute to unexpected fractures and have increased the need for accommodating local plastic strain to prevent failure. The basic concepts of fracture control were developed from the study of these factors. Fracture control consists of controlling the nominal stress and crack size so that the combination always lies below a critical level for the material being used in a given design application.

An important characteristic of fracture behavior is that the magnitude of the nominal applied stress that causes fracture is related to the size of the crack or cracklike flaw within the structure. For example, observations of the behavior of central through-the-thickness cracks in steel and aluminum plates, oriented normal to the applied tensile stress (for illustration, see Fig. 8-15(a)) yielded the results shown in Figure B-12(b) and (c). As the tensile loading on the precracked plates was slowly increased in these tests, the extension of the crack also slowly increased initially until suddenly the crack extended to the point of failure by rapid propagation. Slow, stable growth of cracks is characterized by speeds on the order of fractions of inches per minute, whereas sudden propagation of cracks is characterized by speeds on the order of hundreds of feet per second.

The sudden change from slow to rapid crack propagation, termed *fracture toughness*, is an important property of material. Fracture toughness is used as a design criterion for preventing fracture, just as the yield strength is used as a design criterion for preventing excessive deformation of or irreversible strain on a ductile material under static loading.

Three basic types of stress fields can be defined for crack-tip stress analysis; each mode is associated with a distinct mode of crack deformation (see Fig. 8-15, p. 263). The crack opening mode (Mode I) is associated with local displacement in which the surfaces of the crack move apart in a perpendicular manner (see Fig. 8-15(a)). The edge-sliding or edge-perpendicular-sliding mode (Mode II) is developed when crack surfaces slide over each other in a direction that is perpendicular to the leading edge of the crack (see Fig. 8-15(b)). The tearing or edge-parallel shear mode (Mode III) is characterized by crack surfaces that slide with respect to each other in a direction parallel to the leading edge of the crack (see Fig. 8-15(c)). Superposition of these three modes describes the general three-dimensional local crack-tip deformation and stress field.

For a given cracked plate (see, for example, Fig. B-12(d)), the factor K increases proportionally with gross nominal stress and also is a function of the instantaneous crack length. Thus, K is a single-parameter measure of the stress field around the crack tip. The value of K (associated with the onset of rapid extension of the crack) has been designated as the critical stress intensity, K_c. As noted earlier, the onset of rapid crack propagation for specimens with cracks of different initial lengths occurs at different values of cross-section stress but at a constant value of K_c. Thus, K_c provides a single-parameter fracture criterion that allows the prediction of fracture, that is, fracture is predicted to occur if

$$C\sigma\sqrt{\pi a} \geq K_c \tag{17}$$

A simple adjustment factor is often applied to account for small-scale plasticity at the tip of the crack. The simplest approach is to estimate the extent of yielding by treating the problem as one of plane stress.

When studying the behavior of a given material, one finds that, depending on the state of stress at the crack tip, the critical intensity K_c decreases to a lower limiting value as the state of strain approaches the condition of plane strain. The lower limiting value defines a basic material property K_{Ic}: the plane-strain fracture toughness of the material. Standard testing methods have been established to determine K_{Ic} values. A few data are shown in Table B-3.

For the plane-strain fracture toughness K_{Ic} to be a valid criterion for predicting failure of a specimen of a machine part, plane-strain conditions must exist at the crack tip; that is, the material must be thick enough to ensure plane-strain conditions.

TABLE B-3. Data on the Yield Strength and Plane-Strain Fracture Toughness of Selected Alloys[a]

		Test Temperature		σ_{yp}		K_{Ic}	
Alloy	*Form*	*°F*	*°C*	*ksi*	*MPa*	*ksi* $\sqrt{in.}$	*MPa* $\sqrt{m}$
Steel							
4340 (500°F temper)	Plate	70	21	217–238	1495–1640	45–57	50–63
4340 (800°F temper)	Forged	70	21	197–211	1360–1455	72–83	79–91
D6AC (1000°F temper)	Plate	70	21	217	1495	93	102
D6AC (1000°F temper)	Plate	−65	−54	228	1570	56	62
A 538	—	—	—	250	1722	100	111
Aluminum							
2014-T6	Forged	75	24	64	440	28	31
2024-T351	Plate	80	27	54–56	370–385	28–40	31–44
7075-T6	—	—	—	85	585	30	33
7075-T651	Plate	70	21	75–81	515–560	25–28	27–31
7075-T7351	Plate	70	21	58–66	400–455	28–32	31–35
Titanium							
Ti-6A1-4V	Plate	74	23	119	820	96	106

[a]From R. W. Hertzberg, *Deformation and Fracture Mechanics of Engineering Materials,* Wiley, New York, 1976; also, J. E. Campbell *et al., Damage Tolerant Designs Handbook,* Batelle, Columbus, Ohio, 1973.

For plane-strain conditions, it has been estimated empirically that the minimum material thickness B must be

$$B \geq 2.5 \left(\frac{K_{Ic}}{\sigma_{yp}}\right)^2 \tag{18}$$

If the material thickness does not meet these criteria, plane stress is a likely state of stress at the crack tip, and K_c (the critical stress-intensity factor for predicting failure under plane-stress conditions) can be as much as 10 times higher than the lower limiting value of K_{Ic}.

We must emphasize that fracture toughness is a function not only of metallurgical factors, such as the composition of an alloy and heat treatment, but also of service temperature, loading rate, and state of stress in the vicinity of the crack tip. In many practical applications, the size of the plastic zone ahead of the crack tip becomes so large that the assumption of small-scale yielding is no longer valid and the linear elastic fracture mechanics (LEFM) analyses are no longer appropriate. A large percentage of low- and medium-strength steels used in the fabrication of pressure vessels, ships, bridges, and other structures and machines are used in section sizes that are usually too thin to maintain plane-strain conditions at the crack tip. Using LEFM analysis under such circumstances and using K_{Ic} for the failure criterion are not valid in these cases. Efforts to extend the techniques of fracture mechanics into this elastic–plastic regime have led to procedures such as crack-opening displacement methods, R-curve methods, and J-integral methods. The basic concept of the crack-opening displacement methods is that the fracture behavior in the region of a sharp crack can be described as the opening displacement at the face of the crack.

Effect of Temperature on Tensile Properties (Independent of Time)

As temperature rises, both yield strength and tensile strength decrease and ductility increases. Figure B-13 shows the results of tests on mild steel. For safe operation, temperatures above the transition temperature should be used. The transition temperature is affected by heat treatment, the composition of the alloy, and irradiation. The effects of irradiation are shown in Figure B-14.

Having a structure undergo a loading program above the transition temperature, rather than below it, is desirable because it permits the structure to behave in a ductile manner. In the brittle regime, the danger of a rapid, unstable fracture is present. The study and prevention of

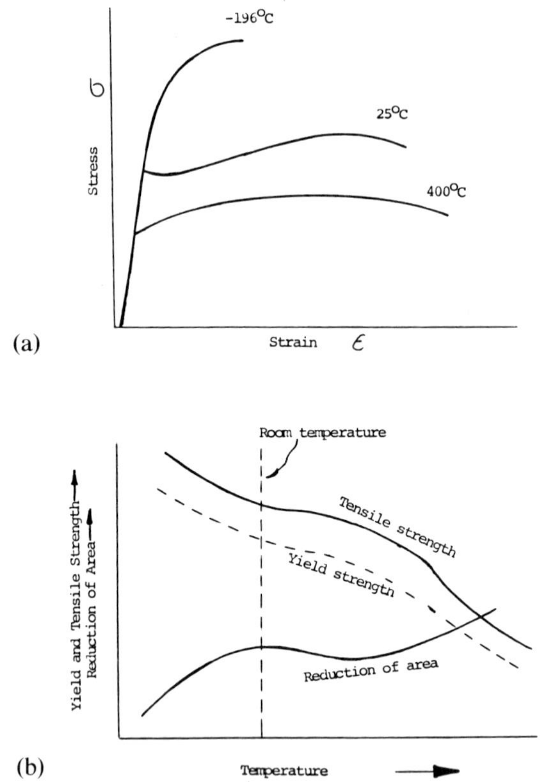

Figure B-13. The effects of temperature on materials. (a) Changes in engineering stress–strain curves of mild steel with temperature. (b) Variation of tensile properties of steel with temperature.

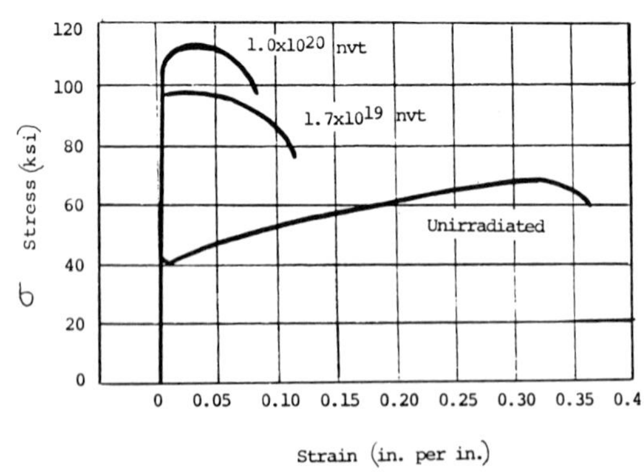

Figure B-14. Stress–strain curves of irradiated carbon steel (ASTM A-212B). Irradiation temperature = 200°F.

this mechanism of material failure are discussed in the section on fracture analysis.

Viscous Effects (Time- and Temperature-Dependent Deformation)

In contrast to solids that exhibit an elastic region in the stress–strain curve, some materials exhibit inelastic deformation from the time that stress is first applied. In many cases, one finds that inelastic deformation is a time-dependent function of stress and temperature. This behavior is usually referred to as the "creep" phenomenon.

To determine the engineering *creep curve* of a metal, a constant load is applied to a tensile specimen that is maintained at a fixed temperature, and the uniaxial strain is calculated as a function of time. The elapse time of such tests can range from several months to more than 10 years. The general procedure for testing creep is described in American Society for Testing and Materials Specification E139-66T.

Figure B-15 illustrates the four stages of behavior that may occur in a constant-load, constant-temperature creep test (D'isa 1968; Lam *et al.* 1983). The first stage (O–A) consists of an initial strain that can be entirely elastic or plastic, or partially elastic and partially plastic. During the second stage (A–B), the rate of creep decreases. During this process, the rate of strain hardening is greater than recovery. The third stage (B–C) is characterized by a linear curve (ε_T versus t), where strain hardening is in equilibrium with recovery. During the fourth stage, tertiary creep (C–D), the creep rate increases rapidly because the cross-sectional area of the specimen is reduced, and therefore the stress and structural

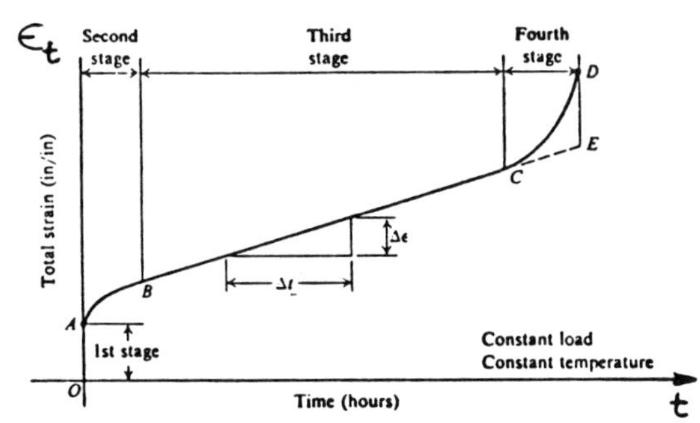

(a)

Figure B-15. Strain (ε) versus time-creep tests. (a) Curve illustrating the four stages of behavior that may occur in a constant-load, constant-temperature creep test. (b) Schematic of test specimen and load in a creep test.

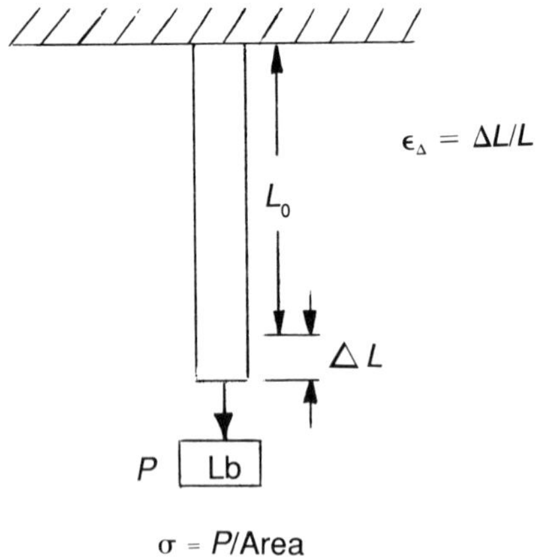

(b)

Figure B-15. (*continued*)

changes possibly occur in the metal (crack and void formation). As the temperature rises, the creep rate increases; for increasing values of stress, a similar family of curves is obtained (see Fig. B-16).

The *stress-rupture* test is similar to the creep test. The difference is that the stress-rupture test can be carried out at significantly higher levels of stress, which results in strains that are several orders higher and a time of test that is several orders lower than the creep test. The basic information obtained from this test is the time required to cause failure at a given nominal stress for a constant temperature. In the creep test,

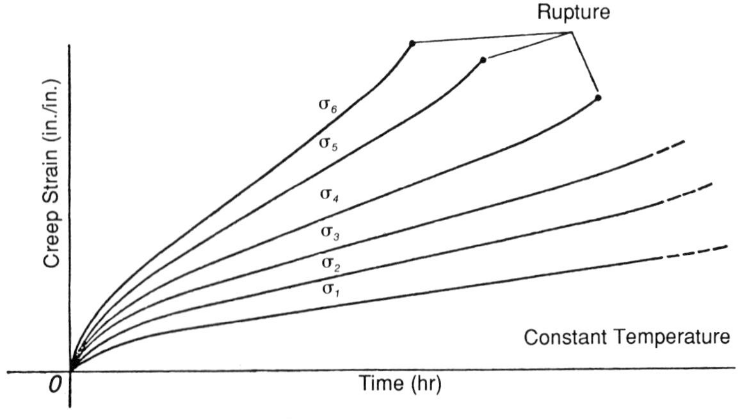

Figure B-16. Comparison of increasing values of creep strain versus time for constant temperature various load curves (see Fig. B-15(b) for test setup).

on the other hand, the recording of the *strain* (ε_T) and *rate of strain* ($\dot{\varepsilon}_T$) is the principal information desired.

So far, we have observed the behavior of materials subjected to an applied stress. Let us consider the *creep-stress relaxation* under an applied fixed strain. The material is evaluated by maintaining the length of the test specimen while measuring the load versus time (see Fig. B-17(a)).

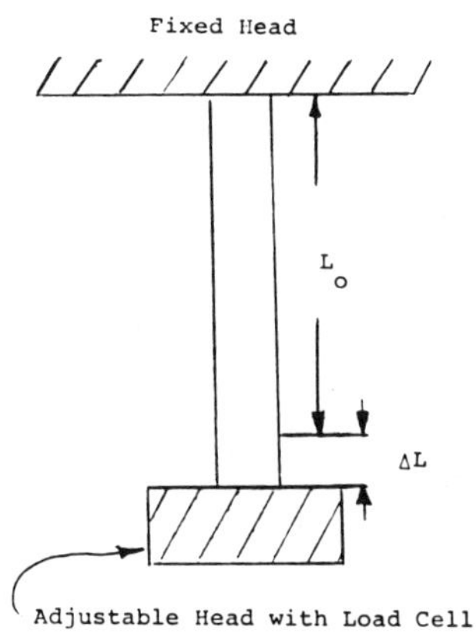

(a)

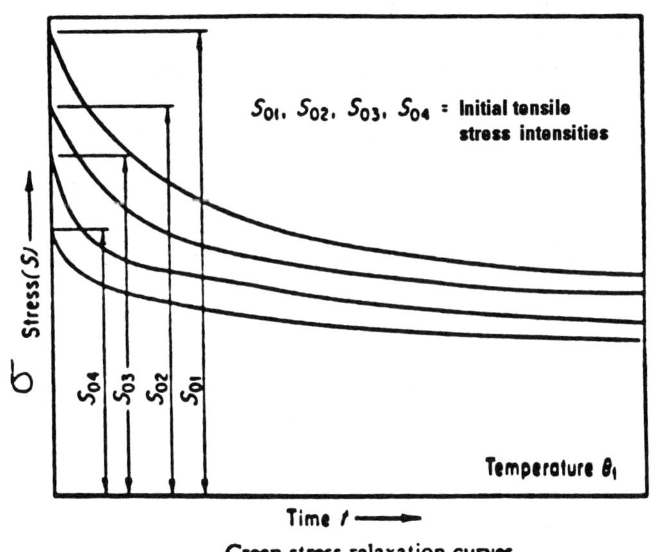

(b)

Figure B-17. Stress (σ) versus time-creep relaxation test. (a) Creep-stress relaxation test specimen and input displacement. (b) Creep-stress relaxation versus time curves.

Figure B-17(b) shows the creep-relaxation curves for the conditions of constant deformation and constant temperature.[1]

The application or occurrence of the time-temperature behavior of materials is important in many industrial operations: long-time–low-strain creep behavior is observed in low-pressure–high-temperature breeder reactors; creep rupture is a matter of concern, particularly in high-pressure tube boilers; and creep relaxation that occurs in bolted closure systems loaded to a fixed strain is of interest.

DYNAMIC/CYCLIC LOADING

Fatigue and Crack Growth

Conventional procedures (Shigley 1977) that have historically been used to design structural components subjected to fluctuating loads (see Fig. B-18(a)–(c)) provide the engineer with (1) a design fatigue curve, the characteristics of the basic unnotched fatigue properties (stress versus number of cycles to failure) of the material (Fig. B-19), and (2) fatigue-strength-reduction factors (Fig. B-20). The fatigue-strength-reduction factor takes into consideration the various effects on the material and its finish (the different parameters characteristic of the specific structural component) that make it more susceptible to fatigue failure than the unnotched specimen (see Fig. B-21). The design fatigue curves are based on the prediction of cyclic life from data on nominal stress (or strain) versus elapsed cycles to failure (S-N curves), as determined from specimens tested in the laboratory. Such data represent both the number of cycles required to *initiate* a crack in the specimen and the number of cycles required to *propagate* the crack from a subcritical to a critical dimension, which often varies from laboratory to laboratory. The dimension of the critical crack required to cause "failure" in the fatigue specimen also depends on the magnitude of the applied stress and on the size of the test specimen.

Figure B-22 illustrates that the endurance limit $S_{e'}$ or $\sigma_{e'}$ is approximately half the ultimate stress (σ_{ut} or S_{ut}). Figure B-23 illustrates a compilation of rotating beam fatigue tests demonstrating that the basis of the construction of a typical S-N curve is the minimum strength curve defined by a fatigue alternating stress, $\sigma_{f'}$ (or stress intensity $S_{f'}$) equal to

[1]For a discussion of the mathematical modeling of these behaviors and cumulative creep under different stresses and strains, refer to Chapters 8 and 10, D'Isa (1968), and Lam *et al.* (1983). For a discussion of structural performance criteria with respect to creep, refer to ASME Case Code N-47 (1986), Dhalla and Gallagher (1982), Manson (1982), Nickell and Yahn (1982), Pugh and Robinson (1982), Snow and Jakub (1982), Swindeman and Brinkman (1982), Wei and Nelson (1982).

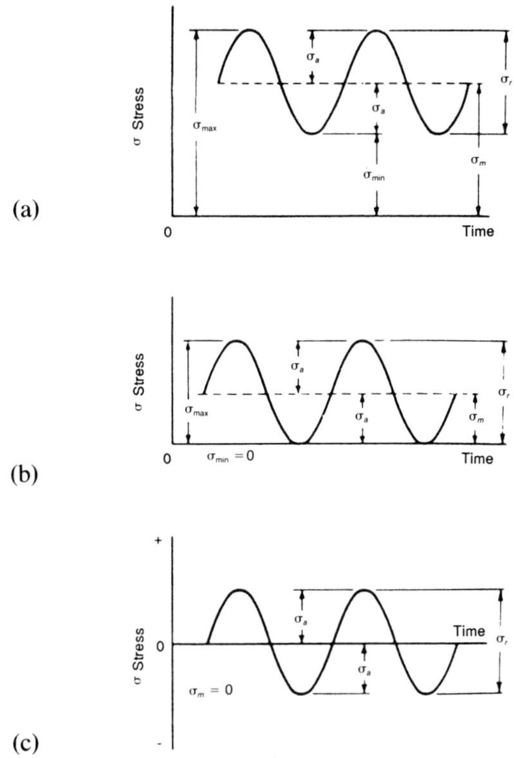

Figure B-18. Types of fatigue stress. (a) Fluctuating stress. (b) Repeated stress. (c) Reversed stress (tension–compression).

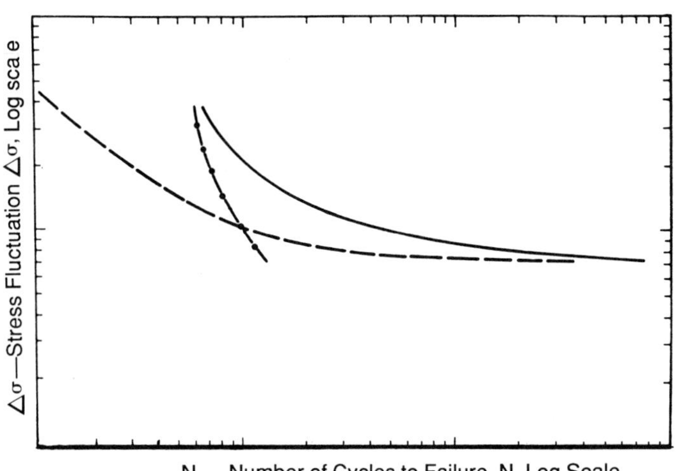

Figure B-19. Schematic S-N curve divided into "initiation" and "propagation" components. —— = initiation life; —·—·— = propagation life; ——— = total life.

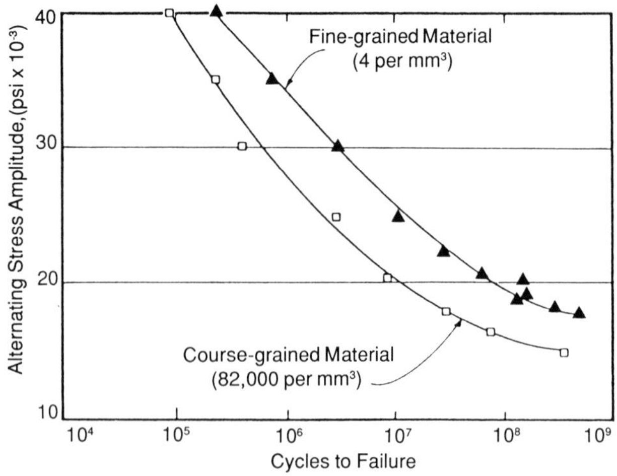

(a)

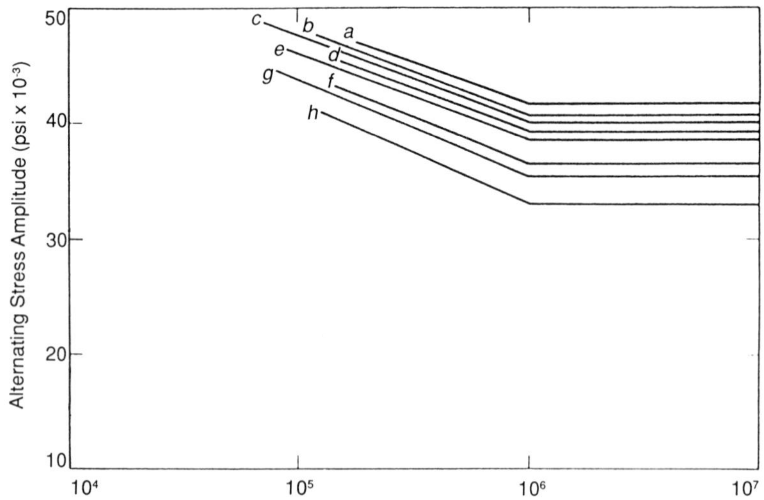

(b)

Figure B-20. Factors that affect fatigue strength reduction. (a) Effect of grain size on the S-N curve for 18S aluminum alloy. Average diameter ratio of coarse to fine grains is approximately 27 to 1. Nominal composition: 4 percent copper, 2 percent nickel, 0.6 percent magnesium. Note that at a life of 10^8 cycles, the mean fatigue strength of the coarse-grained material is about 3000 psi lower than for fine-grained material. (Data from Murray (1952); adapted from *Fatigue and Fracture of Metals,* by W. M. Murray. By permission of the MIT Press, Cambridge, Massachusetts. Copyright 1952) (b) Effect of surface finish on the S-N curve of 0.33 percent carbon-steel specimens, testing in a rotating cantilever beam machine: line a = high polish, longitudinal; line b = FF emery finish; line c = No. 1 emery finish; line d = coarse emery finish; line e = smooth file; line f = as-turned; line g = bastard file; and line h = coarse file. (Data from Thomas (1923))

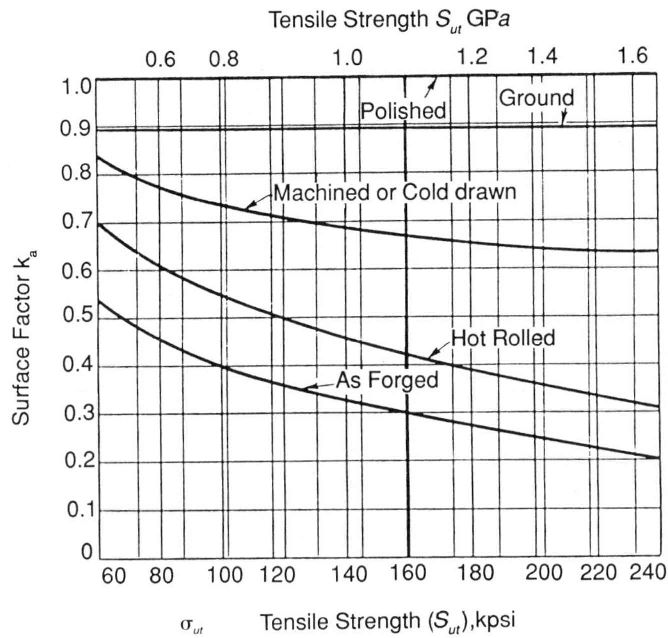

(a)

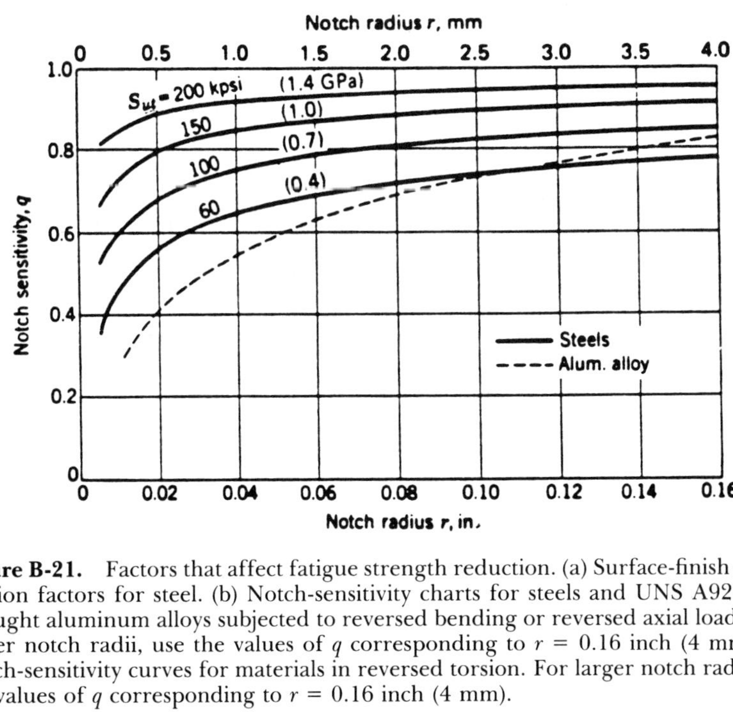

(b)

Figure B-21. Factors that affect fatigue strength reduction. (a) Surface-finish modification factors for steel. (b) Notch-sensitivity charts for steels and UNS A92024-T wrought aluminum alloys subjected to reversed bending or reversed axial loads. For larger notch radii, use the values of q corresponding to r = 0.16 inch (4 mm). (c) Notch-sensitivity curves for materials in reversed torsion. For larger notch radii, use the values of q corresponding to r = 0.16 inch (4 mm).

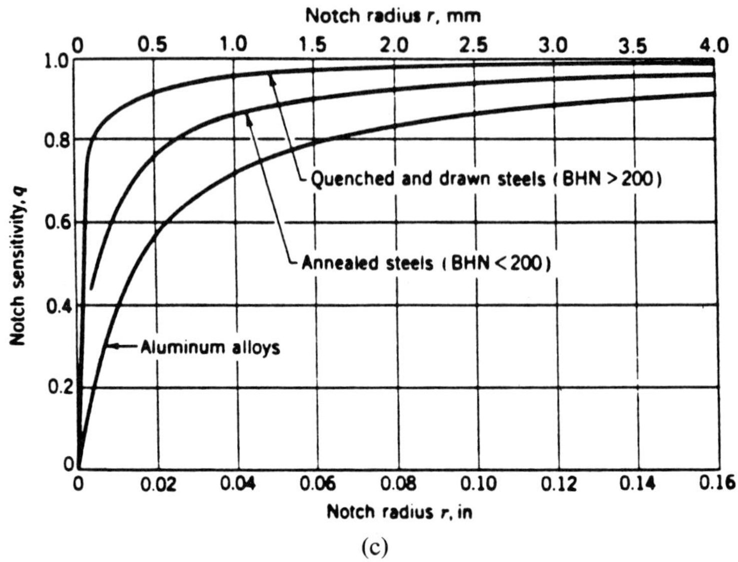

Figure B-21. *(continued)*

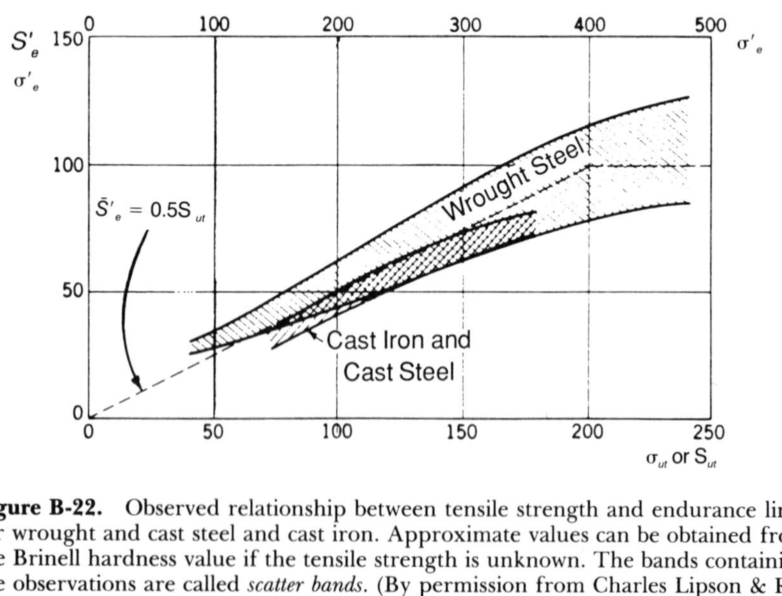

Figure B-22. Observed relationship between tensile strength and endurance limit for wrought and cast steel and cast iron. Approximate values can be obtained from the Brinell hardness value if the tensile strength is unknown. The bands containing the observations are called *scatter bands.* (By permission from Charles Lipson & Robert C. Juvinall [Eds.]. *Application of Stress Analysis to Design and Metallurgy.* The University of Michigan Summer Conference, Ann Arbor, Michigan, 1961)

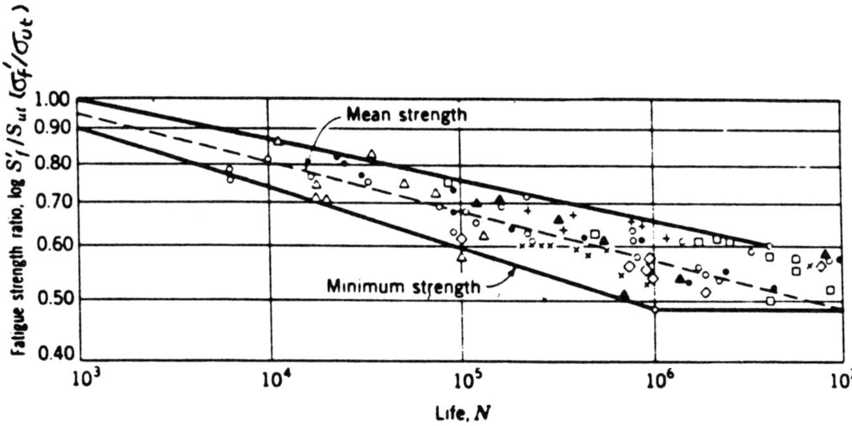

Figure B-23. Use of Miner's rule to predict the endurance limit of a material that has been overstressed for a finite number of cycles. (Shigley 1977)

$0.9\ \sigma_{ult}$ at $N = 10^3$ cycles and $0.5\ \sigma_{ult}$ at $N = 10^6$ cycles. Table B-4 compares the tensile strengths and endurance limits for a sampling of materials.

How does one calculate the life remaining in a structure subjected to stress σ_1 at N_1 cycles? How does one account for the fact that in real life, a product might be subjected to different cyclic stress levels (σ_1, σ_2, and

TABLE B-4. Standard Deviations of Endurance Limit[a]

Material (UNS No.)[b]	*Tensile Strength*		*Endurance Limit*		*Standard Deviation*	
	MPa	*kpsi*	*MPa*	*kpsi*	*kpsi*	*Percentage*
G43400 steel	965	140	489	71	3.5	4.9
	1310	190	586	85	6.7	7.8
	1580	230	620	90	5.3	5.9
	1790	260	668	97	6.3	6.5
G43500 steel	2070	300	689	100	4.4	4.4
R50001-series titanium alloy	1000	145	579	84	5.4	6.4
A97076 aluminum alloy	524	76	186	27	1.6	6.0
C63000 aluminum bronze	806	117	331	48	4.5	9.4
C17200 beryllium alloy	1210	175	248	36	2.7	7.5

[a]Reported by F. B. Stulen, H. N. Cummings, and W. C. Schulte. (June 22, 1961. "Preventing Fatigue Failures," Part 5. *Machine Design,* Vol. 33, p. 166.)

[b]Alloys were heat-treated and hot-worked; specimens were smooth and subjected to long-life, rotating-beam tests.

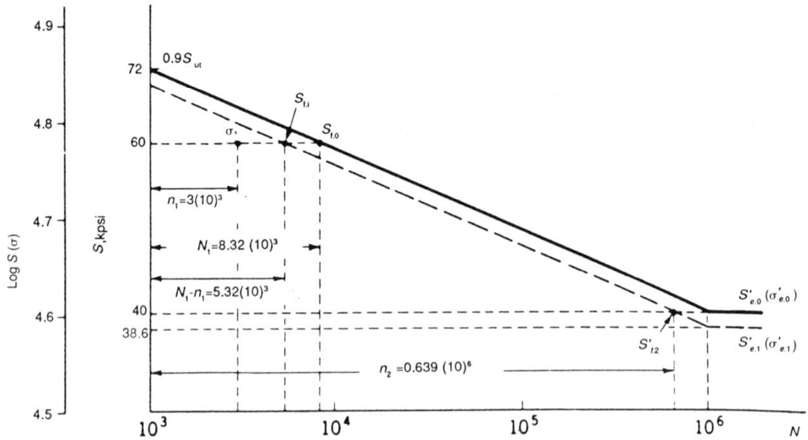

Figure B-24. Use of Miner's rule to predict the endurance limit of a material that has been overstressed for a finite number of cycles. (Shigley 1977)

σ_3) for a different number of cycles (N_1, N_2, N_3, and so on)? A number of methods are available for predicting fatigue life.

Figure B-24 illustrates the use of the Palmgren–Miner's cycle-ratio summation theory (or Miner's Rule) to predict the endurance limit of a material that has been overstressed for a finite number of cycles. An alternate method of calculating cumulative damage by fatigue is the Manson method (see Fig. B-25), which recognizes that the order of

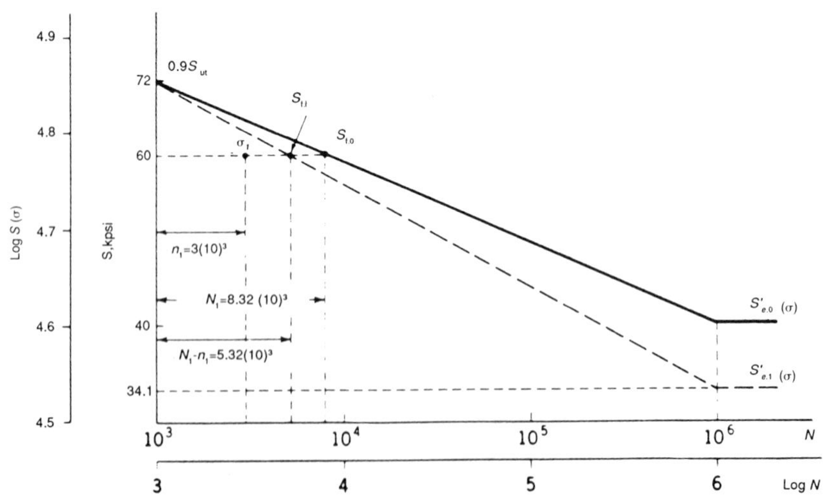

Figure B-25. Use of Manson's method to predict the endurance limit of a material that has been overstressed for a finite number of cycles. (Shigley 1977)

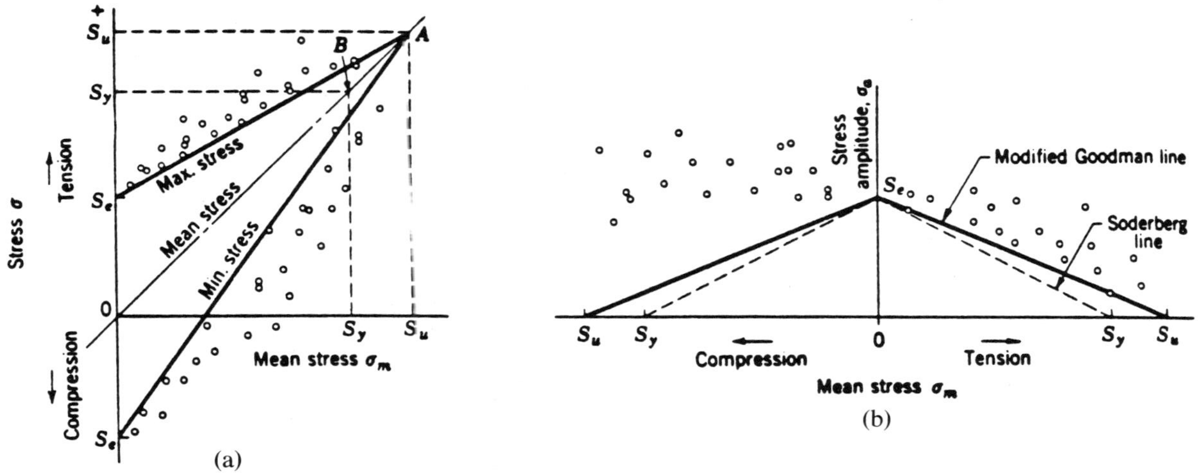

Figure B-26. Two types of fatigue diagrams showing typical points of failure. (a) Modified Goodman diagram. (b) Diagram showing the modified Goodman line.

different cycles of loading affects the fatigue life of the part. Figure B-26 compares alternating, ultimate, yield, and mean stress on a Goodman diagram. Note the effect of compression alternating stress (σ_a) versus tensile alternating stress (σ_a).

The Miner's Rule is currently used most often to explain cumulative fatigue damage. Mathematically, this theory is stated as

$$\frac{n_1}{N_1} + \frac{n_2}{N_2} + \cdots + \frac{n_i}{N_i} = C, \tag{19}$$

where n = the number of cycles of stress σ_1 applied to the specimen and N_1 = the life corresponding to σ_1. The constant C is determined by experiment and is usually found in the range $0.7 \leq C \leq 2.2$. Many authorities recommend using $C = 1$ so that the formula can be written for several different load cycles as

$$\sum_{J=1}^{i} \left(\frac{n_J}{N_J}\right) = 1 \tag{20}$$

Fluctuating Stresses

Frequently, one must determine the strength of parts corresponding to stress situations other than complete reversals. Many times, the stresses in a design fluctuate without passing through zero. Figure B-18 illustrates some of the various stress-time relationships that may occur. The components of stress we must deal with, some of which are shown in Figure B-18(a), are the following:

$\sigma_{\min}$ = minimum stress	σ_m = mean stress
$\sigma_{\max}$ = maximum stress	σ_r = stress range
σ_a = stress amplitude	σ_s = steady or static stress

Endurance-Limit Modifying Factors

The endurance limit S_e of a machine element may be considerably smaller than the endurance limit $S_{e'}$ of the rotating-beam specimen. This difference can be explained by using a variety of modifying factors, each of which accounts for a separate effect. Using this idea, we can write

$$S_e = k_a k_b k_c k_d k_e k_f S_{e'} \tag{21}$$

where

S_e = endurance of the mechanical element
$S_{e'}$ = endurance limit of the rotating-beam specimen
k_a = surface factor
k_b = size factor
k_c = reliability factor
k_d = temperature factor
k_e = modifying factor for stress concentration
= $1/K_f = 1/(1 + q(K_f - 1))$; q = notch sensitivity; K_t = stress concentration
k_f = miscellaneous-effects factor

These fatigue curve-based methods have proved to be invaluable in evaluating and predicting the fatigue life or "strength" of structural components. They are subject to many limitations, however, primarily because they fail to distinguish adequately between fatigue-crack initiation and propagation.

Figure B-19 is a schematic S-N curve divided into an initiation component and a propagation component. The number of cycles corresponding to the endurance limit (S_e) represents initiation life primarily, whereas the number of cycles expended in crack initiation at a high value of applied stress is negligible. Consequently, S-N-type data do not necessarily provide information about safe-life predictions in structural components. This is the case especially in components having surface irregularities that differ from those of the test specimens and in components containing cracklike discontinuities, because the existence of surface irregularities and cracklike discontinuities reduces, and may eliminate, the crack-initiation portion of the fatigue life of structural components.

As was discussed in the subsection called "Impact Testing," fracture mechanics can be used to determine the initiation life of fatigue cracks from various stress concentration factors. The use of fracture mechanics to evaluate the behavior of propagating cracks will be described in the next section (see also, Rolfe and Barsom 1987 for a discussion of fracture mechanics and Kutz 1986 for an overview).

Although S-N curves have been widely used to analyze the fatigue behavior of steels and weldments, closer inspection of the overall fatigue process in complex welded structures indicates that one can conduct a more rational analysis of fatigue behavior by using fracture mechanics methods. Fabrication discontinuities are invariably present in welded structures, even though the structure has been "inspected." A conservative approach to the prevention of fatigue failure is to assume the presence of an initial flaw and to analyze the fatigue-crack-growth behavior of the structural member. Such a fracture-mechanics analysis can reduce the need for expensive fatigue testing for many different types of structural details.

A schematic diagram showing the general relation between fatigue-crack initiation and propagation is shown in Figure B-27. The question of when a crack "initiates" to become a "propagating" crack depends on the level at which the crack is observed: for example, crystal inperfection, dislocation, microcrack, and lack of penetration. Using the fracture-mechanics approach to fatigue, one assumes the size of the initial flaw on the basis of the quality of fabrication or inspection and then calculates the number of cycles required for a crack to grow to the critical size for brittle fracture. This approach can be used to establish inspection requirements.

In addition to subcritical crack growth by fatigue, small cracks can grow by stress corrosion during the life of structures. Crack growth by either fatigue or stress corrosion does not represent catastrophic failure for structures fabricated from materials having reasonable levels of notch toughness; in both mechanisms, small cracks can become large enough to reduce structural efficiency and thus require repairs. In addition, both mechanisms can operate at the same time, driven by corrosion fatigue. A knowledge of the fatigue, corrosion fatigue, and stress corrosion behavior of materials is required to assess a product (structure) by an overall fracture-control plan that includes inspection specifications.

By testing precracked specimens under static loads in specific environments such as salt water and analyzing the results according to frac-

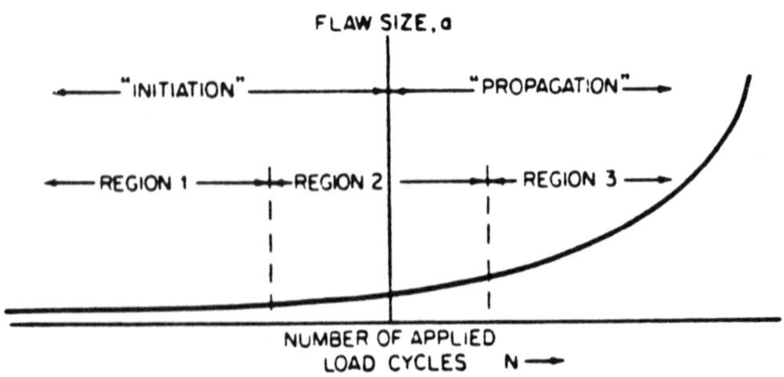

Figure B-27. Schematic showing the relation between flaw or crack-size growth versus N during "initiation" life and "propagation" life. Region 1: Difficulty in defining the size of the flaw (dislocation, microcrack, porosity, and so on). Region 2: Flaws can first be observed in an engineering sense. Region 3. Crack growth can be observed.

ture mechanics concepts, a K_I value below which subcritical crack propagation does not occur can be determined. This threshold value is called K_{Iscc}. The K_{Iscc} value for a particular material and environment is the plane-strain, stress-intensity threshold that describes the value below which subcritical cracks (scc) will not propagate and also have units of ksi $\sqrt{\text{in}}$.

For materials that are susceptible to crack growth in a particular environment, the K_{Iscc} value is used as the failure criterion rather than K_{Ic}. Thus, in a schematic relation between material toughness, design stress, and flaw size shown, K_{Iscc} would replace K_c as the critical value of K_I.

Crack Growth

For most structural materials that are subjected to static or quasi-static loads, the acceptable flaw sizes are usually much larger than any initial undetected flaws (Rolfe and Barsom 1987). For structures subjected to fatigue loading (or stress corrosion cracking), these initial cracks can grow throughout the life of the structure. Thus, an overall approach to preventing fracture or fatigue failures in large welded structures assumes that a small flaw of certain geometry exists after fabrication and that this flaw can either cause brittle fracture or grow by fatigue to the critical size. The initial size of the flaw can be established on the basis of the accuracy and reliability of the *inspection procedure* (see Chapter 8). To ensure that the structure does not fail by fracture, the calculated critical crack size a_{CR} at design load must be large enough and the number of cycles of loading required to make a small crack grow to a critical crack must be greater than the design life of the structure.

The procedure for analyzing the crack-growth behavior of steels and weld metals using fracture mechanics concepts is as follows, where $da/dN = C(\Delta K_I)^n$ and C and n are constants (see Rolfe and Barsom 1987).

1. On the basis of quality of inspection, estimate the maximum initial size of the flaw (a_0) present in the structure and the associated K_I relation for the member being analyzed.
2. Knowing K_c or K_{Ic} and the nominal maximum design stress, calculate the critical flaw size (a_{CR}) that would cause failure by brittle fracture.
3. Obtain an expression relating the fatigue-crack-growth rate of the steel or weld metal being analyzed. The following conservation estimates of the fatigue-crack-growth per cycle of loading, da/dN, have been determined for martensitic steels (e.g., A514/

517) as well as ferrite–pearlite steels in room temperature air:

$$\text{Martensitic steels: } \frac{da}{dN} = 0.66 \times 10^{-8}\, (\Delta K_I)^{2.25} \tag{22}$$

$$\text{Ferrite–pearlite steels: } \frac{da}{dN} = 3.6 \times 10^{-10}\, (\Delta K_I)^3 \tag{23}$$

where da/dN = fatigue-crack growth per cycle of loading, in./cycle, and ΔK_I = stress-intensity-factor range, ksi $\sqrt{\text{in.}}$ ($MN/m^{3/2}$).

4. Determine ΔK_I using the appropriate expression K_I, the estimated initial flaw size (a_0), and the range of live-load stress (cyclic stress range).
5. Integrate the crack-growth-rate expression between the limits of a_0 (at initial K_I) and a_{cr} (at K_{Ic}) to obtain the life of the structure before the failure.

A numerical example of this procedure is shown below:

- Assume the following conditions:

$$\text{A514 steel, } \sigma_{yz} = 100 \text{ ksi} \left(\frac{689\ MN}{m^2}\right)$$

$$K_{Ic} = 150 \text{ ksi } \sqrt{\text{in.}} \left(\frac{165\ MN}{mn^{3/2}}\right)$$

$$a_0 = 0.3 \text{ in. (7.6 mm), edge crack in tension} \tag{24}$$

$$\sigma_{max} = 45 \text{ ksi} \left(\frac{310\ MN}{m^2}\right) \qquad \sigma_{min} = 25 \text{ ksi} \left(\frac{172\ MN}{m^2}\right)$$

$$\Delta\sigma = 20 \text{ ksi} \left(\frac{138\ MN}{m^2}\right) \text{ (live-load stress range)}$$

$$K_I = 1.12\, \Delta\sqrt{\pi}(\sigma)\sqrt{a} \text{ edge crack in tension}$$

- Calculate a_{cr} at $\sigma = 45$ ksi ($310\ MN/m^2$):

$$a_{cr} = \left(\frac{K_{Ic}}{1.12\sqrt{\pi}\,(\sigma)_{max}}\right)^2 = \left(\frac{150}{1.12(1.77)(45)}\right)^2$$
$$= 2.8 \text{ in. (71.1 mm)} \tag{25}$$

- Assume an increment of crack growth, Δa. In this case, assume that $\Delta a = 0.1$ in. (2.5 mm). If smaller increments of crack growth were assumed, the accuracy would be increased slightly.

- Determine the expression for ΔK_I, where a_{avg} represents the average crack size between the two crack increments a_i and a_j:

$$\Delta K_I = 1.12\sqrt{\pi}\Delta\sigma \sqrt{a_{avg}} = 1.98\ (20)\ \sqrt{a_{avg}} \tag{26}$$

- Using the appropriate expression for crack growth rate,

$$\frac{da}{dN} = 0.66 \times 10^{-8}\ (\Delta K_I)^{2.25}$$

  solve for N for each increment of crack growth, replacing da/dN with $\Delta a/\Delta N$.

$$\Delta N = \frac{\Delta a}{0.66 \times 10^{-8}\ [1.98\ (20)\sqrt{a_{avg}}]^{2.25}} \tag{27}$$

- Repeat for $a = 0.4 - 0.5$ in. (10.2 to 12.7 mm), and so on, by numerical integration, as shown in Table B-5.

The flaw size-life results for this example are presented in Figure B-28. If only the desired total life is required, the expression for N can be integrated directly. In this example, direct integration yields a life of 87,600 cycles, whereas the numerical technique gave a life of 86,700 cycles. Note that the total life to propagate a crack from 0.3 to 2.8 in. (7.6 to 71.1 mm) in this example is 86,700 cycles. If the required life is 100,000 cycles, then this design would be inadequate and one or more of the following changes should be made:

- Increase the critical crack size at failure [$a_{cr} = 2.8$ in. (71.1 mm)] by using a material with a higher K_{Ic} value.
- Lower the design stress σ_{max} to increase the critical crack size at failure.
- Lower the stress range ($\Delta\sigma$) to reduce the rate of crack growth, which increases the number of cycles required for the crack to grow to critical size. Because the rate of crack growth is a power function of $\Delta\sigma$, or actually ΔK, lowering the stress range slightly has a significant effect on the life of the material.
- Improve the fabrication quality and inspection capability so that the initial flaw size (a_0) is reduced. Note that according to Table B-5 and Figure B-28, most of the life is taken up in the early stages of crack propagation. To double the initial crack size during the early stages of propagation requires almost half the total number

TABLE B-5. Fatigue-Crack-Growth Calculations[a]

a_0 (in.)[b]	a_f (in.)[b]	a_{avg} (in.)[b]	ΔK (ksi $\sqrt{in.}$)[c]	ΔN (cycles)	ΣN (cycles)
0.3	0.4	0.35	23.5	12,500	12,500
0.4	0.5	0.45	26.7	9,750	22,250
0.5	0.6	0.55	29.4	7,550	29,800
0.6	0.7	0.65	32.2	6,150	35,950
0.7	0.8	0.75	34.6	5,200	41,150
0.8	0.9	0.85	36.6	4,600	45,750
0.9	1.0	0.95	38.8	4,100	49,850
1.0	1.1	1.05	40.5	3,700	53,550
1.1	1.2	1.15	42.5	3,300	56,850
1.2	1.3	1.25	44.5	2,950	59,800
1.3	1.4	1.35	46.1	2,700	62,500
1.4	1.5	1.45	47.7	2,550	65,050
1.5	1.6	1.55	49.3	2,350	67,400
1.6	1.7	1.65	51.0	2,200	69,600
1.7	1.8	1.75	52.5	2,050	71,650
1.8	1.9	1.85	54.0	1,900	73,550
1.9	2.0	1.95	55.6	1,800	75,350
2.0	2.1	2.05	56.8	1,700	77,050
2.1	2.2	2.15	58.5	1,600	78,650
2.2	2.3	2.25	59.6	1,500	80,150
2.3	2.4	2.35	60.8	1,450	81,600
2.4	2.5	2.45	62.5	1,400	83,000
2.5	2.6	2.55	63.5	1,350	84,350
2.6	2.7	2.65	64.8	1,200	85,550
2.7	2.8	2.75	66.0	1,150	86,700

[a] $\Delta N = \dfrac{\Delta a}{0.66 \times 10^{-8}\,[1.98(\Delta\sigma)\sqrt{\sigma_{avg}}]^{2.25}}$,
where $\Delta a = 0.10$ in. (2.54 mm) and $\Delta\sigma = 20$ ksi(138 MN/m^2).

[b] 1 inch = 25.4 mm.

[c] 1 ksi $\sqrt{\text{in.}} = 1.1\ MN/m^{3/2}$.

of cycles. Any reduction in the size of the flaw has a slightly significant effect on the fatigue life of a structural member.

In this example, if a_0 was only 0.2 in. (5.1 mm), the design would be satisfactory. That is, the number of cycles to grow a crack of 0.2 to 0.3 in. (5.1 to 7.6 mm) is about 18,000 cycles, as indicated in Figure B-28, which, added to the 86,700 cycles required to grow the crack from 0.3 to 2.8 in. (7.7 to 71.1 mm), would make the total life equal to 104,000 cycles. As was mentioned earlier, in steels with high levels of toughness, the state of stress ahead of large cracks may be plane stress; thus, larger cracks than are calculated on the basis of plane-strain behavior could be tolerated. Because the rate of crack growth is increasing rapidly for large

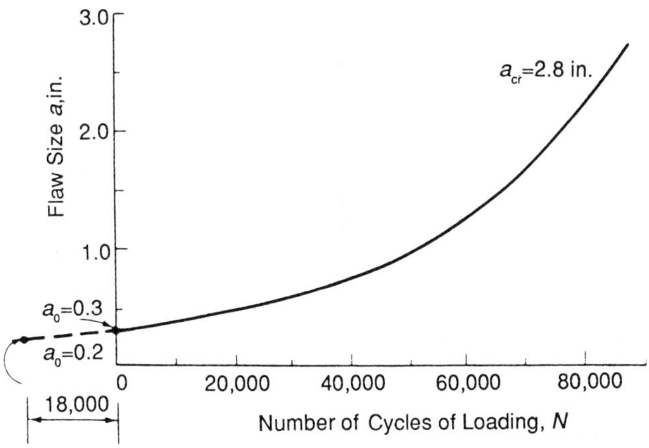

Figure B-28. Illustrative fatigue-crack-growth curve.

cracks, as illustrated in Figure B-28, the life may not be increased significantly. Because the rate of crack growth increases rapidly for large cracks, a significant increase in the fracture toughness of the material—that is, in the size of the tolerable crack at failure—may result in a negligible increase in the fatigue life of the structural member.

FUNDAMENTAL PHYSICAL LAWS

Conservation of Mass

Consider a mass at ambient conditions (Fig. B-29(a)) and zero pressure versus ambient conditions with applied pressure (Fig. B-29(b)): $M_0 = V_0\rho_0 = V_1\rho_1$ and P = pressure, where $V_0;V_1$ = volume, $\rho_0;\rho_1$ = density, and M_0 = mass.

Conservation of Momentum (Linear)

Consider the momentum of two masses before a collision (Fig. B-30(a)) and after a collision (Fig. B-30(b)):

$$M_1\,v_1 + M_2\,v_2 = M_1\,v_3 + M_2\,v_4 \tag{28}$$

where v = velocity and M = mass.

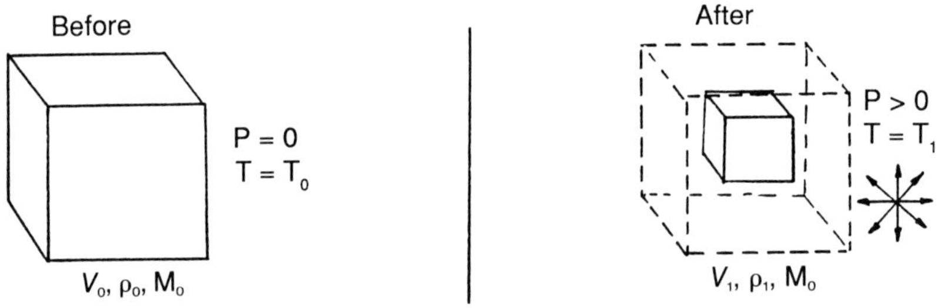

(a) (b)

Figure B-29. Conservation of mass: (a) before and (b) after compression.

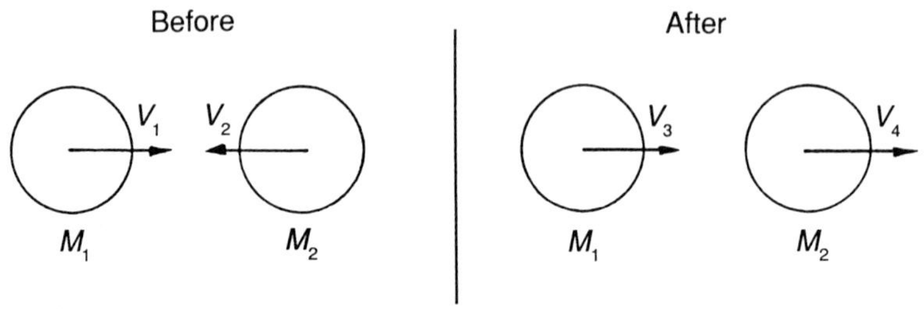

(a) (b)

Figure B-30. Conservation of linear momentum (a) before and (b) after collision.

Conservation of Energy

Consider a container of fluid in an insulated enclosure at a time before (t) and after (t') energy is added to the fluid (Fig. B-31(a) and (b)):

$$Q - W = U_1 - U_0 = (E_1 - E_1') + (E_2 - E_2') + (E_3 - E_3') \qquad (29)$$

where Q = heat energy absorbed, W = work energy extracted from the system, U_1 = energy state at time t', U_0 = energy state at time t, and E and E' = system energy before and after. Consider Fig. B-30 then $\frac{1}{2}(M_1V_1^2 + M_2V_2^2) = \frac{1}{2}(M_1V_3^2 + M_2V_4^2)$.

Irrecoverable Energy (Inelastic Collision)

Consider the change in kinetic energy from before to after a collision represented in Figure B-32(a) and (b):

Coefficient of restitution $$e = -\frac{v_1 - v_3}{v_0 - v_2} \qquad (30)$$

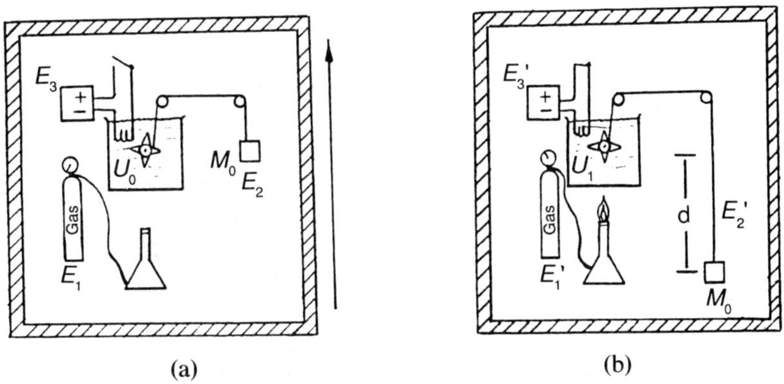

Figure B-31. Conservation of energy before (a) and (b) after conversion of chemical, mechanical, and electrical energy to heat energy.

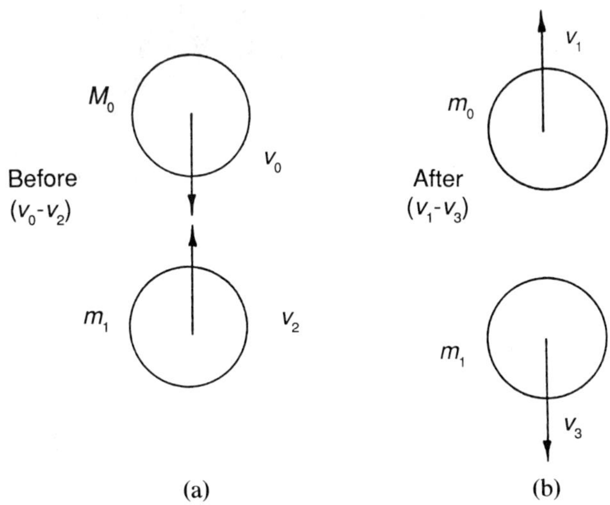

Figure B-32. Nonconservation of kinetic energy (a) before and (b) after a collision of two bodies. Some kinetic energy is converted to heat (1–*e*).

where e = energy and v = velocities. In the case in which $U_3 = 0$ and $U_2 = 0$ (Fig. B-33), then Equation 31 becomes

$$e = -\frac{v_1}{v_0} \tag{31}$$

where $\beta = (1 - e)$ approximate heat loss or deformation work coefficient.

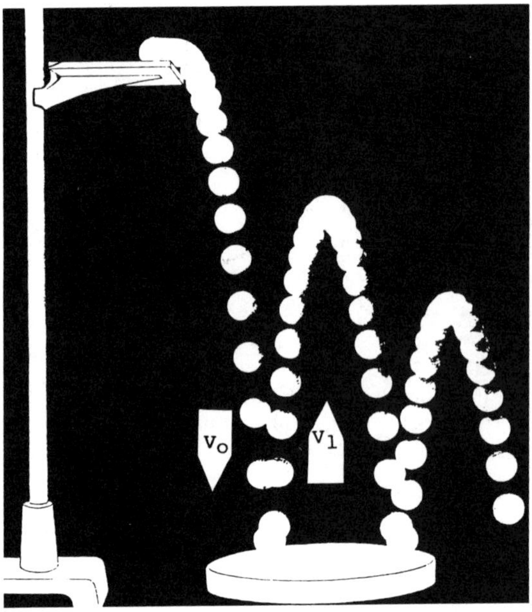

Figure B-33. An illustrative example of nonconservation of kinetic energy, where $v_2 = 0$ and $v_3 = 0$. (From F. W. Sears and M. W. Zemansky, *College Physics,* Addison-Wesley, Reading, MA, 1975)

Friction

Consider two masses sliding relative to one another (Fig. B-34)

$$F = \mu_i N \tag{32}$$

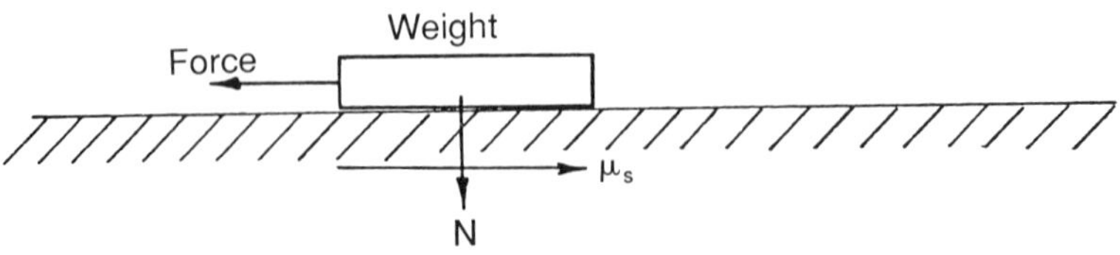

Figure B-34. Illustration of friction. (Conversion of kinetic energy to heat.)

where μ_i = coefficient of friction; $i = S$ and D; μ_S = static, no motion; N = normal force (usually weight); and μ_D = dynamic, motion. If friction energy is equated with kinetic energy,

$$Fd = (\mu_D N)(d) = \left(\frac{1}{2}\right) M(v_1^2 - v_2^2) \tag{33}$$

If $v_2 = 0$, then

$$v_1 = 30\mu_D d \text{ mph} \tag{34}$$

where d = feet and $0 \leq \mu_D \leq 1$.

Note the following examples:

Materials	$\mu_D \cong$
Tire on dry cement	0.8–0.85
Iron on iron	0.15
Brass on brass	0.17
Oak on oak	0.48

This simple formula is used to assess the velocity of a vehicle on the basis of skid marks (see Fig. B-35).

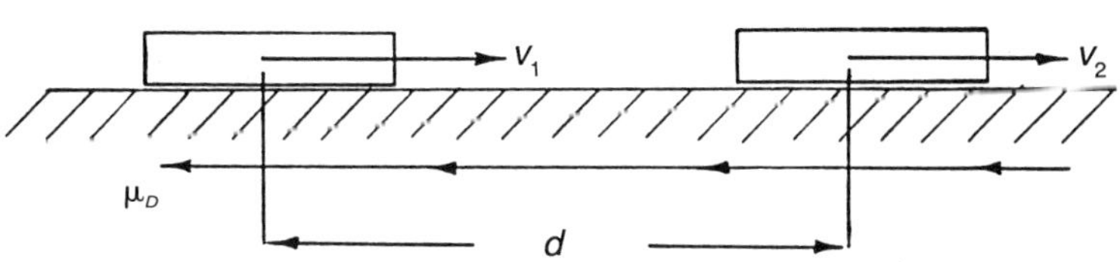

Figure B-35. Special case when friction energy to stop an object is approximately equal to its kinetic energy.

Linear Elasticity

The linear elastic behavior or material (see in Fig. B-4(a)) is analogous to a simple spring, which has the following force/displacement relationship

$$F = kx \tag{35}$$

where k = force/distance, x = distance, and F = force (see Fig. B-36).

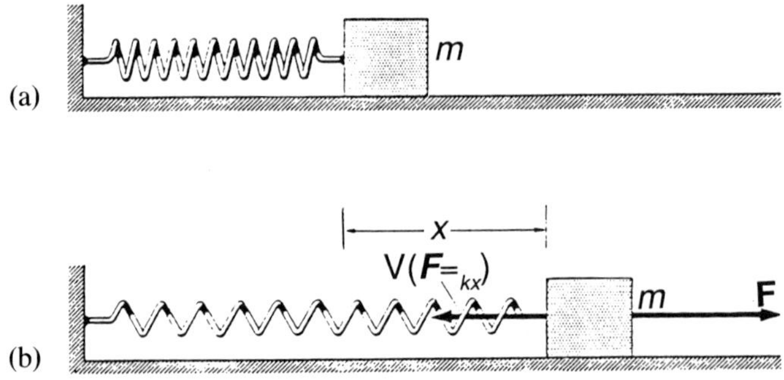

Figure B-36. Concept of elasticity, spring stiffness (k).

Viscoelasticity

The time-dependent behavior of materials (see Fig. B-16 and B-17(b)) is analogous to a dashpot or damper, which has the following force/velocity relationship

$$F = C\dot{x} \tag{36}$$

or $\dot{x} = (dx/dt)$ and

$$C \equiv \frac{\#}{(\text{distance/time})} = \frac{\#}{\text{velocity}}$$

(see Fig. B-37).

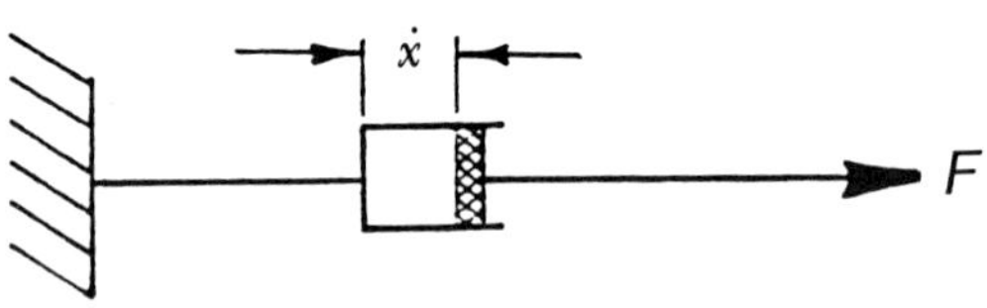

Figure B-37. Concept of viscoelasticity: dashpot or damper. $\dot{x} = dx/dt =$ velocity.

Appendix C

The Organizational Structure and Supporting Members of the American National Standards Institute and the Code of Federal Regulations Titles and Chapters

This appendix gives an illustration of the ANSI voluntary codes and standards and the CFR Statutory Regulations. ANSI is composed of the various industries, societies, and associations that write the various volumes that are incorporated into the overall ANSI collection of codes and standards. The CFR is established by the cognizant government agency that is charged with regulating the various industries. Frequently, the CFR adopts the guidelines, codes, and standards written by the organizations that are members of ANSI.

ORGANIZATION OF THE AMERICAN NATIONAL STANDARDS INSTITUTE

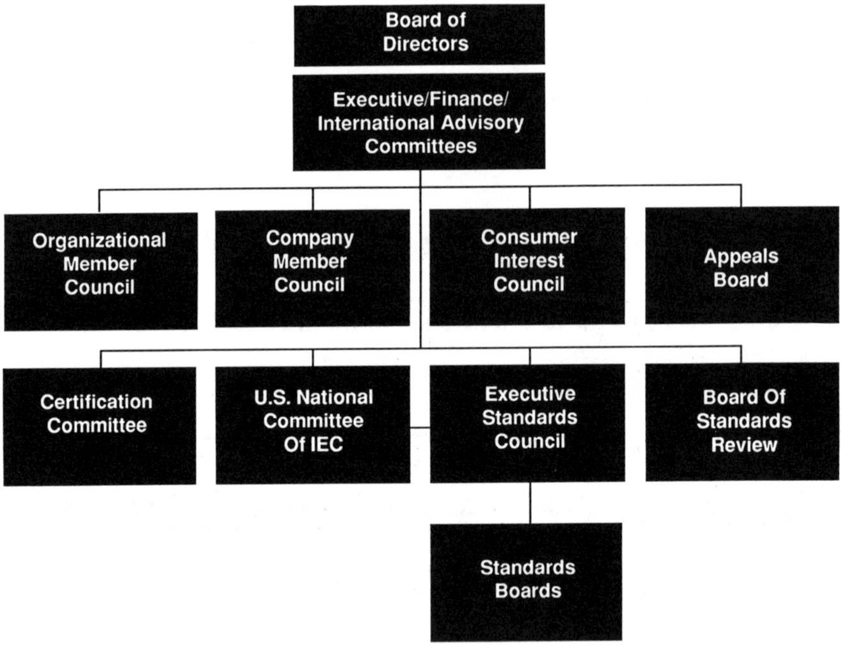

Board of Directors: The governing and policymaking body of the Institute

Executive Committee: Provides guidance for administration of ANSI and acts for the Board of Directors between meetings

Finance Committee: Maintains a continuing review of ANSI financial affairs and makes recommendations to the Board

International Advisory Committee: Advises the Board on international policy and programs

Organizational Member Council and Company Member Council: Ensure participation of their members in ANSI programs and provide a communication channel between their members and constituents and ANSI's Board on programs and policies of the Institute; help determine standards needs

Consumer Interest Council: Responsible for consumer input to standards programs; reviews all consumer

	standards; represents and protects consumer interests in national standards activities
Appeals Board:	Hears complaints of those who believe that they have been adversely affected by the action or inaction of any ANSI board or council
Certification Committee:	Develops and operates programs leading to national accreditation of certification programs
U.S. National Committee of IEC:	Responsible for effective participation in the work of IEC and for operation of the technical advisory groups that develop the U.S. position on international electrotechnical standards
Executive Standards Council:	Manages the standardization activities coordinated by ANSI—promulgates operating procedures; stimulates expeditious completion of standards work; also coordinates U.S. participation in technical work of international organizations
Board of Standards Review:	Approves standards as American National Standards and acts on withdrawal and reaffirmation when it finds that a consensus exists among those substantially concerned with the scope and provisions of the standards under consideration
Standards Boards:	Assist the Executive Standards Council in carrying out its management and coordination functions for standards development in the discipline or homogeneous technical sphere in which the particular SB operates

ORGANIZATIONAL MEMBERS

AAMA
ASTM
Abrasive Grain Association
Acoustical Society of America
Aerospace Industries Association of America
Air-Conditioning & Refrigeration Institute
Air Transport Association of America
Alliance of American Insurers
The Aluminum Association

American Academy of Ophthalmology
American Academy of Otolaryngology
American Association of Nurserymen
American Association of Textile Chemists & Colorists
American Bankers Association
American Bureau of Shipping
American Chain Association
American College of Radiology
American Concrete Institute
American Council of Independent Laboratories
American Dental Association
American Dental Trade Association
American Foundrymen's Society
American Gas Association
American Gear Manufacturers Association
American Hardboard Association
American Hotel & Motel Association
American Industrial Hygiene Association
American Institute of Architects
American Institute of Chemical Engineers
American Institute of Physics
American Institute of Steel Construction
American Institute of Timber Construction
American Insurance Services Group
American Iron & Steel Institute
American Ladder Institute
American Library Association
American Mining Congress
American National Metric Council
American Nuclear Insurers
American Nuclear Society
American Optometric Association
American Paper Institute
American Petroleum Institute
American Plywood Association
American Public Power Association
American Public Works Association
American Society of Agricultural Engineers
American Society for Artificial Internal Organs
American Society of Bakery Engineers
American Society of Civil Engineers
American Society of Heating, Refrigerating & Air-Conditioning Engineers
American Society of Mechanical Engineers
American Society for Quality Control
American Society of Safety Engineers
American Speech-Language-Hearing Association
American Water Works Association
American Welding Society
American Wood Preservers Bureau
Anti-Friction Bearing Manufacturers Association
The Art & Craft Materials Institute
Associated General Contractors of America
Association for the Advancement of Medical Instrumentation
Association of American Railroads
Association for Computing Machinery
Association of Edison Illuminating Companies
Association of Home Appliance Manufacturers
Association for Information and Image Management
Association of Iron & Steel Engineers
Association of Official Analytical Chemists
Association of Records Managers & Administrators
Audio Engineering Society
Automobile Importers of America

Builders Hardware Manufacturers Association

Building Owners & Managers Association International
The Business & Institutional Furniture Manufacturers Association

Cast Iron Soil Pipe Institute
Cemented Carbide Producers Association
Certified Ballast Manufacturers
Chemical Manufacturers Association
Chemical Specialties Manufacturers Association
Coated Abrasives Manufacturers Institute
College of American Pathologists
Compressed Air & Gas Institute
Compressed Gas Association
Computer Aided Manufacturing—International (CAM-I)
Computer & Business Equipment Manufacturers Association
Concrete Reinforcing Steel Institute
Construction Industry Manufacturers Association
Contact Lens Association of Ophthalmologists
Contact Lens Institute
Conveyor Equipment Manufacturers Association
Council of American-Flag Ship Operators
Cultured Marble Institute
CUNA Service Group

Data Processing Management Association
Diving Equipment Manufacturers Association
Door & Hardware Institute
Ductile Iron Pipe Research Association

Edison Electric Institute
Electronic Industries Association
Exchange Carriers Standards Association

Factory Mutual System
Farm & Industrial Equipment Institute
The Fertilizer Institute
Fluid Controls Institute
Fragrance Material Association of the U.S.

Gas Appliance Manufacturers Association
Gas Processors Association
Glass Lens Council
Grinding Wheel Institute

Hand Tools Institute
Hardwood Plywood Manufacturers Association
Health Industry Manufacturers Association
Health Physics Society
Human Factors Society
Hydraulic Institute

Illuminating Engineering Society
Industrial Fasteners Institute
Industrial Safety Equipment Association
The Industrial Truck Association
Institute of Electrical & Electronics Engineers
Institute of Industrial Engineers
Institute for Interconnecting & Packaging Electronic Circuits
Institute of International Container Lessors
Institute of Nuclear Materials Management
Instrument Society of America
Insulated Cable Engineers Association
Insulated Steel Door Systems Institute

International Association of Plumbing & Mechanical Officials
International Brotherhood of Electrical Workers
International Communications Industries Association
International Society of Arboriculture
International Union of Operating Engineers
The Irrigation Association

Laborers' International Union of North America
Laser Institute of America
Lead Industries Association/Zinc Institute

Man-Made Fiber Producers Association
Manufacturers Standardization Society of the Valve & Fittings Industry
Mastercard International
The Material Handling Institute
Mechanical Contractors Association of America
Metal Cutting Tool Institute
Motor Vehicle Manufacturers Association

National Association of Broadcasters
National Association of Corrosion Engineers
National Association of Garage Door Manufacturers
National Association of Home Builders
National Association of Optometrists & Opticians
National Association of Photographic Manufacturers
National Association of Plumbing, Heating, Cooling Contractors
National Association of Printing Ink Manufacturers
National Association of Punch Manufacturers
The National Board of Boiler & Pressure Vessel Inspectors
National Coffee Association
National Committee for Clinical Laboratory Standards
National Computer Graphics Association
National Conference of States on Building Codes & Standards
National Electrical Contractors Association
National Electrical Manufacturers Association
National Elevator Industry
National Fastener Distributors Association
National Fire Protection Association
National Fluid Power Association
National Forest Products Association
The National Golf Car Manufacturers Association
National Information Standards Organization
National Kitchen Cabinet Association
National LP-Gas Association
National Machine Tool Builders Association
National Particleboard Association
National Printing Equipment & Supply Association
National Safety Council
National Ski Areas Association
National Society to Prevent Blindness
National Spa & Pool Institute
National Tooling & Machinery Association
National Wholesale Druggists' Association
National Wood Window & Door Association
Norex U.S.A., Inc.
North American District Heating & Cooling Institute

Optical Laboratories Association
Optical Manufacturers Association
Opticians Association of America
Outdoor Power Equipment Institute

Packaging Machinery Manufacturers Institute
The Painting & Decorating Contractors of America
Paralyzed Veterans of America
Portable Power Equipment Manufacturers Association
Portable Sanitation Association
Portland Cement Association
Powder Actuated Tool Manufacturers Institute
Power Tool Institute
Printing Industries of America

Radiological Society of North America
Recreation Vehicle Industry Association
The Refractories Institute
Robotic Industries Association
Rubber Manufacturers Association

Salt Institute
Scaffold Industry Association
Scientific Apparatus Makers Association
Screen Manufacturers Association
Semiconductor Equipment & Materials Institute
Society of Automotive Engineers
Society of Manufacturing Engineers
The Society of Motion Picture & Television Engineers
Society of Photographic Scientists & Engineers
The Society of the Plastics Industry
Specialty Vehicle Institute of America
Sporting Arms & Ammunition Manufacturers Institute
Steel Door Institute
Steel Service Center Institute
Steel Shipping Container Institute
Sunglass Association of America

Tea Association of the United States of America
Tile Council of America
Toy Manufacturers of America
Tubular Exchanger Manufacturers Association
Tubular Rivet & Machine Institute

Underwriters Laboratories
United Association of Journeymen & Apprentices of the Plumbing & Pipe Fitting Industry of the United States & Canada

United States League of Savings Associations
University of Texas

Virginia Polytechnic Institute & State University
VISA U.S.A.

Wire Rope Technical Board
Wood Machinery Manufacturers of America
World Safety Organization

GOVERNMENTAL MEMBERS

Dallas Water Utilities
Delray Beach, Florida, City of
General Services Administration
Lawrence Livermore Laboratories, University of California
Los Angeles Department of Water & Power

Memphis Light, Gas & Water Division, Memphis, Tennessee
Memphis & Shelby County Office of Construction Code Enforcement
National Archives and Records Administration
National Bureau of Standards
National Communications System
Nebraska Public Power District
New York Power Authority
New York State Public Service Commission
Sacramento Municipal Utility District, Sacramento, California
Shreveport, Louisiana, City of
Somerset County Facilities and Services
Tennessee Valley Authority
U.S. Air Force Engineering & Services Center
U.S. Army Medical Bioengineering Research & Development Laboratory
U.S. Army Armament Research Development Engineering Center
U.S. Department of Agriculture Rural Electrification Administration
U.S. Department of Defense
U.S. Department of Education
U.S. Department of Energy Bonneville Power Administration
U.S. Department of the Interior
U.S. Food & Drug Administration, Center for Devices and Radiological Health
U.S. Navy Naval Construction Battalion Center
U.S. Veterans Administration

CODE OF FEDERAL REGULATIONS TITLES AND CHAPTERS

(Revised as of September 27, 1988)

Title 1—General Provisions

I Administrative Committee of the Federal Register (Parts 1–49)
II Office of the Federal Register (Parts 50–299)
III Administrative Conference of the United States (Parts 300–399)
IV Miscellaneous Agencies (Parts 400–499)

Title 2—[Reserved]

Title 3—The President

I Executive Office of the President (Parts 100–199)

Title 4—Accounts

I General Accounting Office (Parts 1–99)
II Federal Claims Collection Standards (General Accounting Office–Department of Justice) (Parts 100–299)
III General Accounting Office (CASB) (Parts 300–499)

Title 5—Administrative Personnel

I Office of Personnel Management (Parts 1–1199)

II Merit Systems Protection Board (Parts 1200–1299)

III Office of Management and Budget (Parts 1300–1399)

IV Advisory Committee on Federal Pay (Parts 1400–1499)

V The International Organizations Employees Loyalty Board (Parts 1500–1599)

VI Federal Retirement Thrift Investment Board (Parts 1600–1699)

VII Advisory Commission on Intergovernmental Relations (Parts 1700–1799)

IX Appalachian Regional Commission (Parts 1900–1999)

XI United States Soldiers' and Airmen's Home (Parts 2100–2199)

XIV Federal Labor Relations Authority, General Counsel of the Federal Labor Relations Authority and Federal Service Impasses Panel (Parts 2400–2499)

XV Office of Administration, Executive Office of the President (Parts 2500–2599)

Title 6—[Reserved]

Title 7—Agriculture

SUBTITLE A—Office of the Secretary of Agriculture (Parts 0–26)

SUBTITLE B—Regulations of the Department of Agriculture

I Agricultural Marketing Service (Standards, Inspections, Marketing Practices), Department of Agriculture (Parts 27–209)

II Food and Nutrition Service, Department of Agriculture (Parts 210–299)

III Animal and Plant Health Inspection Service, Department of Agriculture (Parts 300–399)

IV Federal Crop Insurance Corporation, Department of Agriculture (Parts 400–499)

V Agricultural Research Service, Department of Agriculture (Parts 500–599)

VI Soil Conservation Service, Department of Agriculture (Parts 600–699)

VII Agricultural Stabilization and Conservation Service (Agricultural Adjustment), Department of Agriculture (Parts 700–799)

Title 7—Agriculture—Continued

VIII Federal Grain Inspection Service, Department of Agriculture (Parts 800–899)

IX Agricultural Marketing Service (Marketing Agreements and Orders; Fruits, Vegetables, Nuts), Department of Agriculture (Parts 900–999)

X Agricultural Marketing Service (Marketing Agreements and Orders: Milk), Department of Agriculture (Parts 1000–1199)

XI Agricultural Marketing Service (Marketing Agreements and Orders; Miscellaneous Commodities), Department of Agriculture (Parts 1200–1299)

XIV Commodity Credit Corporation, Department of Agriculture (Parts 1400–1499)

XV Foreign Agricultural Service, Department of Agriculture (Parts 1500–1599)

XVI Rural Telephone Bank, Department of Agriculture (Parts 1600–1699)

XVII Rural Electrification Administration, Department of Agriculture (Parts 1700–1799)

XVIII Farmers Home Administration, Department of Agriculture (Parts 1800–2099)

XXI Foreign Economic Development Service, Department of Agriculture (Parts 2100–2199)

XXII Office of International Cooperation and Development, Department of Agriculture (Parts 2200–2299)

XXIV Board of Contract Appeals, Department of Agriculture (Parts 2400–2499)

XXV Office of the General Sales Manager, Department of Agriculture (Parts 2500–2599)

XXVI Office of Inspector General, Department of Agriculture (Parts 2600–2699)

XXVII Office of Information Resources Management, Department of Agriculture (Parts 2700–2799)

XXIX Office of Energy, Department of Agriculture (Parts 2900–2999)

XXX Office of Operations and Finance, Department of Agriculture (Parts 3000–3099)

XXXI Office of Environmental Quality, Department of Agriculture (Parts 3100–3199)

XXXII Office of Grants and Program Systems, Department of Agriculture (Parts 3200–3299)

XXXIII Office of Transportation, Department of Agriculture (Parts 3300–3399)

XXXIV Cooperative State Research Service, Department of Agriculture (Parts 3400–3499)

XXXVI National Agricultural Statistics Service, Department of Agriculture (Parts 3600–3699)

XXXVII Economic Research Service, Department of Agriculture (Parts 3700–3799)

XXXVIII World Agricultural Outlook Board, Department of Agriculture (Parts 3800–3899)

XXXIX Economic Analysis Staff, Department of Agriculture (Parts 3900–3999)

XL Economics Management Staff, Department of Agriculture (Parts 4000–4099)

XLI National Agricultural Library, Department of Agriculture (Part 4100)

Title 8—Aliens and Nationality

I Immigration and Naturalization Service, Department of Justice (Parts 1–499)

Title 9—Animals and Animal Products

I Animal and Plant Health Inspection Service, Department of Agriculture (Parts 1–199)

II Packers and Stockyards Administration, Department of Agriculture (Parts 200–299)

III Food Safety and Inspection Service, Meat and Poultry Inspection, Department of Agriculture (Parts 300–399)

Title 10—Energy

I Nuclear Regulatory Commission (Parts 0–199)

II Department of Energy (Parts 200–699)

III Department of Energy (Parts 700–999)

X Department of Energy (General Provisions) (Parts 1000–1099)

XV Office of the Federal Inspector for the Alaska Natural Gas Transportation System (Parts 1500–1599)

Title 11—Federal Elections

I Federal Election Commission (Parts 1–9099)

Title 12—Banks and Banking

I Comptroller of the Currency, Department of the Treasury (Parts 1–199)

II Federal Reserve System (Parts 200–299)

III Federal Deposit Insurance Corporation (Parts 300–399)

IV Export-Import Bank of the United States (Parts 400–499)

V Federal Home Loan Bank Board (Parts 500–599)

VI Farm Credit Administration (Parts 600–699)

VII National Credit Union Administration (Parts 700–799)

VIII Federal Financing Bank (Parts 800–899)

XI Federal Financial Institutions Examination Council (Parts 1100–1199)

Title 13—Business Credit and Assistance

I Small Business Administration (Parts 1–199)

III Economic Development Administration, Department of Commerce (Parts 300–399)

V Regional Action Planning Commissions (Parts 500–599)

Title 14—Aeronautics and Space

I Federal Aviation Administration, Department of Transportation (Parts 1–199)

II Office of the Secretary, Department of Transportation (Aviation Proceedings) (Parts 200–399)

III Office of Commercial Space Transportation, Department of Transportation (Parts 400–499)

V National Aeronautics and Space Administration (Parts 1200–1299)

Title 15—Commerce and Foreign Trade

Subtitle A—Office of the Secretary of Commerce (Parts 0–29)

Subtitle B—Regulations Relating to Commerce and Foreign Trade

I Bureau of the Census, Department of Commerce (Parts 30–199)

II National Bureau of Standards, Department of Commerce (Parts 200–299)

III International Trade Administration, Department of Commerce (Parts 300–399)

IV Foreign-Trade Zones Board (Parts 400–499)

VI Bureau of Industrial Economics, Department of Commerce (Parts 600–699)

VII Bureau of Export Administration Department of Commerce (Parts 700–799)

VIII Bureau of Economic Analysis, Department of Commerce (Parts 800–899)

IX National Oceanic and Atmospheric Administration, Department of Commerce (Parts 900–999)

XII United States Travel Service, Department of Commerce (Parts 1200–1299)

XIII East-West Foreign Trade Board (Parts 1300–1399)

XIV Minority Business Development Agency (Parts 1400–1499)

SUBTITLE C—Regulations Relating to Foreign Trade Agreements

XX Office of the United States Trade Representative (Parts 2000–2099)

SUBTITLE D—Regulations Relating to Telecommunications and Information

XXIII National Telecommunications and Information Administration, Department of Commerce (Parts 2300–2399)

Title 16—Commercial Practices

I Federal Trade Commission (Parts 0–999)

II Consumer Product Safety Commission (Parts 1000–1799)

Title 17—Commodity and Securities Exchanges

I Commodity Futures Trading Commission (Parts 1–199)

II Securities and Exchange Commission (Parts 200–399)

IV Department of the Treasury (Parts 400–499)

Title 18—Conservation of Power and Water Resources

I Federal Energy Regulatory Commission, Department of Energy (Parts 1–399)

III Delaware River Basin Commission (Parts 400–499)

VI Water Resources Council (Parts 700–799)

VIII Susquehanna River Basin Commission (Parts 800–899)

XIII Tennessee Valley Authority (Parts 1300–1399)

Title 19—Customs Duties

I United States Customs Service, Department of the Treasury (Parts 1–199)

II United States International Trade Commission (Parts 200–299)

III International Trade Administration, Department of Commerce (Parts 300–399)

Title 20—Employees' Benefits

I Office of Workers' Compensation Programs, Department of Labor (Parts 1–199)

II Railroad Retirement Board (Parts 200–399)

III Social Security Administration, Department of Health and Human Services (Parts 400–499)

IV Employees' Compensation Appeals Board, Department of Labor (Parts 500–599)

V Employment and Training Administration, Department of Labor (Parts 600–699)

VI Employment Standards Administration, Department of Labor (Parts 700–799)

VII Benefits Review Board, Department of Labor (Parts 800–899)

VIII Joint Board for the Enrollment of Actuaries (Parts 900–999)

Title 21—Food and Drugs

I Food and Drug Administration, Department of Health and Human Services (Parts 1–1299)

II Drug Enforcement Administration, Department of Justice (Parts 1300–1399)

Title 22—Foreign Relations

I Department of State (Parts 1–199)

II Agency for International Development, International Development Cooperation Agency (Parts 200–299)

III Peace Corps (Parts 300–399)

IV International Joint Commission, United States and Canada (Parts 400–499)

V United States Information Agency (Parts 500–599)

VI United States Arms Control and Disarmament Agency (Parts 600–699)

VII Overseas Private Investment Corporation (Parts 700–799)

IX Foreign Service Grievance Board Regulations (Parts 900–999)

X Inter-American Foundation (Parts 1000–1099)

XI International Boundary and Water Commission, United States and Mexico, United States Section (Parts 1100–1199)

XII United States International Development Cooperation Agency (Parts 1200–1299)

XIII Board for International Broadcasting (Parts 1300–1399)

XIV Foreign Service Labor Relations Board; Federal Labor Relations Authority; General Counsel of the Federal Labor Relations Authority; and the Foreign Service Impasse Disputes Panel (Parts 1400–1499)

XV African Development Foundation (Parts 1500–1599)

XVI Japan United States Friendship Commission (Parts 1600–1699)

Title 23—Highways

I Federal Highway Administration, Department of Transportation (Parts 1–999)

II National Highway Traffic Safety Administration and Federal Highway Administration, Department of Transportation (Parts 1200–1299)

III National Highway Traffic Safety Administration, Department of Transportation (Parts 1300–1399)

Title 24—Housing and Urban Development

Subtitle A—Office of the Secretary, Department of Housing and Urban Development (Parts 0–99)

Title 24—Housing and Urban Development—Continued

SUBTITLE B—Regulations Relating to Housing and Urban Development

I Office of Assistant Secretary for Equal Opportunity, Department of Housing and Urban Development (Parts 100–199)

II Office of Assistant Secretary for Housing-Federal Housing Commissioner, Department of Housing and Urban Development (Parts 200–299)

III Government National Mortgage Association, Department of Housing and Urban Development (Parts 300–399)

V Office of Assistant Secretary for Community Planning and Development, Department of Housing and Urban Development (Parts 500–599)

VI Office of Assistant Secretary for Community Planning and Development, Department of Housing and Urban Development (Parts 600–699)

VII Office of the Secretary, Department of Housing and Urban Development (Section 8 Housing Assistance Programs and Public and Indian Housing Programs) (Parts 700–799)

VIII Office of the Assistant Secretary for Housing—Federal Housing Commissioner, Department of Housing and Urban Development (Section 8 Housing Assistance Programs and Section 202 Direct Loan Program) (Parts 800–899)

IX Office of Assistant Secretary for Public and Indian Housing, Department of Housing and Urban Development (Parts 900–999)

X Office of Assistant Secretary for Housing—Federal Housing Commissioner, Department of Housing and Urban Development (Interstate Land Sales Registration Program) (Parts 1700–1799)

XI Solar Energy and Energy Conservation Bank, Department of Housing and Urban Development (Parts 1800–1899)

XII Office of Inspector General, Department of Housing and Urban Development (Parts 2000–2099)

XV Mortgage Insurance and Loan Programs under the Emergency Homeowners' Relief Act, Department of Housing and Urban Development (Parts 2700–2799)

XX Office of Assistant Secretary for Housing—Federal Housing

Commissioner, Department of Housing and Urban Development (Parts 3200–3699)

XXV Neighborhood Reinvestment Corporation (Parts 4100–4199)

Title 25—Indians

I Bureau of Indian Affairs, Department of the Interior (Parts 1–299)

II Indian Arts and Crafts Board, Department of the Interior (Parts 300–399)

IV Navajo and Hopi Indian Relocation Commission (Parts 700–799)

Title 26—Internal Revenue

I Internal Revenue Service, Department of the Treasury (Parts 1–699)

Title 27—Alcohol, Tobacco Products and Firearms

I Bureau of Alcohol, Tobacco and Firearms, Department of the Treasury (Parts 1–299)

Title 28—Judicial Administration

I Department of Justice (Parts 0–199)

III Federal Prison Industries, Department of Justice (Parts 300–399)

V Bureau of Prisons, Department of Justice (Parts 500–599)

VI Office of Independent Counsel, Department of Justice (Parts 600–699)

VII Office of Independent Council (Parts 700–799)

Title 29—Labor

SUBTITLE A—Office of the Secretary of Labor (Parts 0–99)

SUBTITLE B—Regulations Relating to Labor

I National Labor Relations Board (Parts 100–199)

II Bureau of Labor-Management Relations and Cooperative Programs, Department of Labor (Parts 200–299)

III National Railroad Adjustment Board (Parts 300–399)

Title 29—Labor—Continued

IV Office of Labor-Management Standards, Department of Labor (Parts 400–499)

V Wage and Hour Division, Department of Labor (Parts 500–899)

IX Construction Industry Collective Bargaining Commission (Parts 900–999)

X National Mediation Board (Parts 1200–1299)

XII Federal Mediation and Conciliation Service (Parts 1400–1499)

XIV Equal Employment Opportunity Commission (Parts 1600–1699)

XVII Occupational Safety and Health Administration, Department of Labor (Parts 1900–1999)

XX Occupational Safety and Health Review Commission (Parts 2200–2499)

XXV Pension and Welfare Benefits Administration, Department of Labor (Parts 2500–2599)

XXVI Pension Benefit Guaranty Corporation (Parts 2600–2699)

XXVII Federal Mine Safety and Health Review Commission (Parts 2700–2799)

Title 30—Mineral Resources

I Mine Safety and Health Administration, Department of Labor (Parts 1–199)

II Minerals Management Service, Department of the Interior (Parts 200–299)

III Board of Surface Mining and Reclamation Appeals, Department of the Interior (Parts 300–399)

IV Geological Survey, Department of the Interior (Parts 400–499)

VI Bureau of Mines, Department of the Interior (Parts 600–699)

VII Office of Surface Mining Reclamation and Enforcement, Department of the Interior (Parts 700–999)

Title 31—Money and Finance: Treasury

SUBTITLE A—Office of the Secretary of the Treasury (Parts 0–50)

SUBTITLE B—Regulations Relating to Money and Finance

I Monetary Offices, Department of the Treasury (Parts 51–199)

II Fiscal Service, Department of the Treasury (Parts 200–399)

IV Secret Service, Department of the Treasury (Parts 400–499)

V Office of Foreign Assets Control, Department of the Treasury (Parts 500–599)

VI Bureau of Engraving and Printing, Department of the Treasury (Parts 600–699)

VII Federal Law Enforcement Training Center, Department of the Treasury (Parts 700–799)

Title 32—National Defense

Subtitle A—Department of Defense

I Office of the Secretary of Defense (Parts 1–399)

V Department of the Army (Parts 400–699)

VI Department of the Navy (Parts 700–799)

VII Department of the Air Force (Parts 800–1099)

Subtitle B—Other Regulations Relating to National Defense

XII Defense Logistics Agency (Parts 1200–1299)

XVI Selective Service System (Parts 1600–1699)

XIX Central Intelligence Agency (Parts 1900–1999)

XX Information Security Oversight Office (Parts 2000–2099)

XXI National Security Council (Parts 2100–2199)

XXIV Office of Science and Technology Policy (Parts 2400–2499)

XXVII Office for Micronesian Status Negotiations (Parts 2700–2799)

XXVIII Office of the Vice President of the United States (Parts 2800–2899)

Title 33—Navigation and Navigable Waters

I Coast Guard, Department of Transportation (Parts 1–199)

II Corps of Engineers, Department of the Army (Parts 200–399)

IV Saint Lawrence Seaway Development Corporation, Department of Transportation (Parts 400–499)

Title 34—Education

Subtitle A—Office of the Secretary, Department of Education (Parts 1–99)

Title 34—Education—Continued

Subtitle B—Regulations of the Offices of the Department of Education

I Office for Civil Rights, Department of Education (Parts 100–199)

II Office of Elementary and Secondary Education, Department of Education (Parts 200–299)

III Office of Special Education and Rehabilitative Services, Department of Education (Parts 300–399)

IV Office of Vocational and Adult Education, Department of Education (Parts 400–499)

V Office of Bilingual Education and Minority Languages Affairs, Department of Education (Parts 500–599)

VI Office of Postsecondary Education, Department of Education (Parts 600–699)

VII Office of Educational Research and Improvement, Department of Education (Parts 700–799)

Title 35—Panama Canal

I Panama Canal Regulations (Parts 1–299)

Title 36—Parks, Forests, and Public Property

I National Park Service, Department of the Interior (Parts 1–199)

II Forest Service, Department of Agriculture (Parts 200–299)

III Corps of Engineers, Department of the Army (Parts 300–399)

IV American Battle Monuments Commission (Parts 400–499)

V Smithsonian Institution (Parts 500–599)

VII Library of Congress (Parts 700–799)

VIII Advisory Council on Historic Preservation (Parts 800–899)

IX Pennsylvania Avenue Development Corporation (Parts 900–999)

XI Architectural and Transportation Barriers Compliance Board (Parts 1100–1199)

XII National Archives and Records Administration (Parts 1200–1299)

Title 37—Patents, Trademarks, and Copyrights

I Patent and Trademark Office, Department of Commerce (Parts 1–199)

II Copyright Office, Library of Congress (Parts 200–299)

III Copyright Royalty Tribunal (Parts 300–399)

IV Assistant Secretary for Productivity, Technology and Innovation, U.S. Department of Commerce (Parts 400–499)

Title 38—Pensions, Bonuses, and Veterans' Relief

I Veterans Administration (Parts 0–99)

Title 39—Postal Service

I United States Postal Service (Parts 1–999)

III Postal Rate Commission (Parts 3000–3099)

Title 40—Protection of Environment

I Environmental Protection Agency (Parts 1–799)

V Council on Environmental Quality (Parts 1500–1599)

Title 41—Public Contracts and Property Management

Subtitle B—Other Provisions Relating to Public Contracts

50 Public Contracts, Department of Labor (Parts 50-1–50-999)

51 Committee for Purchase from the Blind and Other Severely Handicapped (Parts 51-1–51-99)

60 Office of Federal Contract Compliance Programs, Equal Employment Opportunity, Department of Labor (Parts 60-1–60-999)

61 Office of the Assistant Secretary for Veterans Employment and Training, Department of Labor (Parts 61-1–61-999)

Subtitle C—Federal Property Management Regulations System

101 Federal Property Management Regulations (Parts 101-1–101-99)

105 General Services Administration (Parts 105-1–105-999)

Title 41—Public Contracts and Property Management—Continued

109 Department of Energy Property Management Regulations (Parts 109-1–109-99)

114 Department of the Interior (Parts 114-1–114-99)

115 Environmental Protection Agency (Parts 115-1–115-99)

128 Department of Justice (Parts 128-1–128-99)

SUBTITLE D—Other Provisions Relating to Property Management (Reserved)

SUBTITLE E—Federal Information Resources Management Regulations System

201 Federal Information Resources Management Regulations (Parts 201-1–201-99)

Title 42—Public Health

I Public Health Service, Department of Health and Human Services (Parts 1–199)

III Saint Elizabeths Hospital, Department of Health and Human Services (Parts 300–399)

IV Health Care Financing Administration, Department of Health and Human Services (Parts 400–499)

V Office of Inspector General-Health Care, Department of Health and Human Services (Parts 1000–1999)

Title 43—Public Lands: Interior

SUBTITLE A—Office of the Secretary of the Interior (Parts 1–199)

SUBTITLE B—Regulations Relating to Public Lands

I Bureau of Reclamation, Department of the Interior (Parts 200–499)

II Bureau of Land Management, Department of the Interior (Parts 1000–9999)

Title 44—Emergency Management and Assistance

I Federal Emergency Management Agency (Parts 0–399)

IV Department of Commerce and Department of Transportation (Parts 400–499)

Title 45—Public Welfare

SUBTITLE A—Department of Health and Human Services, General Administration (Parts 1–199)

SUBTITLE B—Regulations Relating to Public Welfare

II Office of Family Assistance (Assistance Programs), Family Support Administration, Department of Health and Human Services (Parts 200–299)

III Office of Child Support Enforcement (Child Support Enforcement Program), Family Support Administration, Department of Health and Human Services (Parts 300–399)

IV Office of Refugee Resettlement, Family Support Administration, Department of Health and Human Services (Parts 400–499)

V Foreign Claims Settlement Commission of the United States, Department of Justice (Parts 500–599)

VI National Science Foundation (Parts 600–699)

VII Commission on Civil Rights (Parts 700–799)

VIII Office of Personnel Management (Parts 800–899)

X Office of Community Services, Family Support Administration, Department of Health and Human Services (Parts 1000–1099)

XI National Foundation on the Arts and the Humanities (Parts 1100–1199)

XII ACTION (Parts 1200–1299)

XIII Office of Human Development Services, Department of Health and Human Services (Parts 1300–1399)

XVI Legal Services Corporation (Parts 1600–1699)

XVII National Commission on Libraries and Information Science (Parts 1700–1799)

XVIII Harry S. Truman Scholarship Foundation (Parts 1800–1899)

XX Commission on the Bicentennial of the United States Constitution (Parts 2000–2099)

XXI Commission on Fine Arts (Parts 2100–2199)

XXII Christopher Columbus Quincentenary Jubilee Commission (2200–2299)

Title 46—Shipping

I Coast Guard, Department of Transportation (Parts 1–199)

II Maritime Administration, Department of Transportation (Parts 200–399)

III Coast Guard (Great Lakes Pilotage), Department of Transportation (Parts 400–499)

IV Federal Maritime Commission (Parts 500–599)

Title 47—Telecommunication

I Federal Communications Commission (Parts 0–199)

II Office of Science and Technology Policy and National Security Council (Parts 200–299)

III National Telecommunications and Information Administration, Department of Commerce (Parts 300–399)

Title 48—Federal Acquisition Regulations System

1 Federal Acquisition Regulation (Parts 1–99)

2 Department of Defense (Parts 200–299)

3 Department of Health and Human Services (Parts 300–399)

4 Department of Agriculture (Parts 400–499)

5 General Services Administration (Parts 500–599)

7 Agency for International Development (Parts 700–799)

8 Veterans Administration (Parts 800–899)

9 Department of Energy (Parts 900–999)

10 Department of the Treasury (Parts 1000–1099)

12 Department of Transportation (Parts 1200–1299)

13 Department of Commerce (Parts 1300–1399)

14 Department of the Interior (Parts 1400–1499)

15 Environmental Protection Agency (Parts 1500–1599)

16 Office of Personnel Management Federal Employees Health Benefits Acquisition Regulation (Parts 1600–1699)

17 Office of Personnel Management (Parts 1700–1799)

18 National Aeronautics and Space Administration (Parts 1800–1899)

19 United States Information Agency (Parts 1900–1999)

22 Small Business Administration (Parts 2200–2299)

24 Department of Housing and Urban Development (Parts 2400–2499)

25 National Science Foundation (Parts 2500–2599)

28 Department of Justice (Parts 2800–2899)

29 Department of Labor (Parts 2900–2999)

34 Department of Education Aquisition Regulation

44 Federal Emergency Management Agency (Parts 4400–4499)

52 Department of the Navy acquisition regulations (Parts 5200–5299)

53 Department of the Air Force Federal acquisition regulation supplement (Parts 5300–5399)

57 African Development Foundation (Parts 5700–5799)

61 General Services Administration Board of Contract Appeals (Parts 6100–6199)

63 Department of Transportation Board Contract Appeals (Parts 6300–6399)

Title 49—Transportation

Subtitle A—Office of the Secretary of Transportation (Parts 1–99)

Subtitle B—Other Regulations Relating to Transportation

I Research and Special Programs Administration, Department of Transportation (Parts 100–199)

II Federal Railroad Administration, Department of Transportation (Parts 200–299)

III Federal Highway Administration, Department of Transportation (Parts 300–399)

IV Coast Guard, Department of Transportation (Parts 400–499)

V National Highway Traffic Safety Administration, Department of Transportation (Parts 500–599)

VI Urban Mass Transportation Administration, Department of Transportation (Parts 600–699)

VII National Railroad Passenger Corporation (AMTRAK) (Parts 700–799)

VIII National Transportation Safety Board (Parts 800–899)

X Interstate Commerce Commission (Parts 1000–1399)

Title 50—Wildlife and Fisheries

I United States Fish and Wildlife Service, Department of the Interior (Parts 1–199)

II National Marine Fisheries Service, National Oceanic and Atmospheric Administration, Department of Commerce (Parts 200–299)

III International Regulatory Agencies (Fishing and Whaling) (Parts 300–399)

IV Joint Regulations (United States Fish and Wildlife Service, Department of the Interior and National Marine Fisheries Service, National Oceanic and Atmospheric Administration, Department of Commerce); Endangered Species Committee Regulations (Parts 400–499)

V Marine Mammal Commission (Parts 500–599)

VI Fishery Conservation and Management, National Oceanic and Atmospheric Administration, Department of Commerce (Parts 600–699)

Subject/Agency Index

List of Agency Prepared Indexes

Parallel Tables of Statutory Authorities and Rules

Acts Requiring Publication in the Federal Register

List of CFR Titles, Chapters, Subchapters, and Parts

Appendix D

Chronology and Cost of a Typical Forensic Investigation

ITEM 1. Preliminary Evaluation of Grinding Wheel Failure (Refer to Fig. 10-14)

1/7–1/10/86	*$510.00*

A. Disassembled the grinder and photographed the process to determine the parts, defects, and particularly the governor that controls rpm.

B. Conducted micro- and macroexaminations of the failed grinder disk. Examined and photographed the fracture site.

C. Performed preliminary theoretical calculations using simplified formulas to determine the stress in the disk during grinding (operator-applied force and centrifugal force).

D. Interviewed the plant supervisor (plaintiff's employer) by telephone.

E. Telephoned the plaintiff's attorney on January 8 to advise him about the preliminary results of examination. Identified the mode of failure as crack propagation and growth to rupture, commonly referred to as a fatigue mode. Indicated that the mechanism of failure was likely to be alternating stress in the disk caused by rotation and pressure applied by the user. The cause of failure could be *deficiencies in design, defects in the material, improper fabrication or use,* or *both.* Identified eight organizations as having a possible role in the failure of the product.

F. Outlined the following forensic investigation strategy or actions to be taken:

1. Perform background research into the eight organizations that may have affected the failure of the product: for example, grinder, codes and standards, and so on.
2. Perform a tensile test and fatigue analysis (S-N curve) of the material in the failed disk and compare the results with the tensile strength and fatigue strength of material from new disks.
3. Obtain a random sample of new grinder disks to examine the ply orientation of the composite and to obtain samples for tensile tests and fatigue tests (S-N).
4. Build a theoretical finite-element computer simulation of a grinder disk to determine the relationship of *stress* (σ or stress intensity, S) in the disk as a result of *load* (P) applied during the grinding operation.
5. Use the experimentally determined data on fatigue (number of cycles [N] before failure at alternating stress [S] value in disk value) and the theoretical finite-element computer results (P vs. S) to predict load (P) versus the number of cycles to failure (N) for the disk. Assess the risk of disk consumption during grinding before its disintegration.
6. Reduce the cost to the plaintiff's attorney of testing the material by obtaining data from the manufacturer on the nominal tensile strength and fatigue for the material in the failed disk.
7. Investigate the possibility of using applicable tensile and fatigue strength data on composite material in the public domain if the manufacturer doesn't provide data on the material or if the plaintiff's attorney can't afford the material tests.

G. Indicated to the attorney that a letter of authorization will be needed before fatigue tests can begin at Penn State. Quoted price estimates for each task. The attorney decided not to have me take action at this time.

ITEM 2. Engineering Evaluation and Analysis, Telecons, and Meetings

1/11–3/28/86	*$1785.00*

A. 1/23/86 Telecon discussions with the plaintiff's attorney. The attorney asked me to proceed with the forensic analysis. I indicated that Penn State University labs quoted a price of $4000 to $5000 to perform the tensile and fatigue tests on material from two sets of specimens (failed disk and new samples) and that written authorization would be needed to proceed with the tests. Advised that the costs could be reduced if data on tensile strength and fatigue is in the open literature or in the manufacturer's possession. Plaintiff's attorney did not authorize the material tests at Penn State.

B. 2/24–3/8/86 Forensic expert's activities:

1. Searched the literature for tensile strength and fatigue properties of composite material at in-house library, at University of Houston science library, at Rice University, and through database network system.
2. Evaluated the applicability of data on composite material available in the open literature for use in computer analysis of the grinder disk.
3. Began developmental basis of a computer simulation model by *finite-element method* that will determine user-applied load (P) versus stress (S) in the disk (this activity is reported in Item 6).
4. Telecon interview with the fabrication shop maintenance employee and plaintiff.

C. 3/17/86 Discussion with plaintiff's attorney regarding status of project and meeting with manufacturer's in-house expert.

D. 3/18/86 Travel and meeting with defendant's in-house expert (3 hours).

E. 3/24/86 Used data from open literature geared to structural applications of fiberglass and resin-based composites to calculate stress levels with computer simulation. (This work activity is reported in Item 8.) Reported to the plaintiff's attorney that because the test data in the open literature on the tensile strength and fatigue properties of structural composite materials are abnormally high, we need to test the actual material. Advised that the manufacturer's data would reduce the cost of the tests.

F. 3/27/86 Meeting with plaintiff's attorney, defense attorney, and manufacturer's in-house expert. Manufacturer provided substantial data on manufacturing procedures, material blend and compositions, quality control, and some tests, but did not provide data on tensile tests and fatigue tests. Time and travel (5 hours).

G. 3/28/86 Read and evaluated the manufacturer's data obtained at the meeting. Read and evaluated the grinder manufacturer's specifications, service manuals, and shop manuals.

H. 4/2/86 Performed grinder speed tests and removed additional sample from disk to determine fabric or ply orientation in new disks. (This activity and charge is reported in Item 3.)

ITEM 3. Testing of Grinder

4/28/86	*$358.00*

Performed grinder speed tests and removed 10 samples from new disks.

ITEM 4. Engineering Evaluation and Analysis

4/3–4/4/86 *$1360.00*

A. 4/3/86 Plaintiff's attorney asked me to give a deposition on work activity thus far, as requested by defense attorney. Advised the plaintiff's attorney that a deposition "is a bit premature" (the attorney's correspondence of July 28, 1986) because he had not authorized completion of the analysis of data on composite material nor the performance of tensile and fatigue tests, and the defense attorney had not provided it. Asked plaintiff's attorney if he anticipated that the case would go to trial and if he intended to wait until after the deposition to perform the tests. Suggested that the trip to give my deposition would be an opportunity to examine the accident site, take some measurements, photograph the site of disk fragmentation, interview personnel at the plant, and interview the plaintiff.

B. 4/4/86 Actions taken for plaintiff's attorney:

1. Removed sections from the failed disk with a special abrasive cutting wheel. Undamaged sections will be used for possible tensile and fatigue testing specimens at Penn State University. Fractured sections will be sent to the manufacturer's in-house expert.
2. Performed final examination, measurements, and photographic documentation of the grinder and disk parts that will be shipped to manufacturer's in-house expert.
3. Read two depositions supplied by the plaintiff's attorney: one by plaintiff; the other by an employee in the fabricating shop, Maintenance Department.
4. Reviewed applicable codes and standards (American National Standards Institute B7.1 and Occupational Safety and Health Agency). (CJR 29, Chapter XVII, Subpart P-1910.241 to 1910.247).
5. Reviewed my computer analysis of the grinder wheel and my fatigue analysis based on conservative data in the open literature.
6. Reviewed the data in the file regarding the case and reviewed the data on protective barrier design and on injuries to personnel caused by projectile impact.

ITEM 5. Deposition on the Forensic Expert's Activities

4/5/86 *$1190.00*

A. Traveled from Houston to Midland, leaving Houston office at 7:00 P.M. Friday and arriving at Midland at 2:00 A.M. Saturday (plane delays).

B. Arrived at plaintiff's attorney's office on Saturday morning. Attorney canceled the visit to the plant and the scene of accident.

C. Interviewed the plaintiff.

D. Reviewed the case with the plaintiff's attorney and prepared for the deposition.

E. Gave my deposition regarding activities on the case regarding the following:

1. Research on background information.
2. Examination of the failed disk and photomicrographs and planned tests.
3. Examination of the failed grinder unit (disassembly and tests on rpm versus line pressure). (See Fig. 10-4.)
4. Computer simulation of the disk by finite element computer method.
5. Samples taken from the failed disk and a number of new disks.
6. Review of depositions provided by the plaintiff's attorney.

Noted in the deposition that calculations have not been completed because I was not authorized to perform tests on material from the failed composite disk and new disks and that the defense has not provided this data, as requested in previous meetings.

F. Travel from Midland to Houston (left attorney's office at 4:45 P.M. and arrived at Houston office at 10:00 P.M.).

ITEM 6. Travel

4/4/86 *$208.00*

Round-trip air fare from Houston to Midland, ground transportation, parking, and meals.

ITEM 7. Testing of Tensile Properties and Fatigue Curve (S-N) Data

4/24/86 *$4000.00*

Applied mechanics testing lab at Penn State University performed three tensile test points and seven fatigue bend test points on samples of material from the failed disk and new disks. Two research associates assisted Dr. Sam Zamrik, project leader. I assisted in establishing data points and procedures required and in evaluating progress.

ITEM 8. Computer Analysis by Intertech Services, Inc.

4/86 *$1710.00*

A. Used the finite-element model of the grinder wheel to predict stresses (S) versus load (P) applied by the user.

B. Developed and outlined a theory for modeling composite materials by the finite-element method.

C. Calculate stresses caused by centrifugal force.

D. Conducted computer-simulated loads using (1) data on fiberglass material from the open literature and (2) data on material properties from the tensile and fatigue tests performed at Penn State University.

ITEM 9. Engineering Evaluation and Analysis: Reconstruction and Causality

5/3/86	*$1317.50*

A. Evaluated the procedure for testing tensile and fatigue strength and did an independent check of test results, examination of specimen, and photograph of specimen.

B. Evaluated fatigue data supplied by the manufacturer for new, similar, but not identical grinder disks (thickness is smaller and shape is different). The defense attorney gave these data to the plaintiff's attorney to forward to me at this time.

C. Compared the fatigue data recently forwarded from Penn State and the manufacturer. Integrated the data to develop a fatigue (S-N) curve. Used the computer simulation of the grinder wheel and experimental fatigue data from Penn State to produce curves of grinder-wheel force versus time to failure.

D. Performed an independent review of Intertech's computer simulation model.

E. Assessed the adequacy of the grinder guard against fragmentation of disk.

ITEM 10. Technical Report

4/86	*$1746.50*

Prepared a 60-page report with the following table of contents:

BACKGROUND
- Applicable Codes and Standards
- Manufacturer's Specifications

EXAMINATION AND INSPECTION
- Grinding Wheel
- Grinder Unit

MODES OF FAILURE

SIMULATION
- Test Data
- Mathematical Simulation of Disk Behavior
  - Composite Properties
  - Stress Analysis
  - Centrifugal Force
  - Concentrated Force
  - Summary

ACCIDENT RECONSTRUCTION

CAUSE OF FAILURE AND OPINION

RECOMMENDATIONS

REFERENCES

APPENDIXES

A. Properties of the Composite

B. Stress Caused by Centrifugal Force

C. Finite-Element Analysis of the Disk Subjected to a Concentrated Force

D. Test Tensile and Fatigue Data on S-N Curve

ITEM 11. Preparation for Trial

5/12–5/13/86	*$255.00*

Reviewed the following with Dr. Zamrik of Penn State University: deposition by the manufacturer's expert regarding the Penn State tests, my report, Intertech's analysis, and so on (3 hours)

ITEM 12. Miscellaneous Expenses

	$40.00

Miscellaneous clerical assistance and photocopies.

ITEM 13. Deliveries

4/8/86	*$119.75*

A. Shipment of grinder to manufacturer.

B. Shipment of samples to Penn State University.

C. Shipment or reports to plaintiff's attorney.

Appendix E

Checklist to Use When Questioning Technical Experts During a Deposition

A checklist such as the one included in this appendix is useful for attorneys when questioning expert witnesses during a deposition or trial (see also Brenton and Lovett 1987). Some questions may be unnecessary if a witness has provided an extensive and current curriculum vitae, which can be attached to his or her deposition as an exhibit.

Before the deposition begins, identify and introduce all persons who are present. Agree on stipulations regarding objections: (1) should they be made during the deposition, (2) should they be reserved until the time of the trial, or (3) should they be reserved except as to form of question and responsiveness of answer. In the absence of agreement, all objections as to the form of the question and the responsiveness of the answers are waived unless stated at the time of the deposition. Also note that if objections need to be stated at the deposition, both the examiner and cross-examiner must state their objections to the nonresponsiveness of answers, (4) the court reporter is qualified to swear in the witness, (5) objections to the formalities of notice and return are waived, (6) the deposition can be used at trial without the witness's signature. The form for stipulations can be obtained from various sources, such as Branton and Lovett (1987), federal Form 54, and state forms.

1. *Preliminary Questions*
   a. What is your name and address?
   b. What is your *occupation*—for example, consulting engineer, widget engineer, engineering manager, safety engineer?
   c. What is your *profession*—for example, professional engineer, educator, administration, doctor of jurisprudence?
2. *Personal Data*
   a. What is your educational background?
      - What university did you attend?
      - What is your degree?
      - From what college in the university did you graduate—for example, engineering, arts and sciences?
      - In what department did you receive your education—for example, mechanical, civil, electrical, chemical engineering?
      - What was your major course of study or specialization—for example, general mechanical engineering, heat transfer, dynamics, structural mechanics?
      - Was the university's baccalaureate program in engineering approved by the Accrediting Board of Engineering and Technology (ABET)?
      - In what year did you graduate?
      - What courses did you take that are relevant to your opinion in the case?
      - What courses were available in your department that had the following words in their title: design, failure analysis and accident investigation, human factors, industrial safety, maintenance, manufacturing, marketing, materials, process, product support, reliability, sales, testing, training?
      - Which of these courses did you take?
   b. Do you have any advanced degrees?
      - From what university?
      - Which degrees did you earn (M.S., M.E., D.E., E.D., Ph.D., D.Sc.)?
      - From which colleges?
      - From which departments?
      - What were your major courses of study?
      - What was the title of your thesis or report?
      - Was the graduate engineering program approved by the ABET?
      - In what year (or years) did you graduate?

- What courses of study were relevant to your opinion in this case?
- In what areas of specialization were your courses and research?

c. Did you have any additional education?
  - Have you attended any seminars, workshops, or short courses that are relevant to your opinion in this case?
  - On what dates?
  - Where?
  - What were the topics of discussion?
  - Who sponsored the program?

3. *Employment History*

a. Who is your current employer?
  - What is your current position?
  - What is your job title?
  - What education background does the job require?
  - What experience does the job require?
  - What is the nature of duties (e.g., what percentage of your time is spent in administration, theoretical calculations, testing, supervision, marketing or sales)?
  - Have you written or cowritten any technical reports?
  - How many of these reports are relevant to your opinion in this case?
  - What are the titles of the reports?
  - When were they written?
  - Who issued the reports (are they in the public domain)?
  - How extensive were these reports (did the projects cover a span of months or years)?

b. What is your previous employment (from immediately after graduation until present job)?
  - What were the inclusive dates?
  - What were your positions and job titles?
  - What were your duties and responsibilities?
  - What degrees were required?
  - What experience was required?
  - Were you the author or coauthor of any technical reports?
  - What were the titles of these reports?
  - When were they written?
  - Who issued them (are they in the public domain)?
  - How extensive were the reports (did the projects cover a span of months or years)?

c. What general and specific employment experience is relevant to your opinions in this case?
d. Do you have any teaching experience (university or college, continuing education, technical society, corporate employer, government agency)?

4. *Professional Activities*

a. Do you have a professional license?
- Did you take the National Council of Engineering Examiners's (NCEE) eight-hour Engineer-in-Training (EIT) examination? State and date?
- Did you take the eight-hour professional practice examination? Did you pass? State, date, designated branch, reciprocity with what states?
- Do you practice engineering under the corporate exemption clause?
- Do you practice engineering under the grandfather clause?
- Does the state where you practice certify your expertise as a (mechanical, civil, electrical, chemical, widget) engineer?

b. Are you a member of any technical societies?
- Are these societies members of ABET (e.g., American Society of Mechanical Engineers, American Institute of Chemical Engineers, Institute of Electrical and Electronics Engineers, American Society of Civil Engineers)?
- What degree and experience are required for membership?
- Have you held any offices?
- Have you been appointed to any committees?
- What is the organization's principal function?
- Are you an active or inactive member?
- How long have you been a member (inclusive dates)?
- Have you developed any of the following technical forums: symposia, workshops, conferences? Were any of these relevant to your opinions in this case? Titles of these forums? Where were they held and when?

c. What journals do you subscribe to in your discipline, industry, and product?

d. Have your published any technical articles or books?
- Are any of them relevant to this case?
- What are the titles?
- What are the subjects?
- When and where were they published?

e. Have you made any technical presentations in journals, magazines, pamphlets, or proceedings of symposia?
f. Are you a member of any committee that write codes and standards?
   - Which organization do you represent on the committee?
   - Does the committee permit membership by representatives of governmental agencies, insurance companies, various industries, technical societies, professional societies, and consumer-user organizations?
   - What degrees are required for membership?
   - What professional engineering registrations are required for membership?
   - Are you an officer of the committee?
   - If membership is open to different interest organizations or individuals is election or appointment to the committee weighted in favor of any specific category?
   - Which memberships in codes and standards committees are relevant to this case?

5. *Patents and Copyrights*
   a. Do you hold any patents? How many?
   b. What is the nature and description of these patents?
   c. What are their titles?
   d. What are the patent numbers?
   e. When were they issued?
   f. When were they applied for?
   g. Did you contribute to any patents that are not held in your name?
   h. Do you hold any copyrights?
   i. Did you assign all copyrights to your employer?
   j. Are any of your patents or copyrights relevant to your opinions in this case?

6. *Field of Specialization or Expertise*
   a. In this case, do you rely on your expertise by your formal educational training (degrees), your occupational experience (employment), or your research (funded, personal development)?
      - What areas did you specialize in after completing your formal education that are relevant to this case: as a result of your occupational experience, your research, or past formal training?
      - Is this specialization or expertise certified or qualified by

professional registration or certification by the Accrediting Board for Engineering and Technology?
- Has your specialization been registered or certified by these or some organization?
- Is your expertise or specialization relevant to your opinion in this case?
- Which areas of expertise does your company or you advertise, promote, or identify to clients for your services or products?

b. What percentage of your time do you devote to consulting versus other occupational duties?
- What percentage of time do you spend consulting—company versus private consulting?
- What is an engineer? ("One who is knowledgeable in mathematics and scientific principles and applies them to the development and use of products in an efficient and safe manner.")
- What percentage of your consulting time is devoted to accident investigations and failure analyses?
- What percentage of your consulting time is devoted to incidents that result in litigation?

c. To what engineering or scientific functions do you devote 50 percent or more of your time and at least 5 percent of your time? To what management functions do you expect to devote 50 percent or more of your time and at least 5 percent of your time?

d. What other persons do you consider to be authoritative in terms of expertise, specialization, and training to assess the technical incident involving the product in question?

e. What type of experience have you had with the product involved in this case?
- Design and analysis?
- Selection of materials?
- Sale?
- Testing?
- Consequences of use and/or failure?
- Accident investigation and failure analysis?
- Personal contact with or use of the product?
- Human factors?
- Reliability?
- Maintainability or inspection?
- Safety?
- Training?

- Material processing?
- Development of predictive methods (experimental and theoretical)?
- Participation in writing codes and standards?
- Research and development?

7. *Experience as an Expert Witness* (in-house versus public service)
   a. Have you done any consulting in the past regarding the subject product or a similar product (if we agree that consulting means temporary professional advisory services)?
   b. For whom have you consulted?
      - What was the client's name?
      - What was the nature of the product or issue involved in the case?
      - Who was the plaintiff (injured consumer)?
      - Who was the defendant (manufacturer, for example)?
   c. Have you had any past experience with lawsuits (currently pending or closed)? Were any of these suits relevant to this case?
   d. What hazard, mechanism, or alleged design defect was involved?
      - What were the names of the parties?
      - What were the facts of the suit?
      - Where was the site of the accident?
      - What was the venue of the suit?
   e. Have you previously given testimony as an expert witness?
      - How many times at trial?
      - How many times at depositions?
      - What were the names of the attorneys involved?
      - What were the names of the parties to the suit?
      - What were the issues involved?
      - What percentage of your time was devoted to testifying?
      - What percentage of your time was devoted to engineering consulting and to expert testimony for industry only?
      - What percentage was devoted to engineering consulting and expert testimony for a personal injury plaintiff?

8. *Consultation in the Present Case*
   a. For what purpose were you hired in this lawsuit?
   b. When were you retained?
   c. Do you have a written agreement with your client?
   d. Were your expenses paid up before the start of the deposition or trial? How much is outstanding?
   e. What information has been furnished to you?

f. What exhibits, photographs, or documents have been furnished to you?
g. Have you inspected the product?
h. Have you inspected the scene of the accident?
i. What depositions have you read?
j. What tests have you performed?
k. Have you attempted to reconstruct the product failure or accident?
l. Have you run any theoretical or experimental simulations to determine the mode or mechanism of failure or both?
m. Have you interviewed eyewitnesses?
n. Have you reviewed relevant background information?
   - Specifications?
   - Reports?
   - Manuals?
   - Codes and standards?
   - Current practices?

9. *Standards*

a. What is meant by voluntary industry standards?
b. What organizations propose voluntary industry standards?
   - American National Standards Institute (ANSI)?
   - National Bureau of Standards?
   - National Safety Council?
   - American Society of Agricultural Engineers (ASAE)?
   - American Society of Mechanical Engineers (ASME)?
   - American Society for Testing and Materials (ASTM)?
   - Federal Drug Administration (FDA)?
   - National Fire Protection Association (NFPA)?
   - Society of Automotive Engineers (SAE)?
   - Underwriters Laboratory (UL)?
c. What is the purpose of a (fill in the applicable) standard from a design, manufacturing, and user standpoint?
d. What is a design standard?
e. What is a performance standard?
f. What is a protection standard?
g. What is a warning standard?
h. What applicable mandatory and voluntary standards are you citing in this case?
   - Which standards are relevant in this case?
   - What type of standards are they (design, performance, protection, warning)?
   - What are the membership requirements for the committee that wrote the applicable standard: degree, professional

registration, industry employee, individual, manager versus practicing professional engineer?

- Why do engineers prefer performance standards? (They do not limit product design and are user-passive compared with protection and warnings.)
- Which one should be correct and accurate if there is an applicable engineering design by analysis standard that is in disagreement with the design standard?

i. If no voluntary or mandatory standards exist for a product, does it mean that nothing needs to be done from a safety standpoint concerning the foreseeable use of the product and the hazards and risks in connection with its use?

j. If the standard does not contain any applicable guidance regarding design, performance, protection, warning of a system, component, part, or feature of a product, does this imply that nothing needs to be done or considered from a safety standpoint regarding the foreseeable use of the system, component, part, or feature and the hazards and risks in connection with its use?

k. Does the compliance of a product with a standard necessarily mean that the product is safe?

l. What are in-house or organizational standards (e.g., Navy, General Motors, NASA)?

m. If mandatory standards are relevant to the case, are they based on the voluntary standards?

n. What weight would be given to a standard that is proposed but not yet approved?
   - Does it demonstrate knowledge of potential hazards and risks?
   - Does it help to identify risks and hazards?
   - Is it a potential means of eliminating or minimizing risks and hazards?

o. Does industry ever consult or use standards before they are actually approved?

10. *State of the Art*

a. What is meant by "state of the art" in design engineering, selection of material, processing of material, manufacturing, operation, maintenance, or inspection?
   - Does it include what is being done in a particular field?
   - Does it include what is known regardless of whether it is being implemented in the particular field and may be common in other industries or professions?
   - Does it include all the literature in the field?

- Does it include patented concepts?
- Does it include that which is technologically feasible?

b. Are there any alternative concepts, principles, methods, and procedures with respect to design engineers, selection of material, processing of material, manufacturer, operations, or maintenance and inspection that could have prevented the incident?

c. Is anyone in the industry using these concepts, principles, methods, and procedures as an alternative? Which manufacturers? For how long? What is the purpose of such a design engineering alternative?

d. Does the mere fact that no one in the industry is using these concepts, principles, methods, or procedures demonstrate that they are not state of the art?

e. Does the mere fact that no one in the industry is using them mean that they are unnecessary from a safety standpoint?

f. Is any industry or organization using these concepts, principles, methods, or procedures?

11. *Engineering Considerations*

a. From a design engineering standpoint, does the fact that a hazard may be obvious and known to the foreseeable user indicate that the design safety does not need to be considered with respect to the particular hazard?

b. When considering how the product should be design engineered, is it of prime importance to foresee the environment in which the product will be fabricated, used, maintained, and inspected?

c. Is the fact that the product will be fabricated, operated, and maintained and inspected by humans the number one environmental consideration in its design engineering?

d. Is the fact that a human could be in close proximity to the product a reasonably foreseeable environmental consideration?

e. Is it reasonably foreseeable that a human could come into actual contact with the product when using, maintaining, or inspecting the product?

f. Do design engineers recognize that humans are not perfect and will make mistakes and errors of judgment when using, maintaining, or inspecting the product?

g. Is the basic purpose of product protection (e.g., product guarding) to prevent injuries caused by human mistakes and errors of judgment?

h. Given these factors, what safety features have been incorporated into the design engineering of this product to account for the human as an environmental element?

i. What safety features have been incorporated specifically to mitigate the hazard (contain, reduce, or eliminate it)?

j. When design engineering a product, is it unwise to rely entirely on operator training, warnings, or the operator's constant attention (inasmuch as humans will make mistakes)?

k. Is the safety of humans and their foreseeable use of a product the primary concern when design engineering the product?

l. Do you agree that the concept of protecting or guarding those parts of the product that move or may move when humans are in close proximity use good safety practices? Explain.

m. What roles do patents play in design engineering?
   - Do they make inventions and design improvements public knowledge?
   - Do they identify hazards, risks, and dangers?
   - Do they suggest ways of handling the hazards?
   - Do they demonstrate a technologically feasible means of eliminating or minimizing the particular hazard or risk?
   - Do they demonstrate an economically feasible means of eliminating or minimizing the particular hazard or risk?

n. Is performance commonly sacrificed on behalf of a safety factor (for example, a shield on a grinder or fan)?

o. Do you agree that if using a product poses a foreseeable risk of injury, an injury will occur sooner or later (Murphy's Law)?

p. To be beneficial, must a protective system minimize or eliminate the risk of injury or failure under (1) all circumstances, (2) circumstances in which injury or failure is highly probable, (3) circumstances in which probability of failure is low (unlikely) but the consequences are severe (e.g., a nuclear or chemical accident), or (4) all circumstances with high consequence?

q. What is the role of design engineering with respect to the following? (Refer to Box 4–4 and Box 7–2 for supplementary questions.)
   - Selection or specification of material with respect to performance, allowable number of defects in raw materials, or the environment.
   - Manufacturing specifications with respect to tolerances, processing, testing, inspection, and defects.

- Shipping requirements.
- Maintenance specifications.
- Inspection requirements.
- Specified responsibilities or requirements regarding users.

r. What is the role of the manufacturer or processor of intermediate-to-finished products with respect to the following? (Refer to Chapter 4.)
  - Selection or specification of material with respect to performance, the environment, or allowable number of defects.
  - Manufacturing specifications with respect to tolerances, processing, testing, inspection, criteria regarding defects, case histories, current practices, sales, and marketing.

s. What is the role of the owner-user responsibilities with respect to specifications or requirements provided to the supplier or vendor (e.g., specifications related to design engineering, materials, defects, tolerances, manufacturing, operation, maintenance, and inspection)? (Refer to Chapter 7, user responsibilities.)

t. What is the role of the project administrator or engineering department administrator?

12. *Findings and Opinions*

a. Has the product been altered or modified from its original design?
b. What is the history or condition of the product with respect to manufacture, transportation, use, and maintenance?
c. Has the product been damaged as the result of an accident?
d. Are there product defects as a result of processing, manufacturing, or use?
e. Was the product used in the environment in which it was expected to be used?
f. What was the user's level of skill and experience?
g. Did your review of background data and your physical examination, physical tests, theoretical and experimental simulations, and reconstruction reveal any significant information?
h. What was the mode of failure? Describe it.
i. What was the mechanism of the failure, the injury, or the accident? Describe it.
j. In your opinion, who is responsible for or caused the product failure or accident?
k. Was there a causal connection between product design, defect, use, and injury? Describe.
l. In your opinion, who caused the injury (who is responsible)?

m. In your opinion, did the product comply with applicable codes and standards?

13. *Reconstruction of the Incident*

a. On the basis of your findings, can you explain how this incident occurred?
   - Was it foreseeable that a person could be injured in this manner?
   - Is the fact that people are being injured because of this hazard a matter of concern to the industry or engineers?
   - What has been done to eliminate or minimize that hazard with respect to this product?
   - What testing has been done with respect to this particular hazard in an attempt to eliminate or minimize the hazard? On what dates were the tests performed? What were the results? Who performed the tests? What methods were used? What were the findings and conclusions?

b. Considering the manner in which you have explained your understanding of how this accident occurred, would a (alleged safe design, material selection, material process, manufacture, operation and maintenance/inspection feature or procedure) have prevented this accident or injury?

c. Who is responsible for this incident?

d. Have you reconstructed the incident or accident, (the sequence of events that led to the failure and/or injury)?

e. What in your opinion caused this accident?

f. Are the (alleged safe design material selection, material process, manufacture, operation, and/or maintenance/inspection fracture or procedure alternatives) offered on the market by any competitors?
   - Name of company?
   - Length of time they have been offered on the market?

g. Are you aware of any testing by competitors in which the (design material selection, material process, manufacture, operation, and/or maintenance/inspection fracture or procedure alternative) was involved?

h. Has there ever been any discussion or recommendation from anybody in the company or outside the company pertaining to the incorporation of a ____________ feature or procedure into the product (that would have prevented this injury)?

14. *Background Research*

a. What technical literature have you reviewed that is relevant to this case and to your opinion?

b. What other information have you reviewed that is relevant to this case and to your opinion (e.g., case histories, statistical data, databases, specifications, drawings, reports, manuals, service history data, current practices, and records of design, testing, material, fabrication, installation, inspection, operation, and maintenance)?
c. What safety standards have you reviewed regarding the product?
d. What safety codes have your reviewed regarding the product?
e. What codes and standards are applicable to the product?

15. *Protection*

a. Would you agree that there are two basic protection techniques: (1) active (requires an intermediate action, as, for example, a water sprinkler depends upon sensor and activation block) and (2) passive (does not depend upon intermediate action, as, for example, a protection guard, a wall, a barrier)?
b. Would you agree that the requirements of interlocking protection are: (1) when protection is opened or removed, the machine, equipment, or system cannot operate? (2) when protection is in the proper operating position it cannot be removed until the equipment is stopped or shut down? and (3) the device must be failsafe?
c. Do you agree that according to good safety engineering, a product must fail in a safe mode: that is, if a protective system is removed, the product inadvertently shuts down, or a hazard is otherwise exposed, further operation is impossible?
d. Do you agree that the state of the art of equipment or product or system protection has always been to design out the hazard, where feasible?
e. Do you agree that if a hazard cannot be designed out, the first alternative is always to seek to protect against the hazard?

16. *Warnings*

a. What is the purpose of a warning? Do you agree that:
   - It serves as a constant reminder of a danger?
   - It warns of a danger that cannot be designed out of the product?
   - It reduces the risk?
b. Do you believe, from a safety design standpoint, that simply

because a danger is known and obvious, no warning is necessary?

c. Do you believe that instructions or warnings provided by a manufacturer should explain the danger of using the product in an unintended but nonetheless foreseeable manner?

d. What warnings were provided with the product?
   - Were they affixed to the product?
   - Were they included in the operator's manual?
   - Were they provided elsewhere?

e. Should the designer, manufacturer, user, regulatory organizations, or insurance companies require that operators be trained to use the product?

f. Were the warning labels or instructions permanently affixed to the product?

g. Do you have expertise in human factors, warnings, or ergonomics?

17. *Hazard or Risk Analysis*

a. Did you evaluate the product from a safety standpoint; that is, did you use fault tree or consequence analysis?
   - What hazards did you identify?
   - What risks in using the product did you identify?

b. Was it technologically and economically possible to eliminate the specific hazards at the time the product was produced or involved in the incident?

c. Should warnings about the hazards have been included?

d. Which of the following options is preferable?
   - Eliminate the hazard.
   - Protect against the hazard.
   - Provide readily recognizable (by any one or all of the senses) warnings.
   - Instruct or train the operator regarding the hazard.

e. How would the following affect the usefulness of the product?
   - Eliminating the hazard.
   - Guarding against the hazard.
   - Warning of the hazard.
   - Instructing or training the operator regarding the hazard.

18. *Feasibility of Preventing the Hazard* (Refer to Fig. 4-1 and Chapter 4.)

a. Was the alternative that you believe would have prevented the accident technologically and economically feasible when the product was produced or the accident occurred?

b. On what dates was the product designed, manufactured, purchased, and resold?

19. *Opinions Regarding Unreasonable Danger* (Refer to Chapters 7 and 11.)

a. Do you agree that any product that poses the risk of serious injury or death and that risk can be eliminated or protected, warned, or instructed against by technologically and economically feasible means, is an unreasonably dangerous product by design?
b. Do you agree that any product that is modified without the designer's, manufacturer's, or cognizant organization's approval poses the risk of economic loss or injury?

20. *Witness's Expertise Regarding Human Factors, Warnings, and Ergonomics*

a. Can you define the terms ergonomics and human factors?
b. Can you provide a brief history of the field of human factors and human-factors psychology?
c. What part of your education was relevant to human factors, warnings, and ergonomics?
d. Do you have teaching experience that is relevant to human factors, warnings, and ergonomics?
e. Do you have any other work experience that is relevant to product labeling?
f. Have you written any articles, given any lectures, or participated in any panels that are relevant to human factors, warnings, or ergonomics?
g. Do you have any current professional involvements that are relevant to human factors, warnings, and ergonomics?
   - Are you a member of any professional or technical societies that are involved in these three subjects?
   - Do you hold any offices in these societies?
   - Do you serve on any committee of these societies?
   - Do you serve on any national committees?
   - Which committees or organizations have some involvement with human-factors psychology as it relates to warnings?
   - Have you received any honors regarding the three subjects?
h. Can you provide examples of effective warnings? For this product?
   - What makes a warning attract attention?

- When is a warning given in relation to when a potentially dangerous activity is to be performed?
- Where is the warning given in relation to the dangerous product or activity?
- What human need does the warning affect or human response does it motivate?
- What instruction does the warning contain concerning how to avoid harm?

i. What dangers are associated with the product?
- What is the danger associated with the product in this case?
- What is the likelihood of serious injury or death?

j. What are your findings and opinions regarding warnings?
- Did this product require a warning?
- Why (or why not)?
- Did the product have a warning on it?
- If so, was the warning adequate?
- What instructions came with the product?
- Did these instructions constitute a warning?
- Did the owner's manual contain any warnings?
- Did the packaging contain any warnings?
- Did anything associated with the product carry a warning about the specific hazard or danger?

k. Do you have any suggestions with respect to warnings?
- What should the warning say?
- Why?
- Why have you not formulated precise language to use in the warning?
- Where would you place the warning?
- Why would you place the warning there?

l. Who bears the responsibility to warn?
- Does the manufacturer's responsibility to warn increase if the manufacturer is aware of the danger?
- Does that responsibility increase if the manufacturer knows why people have been injured by the product?

m. What is your opinion concerning whether the product is unreasonably dangerous?
- What is the basis for your opinion?
- Is the product unreasonably dangerous because it lacks warnings?
- Is it unreasonably dangerous because the warnings or instructions are inadequate?
- Could this kind of accident have been prevented if it contained a warning or if the warning had been effective?

- What is the basis of your opinion regarding causal relationships?

21. *Concluding Checklist*
    a. Has the expert witness signed and dated all diagrams or drawings made during the deposition?
    b. Have all exhibits been marked and given to the court reporter to attach to the deposition?
    c. Has the court reporter been advised if the party wants to buy a copy of the deposition?
    d. When does the court reporter expect the deposition to be ready?
    e. Have the court reporter's name, address, and employer (if any) been obtained for future reference?

22. *Definitions*
    a. Hazard—The injury potential (e.g., a pinch point).
    b. Risk—The probability of a potential injury from the hazard that is affected by the proximity to the hazard and the amount of exposure to the hazard.
    c. Danger—Unreasonable or unacceptable combination of hazard and risk.
    d. Consultant—If we agree that consulting refers to temporary professional advisory services.
    e. ABET (Accrediting Board for Engineering and Technology).
    f. NCEE (National Council of Engineering Examiners).

Appendix F

About Legal Citations

SOURCES CITED IN NOTES

While legal citations often appear complicated, they are, in reality, quite simple. There are four elements to a citation (Table F-1 defines the abbreviations used in legal citations). A typical citation to a decision of the U.S. Supreme Court would be *American Textile Mfgrs. Inst., Inc.* v. *Donovan*, 452 U.S. 490, 101 Sup. Ct. 2478, 69 L.Ed. 185 (1981). The name of the case is given first. In many state court systems the first name in the case is the party that commenced the action. But in the U.S. Supreme Court and in some state courts the first name is the party that has taken the appeal to the court that is issuing the opinion. The second name is usually the party against which the action was originally brought. In citations to opinions given by the U.S. Supreme Court, the second party has won in the next lower court. U.S. is the abbreviation of the case reporter from which the case is taken, in this case the U.S. Supreme Court Reporter. The number preceding the abbreviation of the reporter system (452) is the volume in which the case appears. The number following the abbreviation of the reporter system (490) is the page on which the opinion commences. Finally, the citation concludes with the year that the court announced the decision (1981). The opinions of the U.S. Supreme Court are also reported in the supreme court where the case is taken, in this case the U.S. Supreme Court Reporter, published by the West Publishing Company.

A typical state court citation would be *Henningsen* v. *Bloomfield Motors, Inc.*, 32 N.J. 358, 161 A.2d 69 (1960). N.J. is an abbreviation for the New Jersey Official reports. This abbreviation is preceded by the volume of the reporter (32) in which the opinion is found, and followed by the

TABLE F-1. Table of Abbreviations

A.2d	Atlantic Reporter, Second Series
Cal.2d	California Reports, Second Series
Cal.3d	California Reports, Third Series
Cal.App.3d	California Appellate, Third Series
Cal.Rptr.	California Reporter
F. Supp.	Federal Supplement (Federal Trial Court Decisions)
L.Ed.2d	Lawyer's Edition, Second (U.S. Supreme Court Cases)
N.J.	New Jersey Supreme Court Reports
N.Y.2d	New York Reports, Second Series
P.2d	Pacific Reporter, Second Series
Pa.	Pennsylvania Supreme Court Reports
Pub. L.	Public Law
S.D.N.Y.	Southern District, New York (Federal Trial Court)
Stat.	United States Statutes-At-Large
Sup. Ct.	Supreme Court Reporter (U.S. Supreme Court cases)
U.S.	United States Supreme Court Reports
U.S.C.	United States Code

page number on which the case begins (358). This is followed by the citation to the regional reports, Atlantic, Second (A.2d), with the volume number preceding (161), and the beginning page number following (69). Finally, the citation is concluded with the year of the decision.

New York and California have separate unofficial reports, New York Supplement and California Reporter, that collect in one volume all published opinions of both their highest and their immediate appellate courts.

Federal statutes can be found in a number of different sources. The official source is the United States Code (U.S.C.). Statutes found there, along with case annotations interpreting them, are listed in United States Code Annotated (U.S.C.A.). These statutes are also found in books that give the laws enacted by the Congress each year, but which do not list them according to a subject matter system, as do the U.S.C. and U.S.C.A. These yearly collections are in what is called Statutes-At-Large (Stat.). As with case decisions, the first number is the volume number, and the subsequent number is the number of the page on which the statute begins. These yearly collections of statutes list the cases by Public Law number.

There are a few references to legal periodicals. The volume number precedes the name of the periodical, the page number on which the article begins follows the name of the periodical, and the citation concludes with the year of publication.

Bibliography

Abney, D. L. 1986. Expert testimony and eyewitness identification. *Case & Comment*, 91(2).

ABET. 1987a. *Accredited programs leading to degrees in engineering* (AB-20 12/87). New York: Accreditation Board for Engineering and Technology.

ABET. 1987b. *Accredited programs leading to degrees in engineering technology* (AB-21 12/87). New York: Accreditation Board for Engineering and Technology.

ABET. 1987c. *Accreditation Board for Engineering and Technology* (AB-1 10/87SD). New York: Accreditation Board for Engineering and Technology.

Adler, S. C., and Pierman, B. C. 1975. *History of walking slip-research at the National Bureau of Standards* (NBS Special publication). Washington, DC.

American Bar Association Journal. 1984, February, pp. 12–16.

American National Standards Institute. ANSI-B1-7, 1978.

ASME. 1986. *Guide to writing ASME codes and standards*. New York: American Society of Mechanical Engineers.

ASME. 1968. *Boiler and Pressure Vessel Code*. New York: American Society of Mechanical Engineers.

American Society of Metals, "Case Histories in Failure Analysis," ASM, Metals Park, OH, 1979.

ASNT. 1988. *Nondestructive testing handbook* (4 vols). Columbus, OH. American Society of Nondestructive Testing.

ASTM. 1975. Part 10. Philadelphia: American Society for Testing and Materials.

ASTM. 1977. *A standard specification for special requirements for steel plates for nuclear and other special applications* (Specification A647-77). Philadelphia: American Society for Testing and Materials.

ASTM. 1981. *Standard test method for J_{Ic}, a measure of fracture toughness* (Standard E813-81). Philadelphia: American Society for Testing and Materials.

American Textile Mfrs Inst., Inc. v. Donovan. 452 U.S. 490, 101 Sup. Ct. 2478, 69 L.Ed. 2d 185 (1985).

Amsbartsumyan, S. A. 1964. *Theory of anisotropic shells.* Moscow: State Publishing House for Physical and Mathematical literature. NASA Technical trans. NASA TT F-118.

Ando, K. 1975. Analysis for cracks in ship structures using the finite element method with the concept of sub-structure. In E. F. Rybicy and S. E. Benzle (Eds.), *Computational fracture mechanics.* New York: American Society of Mechanical Engineers.

Anthony v. Abbott Laboratories, 106 F.R.D. 461 (1985).

Apostolakis, G. E., *et al.* (Eds.). 1989. *Reliability engineering and system safety.* Essex, U.K.: Elsevier Applied Science.

Archea, J. C. 1983. *History of stairs.* Georgia Institute of Technology.

Archea, J. C., Collins, B., and Stahal, F. 1979. *Guidelines for stair safety.* "Science series 120." Washington, DC.: National Bureau of Standards, U.S. Department of Commerce.

Argo Industries Inc. v. Heat-Timer Corp., New York Law Journal, December 15, 1972.

Astrand, P. O., and Rodahl, K. 1986. *Textbook of work physiology.* New York: McGraw-Hill.

Atkeson, T. B., and Neidich, G. 1983, February. The business lawyer. American Bar Association. Section of Corporation, Banking and Business Law. Vol. 38, No. 2, p. 623.

Automobile recalls: Caveat vendor. 1972, June 25. *New York Times,* Sec. 3.

Au-Yang, M. K., and Brown, S. J. 1979. Dynamics of fluid pressure systems in the energy industry. ASME. Vol. PVP-39.

Au-Yang, M. K., and Galford, J. E. 1981. Fluid-structure interaction—A survey with emphasis on its application to nuclear steam system design. ASME. Vol. PVP-46.

Baker, W. E. 1983. *Explosions in air.* San Antonio, TX: W. E. Baker Engineering.

Baker, W. E. 1985. Post-test assessment of blast and fragment effects of explosive failure of a large steel containment shell model during pneumatic testing. In S. J. Brown (Ed.), *Proceedings of the 1985 Pressure Vessel and Piping Conference* (PVP–98-6). New York: American Society of Mechanical Engineers.

Baker, W. E., *et al.* 1983. *Explosions hazards and evaluation.* Amsterdam: Elsevier Scientific Publishing Co.

Baldwin, S. 1986. American Trial Lawyers Association Mid-winter Convention.

Barker v. Lull Eng. Co. 20 California Reports, 3rd series, 413, 143 California Reporter 225, 573 Pacific Reporter, 2d series 443 (1978).

Barry v. Manglass. 389 N.Y. Supplement, 2d series 870 (1976).

Barsom, J. M., and Rolfe, S. T. 1977. *Fracture and fatigue control in structures.* Englewood Cliffs, NJ: Prentice-Hall.

Bartknecht, W. 1981. *Explosions.* Berlin, FRG: Springer-Verlag.

Bathe, K. J., Ozdemir, H., and Wilson, E. L. 1974. Static and dynamic geometric and material nonlinear analysis. Report UCSESM 74-4, Department of Civil Engineering, Structural Engineering Lab, University of California, Berkeley, CA.

Bathe, K. J., and Wilson, E. L. 1976. *Numerical methods in finite element analysis.* Englewood Cliffs, NJ: Prentice-Hall.

Baum, M. R. 1984. Large missiles generated by the rupture of gas pressured vessels: A summary of current data. In S. J. Brown (Ed.), *Impact, fragmentation, and blast (vessels, pipes, tubes, equipment).* (Vol. PVP-82). New York: American Society of Mechanical Engineers.

Baum, M. R. 1989, July. Rupture of a gas pressurized cylindrical pressure vessel: The dynamics of end-cap missiles. Institute of Mechanical Engineers, U.K. *Chemical Engineering Research Design*, Vol. 67.

Beachem, C. D. 1972. A new model for hydrogen-assisted cracking (hydrogen "embrittlement"). *Met. Trans*, 3: 437–451.

Beckwith, T. G., and Buck, N. L. 1961. *Mechanical measurements.* Reading, MA: Addison-Wesley.

Bednar, H. H. 1986. *Pressure vessel design handbook.* New York: Van Nostrand Reinhold.

Bellavita, P. 1978. *Advances in composite materials* (pp. 357–381). London, England: Applied Science Publishers.

Belytschko, T., and Greer, T. L. 1977. *Computational methods for fluid-structure interaction problems.* ASME AMD, Vol. 26.

Belytschko, T. B., and Kennedy, J. M. 1975. Thermal and fluid/structure dynamics analysis. In *Structural mechanics in reactor technology* (Vol. B). Berlin, FRG.

Bennett, G. F., Feates, F. S., and Wilder, I. 1982. *Hazardous material spills handbook.* New York: McGraw-Hill Book Co.

Berkebile v. Brantley Helicopter Corp. 219 Pennsylvania Superior Court Reports 479, 281 Atlantic Reporter, 2d series 707 (1971).

Beshada v. Johns Manville Products Corp., 90 NJ, 191, 447 2nd 539, 1982.

Blessington v. McCrory's Stores, 305 NY 140.

Blundell, J. 1985. *Machine guarding accidents.* "Nuts and bolts series." Columbia, MD: Hanrow Press.

Boley, J. W. 1977. *A guide to effective industrial safety*. Houston, TX: Gulf Publishing.

Bowden, V. P., and Tabor, D. 1964. *Friction and lubrication of solids* (Parts I and II). Oxford, England: Clarendon Press.

Boyer, J. E. (Ed.). 1975. *Metals handbook* (10 vols). Metals Park, OH: American Society for Metals.

ASM. 1986. *Metals handbook* (9th ed. 13 vols; Vol. 12. *Fractography*. Metals Park, OH: American Society for Metals.

ASM. 1986. *Metals handbook* (9th ed. 13 vols; Vol. 11. *Failure analysis and prevention*. Metals Park, OH: American Society for Metals.

ASM. 1979. *Case histories in failure analysis*. Metals Park, OH: American Society for Metals.

Brandon v. Yale & Towne Mfg. Co., (DC Pa. 1963). Commerce Clearing House, Products Liability Reports, Par. 5392, 220F, Supp. 855.

Branton, J. L., and Lovett, J. D. 1987. *Trial lawyers series*. Fort Worth, TX: Knowles Law Books.

Briem, J. 1987. Warning defect cases: An engineer's involvement. Paper delivered St. Louis Engineers Club, St. Louis, MO, January 30, 1987.

Brody v. Overlook Hospital, 121 N.J. Super 299, 296 A 2d 668 (1972).

Brown, D. B. 1976. *System analysis and design for safety*. International series in industrial and systems engineering. Englewood Cliffs, NJ: Prentice-Hall.

Brown, M. W., and Miller, K. J. 1973. A theory for fatigue failure under multiaxial stress-strain conditions. *Proceedings of the Institute of Mechanical Engineers*, 187, 745–755 and D229–D244.

Brown, S. J. 1976. Pressure vessel failure. In Waite, H. *Pressure vessel and piping: Design and analysis* (Chap. 5). "Decade of progress." New York: American Society of Mechanical Engineers.

Brown, S. J. 1979. On the mechanical and thermal transient analysis of cylinder-to-cylinder vessels by a finite plate method. *International Journal of Pressure Vessel Piping*, 7.

Brown, S. J. 1980. Nonlinear analysis of LWR components: Areas of investigation-benefits-recommendation (1AO-76-1). Albuquerque, N.M. Sandia National Laboratory, U.S. Department of Energy.

Brown, S. J. 1982. Hydrodynamic response of fluid coupled cylinders: Simplified damping and inertia coefficients. Welding Research Council Bulletin No. 281. New York, N.Y.

Brown, S. J. 1983a. Fluid component analysis and the finite element method. Seventh Annual Modern Valve Technology Conference. University of Wisconsin, Milwaukee.

Brown, S. J. 1983b. Stress-strain (constitutive) relationships. In *ASM handbook of engineering mathematics* (Chap. 19). Metals Park, OH: American Society of Metals.

Brown, S. J. (Ed.). 1984. *Impact, fragmentation and blast (vessels, pipes, tubes, equipment)*. Vol. PVP-82. New York: American Society of Mechanical Engineers.

Brown, S. J. 1985a. Energy release protection of pressurized systems: Part I. A review of studies into blast and fragmentation. *Applied Mechanics Review*, 38(12), pp 1625–51.

Brown, S. J. 1985b. Fracture, fatigue and advanced mechanics. In Brown (Ed.), *Proceedings of the Pressure Vessel and Piping Conference* (Vol. PVP–98-8). New York: American Society of Mechanical Engineers.

Brown, S. J. (Ed.). 1985c. Structural dynamics. ASME Proceedings of the ASME Pressure Vessel and Piping Conference (Vol. PVP–98-6). New York: American Society of Mechanical Engineers.

Brown, S. J. 1985d. Fluid Structure Interaction (Chap. 4 Sec. 7). *Decade of progress*. New York: American Society of Mechanical Engineers.

Brown, S. J. 1986a. Energy release protection for pressurized systems: Part II. Review of studies into impact/terminal ballistics. *Applied Mechanics Reviews*, 39(2), pp. 177–201.

Brown, S. J. 1986b. Instrumentalities of engineering and the role of engineering education. In Grayson, L. P., Biedenbach, J. M., Proceedings of the 1986 Annual Conference (pp. 1489–95). University of Cincinnati, June 23–6, 1986. American Society of Engineering Education.

Brown, S. J. 1986c, May. *Pressure systems energy release protection* (NASA Contract Report No. 178090). NASA Langley Research Center, Hampton, VA.

Brown, S. J. 1987a, January. The effects of system dynamics on failure: Illustrative examples. *AIChE Plant/Operations Progress*, Vol. 6, No. 1, p. 20.

Brown, S. J. 1987b. Product liability and the use of videotape at trial. Session on the use of videotape evidence at trial. Chu, M., and Martindale, J. E. Section of litigation committee on trial practices, mid-year meeting, J. R. Archibald, Program chairman. April 24–25.

Brown, S. J. 1989. Flow induced vibration in pump impellers: A case study. In *Proceedings of the Tenth Conference of Structural Mechanics in Reactor Technology*, Los Angeles, California, Vol. T, pp. 13–18.

Brown, S. J. 1990. On the use of a flange and stud array to affect a component post-tension stress field: A case study. In *The analysis of pressure vessel and heat exchanger components*. (Vol. PVP-194). New York: American Society of Mechanical Engineers, p. 103.

Brown, S. J., and Brown, T. J. 1990a. An analysis of thickwalled 90 and 45 degree pipe fittings. In *Design and analysis of piping components*. (Vol. PVP-188), New York: American Society of Mechanical Engineers, p. 25.

Brown, S. J., and Brown, T. J. 1990b. A computer simulation of a bolted flange with a spiral-wound gasket configuration in a tee shell. In *Finite element analysis, computer applications and data management.* (Vol. PVP-185). New York: American Society of Mechanical Engineers.

Brown, S. J., and Baker, W. E. 1992. *Handbook of explosions and hazardous release.* New York: Butterworth.

Brown, S. J., and Chu, M. 1983. A case study of the seismic response of fluid-coupled-flexible cylinders. *Journal of Experimental Mechanics,* Vol. 23, No. 3, pp. 270–81.

Brown, S. J., and Hsu, K. H. 1977. *On the verification of finite element transient temperature analysis with thermally orthotropic material* (Vol. PVP-PB-024). New York: American Society of Mechanical Engineers.

Brown, S. J., and Hsu, K. H. 1978. On the use of the finite element displacement method to solve solid-fluid interaction vibration problems. In *Fluid transients and acoustics in the power industry,* C. Papadakas and H. Scarton (Eds.) New York: Fluids Engineering Division, American Society of Mechanical Engineers.

Brown, S. J., and Moody, F. W. *Fluid structure dynamics.* New York: Van Nostrand Reinhold. In press.

Brown, S. J., and Reimus, W. S. 1984. Metallic bellows and expansion joints (PVP, Vol. 83, Part 2). New York: American Society of Mechanical Engineers.

Brown, S. J., and Wu, J. N. C. 1977. The influence of boundary conditions upon the axisymmetric vibration of a shallow spherical shell. (Vol. PVP-PB-022). New York: American Society of Mechanical Engineers.

Brown, S. J., Hailzlip, L., Nielson, J., and Reed, S. Analytical and experimental stress analysis of a cylinder-to-cylinder structure. *ASME Journal of Pressure Vessel Technology,* Nov. 1977. (Published also in 4th Structural Mechanics in Reactor Technology Conference: Finite element, photoelastic, and strain gage stress analysis of a cylinder-to-cylinder structure.)

Buck, T. H., and Buck, N. L. 1961. *Mechanical measurements.* Reading, MA: Addison-Wesley.

Bunton, L. D., Judge (U.S. District Court, Western District of Texas). Personal correspondence to Dr. S. J. Brown, Jr., September 19, 1986.

Burbage v. Boiler Engineering 7 Supply Co. Inc., 1967 16 Ches Co. Rep. 111, aff'd. 433 Pa. 319, 249 A 2d 563.

Burgher, P. H. 1988. Timing is important in selecting an expert witness. *The Expert and the Law,* 8(3): p. 3.

Burgoyne, J. H. 1985. The assessment and control of major hazards. Three-day symposium on Integrity and Reliability of Plants, Consequence Assessment and Risk Analysis in Case Studies, Institute of Chemical Engineering Symposium Series No. 93. Rugby, Warks, CB-21 CHQ.

Bushnell, D. 1969. Computer analysis of shell structures. ASME paper No. 69-WA/PVP-13.

Cantor, B. J. 1985. *The expert witness*. Belmont, MA: Civil Evidence Photography Seminars.

Carson, D. H., Archea, J. C., Margulis, S. T., and Carson, F. E. 1978. *Safety on stairs*. (NBS Building Science Series no. 108). Washington, DC.: Center for Building Technology, National Engineering Laboratory, National Bureau of Standards.

Carlson, D. 1972. *How man moves, kinesiological methods and studies*. London, England: Heinemann.

CEP Manual. 1987, May. Dow's fire and explosion index hazard classification guide, sixth edition. AIChe for *Chemical Engineering Progress*, New York.

Chen, S. S. 1985. Flow induced vibration of circular cylindrical structures. Argonne National Laboratory ANL 85-51.

Chu, M. L., and Brown, S. J. Experiments on the dynamic behavior of fluid-coupled co-axial cylinders. *Journal of Experimental Mechanics*, April 1981.

Chu, M. L., and Brown, S. J. 1985. An experimental study of vibrational viscous effects in fluid coupled concentric cylinders (pipes and vessels). In *Seismic performance of pipelines and storage tanks* (Vol. PVP-98-4). New York: American Society of Mechanical Engineers.

Chu, M. L., Brown, S. J., Lieb, B., and Lestingi, J. 1979. An experimental study of gap and thickness influence on the vibration response and damping of flexible fluid coupled co-axial cylinders. In *The dynamics of fluid-structure systems in the energy industry* (Vol. PVP-39). New York: American Society of Mechanical Engineers.

Chu, M., Lam, P., Gross, R., Chung, B., and Brown, S. 1983. *Handbook of engineering mathematics*. Metals Park, OH: American Society for Metals.

Clare, V. R., Lewis, J. H., Mukiewicz, A. P., and Sturdivan, L. M. 1976. *Handbook of human vulnerability criteria* (EB-SP-76011-9). Edgewood Arsenal, Aberdeen Proving Grounds, Aberdeen, MD.

Clark Equipment Co. 1983. *Operator restraint for the forklift truck: A compendium*. Ann Arbor, MI: Industrial Truck Division.

Clugston, S. F., Weertman, J. R., and Shewmon, P. G. 1983. The enhancement of hydrogen attack in steel by prior deformation. *Met. Trans. A*, 14A, pp. 695–699.

Codling v. Paglia, 327 NYS 2d 978.

Coffin, L. F., Jr. 1954. A study of the effects of cyclic thermal stresses on a ductile metal. *Transactions of the ASME*, 76, 923–949.

Cohen, G. A. 1968. Computer analysis of asymmetric buckling of ring-stiffened orthotropic shells of revolution. *AIAA Journal*, 6, 141–149.

Colangelo, V. J., and Thornton, P. A. 1981. *Engineering aspects of product liability*. Metals Park, OH: American Society for Metals.

Coles, J. 1943. Standardization of consumer's goods, 78, 1932.

Collins, J. A. 1981. *Failure of materials in mechanical design.* New York: John Wiley & Sons.

Colwell, A. B. 1981, August. An evaluation of ultrasonic radiographic inspection of welding technology. Unpublished master's thesis, University of Texas at Austin.

Connor, J. J., and Brebbia, C. A. 1976. *Finite element techniques for fluid flow.* London, England: Newnes-Butterworths.

Constance, J. D. 1988. *How to become a professional engineer,* 4th ed. New York: McGraw-Hill.

Coskuntuna, S., and Clmauro, 1980. Human factors in industrial design in consumer products. In *Proceedings of the symposium of human factors in industrial design in products.* Boston, MA: Tufts University.

Cost-benefit analysis, cost effective analysis and the cotton dust standard. 1982. *Rutgers Law Review* 35: 133.

Criteria for design of elevated temperature class I components. 1976, May. New York: American Society of Mechanical Engineers.

Crocker v. Winthrop Laboratories Division. 502 S.W. 2d 850 (Civil Ct. App. Tex.) (1973).

Crouch, T. D. 1987, Spring. The feud between the Wright brothers and the Smithsonian. *American Heritage of Invention and Technology.*

Crown Cork & Seal Co. Inc. v. Morton Pharmaceutical Inc. (CA-6 Tenn. 1969), *Products Liability Reports,* Sec. 6246, 417 F. 2d 921.

Cunitz, A. J. 1980, May–June. *Psychologically effective warnings: Hazard reduction.*

Dalehitt v. U.S., 346 U.S. 15.

Deeds, W. E., and Dodd, C. V. 1985. *Eddy-current techniques.* "Decade of Progress." New York: American Society of Mechanical Engineers.

Defects in the courtroom: Review and outlook. 1983, June 15. *Wall Street Journal,* 28, p. 142.

Dentzer, S. 1984. The product-liability debate. *Newsweek,* September 10, p. 54.

Dhillon, B. S. 1986. Fault-tree analysis. In J. Kurtz (Ed.), *Mechanical engineer's handbook* (Chap. 20). New York: John Wiley & Sons.

Dieter, G. E. 1961, 1976. *Mechanical metallurgy.* New York: McGraw-Hill.

Dimeo v. Minster Machine Co. Inc., 388 F 2d 18.

D'Isa, F. 1968. *Mechanics of metals.* Reading, MA: Addison-Wesley.

Doelle, B. 1974, Spring. Product liability law has come of age. *Cleveland State Law Review,* Vol. 23, No. 2, p. 201.

Dorey, J., and Gachot, B. 1978. Pump reliability data. Derived from Electrícite de Francé operating experience. (ASME, PVP-PV032). New York: American Society of Mechanical Engineers.

Dorn, J. E. 1957. The spectrum of activation energies for creep. In *Creep and recovery* (pp. 266–283). Metals Park, OH: American Society for Metals.

Dorn, J. E. (Ed.). 1961. *Mechanical behavior of materials at elevated temperatures.* New York: McGraw-Hill.

Dowling, A. R. 1970. *J. Inst. Met.,* Vol. 98, p. 215.

Dow's fire & explosion index hazard classification guide (6th ed.). 1987. New York: American Institute of Chemical Engineers.

Eastman Kodak Co. 1983. *Ergonomic design for people at work* (Vols. 1 and 2). New York: Van Nostrand Reinhold.

Eberhard, J. W. 1982, January. Quantitative imaging in nondestructive evaluation (NDE) by ultrasonic time of flight (TOF) tomography. Materials evaluation.

Edison, T. 1889. *Scientific American.*

Elices, M., Planas, J., and Corres, H. 1984. *Fracture prevention in energy and transport systems* (pp. 501–509). Cradley Heath, Warley, U.K.: Engineering Materials Advisory Services.

Ellenger, H., Buchele, S. F., and Hopkins, F. 1988, April 13. 3-D imaging techniques using industrial computerized tomography. American Society for Industrial Technical Material Evaluation.

Emerging product liability under Section 2-318 UCC. *Business Lawyer,* April 1974 (American Bar Association).

Engle, H. 1987. Fifth NASA Nondestructive Evaluation Workshop. Washington, DC.

Entwisle, F. 1983. *Operator restraint for the lift truck: A compendium by the Clark Equipment Co.* Industrial Truck Division, Ann Arbor, MI.

Environmental Protection Agency. 1972. *Control of hazardous material spills.* Washington, DC.

Epstein, R. 1980. Modern product liability law. In *The expert and the law,* August 1986. National Forensic Center, Princeton, N.J.

The expert and the law. 1987. National Forensic Center, Princeton, NJ.

Factory. 1973. June, p. 42.

Farr, J. R. 1982. Code design and technology. In Zamrik, S. Y., and Dietrich, D. *ASME pressure vessel and piping design technology: A decade of progress* (Chap. 1.) New York: American Society of Mechanical Engineers.

Farrar, J. C. M., Charles, J. A., and Dolby, R. E. (Eds.). 1971. *Effect of second-phase particles on the mechanical properties of steel.* London, England: ISI.

Fault-trees: A state of the art discussion. *IEEE Transactions on Reliability, R-23*(1), 51.

Fauske, H. K., and Lueng, J. C. 1985. New experimental techniques for characterizing runaway chemical reactions. *Journal of Chemical Engineering Progress.*

Federal Register. 1972, October 18. Department of Labor Occupational Safety & Health Standards, Vol. 37, No. 202, Part II.

Federal Register. 1979. 44 2996.

Federal Register. 1979. 44 62714.

Federal Trade Commission. 1983, April. *Final staff report: Federal Trade Commission standards and certification.* 1983, April. Washington, DC.

Federal Trade Commission Task Force on Industry Self-Regulation, Preliminary Staff Study, April 20, 1972.

Fenves, S. J., *et al.* 1973. *Numerical and computer methods in structural mechanics.* New York: Academic Press.

Ferry, T. (Ed.). 1985. *New directions in safety.* DesPlaines, IL: ASSE Florence, *et al.* 1975.

Final report of the Federal Interagency Task Force on Product Liability Law (PB273-220). 1-1. 1977. Springfield, VA: National Technical Information Service. U.S. Department of Commerce.

Final report of the National Commission on Product Safety. 1970, June.

Finn, R. 1987. The experts speak, engineering & science. In *The expert and the law.* National Forensic Center.

Flynn, W. J., Jr. 1972. Products liability requisites. *New York State Bar Journal,* June, 44(4).

Fong, J. T. 1978. *Inservice data reporting and analysis* (PVP-PV-032). New York: American Society of Mechanical Engineers.

Ford, R. D. 1970. *Introduction to acoustics.* New York: Elsevier.

Forli, O. 1979. Reliability of ultrasonic and radiographic weld testing. *Proceedings of the Ninth World Conference on Nondestructive Testing.*

Forster, J. F. 1971. *How states plan job safety and health.* Washington, DC.: U.S. Department of Labor.

Fowler, F. D. 1980. Failure to warn: A product design problem. *Proceedings of the Symposium of Human Factors in Industrial Design in Products.* Boston, MA: Tufts University.

Frank, R. 1981. Microfractographic features of plastic failures. *Praktische Metallographie,* 18: 157–171.

Frank, K. H., Kohutek, T. L., and Hopkins, F. 1983. *Application of photon tomography to inspect bridge weldments* (Scientific measurements systems report). Scientific Measurements Systems, Inc., Austin, TX.

Frederick, D., and Chang, T. S. 1965. *Continuum mechanics.* Boston, MA: Allyn & Bacon.

Frey v. Chrysler Corporation (DC Pa. 1968).

Frumer, L. R., and Friedman, M. I. 1988. *Products liability.* New York: Matthew Bender.

Frumer, L. R., Benoit, R. L., Friedman, M. I. 1988. *Personal injury actions, defenses, damage* (4 vols.). New York: Matthew Bender.

Fund, Y. C., and Anliker, M. 1972. In Fund and Anliker (Eds.), *Biomechanics: Its foundations and objectives*. Englewood Cliffs, NJ: Prentice-Hall.

Fussell, J. B., Powers, G. J., and Bennetts, R. G. 1974. Fault-trees—A state of the art discussion. *IEEE Transactions on Reliability*, R-23(1): 51.

Gallagher, R. H. 1975. *Finite element analysis: Fundamentals*. Englewood Cliffs, NJ: Prentice-Hall.

Gallagher, R. H., *et al.* 1982. *Finite element in fluids* (Vol. 4). New York: John Wiley & Sons.

Gangadharan, A. C. 1976. Reliability engineering. In Waite, H., *Pressure vessel and piping: Design and analysis* (Chap. 6). "Decade of progress" series. New York: American Society of Mechanical Engineers.

Garrett v. Holiday Inns, Inc. 58 N.Y. 2d 253, 460 N.Y. 2d 774, 447 N.E. 2d 717 (1983).

Garrison v. Rohm & Haas, CCH Product liability reports, par. 7127.

Garwood, S. J., and Harrison, J. D. 1985. The use of yielding fracture mechanics in post failure analysis. In *Pressure vessel and piping technology: A decade of progress* (Chap. 9). New York: American Society of Mechanical Engineers.

Gee, E. G. 1975. The American business system can live with change—Technology: Our most important asset. American Society of Agricultural Engineers, June 23, 1975.

George, G. S. (Ed.) 1990. *USGF gymnastics safety manual,* 2nd ed. Indianapolis, IN: USGF Publications.

Glodowski, R. J. (Ed.). 1983. Through-thickness tension testing of steel. American Society of Testing and Materials, STP 794.

Goodyear Tire v. Hammond. 269 F. 2d 501.

Green, A. E. (Ed.). 1982. *High risk safety technology*. New York: John Wiley & Sons.

Griffith, A. A. 1920. The phenomena of rupture and flow in solids. *Philosophical Transactions of the Royal Society*, 221A, 163–198.

Grimshaw v. Ford Motor Co. 119 Cal. App. 3d 757, 174 Cal. Rptr. 248 (1981).

Guidelines for organizing a product safety program. American National Standards Institute. Consumer Council Publication 2.

Hall, G. 1985. *The failure to warn handbook.* "Nuts and bolts series." Columbia, MD: Hanrow Press.

Hamilton. 1978. Role of nongovernmental standards in the development of mandatory standards affecting safety or health. *Texas Law Review* 56: 1329.

Hammer, W. 1972. *Handbook of systems and product safety*. Englewood Cliffs, NJ: Prentice-Hall.

Hazard, I. F. 1968, October. Product liability claims. *American Bar Association Journal*, 54: 981.

Heath v. Sears, Roebuck & Co. 464 A. 2d 288 N.H. (1983).

Helene Curtis Industries Inc. v. Pruitt. 385 F 2d 841, 849 (5th Cir. Ct) (1967).

Henningsen v. Bloomfield Motors, Inc. 32 N.J. 358, 161 A. 2d 69 (1960).

Hertzberg, R. W., Skibo, M. D., and Manson, J. A. 1979. *Fatigue mechanisms* (STP 675, pp. 471–500). Philadelphia: American Society for Testing and Materials.

Hicks, T. G., Tyler, G., and Mueller, J. F. 1986. *Standard handbook of consulting engineering practice.* New York: McGraw-Hill.

Higgins, L. R. 1988. *Maintenance engineering handbook.* New York: McGraw-Hill.

Hodge, F., and Phillip, G. 1971. Plastic analysis and pressure vessel safety. *Applied Mechanics Review*, 24(4).

Hooker, L. 1990. Danger below. *Chemical Engineering Progress,* Vol. 86, No. 5, p. 52.

Hull, D. 1981. *An introduction to composite materials.* Cambridge, England: Cambridge University Press.

Hydraulics and pneumatics. 1973, July. p. 30.

Hydrolevel Corp. v. American Society of Mechanical Engineers (1982).

INS v. Chadha. U.S. 102 S. Ct. 2764, 77 L. Ed. 317 (1983).

Interagency Committee on Standards Policy. 1970, April 17. Memorandum from the acting chairman.

International Harvester Co. v. Sharoff, 202 F 2d 52.

International Institute of Welding. 1956. Pressure vessels, boilers and pipelines. Document XI-2-56 of Commission XI.

Irving, R. R. 1980. The product liability case of the exploding range (Part II). *Iron Age*, August 11.

Irwin, G. R. 1948. *Fracturing of metals* (pp. 147–166). Metals Park, OH: American Society of Metals.

Irwin, G. R. 1957. Analysis of stresses and strains near the end of a crack transversing a plate. *Transactions of the ASME, Journal of Applied Mechanics*, 24, 361–364.

Irwin, G. R., Krafft, J. M., Paris, P. C., and Wells, A. A. 1967, November 21. Basic aspects of crack growth and fracture (NRL Report 6598). Washington, DC.

Isecke, B., and Stichel, W. 1984. *Structural failure, product liability and technical insurance* (pp. 107–113). Amsterdam: North-Holland.

Jackson v. Muhlenberg Hospital, 53 NJ 138, 249, A 2d 65 (1969).

Jawad, M. H., and Farr, J. R. 1984. *Structural analysis and design of process equipment.* New York: John Wiley & Sons.

Jetter, R. I., Brown, S. J., and Pamidi, M. R. 1981. Metallic bellows and expansion joints (PVP, Vol. 51, Pt. 1).

Johnson, J. I., and Johnson, K. L. 1973, May 8. Effective utilization of third party testing. Paper presented at NFPA convention, Boca Raton, FL.

Johnson, L. G. 1984. The ten deadly deposition sins. *American Bar Association Journal,* 70 (September): 62.

Jur, T. H., Peters, L. C., Talbot, T., and Weinstein, A. S. 1981. Engineering, the law, and education in design. American Society of Electrical Engineers, 71:271–274.

Justice, J. T. 1983. *Proceedings of the International Conference on Pipeline Inspection.* Edmonton, Alta., Canada.

Kalins, A. 1964. Analysis of shells of revolution subject to symmetrical and nonsymmetrical loads. *Journal of Mechanics* Transactions of the ASME, 31, 467–476.

Keefer v. C. R. Bard, Inc. 313 N2d 151 155 (Mich. APP. 1981).

Keher v. Armour & Co., 200 Fed. 322, 323 S.D. NY (1912).

Kellerman, O. 1966, October. Present views on recurring inspection of reactor pressure vessels in the Federal Republic of Germany. In Proceedings of Conference on In-Service Inspection, Pilsen, IAEA, Vienna.

Kellerman, O., and Siepel, E. H. 1968. *Analysis of the improvement of safety obtained by a containment and by other safety devices for water cooled reactors* (SM-89, TB 89). IAEA.

Kellogg Corporation News, 1986. The role of the expert—litigation/arbitration. M. C. Wilkinson (Ed.). Kellogg Corp. Littleton, CO.

Ketterer v. Armour & Company, 200 Fed. 322, 323 S.D. NY (1912).

Kincaid, J. F. 1967, November 14. Address before the National Electrical Manufacturers Association, Washington, DC.

Kitagawa, H., and Takahashi, S. 1976. *Proceedings of the Second International Conference on the Mechanical Behavior of Materials* (ICM-II), pp. 627–631. Boston.

Klayman, 1982. Standard setting under the consumer product safety amendments of 1981: A shift of regulatory philosophy. *George Washington University Law Review* 51: 96.

Klesnil, M., and Lukas, P. 1980. *Fatigue of metallic materials.* Amsterdam: Elsevier.

Kobayashi, A. S., *et al.* 1985. Rapidly propagating ductile crack in a zin pressurized pipe. In S. J. Brown (Ed.), *Proceedings of the Pressure Vessel and Piping Conference* (Vol. PVP–98-8). New York: American Society of Mechanical Engineers.

Kobayashi, A. S. (Ed.). 1988. *Experimental mechanics handbook.* Bethel, CT: Society of Experimental Mechanics.

Kolar, M. J. 1985. Documenting failures: A case study of turbine disks cracking. In S. J. Brown (Ed.), *Structural failures: A legal-engineering overview* (Chap. 9). "Decade of Progress" series. New York: American Society of Mechanical Engineers.

Kraus, H. 1967. *Thin elastic shells.* New York: John Wiley & Sons.

Kurtz, M. 1986. *Mechanical engineers' handbook.* New York: John Wiley & Sons.

Labbens, R., 1984. Safety factors calculation and their conservation in pressure vessels. 5th International Conference on Pressure Vessel Technology, Vol. III, American Society of Mechanical Engineers, New York, NY.

Labbens, R. 1985. Linear elastic fracture mechanics calculations in pressure vessels. In S. J. Brown (Ed.), *Decade of progress* (Chap. 9). New York: American Society of Mechanical Engineers.

Ladino, M. J. 1986, December. The expert witness. *Specifying Engineer.*

LaGorga v. Kroger Co. 275, F. Supp. 373, 378 (W.D. Pa. 1967), aff'd 407 F2d 671 (3rd Cir), 1969.

Lam, P., Gross, R., Chung, B., and Brown, S. 1983. *ASM handbook of engineering mathematics.* Metals Park, OH: American Society of Metals.

Lamb, Carrington, and Shields. 1971. Trade association law and practice. p. 76.

LaQue, F. L., and Copson, H. R. (Eds.). 1963. *Corrosion resistance of metals and alloys.* New York: Van Nostrand Reinhold.

LeMay, I. 1978. Failure mechanisms in metallography: A review. *Metallography & failure analysis.* New York: Plenum Press.

LeMay, I. 1978. *Metallography in failure analysis* (pp. 1–31). New York: Plenum Press.

LeMay, I. 1981. *Principles of mechanical metallurgy.* New York: Elsevier.

LeMay, I. 1985. Failure modes. In S. J. Brown (Ed.), *Decade of progress* (Chap. 9). New York: American Society of Mechanical Engineers.

LeMay, I., and Da Silveira, T. L. 1984. *Proceedings of the Second International Conference on Creep and Fracture of Engineering Materials and Structures,* pp. 1117–1133). Swansea, Wales: Pineridge Press.

LeMay, I., Justice, J. T., and Jamieson, R. M. 1984. *Fracture prevention in energy and transport systems* (pp. 13–24). Cradley Heath, Warley, U.K.: Engineering Materials Advisory Services.

Lestingi, J., and Brown, S. J. 1973, September. Comparison of the numerical integration technique and the finite element method in the analysis of thin shells structures. *Proceedings of the Second International Conference on Structural Mechanics in Reactor Technology.* Berlin, FRG.

Liability of certifiers of products for personal injuries to the user or consumer. 56 *Cornell Law Review,* Vol. 56, p. 132.

Lindholm, U. S. 1971. *Techniques in metals research* (Vol. 5, Pt. 1). New York: Wiley–Interscience.

Liu, T. H., *et al.* 1986. *Seismic engineering for piping systems, tanks and power plant equipment* (Vol. PVP-108). New York: American Society of Mechanical Engineers.

Lukasiewicz, S. 1979. *Local loads on plates and shells.* Warsaw: (PWN) Polish Scientific Publishers.

Lynch, S. P. 1979. Hydrogen embrittlement and liquid-metal embrittlement in nickel single crystals. *Scripta Metallurgica,* 13, 1051–1056.

Lyons, J. 1982. *Lyon's valve designer's handbook.* New York: Van Nostrand Reinhold.

MacPherson v. Buick Motor Company. 217 N.Y. 382, 111 N.E. 1050 (1916).

Magic Chef v. Sibley 546 S.W. 2d 851 (Ct. Civ. App. Tex.) (1977).

Mahan v. Stover 679 W.W.2d 707 (Tex. App. [9th Dist.] 1984, writ dismissed.

Malvern, L. E. 1969. *Introduction of mechanics of a continuous medium.* Englewood Cliffs, NJ: Prentice-Hall.

A management guide to product quality and safety. Hartford, CT. Travelers Insurance Co.

Mandel, J. 1974, April. The evaluation of standard test methods. *ASTM Standardization News,* p. 17.

Manjione, J. J. 1944. The influence of rate of strain and temperature on yield stresses of mild steel. *Transactions of the American Society of Mechanical Engineers,* 66, A-211.

Manson, S. S. 1953. Behavior of materials under conditions of thermal stress. NACA TN 2933.

Marano vs. Ogden Marine (1986).

Marin, J. 1968. *Mechanical behavior of engineering materials.* Englewood Cliffs, NJ: Prentice-Hall.

Marsh Wood Products Company v. Babcock and Wilcox. 207 Wis. 209, 240 N.W. 392 (1932).

Maurizio D. Fortunato v. Ford Motor Co., New York Law Journal, July 11, 1972.

McCahill, F. X., Jr. 1971, May–June 1971. Avoid losses thru risk management. *Harvard Business Review.*

McCall, J. L., and French, P. M. (Eds.). 1978. *Metallography and failure analysis.* New York: Plenum Press.

McComish v. Desoi (1963), CCH Products liability reports, par. 5256, 83 NJ Super 505, 200 A 2d 511.

McConville, J. B. 1985. Safer design through dynamic simulations. *Journal of Computer Aided Engineering.* Society of Manufacturing Engineers.

McCormick v. Hankscraft Company. 278 Minn. 322, 154 N.W. 2d 488 (1967).

McGuiness, J. 1977. Human factors in consumer product safety. *Proceedings of Human Factors Society.*

McLean, W. G. 1985. American Society of Mechanical Engineers: Role of professional societies. In Yeater, M. L., *ASME pressure vessel and piping design technology: A decade of progress* (Chap. 10). New York: ASME.

McQuaid, J. Trials on dispersion of heavy gas clouds. *Plant/Operations Progress* (Vol. 4, No. 1). Jan. 1985. AIChE.

Megyesy, E. F. 1983. *Pressure vessel handbook* (6th ed.). Tulsa, OK: Pressure Vessel Handbook Publishers.

Meidl, J. H. 1970. *Explosive and toxic hazardous materials*. "Fire science series." Beverly Hills, CA: Glencoe Press.

Meidl, J. H. 1978. *Flammable hazardous materials*. "Fire science series." Beverly Hills, CA: Glencoe Press.

Mekdeki v. Merrill Nat. Lab. 1983. 711 F. 2d 1510 (11th Cir. Ct.) (1983).

Melvin, J. W., *et al.* 1982. Operator restraint testing program: Phase II. Ann Arbor, MI: Highway Safety Institute, University of Michigan.

Mescall, J. F. 1965. Large deflection spherical shells under concentrated loads. *Journal of Applied Mechanics,* 32, 936–938.

Miller, K. J. 1985. *Proceedings of the Eshelby Memorial Conference*. Cambridge, England: Cambridge University Press.

Miller, R. L. 1986. *Industrial hygiene handbook for safety specialists (toxie torts)*. "Nuts and bolts series." Columbia, MD: Hanrow Press.

Mooring v. Fram Corp., 420 SW 2d 462.

Myer, K. (Ed.). 1986. *Mechanical engineers' handbook*. New York: John Wiley & Sons.

Nadai, A. 1950. *Theory of flow and fracture of solids*. New York: McGraw-Hill.

Nadel, E. R., and Bussolari, S. R. 1988, August. Daedalus project: Physiological problems and solutions. *American Scientist.*

National Academy of Engineers. 1970. Public safety: A growing factor in modern design. Fifth Annual Meeting of the Washington, DC, National Academy of Sciences.

NCCHMS. 1972, March. Hazardous material spills. *Proceedings of the 1972 National Conference on Control of Hazardous Material Spills.* Sponsored by EPA and University of Houston.

Newman, R. P. 1960. Fatigue strength of butt welds in mild steel. *British Welding Journal,* 7: 169–178.

NFPA Reporter. 1973, July–August. Concerning the American Conference of Governmental Industrial Hygienists Meeting.

NFPA recommended standard. Project T2.7.3, Draft No. 1 and T3.9.14, 1971.

Nicholas, T. 1982. Material behavior at high strain rates. In Zukas, J., Nicholas, T., Swift, H., Greszczwick, L., and Curran, D. (Eds.). *Impact dynamics*. New York: John Wiley & Sons.

Nordstrom, Philip, J., v. White Metal Rolling and Stamping Corp. 453, PA. Rep. 2d–619.

Occupational Safety and Health Act. 84 Stat. 1590, 29 USC, Par. 651.

Onoe, M., Yamada, H., and Liu, Z. C. 1983, March. Transmission and reflection-type ultrasonic computed tomography. *Materials Evaluation.*

Oriani, R. A. 1972. A mechanistic theory of hydrogen embrittlement of steels. *Bericht der Bunsen-Gessellschaft fur Physikalische Chemie,* 76, 848–857.

Oriani, R. A. 1981. *Environmental degradation of engineering materials in hydrogen* (pp. 3–10). Blacksburg, VA: Virginia Polytechnic Institute.

Orowan, E. 1949. Fracture and strength of solids. *Reports on Progress in Physics,* 12, 185–232.

Oxendine v. Merrell Dow Pharm. Inc. 506 A. 2d 110 (D.C. Cir. Ct.) (1986).

Pankow, D. H., and Finnie, I. 1980. *Mechanical behavior of materials* (ICM 3, Vol. 3, pp. 67–76). Oxford, England: Pergamon.

Padovan, J., and Tovichakchaikul, S. 1982a. Self-adaptive predictor corrector algorithms for static non-linear structural analysis. *Journal of Computers and Structures,* Vol. 15, No. 379.

Padovan, J., and Tovichakchaikul, S. 1982b. On the solution of creep-induced buckling in general structures. *Journal of Computers and Structures,* Vol. 15, No. 379.

Padovan, J., and Tovichakchaikul, S. 1983. On the solution of elastic-plastic static and dynamic post-buckling collapse of general structures. *Journal of Computers and Structures,* Vol. 16, No. 199.

Padovan, J., and Tovichakchaikul, S. 1984. Algorithms for elasto-plastic creep post-buckling. *ASCE Journal of Engineering Mechanics,* Vol. 110, No. 6, June.

Papadakas, C., and Scarton, H. (Eds.) 1978. *Fluid transients and acoustics in the power industry.* New York: Fluids Engineering Division, American Society of Mechanical Engineers.

Paris, P. C., and Erdogan, F. 1963. A critical analysis of crack propagation laws. *Transactions of the ASME, Journal of Basic Engineering,* 85, 528–534.

Park, K. C., and Gartling, D. K. 1978. Computational techniques for interface problems. AMD, Vol. 30. New York: American Society of Mechanical Engineers.

Pellini, W. S. 1971, March. Principles of fracture-safe design Part I and II. Welding Journal Research Supplement. American Society, Adams Lecture.

Pellini, W. S. 1973, March 13–15. Design options for selection of fracture control procedures in the modernization of codes, rules and standards. *Proceedings of the Joint United States and Japan Symposium on Application of Pressure Components Codes.* Tokyo, Japan.

Perfecting Service Co. v. Product Development & Sales Co., (1963), *Product Liability Reports,* Sec. 5065, 259 N.C. 400, 131 S.E. 2d 9.

Peters, G. A., and Peters, B. J. 1984. *Automotive engineering and litigation.* New York: Garland Publishing.

Peters, L. C., Sandor, G. N., Thorpe, A. S., Talbot, T., Weinstein, A. S., and Wolosewicz, R. M. Design engineering and the law. *Mechanical Engineering Journal,* October, 1977.

Petroski, H. 1982, August–September. When cracks become breakthroughs. *Technology Review,* 19–30.

Peyton, D. L. 1973a. The American National Standards Institute: Contribution of standards to products liability prevention. Products Liability Prevention Conference, August 22, 1973.

Peyton, D. L. 1973b. The American National Standards Institute: The contribution of standards to products liability prevention. Products Liability Prevention Conference, August 22, 1973.

Phillips, C. A., and Warwick, R. G. 1968. A survey of defects in pressure vessels built to high standards of construction and its relevance to nuclear primary circuit envelope (AHSB (S) R. 162. United Kingdom, Atomic Energy Association.

Piggott, M. R. 1980. *Load-bearing fibre composites.* Oxford, England: Pergamon Press.

Pletta, D. H. 1984. *The engineering profession: Its heritage and its emerging public purpose.* Lanham, NY: University Press of America.

Pletta, D. H. 1985a. Professional support for safety conscious whistleblowers. Seminar on whistleblowing, Association of Professional Engineers of Ontario, Toronto, October 24.

Posner, R. 1977. *Economic analysis of law.* 2d ed. Boston: Little, Brown.

Post Tensioning Institute. 1984. *Design of post tension slabs.* Phoenix, AZ.

Post Tensioning Institute. 1985. *Post-tensioning manual* (4th ed.). Phoenix, AZ.

Post Tensioning Institute. 1986. *Design and construction of post tension slab-on-construction.* Phoenix, AZ.

Powers, G. J., and Thompkins, F. C. 1974. Fault-tree synthesis for chemical processes. *American Institute of Chemical Engineers Journal,* 20(2), 376.

Poynter, Dan. 1988. *The expert witness handbook, tips & techniques for the litigation consultant.* Lawrenceville, NJ, National Forensic Center. Santa Barbara, CA: Para Publishing.

Preece, C. M. 1979. *Treatise on materials science and technology.* Vol. 16. *Erosion* (pp. 249–308). New York: Academic Press, 1979.

Product Liability Reports. 1974.

Production Magazine. 1972, December. What you can do about noise. p. 85.

Prosser, W. 1971. *Handbook of the law of torts.* 4th ed. St. Paul, Minn.: West Publishing.

Pruppecher, H. R., *et al.* 1983. *Precipitation scavenging, deposition and resuspension function.* New York: Elsevier.

Public Law 91–596. 84 Stat. 1590 (1970).

Public Law 91–605. 84 Stat. 1739 (1970).

Public Law 92–573. 86 Stat. 1207 (1972).

Public Law 97–35. S 1202(a), 95 Stat. 703 (1981).

Public Law 97–35. S 1210, 95 Stat. 703, 721 (1981).

Public Law 97–45, 95 Stat. 949 Codified at 15 USC, Secs. 3901–3904, 1981.

Pugh, C. E. 1982. Elevated temperature design technology, Chapter 3. In S. Y. Zamrick and D. Dretrich (Eds.) *ASME—A Decade of Progress.* American Society of Mechanical Engineers, New York, NY.

Pugh, C. E. 1989. Fracture mechanics applications in the nuclear industry. Oak Ridge National Lab paper, 12th Energy Technology Conference, Houston, TX.

Rech, 1956. *National standards in a modern economy.*

Rechlin, C. F. 1970. Liability of certifiers of products for personal injuries through the user or consumer. *Cornell Law Review,* 56(1).

Reimers, P., and Goebbels, J. 1983, May. New possibilities of nondestructive evaluation by x-ray computed tomography. *Materials Evaluation.*

Reinhart, E. R. 1984. Improved fossil plant availability with NDE. *Mechanical Engineering,* 106(2): 46–49.

Restatement [second] of torts. 1965. 291, 395, 402a, American Law Institute.

ReVelle, J. E. 1986. Safety engineering. In J. Kurtz (Ed.), *Mechanical engineer's handbook* (Chap. 14). New York: John Wiley & Sons.

Richard, F. E., Hall, J., and Woods, R. 1970. *Vibration of soils and foundations.* Englewood Cliffs, NJ: Prentice-Hall.

Richardson, R. C. D. 1968. Wear of metals by relatively soft abrasives. *Wear,* 11, 245–275.

Rider, D. A., *et al.* 1975. *General practice in failure analysis.* "Failure analysis and prevention," Vol. 10. Metals handbook. Metals Park, OH: American Society for Metals.

Ritter v. Narragansett Electric Co. 109 R.I. 176, 283 Atlantic 2d 255 (1971).

Rockwell, W. H. 1970. Legal aspects of standardization. Annual meeting of the International Society of America, Philadelphia, October 29, 1970.

Rockwell, W. H. 1972, April 15. Quality and the consumer. Conference of the American Society for Quality Control. Stonybrook, New York.

Rogers v. Toni. 167 OH. 244.

Rolfe, S. T., and Barsom, J. M. 1987. *Fracture and fatigue control in structures: Applications of fracture mechanics.* Englewood Cliffs, NJ: Prentice-Hall.

Roofire Alarm Co. v. Underwriters Laboratories Inc., 188 F, Supp. 753 (E.D. Tenn. 1959), aff'd 284 F 2d 360 (6 Cir 1960).

Rooke, D. P., and Cartwright, D. J. 1976. *Compendium of stress intensity factors.* London, England: IIMSO.

Rosen, S. I. 1984. *The slip and fall handbook.* "Nuts and bolts series." Columbia, MD: Hanrow Press.

Sances, A., Thomas, D. J., Ewing, C. L., Larson, S. J., and Unterharnscheidl, F. 1986. *Mechanism of head and spine trauma.* Deer Park, NY: Aloray Publisher.

Sax, N. I., and Lewis, R. J. 1986. *Rapid guide to hazardous chemicals in the workplace.* New York: Van Nostrand Reinhold.

Schilling, E. G. 1982. *Acceptance sampling in quality control.* New York: Marcel Dekker.

Schlapper, G. A., Brugger, R. M., Koeppe, R. A., Larson, G. H., and Jost, R. J. 1981, November. A demonstration of filtered neutron beam computed tomography: Evaluation of a 37 pin fuel assembly phantom. Argonne, IL: Argonne National Laboratory and U.S. Department of Energy.

Schroeder, J. R. 1980, September. Limit and blast pressure of fabricated branch connection. Pressure Vessel Research Council (PVRC) of the Welding Research Institute Subcommittee on Reinforced Opening and External Load, University of Waterloo, Waterloo, Ont., Canada.

Schwartz, 1982. The Consumer Product Safety Commission: A flawed product of the consumer decade. *George Washington Law Review* 51, 32.

Scott, R. A. 1979. Toxic chemical and explosions facilities in safety and engineering design. *American Chemical Society, Symposium Series 96.* Washington, DC.

Scott, Seifert, and Wescott. 1974, May. The particles of wear. *Scientific American,* p. 88. (Bibliography, p. 144.) (See also Richie, 1974; and Snowden, 1975, in Additional Reading).

Sears, F. W., and Zemansky, M. W. 1975. *College physics.* Reading, MA: Addison-Wesley.

Seeley, F. B., and Smith, J. O. 1963. *Advanced mechanics of materials.* New York: John Wiley & Sons.

Segal, E., Notea, A., and Segal, Y. 1982, November. Dimensional information through industrial computerized tomography. *Materials Evaluation.*

SEM. 1988. *Handbook on experimental mechanics.* Bethel, CT: Society of Experimental Mechanics.

Sendeckyj, G. P. 1983. *Characterization, analysis and significance of defects in composite materials.* AGARD Conference Proceedings, No. 355, pp. 2.1–2.22. Paris, France: NATO/AGARD.

Severy, D. M., Brink, H. M., and Baird, J. D. 1968, March. Full scale crash test shows Liberty Capsule Seat protects passenger best. *Society of Automotive Engineers Journal.*

Sherby, O. P., Orr, R. L., and Dorn, J. E. 1954. *Transactions of the American Society of Mechanical Engineers*, 200, 71–80.

Shigley, J. E. 1977. *Mechanical engineering design*. New York: McGraw-Hill.

Shoemaker, A. K., and Rolfe, T. S. 1969, September. Static and dynamic low temperature K_{Ic} behavior of steels. *ASME Transactions*.

Shroder, J. R. 1979. Burst test on full size pipe. PVRC, Dynamics Design Committee Status Report, University of Waterloo, Waterloo, Ont., Canada.

Silva, T. C. V., Pascual, R., and de Miranda, P. E. V. 1984. *Fracture prevention in energy and transport systems* (pp. 511–520). Cradley Heath, Warley, U.K.: Engineering Materials Advisory Services.

Simpson v. American Oil. 217 N.C. 542.

Simpson v. Wellington Service Corp. 355 Mass. 685 246 N.E. 2d 801 (1969).

Sindell v. Abbott Laboratories. 246 Cal. 2d 588, 163 Cal. Rptr. 132, 607 P. 2d 924 (1980).

Smialowski, M. 1962. *Hydrogen in steel*. London, England: Pergamon Press.

Smith, E. 1966. *The physical basis of yield and fracture* (pp. 36–46). London, England: Institute of Physics and the Physical Society.

Smith, T. A., and Warwick, R. G. 1974. Second survey of defects in pressure vessels built to high standards of construction and its relevance to nuclear primary circuit envelope (SRD-R-30-1974). United Kingdom, Atomic Energy Commission.

Society of engineers must pay $2.5 million damages, court finds. 1983. *Wall Street Journal*, June 27, 16.

Spanner, J. C. 1985. Acoustic emission. *Nondestructive evaluation* (Chap. 6). *ASME Decade of progress*. New York: American Society of Mechanical Engineers.

Spindle v. Travelers Insurance Co. 66 Cal. App. 3d, 136 Cal. Rptr. 404 (1977).

Steele, L. E. 1984. *Fracture prevention in energy and transport systems* (pp. 109–118). Cradley Heath, Warley, U.K.: Engineering Materials Advisory Services.

Strang, G., and Fix, G. J. 1973. *An analysis of the finite element method*. Englewood Cliffs, NJ: Prentice-Hall.

Stricklin, J. A. 1970, August. Geometrically nonlinear static and dynamic analysis of shells of revolution. *Proceedings of the Symposium of High-speed Computing of Elastic Structures*, pp. 384–411. Liege, Belgium. IUTAM

Stulen, F. B., Cummings, H. N., and Schulte, W. C. 1961, June 22. Preventing fatigue failures, Part 5. *Machine Design*, Vol. 33, p. 166.

Suarez, M. G., and Schupack, M. 1988. *Evaluation of performance of single strand unbounded tendons in the collapse of the L'Ambiance Plaza brick slab building*. Phoenix, AZ: Post Tension Institute.

Sundararajan, C. 1985. Reliability. In *Decade of progress: Pressure vessel and piping technology*. New York: American Society of Mechanical Engineers.

Svalbonas, V. 1983. Failure analysis of the world's largest grinding mill. Paper No. 83-PVP-37. New York: American Society of Mechanical Engineers.

Sweeney v. Matthews, 2968, 94 Ill. App. 2d 6, 236, NE 2d 439.

Sweet, J. 1977. *Legal aspects of architecture, engineering and the construction process*. 2d ed. S 35.06. St. Paul, Minn.: West Publishing.

Szilard, R. 1974. *Theory and analysis of plates: Theoretical and classical methods*. Englewood Cliffs, NJ: Prentice-Hall.

Tada, H., Paris, P. C., and Irwin, G. R. 1973. *The stress analysis of cracks handbook*. Hellertown, PA: Del Research Corp.

Tagart, S. W. 1985. Decision analysis applied to structural mechanics, Chap. 8, Reliability. In Sundararajan, C. *Decade of progress: Pressure vessel and piping technology*. New York: American Society of Mechanical Engineers.

Talbot, R. F. 1975. Your day in court. *Machine Design*, 47(3): 68–72.

Target health hazards. 1983. OSHA-2051.

Tarkov, J. 1986. A disaster in the making. *American Heritage Invention Technology* 1(3).

Taylor v. The Carborundum Co., 107, Ill. App. 2d 12, 246 NE 2d 898.

Terry, 1915. Negligence. *Harvard Law Review* 40.

Texas & Pacific Ry. v. Behymer, 189 U.S. 468, 470 (1903).

Thielsch, H. 1965. *Defects and failures in pressure vessels*. New York: Reinhold Publishing Corp.

Timoshenko, S. 1976. *Strength in materials. Part II. Advanced theories in problems*. Huntington, NY: Robert E. Krieger Publishing.

Timoshenko, S. P., and Woinowsky-Krieger, S. 1959. *Theory of plates and shells*. New York: McGraw-Hill.

Tobias, T. A., and Trindad, D. 1988. *Applied reliability*. New York: Van Nostrand Reinhold.

Tomlinson, G. R. "Use of Video Tape Evidence at Trial." ABA Proceedings, April 1987. Litigation Committee on Trial Practices, Midyear Meeting.

Tong, P. 1970. New displacement hybrid finite element models for solid continuum. *International Journal of Numerical Methods Engineering*, 2.

Tonner, P. D., and Tosello, G. 1986, February. Computed tomography scanning for location and sizing of cavities in valve casings. *Materials Evaluation*. Atomic Energy of Canada, Limited. Chalk River, Ontario.

Tort liability of independent agencies. 1968. *Rutgers Law Review*, 299, 317.

Troiano, A. R. 1960. The role of hydrogen and other interstitials in the mechanical behavior of metals. *Transactions of the American Society of Metals*, 52, 54–80.

Tronsberg, W. 1982. *Synopsis of results of the first programme on the production and utilisation of hydrogen* (pp. 452–455). Luxembourg, European Economic Community (EEC).

Tsao, M. D. 1983, October. Industrial ultrasonic tomography: Principle, practice and limitation. *Materials Evaluation.*

Uloth vs. Cititak Corp. 1970.

U.S. Atomic Energy Commission. 1974. *Integrity of reactive vessels for light-water power reactors* (1285, VC-78). Washington, DC.

U.S. Department of Defense. 1988. Minutes of the Twenty-Third Explosive Safety Seminar, two volumes, pg. 1-2443. U.S. Department of Defense (approved for public release), Atlanta, GA, Aug. 9–11, 1988.

U.S. Department of Energy. 1985. *U.S. Department of Energy Manual for Prediction of Blast and Fragment Loads on Structures.* DOETIC11268. Albuquerque, NM: Albuquerque Operations Office.

U.S. Nuclear Regulatory Commission. 1975, July. *Regulatory Guide 1.99.*

U.S. v. American Society of Civil Engineers. 446 F. Supp. 803 NY (1977).

Vance, J. M. 1988. *Rotordynamics of turbomachinery.* New York: John Wiley & Sons.

Vance, J. M., and Gould, R. W. 1985. Failure analysis and the expert witness for products liability: A responsibility for the engineering profession. In S. J. Brown (Ed.), *Decade of progress* (Chap. 9). New York: American Society of Mechanical Engineers.

Vander Voort, G. F. 1978. Macroscopic examination procedure for failure analysis. In J. L. McCall and P. M. French (Eds.), *Metallography and failure analysis.* New York: Plenum Press.

Verman, L. 1973. *Standardization,* Vol. 4.

Vernon v. Lake Motors. 26 Utah 2d 269, 488 Pacific 2d 302 (1971).

Vinson, J. R. 1980. Emerging technologies in aerospace structures, design, structural dynamics and materials. ASME, Aerospace Division.

Vlasov, V. Z. 1964. *General theory of shells and its application of engineering.* GITTL. Moscow-Leningrad. Trans. NASA, N64-19883.

Vroman v. Sears Roebuck & Co., Ca 6 Mich, 1969, CCH products liability reports, Par. 5873, 387 F 2d 732.

Wade, D. 1965. Strict liability of manufacturers. *Southwestern Law Journal* 19(5).

Wah, T. H., and Calcote, L. R. 1970. *Structural analysis by finite difference calculus.* New York: Van Nostrand Reinhold.

Walsh v. Ford Motor Car Co., 1969 59 Misc. 2d 241, 298 NYS 2d 538.

Watanabe, J. 1984. *Fracture prevention in energy and transport systems* (pp. 71–89). Cradley Heath, Warley, U.K.: Engineering Materials Advisory Services.

Waterhouse, R. B. 1972. *Fretting corrosion.* Oxford, England: Pergamon Press.

Waterman, T. E. 1985, December 26. Examination of seat belt acceptance by forklift operators. Institute for Advanced Safety Studies for Allis-Chalmers.

Weinstein, A. S., et al. 1970. *Products liability and the reasonably safe product.* New York: John Wiley & Sons.

Weinstein, A. S., Twerski, A., Prehler, H., Donaher, W. A. 1978. *Product liability and the reasonably safe product: Guide for management, design and marketing*. New York: John Wiley & Sons.

White, C. S., Bowen, I. G., and Richmond, D. R. 1965. Biological tolerance to air blast and related biomedical criteria. CEX-65.4. Washington, DC.: U.S. Atomic Energy Commission.

Whitfield v. Earl E. Knox Co., (D.C. Pa. 1956), 6 Negligence Case (2d) 238, 144 F. Supp. 244.

Whitmal, J., and Playter, K. M. 1987. *Forensic photogrammeter seeking accident cause and effect*. Yakima, WA: Photo-Graphics, Inc.

Wilson, A. D. 1981. Fractographic characterization of the effect of inclusions on fatigue crack propagation. Philadelphia: American Society of Testing and Materials, STP 733: 166–186.

Wilson, E. L., and Nickell, R. E. 1966. Application of the finite element method to heat conduction analysis. In *Nuclear engineering and design*. Amsterdam: North-Holland Publishing Co.

Woodford, D. A. 1980. *Proceedings of the International Conference on Engineering Aspects of Creep* (Paper No. C223/80). London, England: Institute of Mechanical Engineers.

Woodson, W. 1981. *Human factors design handbook*. New York: McGraw-Hill.

Work fatalities involving overturning forklifts, by accident cause. 1977. San Francisco: Division of Labor Statistics and Research, State of California.

The yearbook of the Tire and Rim Association. 1986. Akron, OH: Tire and Rim Association.

Yan, M. J. 1982. Dynamic and seismic analysis of systems and components. ASME Symposium, Vol. PVP-65. New York: American Society of Mechanical Engineers.

Zajic, J. E., and Himmelman, W. A. 1978. *Highly hazardous materials spills and emergency planning*. New York: Marcel Dekker.

Zamrik, S. Y. 1990. An interpretation of axial creep–fatigue damage interaction in type 316 stainless steel. *Journal of Pressure Vessel Technology*, Vol. 112.

Zamrik, S. Y., and Pangborn, R. N. 1989. Fatigue damage assessment using x-ray diffraction and life prediction methodology, *Nuclear Engineering and Design*, Vol. 116, pp. 407–413, North Holland Physics Publishing, Amsterdam.

Zienkiewicz, O. C. 1977. *The finite element method*. New York: McGraw-Hill.

Zukas, J. A. 1980. Impact dynamics. In *Emerging technologies in aerospace structures, design, structural dynamics and materials* Century 2. J. R. Vinson (Ed.). Emerging Technology Conference, San Francisco, CA.

Zukas, J., Nicholas, T., Swift, H., Greszczwick, L., and Curran, D. (Eds.). 1982. *Impact dynamics*. New York: John Wiley & Sons.

Additional Readings

Alen, N. N. 1985. *Effects of seat wings and seat belts on the response of forklift truck operating during tip overs*. Ann Arbor, MI: Institute of Advanced Safety Studies.

ASM. 1989. *Metals handbook* (9th ed., Vols. 1–17). Metals Park, OH: American Society for Metals.

ASME. 1976. Criteria for design of elevated temperature Class 1 components. Boiler and Pressure vessel code for design by analysis (Sec. 3. Components, Division 1), New York: American Society of Mechanical Engineers.

American Textile Manufacturers Institute, Inc. v. Donovan, 452 U.S. 490, 101 Superior Court 2372, 65 L Ed. 2d 1010 (1980). (Regulation setting permissible levels of benzene invalid.)

Anderson, V. L., and McLean, R. A. *Design of experiments: A realistic approach*. New York: Marcel Dekker.

Ascher, H., and Feingold, H. 1984. *Repairable systems reliability: Modeling, interference, misconceptions and their causes*. New York: Marcel Dekker.

Assheton, R. 1930. History of explosions (on which the American Table of Distances was based). Institute of Makers of Explosions, USAF Academy Library. Colorado Springs, CO.

Azzarello v. Block Bros. Corp. 480 Pa. 547, 391 A. 2d 1020 (1978).

Beecher, N., and Rappaport, A. 1990. Hazardous waste management policies overseas. *Chemical Engineering Progress,* Vol. 86, No. 5, p. 30.

Blevins, R. D. 1977. *Flow-induced vibration*. New York: Van Nostrand Reinhold.

Bodhurtha, F. T. 1980. *Industrial explosions prevention and protection*. New York: McGraw-Hill.

Boesebeck, K., and Homke, P. Failure data collection and analysis in the Republic of Germany (ASME-PVP-PV-023). New York: American Society of Mechanical Engineers.

Boisjoly. 1987. *Ethical decisions: Morton Thiokol and the space shuttle Challenger disaster* (Pub. No. 87-WA/TS-4). New York: American Society of Mechanical Engineers.

Bota, K. J., Ozdemir, H., and Wilson, D. L. 1974, February. Static and dynamic geometric and material nonlinear analysis. Report UC SESM 74-4. Structure and Material Research Department of Civil Engineering, University of California, Berkeley.

Buchannan, J. C., Bos, C., and Heller, F. I. 1986. *How to use video tape in litigation.* New York: Prentice-Hall.

Burck, C. G. 1975. The intricate politics of the corporation. *Fortune Magazine,* April.

Cast iron maintenance optimization system. 1988. *Gas Institute Magazine,* Forecasting cast-iron failures. Gas Research Institute, Park Ridge, IL.

CCH Products liability reports. p. 4756. Preventive management.

Chaney, E. M. 1986. Computer stimulations: How they can be used at trial and argument for admissibility. *Indiana Law Review,* 19: 735.

Chang, T. Y. 1980. A nonlinear finite element analysis program: NFAP. Vols. 1 and 2. The University of Akron, Akron, Ohio: College of Engineering, Civil Engineering Department.

Char, W. T. 1977. *Suppressive shields: Structural design and analysis handbook.* Huntsville, AL: U.S. Army Corps of Engineers. HNDM-1110-1-2.

Code of Federal Regulations, U.S. Department of Labor, Vol. 29.

Control of hazardous material spills. 1972. *Proceedings of the 1972 National Conference on Control of Hazardous Material Spills.* Environmental Protection Agency, University of Houston, March.

Cox, D. O., Lund, R. A., and Roman, I. 1985. Developments in fractography for failure analysis. In S. J. Brown (Ed.), *Decade of progress* (Chap. 9). New York: American Society of Mechanical Engineers.

CRC Handbook on Biology, Chemistry, Zoology, 70th ed. 1989. Boca Raton, FL: CRC Press.

Davis v. The Coats Co., 119 NW 2d 1980.

Design Institute for Emergency Relief Systems. 1988. *Project manual.* New York: DIERS User's Group, American Institute of Chemical Engineers.

Dhalla, A. K., and Gallagher, R. H. 1982. Computational methods for nonlinear structural analysis. In *Pressure vessel & piping design and technology: A decade of progress.* New York: American Society of Mechanical Engineers.

Diamond. 1983. Eliminating the "defect in design" strict liability theory. *Hastings Law Journal* 34: 529.

Dobbs, N., and Caltagirone, J. P. 1987. Structures to resist the effects of accidental explosions. U.S. Army Armament, Research, Development, and Engineering Center (ARLCD-SP-84001, Vols. 1–6). Picatinny Arsenal.

Department of Energy. 1982. *Manual for prediction of blast and fragmentation loads on structures.* Albuquerque, NM: Albuquerque Office (DOETIC11268).

E. I. Dupont de Nemours and Company v. McCain, 414 F 2d 369, 5th Circuit (1969).

Dusinberre, G. M., and Lester, J. C. 1962. *Gas turbine power.* Scranton, PA: International Text Book Publishing.

Eginton, W. W. 1970. The manufacturer and product liability. *Quality Progress,* August.

Erb, J., Ortiz, E., and Woodside, G. 1990. On-line characterization of stack emissions. *Chemical Engineering Progress,* Vol. 86, No. 5, p. 40.

Federal Register. Department of Labor—Occupational Health Administration—Occupational Safety & Health Standards. Sec. 1910, Vol. 37, No. 202, Part II.

Federal Hazardous Substances Act, 15 USC, Para. 1261.

Ferry, T. S. 1988. *Modern accident investigation and analysis.* New York: John Wiley & Sons.

Fogel, I. M. 1987. Fail safe management. *Civil Engineering,* September. New York: American Society of Civil Engineers.

Fromson, D. 1973, November 12. A legal review of selected operational and enforcement problems—When is enough enough? Speech delivered at annual fall meeting of the NFPA, Minneapolis, MN.

Fromson, D. 1985. Product liability and the use of codes and standards. In S. J. Brown (Ed.), *Decade of progress* (Chap. 9). New York: American Society of Mechanical Engineers.

Gaillard, J. 1943. Industrial principles and application.

Garcia v. Halsett, 3 Cal. App. 3rd 319, 82 Cal. Rptr. 420 (1970).

Garg, S., and Garg, D. 1990. Genetic engineering and pollution control. *Chemical Engineering Progress,* Vol. 86, No. 5, p. 46.

Gelmusin v. E. W. Bliss Co., 295 NE 2d 110.

George, G. S. 1980. *Biomechanics of womens' gymnastics.* Englewood Cliffs, NJ: Prentice-Hall.

Gilbert, R. O. 1988. *Statistical methods for environmental pollution monitoring.* New York: Van Nostrand Reinhold.

Griffith v. Chevrolet Motor Div., 1962, 125 SE 2d 525.

Haines, R. W. 1988. *Control systems for heating, air conditioning and ventilating.* New York: Van Nostrand Reinhold.

Handbook on experimental mechanics. 1988. Bethel, CT: Society of Experimental Mechanics.

Hasegawa, K., and Sato, K. 1978. Experimental investigation of uncoupled vapor—Cloud explosions of hydrocarbons. Technical memorandum of Firi Research Institute, No. 12. Japan: Firi Defense Agency.

Hazardous material spills. 1972. *Proceedings of the 1972 National Conference on Control of Hazardous Material Spills.* (NCCHMS) Sponsored by the Environmental Protection Agency and the University of Houston, Houston, TX.

H.R. 7284 and S 2631 (earlier citation) as reported by the Senate on Commerce, Science and Transportation.

Hawkins Construction Co. v. Matthews Company Inc., Supreme Court of Nebraska, No. 38627, July 27, 1983.

Henshaw, T. L. 1988. *Reciprocating pumps.* New York: Van Nostrand Reinhold.

Hoffman, J. E. 1968, Summer. Industry-wide codes, advertising, seals of approval and standards as participated in by the Trade Association. *Antitrust Bulletin,* 13.

Homke, P., Lindauer, E., and Meinlschidt, G. 1978. Data collection in a nuclear power plant in a pilot system in a lignite power station (PVP-PV-032). New York: American Society of Mechanical Engineers.

Hooper, T. J., 60 Federal 2d 737 (Second Circuit Court of Appeals, 1932).

Hsu, K. H. (Ed.). 1985. *Computer-aided engineering.* Proceedings of the ASME-PVP, Vol. 98-5. New York: American Society of Mechanical Engineers.

Ibach v. Grant Donaldson Service, Inc., 38 App Div. 2d 39, 326 NYS 2d 720 (rth Dept.) (1971).

Industrial Union v. American Petroleum Inst. 448 U.S. 607, 100 Sup. Ct. 2372, 65 L. Ed. 2d 1010 (1980).

Institute of Advanced Safety Studies. 1986. *Effects of operator restraints on measured head impact during overturn of forklift trucks.* Niles, IL.

ISO/TC 131/SC8 USA-7. 1971, September 13.

Jennings, B. H., and Lewis. 1965. *Air-conditioning & refrigeration.* International Text Book Publishing Co.

Job Safety and Health Act of 1970.

Joseph, G. P. 1985. Video tape evidence in courts. *South Texas Law Journal,* 26(3).

Judge Etheridge gives advice to experts. 1988. *The Expert and the Law,* Vol. 8, No. 1, January.

Kenny, B., and Patterson, E. A. 1985. Load and stress distribution in screw threads. *Experimental Mechanics.*

Klein, J. L. 1981, March. Repair welding of stainless steel and bronze propellers: Theory and practice, welding and shipbuilding symposium.

Kletz, T. A. 1987. *Why are fewer case histories being published?* Twenty-first annual "Loss Prevention Symposium." New York: American Institute of Chemical Engineers, August.

Kumana, J. D. 1990. The impact of excess boiling point rise on evaporators and crystallizers. *Chemical Engineering Progress,* Vol. 86, p. 10.

Leaf, D. A. 1990. Acid rain and the Clean Air Act. *Chemical Engineering Progress,* Vol. 86, No. 5, p. 25.

Lees, F. P. 1980. *Loss prevention in the process industries* (Vols. 1 and 2). London, England: Butterworth Publishing.

LeMay, I. 1982. Failure analysis of a large crude oil storage tank. *Metal Progress,* 122(3), 35–37.

Lofgren, M. J. 1983. *Handbook for the accident reconstructionist.* Jacksonville: Institute for Police Traffic Management, University of North Florida.

Louis Dimeo v. The Minster Machine Co., Inc., Jan. 8, 1968, US Ct. of Appeals 2d Cir.

Lubrication and maintenance. Truckee, CA: Fluid Marketing Corp. (General Source Reference).

Lyons, C. E. 1990. Environmental problem solving: The 1987-88 Metro Denver Brown Cloud Study. *Chemical Engineering Progress,* Vol. 86, No. 5, p. 61.

Lyons, J. 1985, October. America's greatest challenge. *Gateway Engineer Magazine.* St. Louis, MO: Engineers Club of St. Louis.

Mahajan, K. K. 1979. *Design of process equipment.* Tulsa, OK: Pressure Vessel Handbook Publishing Co.

Majundar, S. B. 1990, May. Regulatory requirements and hazardous materials. *Chemical Engineering Progress,* Vol. 86, No. 5.

Manson, S. S. 1982. A critical review of predictive methods for treatment of time-dependent metal fatigue at high temperatures. *Pressure vessel & piping design and technology: A decade of progress.* New York: American Society of Mechanical Engineers.

Mason, P., *et al.* 1965. Nonlinear numerical analysis of axis-symmetrically loaded arbitrary shells of revolution. *AIAA Journal,* 3, 1307–1312.

Maurizio D. Fortunato v. Ford Motor Co. 1972, July 11. *New York Law Journal.*

McLean, W. G. Role of professional societies. 1985. In M. L. Yeater (Ed.), *Decade of progress* (Chap. 10). New York: American Society of Mechanical Engineers.

Memdel v. Pittsburgh Plate Glass Co., 25 NY 2d 340, 345; 253 NE 2d 207, 210 (1969).

Moody, F., and Brown, S. *Fluid-structure dynamics.* New York: Van Nostrand Reinhold. To be published in 1991.

Murray, W. M. (Ed.). 1952. *Fatigue and fracture of metals.* New York: John Wiley & Sons.

National Fluid Power Association. 1973. *Fluid power standards* (3rd ed.). Milwaukee, WI.

Negotia, C. V., and Ralescu, D. 1988. *Simulation, knowledge-based computing and fuzzy statistics.* New York: Van Nostrand Reinhold.

New York State Bar Association Journal. 1979, October. Note 31, p. 499.

NFPA recommended standard. 1971, May 9. T3.9.14.

NFPA recommended standard. 1970, April 19. T39.70.12.

Nickell, R. E., and Yahn, G. T. 1982. Simplified methods for design. *Pressure vessel & piping design and technology: A decade of progress.* New York: American Society of Mechanical Engineers.

Noise, the environmental problem—A guide to OSHA standards. Washington, DC.: U.S. Department of Labor.

Odeh, R., and Owen, D. B. 1983. *Attribute sampling plans, tables of tests and confidence limits for proportions.* New York: Marcel Dekker.

Occupational Safety and Health Act of 1970—A legal review of selected operational and enforcement problems—When enough is enough. 1973, November 12. Speech by David Fromson before the Annual Fall Meeting of the National Fire Prevention Association, Minneapolis, MN.

OSHA, Sec. 5. 1972, October 1. An index to the hazardous materials regulations, Title 49, CFR. Washington, DC.: Office of the Secretary, Office of Hazardous Materials, U.S. Department of Transportation.

OSHA. Machinery and machine guarding. Subpart O. Sec. 1910.212.

OSHA concerning asbestos. *Wall Street Journal,* April 10, 1984, p. 7.

Pennwalt Corporation. 1972. *Annual report,* p. 26.

Peters, G. A. 1986. Warnings and instructions. *Hazard Prevention.*

Pickett, A. G., and Rinehart, E. R. 1973. Review of recent utility experience with power reactor cooled and pressure inspection service conditions with respect to defect detection capability, defect size, and defect orientation. U.S. Atomic Energy Commission, WASH-1285. South West Research, Washington, DC. Institute Report EEI-TVA.

Pinsker v. Pac. Coast Soc. of Orthodontists. 12 Cal. 3d 541, 116 California Reporter 245, 526 P. 2d 253 (1974).

Pletta, D. H. 1985, July. When law makers set standards, who protects the public? *Engineering Times.*

Pollack, A. 1983, March 11. Engineers: A drop-out problem for the U.S. *New York Times,* Section 12.

Pressure vessels, boilers and pipelines. 1956. Document XI-2-56. International Institute of Welding.

Product liability report. CCH, Design, p. 4735.

Progress in measuring fracture toughness and using fracture mechanics. 1964. *Materials Research and Standards,* 103–119, March.

Pugh, C. E., and Robinson, D. N. 1982. Constitutive equations for meeting elevated temperature design needs. In *Pressure vessel & piping design and tech-*

nology: A decade of progress. New York: American Society of Mechanical Engineers.

Recommended practices for design and installation of pressure-relieving systems in refineries. 1973. API-RP-520. Washington, DC.: American Petroleum Institute.

Regulatory Guide 1.99. 1975. U.S. Nuclear Regulatory Commission, July.

The relationship between federal standards and litigation in the control of automobile design. 1982. *New York University Law Review*, Vol. 57, No. 804.

Ritchie, D. 1974, July. Oil analysis—A window in every engine. *Commercial Car Journal*.

Roark, R. J., and Young, W. C. 1975. *Formulas for stress and strains* (5th ed.). New York: McGraw-Hill.

Robinson, E. L. 1947. Steam-piping design to minimize creep concentrations. *ASME Transactions*, 1147–1162.

Rybicy, E. F., and Benzle, S. E. (Eds.). 1975. *Computation fracture mechanics*. New York: American Society of Mechanical Engineers.

Salone v. Hyster, U.S. District Court, Shreveport, LA 1987.

Sax, N. I. 1979. *Dangerous properties of industrial materials*. New York: Van Nostrand Reinhold.

Scerbo, F. A. 1974, February. Product liability prevention techniques for small manufacturers. *Quality Progress*, p. 23.

SAE Journal. 1970. 78(6): 35.

Schaum. AAUM's Schaum's outline series of science handbooks (separate volumes and summaries). New York: Schaum Publishing Co.

Schroeder, J. 1980. Limit and blast pressure of fabricated branch connection. Pressure Vessel Research Council, Subcommittee on Reinforced Openings and External Load, University of Waterloo, Waterloo, Ont., Canada.

Shaver, D. K., and Berkowitz, R. L. 1984. Post-accident procedures for chemicals and propellants. In *Pollution technology review*. Park Ridge, NJ: Noyer Publications.

Smith, A. L. 1984. Reliability of engineering materials. *Proceedings of a symposia on materials reliability* in Baden, Switzerland. Boston: Butterworth Publishers.

Snow, A., and Jakub, M. T. Developments in elevated temperature structural design criteria. In *Pressure vessel & piping design and technology: A decade of progress*. New York: American Society of Mechanical Engineers.

Snowden, J. E., and Conway, J. 1975. Oil analysis as a significant factor in oil and equipment maintenance. Presented at the 30th Annual Meeting, Atlanta, GA.

Solovy, J. S. (Ed.). 1987. Minutes of the mid-year meeting of the Committee on Trial Practice, Section of Litigation. Chicago, IL: American Bar Association.

Sound advice—How to become an instant noise expert, what you can do about noise. 1972, December. *Production Magazine*, p. 85.

Spotts, M. F. 1964. *Design of machine elements.* Englewood Cliffs, NJ: Prentice-Hall.

Standard specification for special requirements for steel plates for nuclear and other special applications. 1977. Specification A647-77. Philadelphia: American Society for Testing and Materials.

Standard test for J_{Ic}, a measure of fracture toughness. 1981. Standard E813-81. Philadelphia: American Society for Testing and Materials.

Stroud, R. C. 1987, August. Excitation, measurements and analysis methods for modal testing. *Journal of Sound and Vibration.*

Svalbonas, V. 1985. Cause of failures. In S. J. Brown (Ed.), *Decade of progress* (Chap. 9). New York: American Society of Mechanical Engineers.

Sweet, J. 1985. Design quality: Standards & liability. In S. J. Brown (Ed.), *Decade of progress* (Chap. 9). New York: American Society of Mechanical Engineers.

Sweet, J. 1987. *Construction industry contracts.* New York: John Wiley & Sons.

Sweet, J. 1989. *Legal aspects of architectural, engineering, and contracting process* (4th ed.). St. Paul, MN: West Publishing.

Swindeman, R. W., and Brinkman, C. R. 1982. Progress in understanding the mechanical behavior of pressure vessel materials at elevated temperatures. In *Pressure vessel & piping design and technology: A decade of progress.* New York: American Society of Mechanical Engineers.

Thermal and fluid/structure dynamics analysis. 1975. Proceedings of Structural Mechanics in Reactor Technology (SMIRT), Vol. B. Berlin, West Germany.

Thomas, W. N. 1923. Effect of scratches and various workshop finishes upon the fatigue strength of steel. *Engineering* 116, ff 449.

Traynor, R. H. Speech delivered to the 75th Anniversary Program of the University of Tennessee College of Law.

UCC Sec. 2-725. Rev. statute of limitations.

U.S. Department of Defense, 1988. Minutes of the Twenty-Third Explosives Safety Seminar, Vols. I and II, pp. 1-2429. Explosives Safety Board, NTIS.

Unterweiser, P. M. (Ed.). *Case histories in failure analysis.* Metals Park, OH: American Society for Metals.

Wall Street Journal. 1984, June 18, p. 2. Per exposure to Ethylene Oxide.

Ward, F. Images for the computer age. *National Geographic,* 175(6).

Wei, B. C., and Nelson, D. V. 1982. Structural design criteria for highly irradiated core components. In *Pressure vessel & piping design and technology: A decade of progress.* New York: American Society of Mechanical Engineers.

Weinstein, A. S. 1985. The legal process: What governs liability? In S. J. Brown (Ed.), *Decade of progress* (Chap. 9). New York: American Society of Mechanical Engineers.

Wells, A. A. 1981. The meaning of fitness for purpose and the concept of defect tolerance. International Conference on Fitness for Purpose Validation of Welded Constructions (Paper 33).

Wilson v. Savage Arms Corp., DC Pa. 1969. Product Liability Reports, Par. 6328, 305 F Supp. 1163.

Zohdi and Biles. 1986. Production processes and equipment. In Kutz (Ed.), *Mechanical engineers handbook* (Chap. 1). New York: John Wiley & Sons.

16 CFR 1512. Requirements for bicycles.

16 CFR 1500. Hazardous substances and articles, administration and enforcement regulations.

22 AM Jr. 2d Damages section 18 (1965).

29 CFR 1910. Secs. 218 (f)(2), Forging machines; 244 (a), Other portable tools and equipment; 217, Mechanical power presses(b)(11)(12); 265, Sawmills(c)(13).

29 CFR 1910.95. Occupational noise exposure.

29 CFR 1910, Sec. 218(e)(1), Forging machines; 169, Air receivers; 242 (b) Hand and portable powered tools; 243(b), Guarding of portable power tools; 217(b), (9)(10)(12), Mechanical power presses; and 107(e)(5), Spray finishing using flammable and combustible materials.

29 CFR 1910. Occupational safety and health hazards.

29 CFR, Subpart H, Hazardous materials.

29 CFR 1910, 177, Servicing multi-piece and single-piece rims.

29 CFR 1910, Subpart P, Hand and portable powered tools and other hand-held equipment.

29 CFR 1910, Subpart O, Machinery and machine guarding.

49 CFR Subchapter V. Hazardous materials transportation and pipeline safety.

75 ALR 2d 778 (1961).

81 Stat. 466.

Index